INTRODUCTION TO ENGINEERING
AND THE ENVIRONMENT

INTRODUCTION TO ENGINEERING AND THE ENVIRONMENT

Edward S. Rubin

Carnegie Mellon University

with Cliff I. Davidson

and other contributors

Boston   Burr Ridge, IL   Dubuque, IA   Madison, WI   New York
San Francisco   St. Louis   Bangkok   Bogotá   Caracas   Kuala Lumpur
Lisbon   London   Madrid   Mexico City   Milan   Montreal   New Delhi
Santiago   Seoul   Singapore   Sydney   Taipei   Toronto

McGraw-Hill Higher Education

A Division of The McGraw-Hill Companies

INTRODUCTION TO ENGINEERING AND THE ENVIRONMENT

Published by McGraw-Hill, an imprint of The McGraw-Hill Companies, Inc., 1221 Avenue of the Americas, New York, NY 10020. Copyright © 2001 by The McGraw-Hill Companies, Inc. All rights reserved. No part of this publication may be reproduced or distributed in any form or by any means, or stored in a database or retrieval system, without the prior written consent of The McGraw-Hill Companies, Inc., including, but not limited to, in any network or other electronic storage or transmission, or broadcast for distance learning.

Some ancillaries, including electronic and print components, may not be available to customers outside the United States.

This book is printed on acid-free paper.

3 4 5 6 7 8 9 0 DOC/DOC 0 9 8 7

ISBN-13: 978-0-07-235467-6
ISBN-10: 0-07-235467-4

Vice president and editor-in-chief: *Kevin T. Kane*
Publisher: *Thomas E. Casson*
Executive editor: *Eric M. Munson*
Editorial coordinator: *Zuzanna Borciuch*
Senior marketing manager: *John Wannemacher*
Senior project manager: *Marilyn Rothenberger*
Production supervisor: *Laura Fuller*
Designer: *K. Wayne Harms*
Cover designer: *Denise Rubin*
Top cover photo: *Allan Montaine/Photonica*
Bottom cover photo: *Gary Randall/FPG International*
Senior photo research coordinator: *Carrie K. Burger*
Photo research: *LouAnn K. Wilson*
Senior supplement producer: *David A. Welsh*
Compositor: *LaChina Publishing Services*
Typeface: *10/12 Times Roman*
Printer: *R. R. Donnelley & Sons Company/Crawfordsville, IN*

Library of Congress Cataloging-in-Publication Data

Rubin, Edward S.
    Introduction to engineering and the environment / Edward S. Rubin. — 1st ed.
        p.   cm. — (McGraw-Hill water resources and environmental engineering series.)
    ISBN 0–07–235467–4

TA170 .R83 2001
628—dc21
                                                                              00–058734
                                                                                    CIP

www.mhhe.com

In memory of
Hy and Esther Rubin,
who made everything possible

CONTENTS

PREFACE

Environmental concerns today profoundly influence all aspects of modern engineering design and practice. Yet, most colleges and universities have been slow at integrating environmental considerations into the fabric of engineering curricula. Although elective courses and degree programs in environmental engineering can be found at many engineering schools, students who do not major in environmental engineering often receive little or no exposure to environmental issues relevant to their profession. At the same time, concern for the environment has been recently highlighted by the Accrediting Board for Engineering and Technology (ABET), and by major U.S. technical societies, as part of the basic ethical responsibility of all engineers.

This text was designed to introduce environmental issues and problem-solving methods to engineering students in all disciplines, primarily at the freshman and sophomore levels. Many of the chapters also have been used successfully in upper-division courses, especially for students with little or no prior background in environmental studies.

The book uses a case study approach to environmental education, drawing on basic science and engineering principles to assess a particular problem, and to design solutions that reduce or eliminate environmental impacts. The case studies thus demonstrate how environmental considerations can be an integral part of good engineering practice. Through applications in different disciplinary domains, students develop and apply the fundamental skills and insights needed to recognize and address a variety of environmental problems. They also gain an appreciation of the interdisciplinary nature of environmental issues and solutions. This case study approach has proved an effective method of introducing environmental subjects in pilot applications at Carnegie Mellon University.

The pedagogical emphasis of this book is on principles of green design and pollution prevention. Thus, the primary thrust is on ways to avoid creating environmental problems in the first place. In many applications, of course, the design and analysis of waste treatment and remediation processes are still very important, and examples of such technologies are also presented. Nonetheless, we believe the green design perspective best reflects the future direction of the environmental field, and of engineering education in general. Although there are limitations on how far this approach can be developed in an introductory textbook, there are substantial opportunities even at the freshman and sophomore levels for students to apply their technical skills to solve environmental problems through improved design. Detailed illustrative examples are included throughout the text to assist students toward this end, along with a set of homework problems at the end of each chapter. More advanced students can be challenged by supplemental projects and assignments that are more open-ended, and that emphasize engineering judgment and the integration of disciplinary knowledge to address particular problems.

Organization and Use of This Book

The text is organized into four parts. Part 1 includes a brief introductory chapter that lays out the relationships between the things engineers do and their environmental consequences. Environmental impacts are seen to arise from both the design and the deployment of technology. The selection of materials, the design of products and manufacturing processes, and the use of energy in its various forms, are identified as areas where engineers play a key role in influencing environmental outcomes. A life cycle perspective and the principles of industrial ecology and sustainable development are also introduced to lay the groundwork for elaboration in later chapters.

The second chapter gives an overview of current environmental issues, including problems related to atmospheric emissions, water pollution, solid wastes, resource depletion, land use, and ecological impacts. This chapter motivates the need for cleaner technologies and better methods to understand and address environmental concerns. Its coverage is broad, but its primary emphasis is on problems and issues most relevant to engineering design, analysis, and practice. This chapter provides the principal background material for the subsequent chapters. Relevant sections of Chapter 2 can be discussed in class, or assigned to students as background reading on selected topics.

Part 2 of the book is a set of case studies focused on the environmental design of technology. Each chapter begins with an overview of the technology, its societal benefits, and its environmental concerns. For example, Chapter 3, "Automobiles and the Environment," discusses the problems of automotive emissions, energy use, materials consumption, and the disposal of used cars. Science and engineering fundamentals are then introduced and used to explore in greater detail the engineering design variables that can alter environmental outcomes. Armed with these insights, students can propose and analyze alternative technology designs that reduce or eliminate environmental problems. The choice of case studies in Part 2 reflects a spectrum of major environmental concerns, as well as a variety of disciplinary approaches to environmental problem solving. Part 2 concludes with a chapter on life cycle assessments that illustrates how different technologies are linked from an environmental perspective. This chapter reinforces an industrial ecology perspective by looking beyond the immediate boundaries of an engineering design problem to consider overall environmental impacts, including upstream and downstream processes.

The case studies in Part 3 focus on environmental modeling. Again, science and engineering fundamentals are employed to understand and predict how various types of pollutants (such as air emissions from power plants, wastewater discharges from manufacturing processes, or CFCs from household refrigerators) are transported and transformed in the environment. The understanding and insights obtained in each case are applied to identify strategies that engineers and society can adopt to control adverse environmental effects. The topics chosen for Part 3 span a range of local, regional, and global environmental concerns involving all environmental media (air, water, and land). Greater emphasis is given to regional and global issues—such as urban air pollution, bioaccumulative chemicals, and global warming—that are likely to dominate the environmental agenda in coming decades.

Finally, Part 4 addresses selected topics in environmental policy analysis. These topics include engineering economics, cost–benefit analysis, risk assessment, decision analysis, and environmental forecasting. Most engineering students are not usually exposed to these topics, although some universities, like Carnegie Mellon, do incorporate policy-related studies as an option for undergraduate engineering and science students. These subjects are especially important in the context of environmental issues, which are seldom purely technical. Accordingly, the topics introduced in Part 4 provide the basic tools needed to extend the technical analyses of Parts 2 and 3 to also consider the costs, risks, and benefits of environmental control strategies and policy options. Chapter 15, "Environmental Forecasting," also includes introductory treatments of population dynamics, economic development, and technological change as they relate to future environmental quality. Some instructors may wish to introduce one or more of these topics at the outset in order to motivate environmental discussions.

Following the first two chapters, students and instructors should feel free to visit the remaining topics of this book in whatever order they desire. Each chapter was designed as a stand-alone module, relying mainly on Chapter 2 for a background discussion of the environmental concerns and impacts of the topic at hand. Although all of the technology design cases in Part 2 are also related to topics in environmental modeling (Part 3) and policy analysis (Part 4), instructors may wish to select (or vary) the subjects presented. The range of topics covered is sufficiently broad to support an introductory course tailored to the particular needs and interests of faculty and students. Thus a selection of chapters from Parts 2, 3, and 4 can be combined to explore certain topics in depth, or the introductory sections of a chapter can be used to obtain a brief overview of the subject. A more detailed guide for instructors is available that includes further suggestions on the use of this text.

ACKNOWLEDGMENTS

Several Carnegie Mellon colleagues played an especially important role as contributors to this text. Professor Cliff Davidson deserves special recognition as a contributor in this effort. He is particularly acknowledged as the author of Chapters 7 and 10, and principal contributor to Chapter 11. He also has strongly supported our joint efforts to enhance the environmental content of undergraduate education at Carnegie Mellon, especially a recent project supported by the National Science Foundation, which was the genesis for several chapters of this book. Another colleague, Professor Spyros Pandis, contributed his special insights as the principal author of Chapter 8. Professors Dave Dzombak and Dick Luthy (now at Stanford University) lent their expertise as the principal authors of Chapter 9.

Research assistants Patricia Bruno, Miles Levin, Laurie McNair, and Jeff Rosenblum helped to develop much of the raw material for this text. Jeff Rosenblum is also credited for his substantial contribution to the writing of Chapter 4. Many other colleagues provided invaluable feedback and assistance on earlier drafts of individual chapters. They include Alex Farrell, Paul Fischbeck, Scott Farrow, Baruch Fischhoff, Don Hanson, Chris Hendrickson, Arpad Horvath, Milind Kandlikar, Lester Lave, Leonard Levin, Deana Matthews, Scott Matthews, Granger Morgan, Indira Nair, Peter Noymer, James Risby, Mitchell Small, and Ross Strader. In addition, five reviewers selected by McGraw-Hill provided valuable comments and suggestions on the overall manuscript. They were Professors Robin L. Autenrieth (Texas A & M University), Brian A. Dempsey (Penn State University), Mel S. Manalis (University of California, Santa Barbara), John T. Novak (Virginia Tech), and Jae K. Park (University of Wisconsin–Madison).

Special debts of thanks are owed to Gloria Rogulin-Blake, who typed (and retyped) the entire manuscript and gave generously of her time under difficult circumstances; and to Mike Berkenpas, whose computer skills and boundless patience repeatedly turned chaos into finished products. Thanks also go to Megan Davidson for her splendid drawings, and to Terrea Cinkovic, Jill Hatch, Cathy Ribarchak, and Sara Schultzer for their secretarial assistance. Executive Editor Eric Munson and the entire McGraw-Hill production team draw special thanks for making this venture into the world of textbook writing as painless and enjoyable as one could hope for. They are a terrific team. I am especially grateful to New York's top graphic designer, Denise Rubin, for contributing her cover design for the book. The unfailing support of my wife, Maria, and my daughters Lisa and Denise, plus the encouragement of close friends and family members, is what truly brought this effort to fruition. The Marsha and Philip A. Dowd Fellowship Award from Carnegie Mellon University made the final stretch all the more rewarding.

Finally, to the many students who provided feedback and comments on earlier drafts of this text, I have listened to you more carefully than anyone. The message that rang especially clear was, "write this book for *us* so *we* can understand it, not just our professors." Together with my colleagues, I invite all readers to let us know how well we did, what you like, and what still needs to be fixed. Your suggestions on other topics of interest are also most welcomed.

ESR
Pittsburgh, PA
June 2000

ABOUT THE AUTHOR
AND CONTRIBUTORS

Edward S. Rubin is the Alumni Professor of Environmental Engineering and Science at Carnegie Mellon University, holding joint appointments in the departments of Engineering & Public Policy and Mechanical Engineering. He is also director of the Center for Energy and Environmental Studies, and was founding director of CMU's Environmental Institute. His teaching and research activities are in the areas of energy systems, environmental control, and the interactions of technology and policy, and he has over 200 publications in these areas. Dr. Rubin is a Fellow of ASME and past chairman of its Environmental Control Division. He has served as a consultant to industry and adviser to governmental organizations in the United States and abroad, including the U.S. Environmental Protection Agency, the U.S. Department of Energy, the National Academies of Science and Engineering, and the International Organization for Economic Cooperation and Development. He currently serves on the Board on Energy and Environmental Systems of the National Research Council, and is a recent recipient of the Lyman A. Ripperton Award of the Air & Waste Management Association for distinguished achievements as an educator.

Cliff I. Davidson is a professor in the departments of Civil & Environmental Engineering and Engineering & Public Policy at Carnegie Mellon. He is also the director of the Environmental Institute. He has taught and conducted research in the environmental field for the past 23 years, and has published over 100 technical papers and several books. He has served on committees of the National Academy of Sciences, the U.S. Environmental Protection Agency, the Department of Energy, the National Science Foundation, and other agencies. He is active in several professional societies, and he recently completed a term as president of the American Association for Aerosol Research.

David A. Dzombak is a professor of Civil and Environmental Engineering at Carnegie Mellon University, and ***Richard G. Luthy*** is the Silas H. Palmer Professor of Civil and Environmental Engineering at Stanford University. Their teaching and research activities are focused on water quality engineering, including problems related to soils, sediments, and the fate and transport of organic compounds in aquatic systems. Recent efforts include a collaborative study of PCB releases from contaminated river sediments, the subject of Chapter 9. Dr. Dzombak and Dr. Luthy both hold the position of Diplomate of the American Academy of Environmental Engineering. Dr. Luthy also is a member of National Academy of Engineering.

Spyros N. Pandis is the Gerard G. Elia Professor of Engineering at Carnegie Mellon University, holding joint appointments in the departments of Chemical Engineering and Engineering & Public Policy. His primary teaching and research interests include the study of atmospheric chemistry as it relates to photochemical smog, acid rain, and global climate change. He is the author of numerous technical publications as well as a leading textbook on air pollution and atmospheric science.

MOTIVATION AND FRAMEWORK

Engineering and the Environment

1.1 INTRODUCTION

Traditionally, the study of engineering has divided along disciplinary lines, most commonly electrical, chemical, civil, and mechanical engineering. From time to time, more specialized topics of popular interest have emerged, giving rise to new fields such as aerospace engineering (an outgrowth of mechanical engineering, popular in the 1960s) and computer engineering (an outgrowth of electrical engineering, popular in the 1980s and 1990s). Environmental engineering—the study of engineering solutions to environmental problems—is also widely seen as a specialized area gaining in popularity.

Historically, the field has been largely identified with civil engineering, where environmental studies have focused primarily on the problems of water pollution and solid waste disposal. Today, however, environmental problems extend far beyond the traditional domains of civil engineering alone. Indeed, the premise of this book is that solutions to modern environmental problems are part of *every* engineering discipline. No engineer can think that environmental problems will be solved by "someone else." As we shall see, nearly everything we do as engineers can have an impact on the environment—impacts that can be part of the solution or part of the problem.

This chapter highlights some of the ways in which engineers of all stripes are involved in both creating and solving environmental problems. Although you may not think of yourself as an environmental engineer, you may be surprised to see the many links between engineering and the environment, no matter what your field of interest.

1.2 WHAT IS "THE ENVIRONMENT"?

Most people care about the environment. But what exactly does that mean? When we talk about "environmental problems" that need to be addressed, what do we have in mind?

The "environment" can mean different things to different people. The unabridged Random House dictionary defines *environment* as "the aggregate of surrounding things, conditions or influences, especially as affecting the existence or development of someone or something." What we have in mind when talking about the environment can make a big difference in the way we approach the subject and in how we formulate or frame environmental issues and problems.

In this book the term *environment* will generally refer to the physical environment that surrounds us. This includes the air we breathe, the water we drink, and the lands, oceans, rivers, and forests that cover the earth. To an increasing extent it also includes the buildings, highways, and modern infrastructure of the urban settings in which a growing proportion of the world's population resides. The state of this physical environment directly and indirectly affects the viability of all living things on the planet—the people, plants, birds, fish, and other animals that we care about. The welfare of these living things motivates most environmental concerns.

This chapter provides a brief introduction to some of the major environmental themes that are discussed later in the text. These themes emphasize areas where engineers play a major role in solving (as well as creating) environmental problems.

For perspective, we begin with a broad overview of environmental impacts and the role of engineering.

1.3   FRAMING ENVIRONMENTAL ISSUES

We know from the study of geology, biology, and other natural sciences that the earth's environment has been changing since time began. What motivates modern concerns about the environment is the expanding role that human activities play in accelerating environmental change.

Figure 1.1 illustrates schematically how human activity influences the environment. At the most basic level, human demands for food and shelter mean that some living things (plants and animals) will be killed or harvested for food, and some natural resources, such as trees, will be used to build structures and provide energy for shelter, cooking, and warmth. These basic demands of human survival thus give rise to an environmental impact. As human populations increase, more and more land is altered to provide settlements and to support activities such as agriculture, industrial processes, and transportation systems. And as people become more affluent, their demand on the earth's resources grows far beyond the basic needs for survival.

In addition to altering the landscape, diverse human activities give rise to various types of waste emissions or *residuals* discarded to the environment. Modern examples include air pollutants from factories and automobiles, water contaminants from manufacturing processes, solid wastes from household and municipal activities, and pesticides used in agriculture.

Many of the wastes or pollutants discharged to the environment are subsequently transported and chemically or biologically transformed over time and dis-

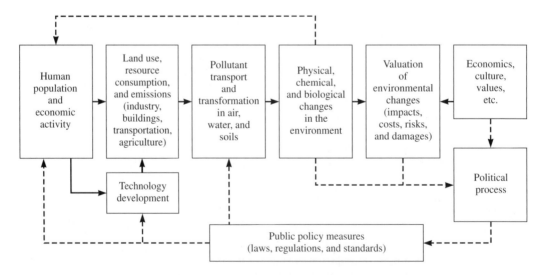

Figure 1.1   A framework for environmental impact assessments. Solid lines show the path of primary or initial impacts; dashed lines show the major feedbacks and responses to these impacts.

tance. For example, gaseous sulfur dioxide emitted from the tall chimneys of coal-burning power plants is partially converted to fine sulfate particles by chemical processes in the atmosphere. This mix of sulfurous gases and particles leads to *acid rain,* which can damage lakes and forests hundreds of miles away. Similarly, many types of organic wastes from factories and households are discharged to rivers and streams, where they are transformed by biological reactions that deplete the oxygen dissolved in water. A deficiency of oxygen directly affects the viability of all aquatic species. Other examples of physical, chemical, and biological changes in the environment due to human activities are shown in Table 1.1.

Historically, the most pervasive environmental changes caused by people have been related to land use—particularly the clearing of forests for agriculture and urbanization. Now, in the modern industrial era, the added impacts of emissions from modern industrial technology are rapidly accelerating the process of human-induced environmental change.

1.3.1   Good Change or Bad?

Are all changes in the environment a reason for concern? Isn't the environment always changing, and aren't many of these changes beneficial?

The answers to such questions are not always simple. In general, our greatest concern is over environmental changes that may harm us or affect our welfare. This is illustrated by the feedback loop in Figure 1.1 between environmental change and human activities. This interaction is complex because some environmental changes that are beneficial in the short run may have adverse consequences later on. In preindustrial

Table 1.1   Some examples of environmental change from human activities.

Human Activity	Physical Changes	Chemical Changes	Biological Changes
Land and water use for housing, agriculture, industry, transportation, and recreation	Deforestation and other alterations of landscapes (e.g., changes in terrain slope, vegetation coverage, pavement); alteration of waterways (e.g., flooding, dams, changes in river channels, drainage of wetlands)	Changes to chemical constituents of soils and sediments (e.g., increased acidity and turbidity of waters, removal of nutrients from soils)	Changes in the viability of plants, fish, animals, and microorganisms due to altered habitat and chemical constituents or concentrations, possibly leading to species succession, extinction, migration, or disease
Emissions or discharge of chemical substances to air, land, and water	Changes to the built environment (structures such as buildings, bridges, monuments, etc.) from deposition and chemical attack caused by emissions such as soot deposits, acid gases, and liquid chemicals	Increases in the concentration of emitted substances in the air, water, and soil; other chemical changes resulting from secondary reactions (e.g., ozone buildup in urban areas)	Injury or illness to people, plants, and animals from exposure to and/or accumulation of chemicals and their derivatives

Land use impacts of human activities. This countryside in southern California originally was transformed for agriculture. Now tree groves have been uprooted to make way for construction of roads and expansion of human settlements.

societies, for instance, the clear-cutting of forests to support agriculture was essential for providing food for a growing population. Over time, however, the continued depletion of forests often led to soil erosion, loss of soil nutrients, and a subsequent inability to sustain agricultural production. As a result, many communities disappeared or were forced to migrate elsewhere.

In today's industrial society, many air pollutants and water contaminants entering the environment as byproducts of modern technology are widely recognized sources of human illness and ecological damage. More subtle or indirect impacts, such as the effects of long-term global warming from anthropogenic emissions of carbon dioxide and other gases, are less clearly defined at the moment but raise new concerns about the longer-term effects of energy use and industrial activities.

Of course, not all people agree on whether a particular environmental change is cause for concern. For example, the clearing of wooded lands to provide space for a new farm, factory, shopping mall, or housing development is commonly seen as a beneficial change that allows people to lead better, more productive lives. Others, however, may view these same changes as destructive of natural habitats for vegetation and wildlife that have intrinsic value and are part of natural ecosystems that affect our long-term welfare and survival. So who's right? And if both views have merit, how do we weigh the tradeoffs and make a decision?

1.3.2   Enter Public Policy

In democratic societies, the political process is how decisions are made as to whether actual or potential changes in the environment are of sufficient concern to adopt *policy measures* (e.g., laws and regulations) to prevent, alter, or reverse these changes. Some notion of the *damages* or *risks* that will be avoided by taking action is one essential element of the decision-making process. Providing such information is an area where science and engineering often play a key role. Ultimately, a host of other factors also influence community and national decisions about environmental policies. As shown in Figure 1.1, such factors include the influences of culture, societal values, and economics.

Because of these factors, priorities and preferences for environmental protection often vary from community to community and from country to country. For instance, a nation struggling to provide its citizens with the basic necessities of life is unlikely to be as worried about wilderness preservation as a wealthy nation. Nonetheless, over time societies tend to address environmental problems in an order roughly equivalent to the relative risks they pose. Often this starts with basic health issues like water quality and sanitation, then broadens (often in tandem with economic development) to include other issues of health and environmental quality.

Environmental policies also are often based on concepts of fairness, or equity, such as the idea that all citizens have a right to breathe clean air. In other cases, the symbolism or basic ethic of environmental protection is most important, as with the protection of endangered species. Because environmental policy often has significant economic implications, it is almost always influenced by private interests as well as the public interest.

At all stages in the policy development process, one important approach is to look for policies whose benefits clearly outweigh the cost of measures taken. An economic analysis of costs and benefits is sometimes employed to evaluate the merits of proposed policy measures. In many cases, however, putting a dollar value on expected environmental and health benefits is especially difficult and controversial. A less controversial approach is to identify measures that are lowest in cost. The adoption of such measures often serves as an initial basis for policy actions.

If environmental protection measures are adopted, their influence might be felt in a number of ways. For example, some measures might alter human behavior by forbidding certain types of activities, such as the dumping of wastes or drilling for oil in pristine areas. Other measures might require the use or development of new technology to abate harmful emissions to the environment. Thus, new cars might have to be equipped with catalytic converters to reduce air pollution and cleaner technologies may be demanded to generate electricity. To an increasing extent, environmental protection measures in the form of standards and regulations have established new constraints on the design and deployment of modern technology. Environmental policies thus shape the development of technology in directions that reflect the goals and preferences of society.

1.4 THE ROLE OF ENGINEERING

Where do engineers fit into all of this? Figure 1.2 suggests how pieces of the environmental puzzle map roughly into the traditional bailiwicks of undergraduate majors or disciplines. Engineers are primarily involved in problems related to technology development and deployment. These include designing, developing, and building the cars, computers, television sets, and other consumer products that people enjoy. Engineers also design and build all the manufacturing processes, industrial technology, and transportation infrastructure needed to extract, transport, and refine raw materials; fabricate products; and distribute the goods and services of

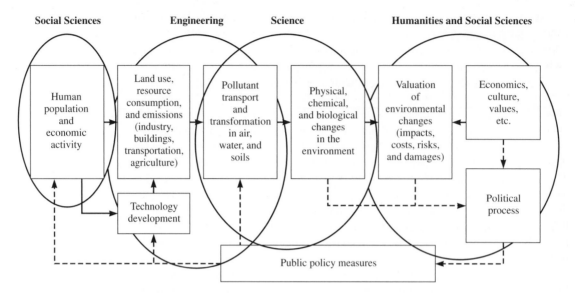

Figure 1.2   Mapping of environmental topics into traditional undergraduate disciplines.

modern societies worldwide. In order to predict the consequences of technology deployment, engineers, along with scientists, also are involved in the study of how pollutants are transported and transformed in the environment. Thus, in the broadest sense, engineers are concerned with—and often responsible for—a wide range of activities that directly or indirectly contribute to environmental change.

From this discussion we see that the sources of anthropogenic environmental change fall into two broad categories: (1) *changes associated with land use* (including depletion of natural resources) and (2) *changes induced by emissions* or residues from products and industrial processes. Engineers have responsibilities for both types of impacts, especially the latter. For example, chemical and mechanical engineers design the petrochemical refineries, paper mills, power plants, and myriad other processes at the heart of industrial societies. The design and operation of these processes directly determine how much air pollution, water pollution, and solid wastes are produced and released to the environment. The same is true for the water treatment plants designed by civil engineers, the microchips developed by electrical and computer engineers, and the steel mills, aluminum smelters, and extraction processes developed by metallurgical engineers.

In addition, all of the consumer products we use (and eventually discard)—such as batteries, toasters, light bulbs, refrigerators, computers, and automobiles—are a result of engineering design decisions that impact the environment. For example, the engineering design of an automobile, including the choice of materials, the type of engine or power plant, and the manner of construction, determines the type and level of air pollutants that are emitted over the life of the car, the amount of fuel required, and the type and quantity of solid wastes that must be disposed of at the end of the car's useful life. Thus, the way engineers design products and processes plays a big role in creating as well as solving environmental problems.

What about other types of environmental concerns, such as species extinction from the clearing of tropical forests, drilling for oil in the Arctic, or the ecosystem impacts of wetlands development? Do engineers also have a professional role in these types of issues?

Yes, they do. These are examples of environmental changes associated with land use. Here engineers also play a key role, but often less directly. The environmental impact of land use decisions is a broader societal issue that involves not only engineers, but a host of other professional and political interests as well. Thus, while the technical design of an automobile, a power plant, or a steel mill is primarily an engineering responsibility with direct consequences for emissions to the environment, decisions about *deploying* that technology—that is, *whether* to use it, or *where* to put it, or *how many* to deploy—typically extend beyond the realm of engineering practice. Often those decisions fall in the domains of city and regional planners, government regulatory agencies, and other groups responsible for review and approval of land use proposals from developers, landowners, corporations, and others. Thus, although engineers frequently play a central role in formulating recommendations with respect to land use—such as where to locate a highway, build a dam, or deploy an oil rig—the process of arriving at a final decision transcends engineering. Throughout this process, however, engineers are intimately involved in defining, collecting, and interpreting the data needed to assess the environmental implications of major land use decisions.[1]

1.5 APPROACHES TO "GREEN" ENGINEERING

Think about something you've thrown away recently. Maybe it was something small like the alkaline batteries from your CD player, or something big like an old refrigerator or an old used car that finally died. Now imagine that you were the engineer responsible for designing and manufacturing that product. The laws have suddenly changed, and now you must take back all of the discarded product and be responsible for its environmentally safe disposal. If you had known this in advance, would you have designed and built the product in the same way?

This scenario is not at all far-fetched. In Germany, manufacturers like IBM are now required to take back all of their discarded computers, which contain quantities of toxic metals used to manufacture circuit boards and other electronic components. The cost of dealing with this enormous problem has caused IBM to completely rethink the way it designs, manufactures, and ships its computers. The result has been a substantial reduction in environmental impacts throughout the product life cycle, without compromising the product's performance or adding to its cost. Indeed, many changes have *saved* money.

This is but one example of how the philosophies of green design, pollution prevention, and industrial ecology are changing the way engineers do their jobs. Here

1 Note that the term *land use* as used here should be thought of broadly to include all of the earth's surface, including oceans as well as land. Indeed, as we enter the 21st century, our sense of the natural environment may expand to include outer space. Over the past 30 years, for example, an increasing array of high-tech debris from satellite launches and other space program ventures has been left in orbit as "space junk." Might space pollution be the next environmental frontier?

we give a brief introduction to these concepts. Chapter 7 goes into greater detail, including several examples of green design applications.

1.5.1   Sources of Environmental Impacts

Over the years, the authors of this text have been asking freshman engineering students to define what engineers do. Invariably, one of the first answers is that engineers "solve problems." True. But doctors, lawyers, mathematicians, and many other professionals also solve problems. So what is it that makes engineering distinctive? Upon further reflection, students conclude that a unique characteristic is that engineers "build things" and "make stuff" that serves society.

Let us use this definition to refine our thinking about the types and sources of environmental impacts that engineers most directly influence. As discussed in Chapter 2, many environmental concerns are related to atmospheric emissions, water pollution, solid wastes, and natural resource depletion. If we step back and ask about the *sources* of these impacts as they relate to the engineers who design, analyze, and "build things," three major categories emerge: materials selection, manufacturing processes, and energy use.

Materials Selection   Anything that engineers design and build—be it a refrigerator, car, chemical plant, or circuit board—has to be made out of something, and the choice of that "something" will directly affect the environment. After all, the material has to come from somewhere, and eventually it all tracks back to the environment.

Not only the choices of materials, but also the quantities needed, are important variables that engineers can influence. Thus, knowing something about how the use of different materials may affect environmental quality is important for all engineering disciplines. Two key questions to keep in mind when selecting materials are "Can I use alternative materials that are environmentally preferable?" and "Can I use less material without compromising function or reliability?"

Manufacturing Processes   This topic refers to the methods that engineers devise to turn raw materials into finished materials and products. It starts with the raw materials dug out of the earth's crust, and continues through the stages of refining, transport, transformation, and assembly into final products. In most cases, every step along this chain releases waste materials to the environment in the form of air pollutants, water pollutants, and solid wastes. Historically, much of what went on in the field of environmental engineering involved developing additional technology and methods for cleaning up the problems created by primary manufacturing and processing technologies.

Many engineering students are not exposed to manufacturing methods because college curricula have become increasingly devoted to the fundamentals of analysis and design. But if you get to actually build something in a project course or design class, stop and think about how the material you're working with was produced, how it got its shape, and what environmental impacts resulted from those processes. The problems at the end of this chapter will introduce you to some of these processes.

Energy Use   This source of environmental impacts is perhaps the most pervasive and most important of any that engineers deal with. Energy is vital for life and for an economy, and the quantities and types of energy that a society uses directly affect environmental quality.

Energy use encompasses everything from the heating and cooling of homes and buildings, to the electricity that runs modern computers and appliances, to the gasoline and other fuels that power our transportation system. The manufacturing processes discussed earlier all require energy to perform their tasks. Most of the world's energy today comes from fossil fuels—oil, coal, and natural gas—and when these fuels are burned or converted to electricity the environment is affected, as you will see in Chapter 2. Nuclear power leaves a different kind of environmental legacy, and even renewable energy sources like hydroelectric, biomass, solar energy, and wind power are not without their adverse environmental consequences.

A good rule of thumb is that any engineering improvement that reduces the energy required for a particular service will be beneficial for the environment. Thus a desktop computer designed to use 400 watts of power instead of 800 watts, or a car that gets 30 miles per gallon (mpg) instead of 20 mpg, will be environmentally friendlier. Switching from one energy form to another—e.g., from gasoline-powered cars to electric-powered cars—is a lot trickier to deal with. As we shall see in later chapters, the environmental consequences may involve trade-offs that are difficult to evaluate.

1.5.2   A Life Cycle Perspective

An *environmental life cycle assessment (LCA)* provides the "big picture" of how engineering decisions in any particular area affect the environment. Figure 1.3 illustrates the connections among the stages commonly involved in building and creating goods to serve society. Environmental impacts occur at each stage as a result of materials consumption and transformation, and the use of energy.

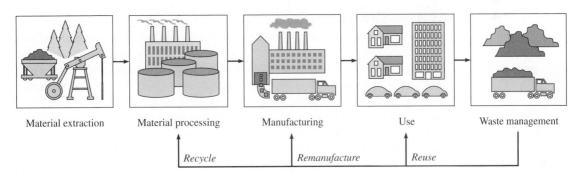

Figure 1.3   Stages of a product life cycle. Environmental impacts occur at all stages. These impacts can be reduced by engineering designs that change the type and/or amount of materials used in a product; by creating more efficient extraction and manufacturing processes; and by improving the recovery and reuse of materials and energy at the end of the product life. (*Source:* Based on OTA, 1992)

Traditionally, most engineers focus on only small pieces of this overall system. But this is starting to change. Environmental awareness has made it clear that all stages of a product's life cycle must be considered in finding ways to reduce environmental impacts. Improvements result from creating cleaner, more efficient manufacturing operations; from reducing the energy and materials needed for use of a product; and from improving the recovery of energy and materials during waste management. Life cycle assessments are thus an important tool in implementing the concepts of green design, pollution prevention, and waste minimization that are becoming fundamental paradigms of good engineering practice.

1.5.3   Industrial Ecology and Sustainable Development

The perspectives of a life cycle assessment provide the foundation for a more comprehensive view of design for the environment that has come to be known as *industrial ecology*. This term has gained popularity in recent years as an embodiment of the environment principles that should be reflected in engineering design. As defined by Graedel and Allenby (1995),

> Industrial ecology is the means by which humanity can deliberately and rationally approach and maintain a desirable carrying capacity, given continued economic, cultural, and technological evolution. The concept requires that an industrial system be viewed not in isolation from its surrounding systems, but in concert with them. It is a systems view in which one seeks to optimize the total materials cycle from virgin material, to finished material, to component, to product, to obsolete product, and to ultimate disposal. Factors to be optimized include resources, energy, and capital.

This view is very much in keeping with the concept of *sustainable development,* which also has been widely popularized in recent years. As originally defined by the World Commission on Environment and Development (WCED, 1987),

> Sustainable development is development that meets the needs of the present without compromising the ability of future generations to meet their own needs.

This statement reflects a strong moral and ethical position about our own responsibility to future generations. But how does one operationalize this goal? Allenby (1999) suggests that the application of industrial ecology principles offers a means by which sustainable development can be approached and maintained. In more practical terms, industrial ecology is based on the concept that natural systems tend to recirculate and reuse materials, thus eliminating or minimizing the production of wastes and the use of energy. This offers a lesson and a path for the development of industrial systems and technology. Another embodiment of industrial ecology principles was offered in a recent National Research Council study (NRC, 1996) that suggests that industrial ecology should include (1) circulating and reusing material flows within the system; (2) reducing the amount of materials used in products to achieve a particular function; (3) protecting living organisms by minimizing or eliminating the flow of harmful substances; and (4) minimizing the use of energy and the flow of waste heat back to the environment.

More detailed discussions of these concepts and their implications may be found in the references cited and in other current literature and technical journals, such as the *Journal of Industrial Ecology.* Later chapters of this text also elaborate on these principles in the context of specific environmental issues. But what engineering analysis tools are needed to implement these concepts? We turn next to a brief review of some of these fundamentals.

1.6 BASIC ENGINEERING PRINCIPLES

So far, we have spoken only in general terms about the relationships between engineering design decisions and their environmental consequences. But engineering design and analysis invariably involve the application of quantitative methods based on fundamental principles and "laws" that are commonly taught in undergraduate science and engineering courses. These fundamentals also apply to the study of environmental impacts and the concepts of green design and industrial ecology. Throughout this book, relevant engineering principles are introduced wherever necessary to address a particular environmental issue or problem. Here we introduce two basic principles that apply to nearly all types of environmental problem solving: conservation of mass and conservation of energy.

1.6.1 Conservation of Mass

Simply put, the law of mass conservation states that mass can be neither created nor destroyed. It is called a "law" because it has been found experimentally to always be true. This relationship can be expressed in equation form as follows:

$$\text{Rate of creation of mass} = 0 \qquad (1.1)$$

This equation is especially important for environmental studies and industrial ecology applications because it provides the basis for a quantitative accounting of "where stuff goes." A common application of the mass conservation principle is a *mass balance* for a specific technology or environmental system defined by an explicit boundary or *control volume* like the one sketched in Figure 1.4.

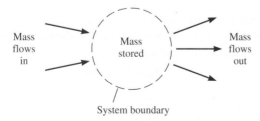

Figure 1.4 Mass flows into and out of a system. The dashed line defines the system boundary.

In general, mass can flow into and out of the system via any number of streams. Mass also can be stored (accumulate) within the system or be removed from storage (mathematically equivalent to a negative accumulation). According to Equation (1.1), the total mass within the system must remain constant. Thus

$$(\text{Total mass flow in}) = (\text{Total mass flow out}) + (\text{Change in mass stored}) \quad (1.2)$$

This equation is a common expression of the mass conservation principle. It also applies to situations where mass flows continuously across the system boundaries. In this case, all quantities are expressed on a rate basis:

$$\begin{pmatrix} \text{Total mass} \\ \text{flow rate in} \end{pmatrix} = \begin{pmatrix} \text{Total mass} \\ \text{flow rate out} \end{pmatrix} + \begin{pmatrix} \text{Rate of mass} \\ \text{storage} \end{pmatrix} \quad (1.3)$$

A special case is the *steady state, steady flow* situation where there is no change in the energy storage term. In this case we simply have

$$(\text{Total mass flow rate in}) = (\text{Total mass flow rate out}) \quad (1.4)$$

Different forms of the mass balance equation are widely used to analyze the behavior of environmental systems and the environmental implications of technology design. A simple example is presented next.

Example 1.1

An application of the mass conservation principle.   An industrial plant discharges 100 kg/day of liquids into a disposal pond. Measurements show that 1 kg/day seeps out of the bottom of the pond into the ground and 2 kg/day evaporates into the air. What is the rate of mass accumulation in the pond?

Solution:

We define the system boundary around the pond. Thus, there is one mass flow rate into the system and two flow rates out of the system. Applying Equation (1.3) we have

$$\text{Total mass flow rate in} = \text{Total mass flow rate out} + \text{Rate of mass storage}$$

$$100 \text{ kg/d} = (1 \text{ kg/d} + 2 \text{ kg/d}) + (\text{Rate of mass storage})$$

$$\text{Rate of mass storage} = 100 \text{ kg/d} - 3 \text{ kg/d} = 97 \text{ kg/d}$$

In this case the two mass flows out of the pond are potentially an environmental concern.

Often in engineering problems a mass (or mass flow rate) is calculated from other variables and properties of matter, such as the density of a material (mass per unit volume). For example, we may write

$$\text{mass} = \text{density} \times \text{volume}$$

or in equation form

$$m = \rho V$$

Although the specific form of the mass conservation equation can vary from one application to another, its basic purpose remains the same: It provides the basis for quantifying the flows of materials and the fate of substances in the environment. This includes situations involving chemical reactions, which occur frequently in nature and in engineered systems. The concept of "balancing" a chemical reaction equation is simply another expression of mass conservation, where the mass of each chemical element is the basic quantity that must be conserved.

Many environmental pollutants are the product of chemical reactions; thus environmental applications of the mass balance principle often involve some chemistry. One of the most pervasive and important types of chemical reaction is the combustion of fuels, such as the gasoline burned in automobile engines or the coal and natural gas burned at electric power plants. Because most of the world's energy is provided by fuel combustion, tracking the flow of energy is another important element of environmental analysis.

1.6.2 Conservation of Energy

Analogous to the law of mass conservation, a second fundamental tenet of engineering analysis is that energy cannot be created or destroyed, but merely transformed from one form to another. Thus the basic principle of energy conservation can be written as follows:

$$\text{Rate of creation of energy} = 0 \qquad (1.5)$$

This equation (or any equivalent expression) is also known as the *First Law of Thermodynamics*. When applied to a well-defined system or control volume, such as the one sketched in Figure 1.5, the governing equation becomes

$$\text{(Total energy flow in)} = \text{(Total energy flow out)} + \text{(Change in energy stored)} \quad (1.6)$$

Although this equation may seem simple, its application to engineering and environmental problems is often complex because of the many different forms of energy that may occur. Later chapters discuss some of the different energy forms relevant to environmental analysis. A key point to emphasize here is that energy flows are directly coupled to mass flows in many types of environmental problems. This is because energy is stored in mass in a variety of ways, kinetic and potential energy being perhaps the most commonly known. So fulfilling requirements for energy usually requires flows of mass—often with environmental consequences.

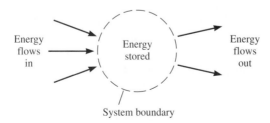

Figure 1.5   Energy flows into and out of a system. The dashed line defines the system boundary.

Example 1.2

An application of the energy conservation principle.   Over a year an electric power plant generated 30 million kilowatt-hours (kW-hr) of useful electrical energy by burning a fuel. The total fuel energy input to the plant was 100 million kW-hr. How much energy was released to the environment in the form of waste heat, assuming no change in energy stored?

Solution:

We define the system boundary around the power plant. Since there is no net storage of energy within the power plant, Equation (1.6) simplifies to

$$\text{Total energy flow in} = \text{Total energy flow out}$$

In this case there is one principal energy flow in (that of the fuel) and two types of energy flows out of the system: (1) the electrical energy generated and (2) the thermal energy released to the environment as waste heat. Thus

$$\text{Fuel energy in} = (\text{Electrical energy out}) + (\text{Heat to environment})$$

$$\text{Heat to environment} = (100 - 30) \times 10^6 \text{ kW-hr} = 70 \text{ million kW-hr}$$

Even simple examples like this can reveal important insights about the environmental impacts of engineering design. For instance, the waste heat in this example amounted to 70 percent of the fuel energy input. An industrial ecology approach would seek ways to improve the power plant design so as to reduce or utilize the waste heat released to the environment.

1.6.3   The Use of Mathematical Models

Engineering analysis relies heavily on the use of mathematical models to make predictions about environmental impacts and to find solutions to environmental problems. Mathematical models consist of one or more equations that describe or approximate the behavior of a system. Many of the equations and formulas you learned in elementary physics are examples of mathematical models.

Each chapter of this text uses some type of mathematical model to address an environmental issue. Many of these models reflect the principles of mass and energy conservation just discussed. Others are based on empirical and/or theoretical approximations to the behavior of natural and engineered systems. If you are not already familiar with the concepts and use of mathematical models in science and engineering, additional readings such as Miller (1996) might be helpful to gain an appreciation of their use and limitations in environmental problem solving.

The mathematical symbols used in each chapter refelct the prevailing conventions in each subject area wherever possible. Thus, a given symbol occasionally may have a different meaning in different chapters. With respect to units, this text relies mainly on the international system of metric (SI) units found in most engineering textbooks. However, readers also will find English (or customary U.S.) units used for specific problems or data sources where such units are commonly employed, especially in the United States. The appendix provides a list of conversion factors relating the different unit systems.

1.7 WHAT LIES AHEAD

For students who wish to learn more about the nature of environmental threats and the status of programs to deal with them, Chapter 2 provides a summary of the major environmental issues facing us today. These are the problems that motivate the remaining chapters of this book.

Chapters 3–12 explore the relationships between engineering practice and its impact on environmental emissions and environmental quality. A series of case studies is used to illustrate applications to problems involving a variety of engineering disciplines. These case studies can be used independently as stand-alone modules, or they can be linked together to provide a more complete picture of engineering and the environment.

The case studies are organized into two general areas. The first set of cases (Chapters 3–7) deals with technology design and its impact on environmental emissions and resource requirements. Commonplace technologies like batteries, power plants, refrigerators, and automobiles are examined with regard to their environmental impacts. Basic science and engineering skills are applied to show how engineers can reduce or eliminate environmental impacts through improved design, including a life cycle analysis perspective.

The second set of case studies (Chapters 8–12) deals with topics in environmental modeling. Here the objective is to understand how pollutants released by modern technology are transported and transformed in the environment. Such understanding is critical to devising effective strategies to avoid harmful effects and achieve a cleaner, sustainable environment.

Finally, for students (and faculty) who wish to explore the realm of engineering and environmental policy, Chapters 13–15 provide an introduction to topics such as engineering economics, cost–benefit analysis, risk analysis, decision analysis, and modeling methods for looking into the future (including models for population growth, economic growth, and technological change). Such methods are an integral part of modern environmental policy analysis. In all chapters, problem sets and numerical examples are heavily emphasized to provide a hands-on experience.

1.8 REFERENCES

Allenby, B.R., 1999. *Industrial Ecology: Policy Framework and Implementation.* Prentice-Hall, Upper Saddle River, NJ.

Graedel, T.E., and B.R. Allenby, 1995. *Industrial Ecology.* Prentice-Hall, Upper Saddle River, NJ.

Miller, G.T., 1996. *Living in the Environment,* 9th ed. Wadsworth, Belmont, CA.

NRC, 1996. *Linking Science and Technology to Society's Environmental Goals.* National Research Council, National Academy Press, Washington, DC.

OTA, 1992. *Green Products by Design: Choices for a Cleaner Environment.* Office of Technology Assessment, U.S. Congress, Washington, DC.

WCED, 1987. *Our Common Future.* World Commission on Environment and Development. Oxford University Press, Oxford, England.

1.9   PROBLEMS

1.1   Figure 1.1 shows a dashed line (feedback loop) connecting "changes in the environment" and "human population and economic activity." Give three specific examples that illustrate this type of interaction.

1.2   With reference to Figure 1.1, give a specific example of how public policy measures can directly affect (a) human population/economic activity, (b) technology development, and (c) pollutant transport/transformation.

1.3   Identify one major product or process associated with the field of engineering you are studying or interested in. Discuss the types of environmental impacts that might arise from either the manufacturing or use of that technology.

1.4   Contact the headquarters of one of the *Fortune 500* companies and ask what measures they have taken over the past decade to reduce the environmental impacts of their activities. Prepare a brief report on your findings.

1.5   How do the concepts of green design, industrial ecology, and sustainable development differ from past approaches to engineering design? Prepare a brief report on this topic, using some specific examples for illustration.

1.6   A field test program is established to measure the environmental releases of mercury from an industrial facility operating at steady state, steady flow conditions. The measurements show releases of 14.3 grams/hour (g/h) to the atmosphere, 0.88 g/h to the wastewater discharge, and 3.19 g/h in the solids leaving the plant. Simultaneous measurements of the mass inputs to the process show a total mass flow rate of 73,700 kg/h of raw materials, of which 0.00002% by weight is mercury. Comment on the accuracy or reliability of the reported mercury test data, and explain the basis for your conclusions.

1.7   Identify three examples of energy use in your daily life. For each case, sketch the system or device that supplies that energy (as well as you understand it). Then identify any mass flows associated with each energy stream. Explain why or how these mass flows might be of environmental concern.

chapter
2

Overview of Environmental Issues

2.1   INTRODUCTION

Many people today take for granted the essentials of life. But stop and think for a moment about your most basic needs. Without air you would not survive for more than a few minutes. Without water, only a few weeks. And without food, your survival would be measured in months. All of these basic needs are supplied by the environment. And if the air, land, or water that sustains you became badly polluted, your health or survival could be jeopardized.

In modern industrial societies most people no longer worry about smoke-laden air, contaminated food, or waterborne diseases. Such life-threatening concerns of the past have been largely eliminated over the last century as economic development has been accompanied by major improvements in public health and a growing concern for the environment. Such concern has been particularly strong in the final decades of the 20th century. In the United States, the number of major federal laws related to environmental protection has skyrocketed since 1970 (see Figure 2.1), spawning a host of environmental regulatory programs and technology innovations to address environmental problems. In many developing and newly industrialized countries, however, severe environmental problems such as air pollution and water pollution remain to be solved. Indeed, on a global basis, waterborne diseases remains an enormous problem. But even as these traditional problems are tackled, new concerns continue to arise, posing new challenges for our technology-intensive societies.

This chapter presents an overview of the major environmental issues and concerns associated with modern technology. Brief summaries of key environmental regulatory actions and trends in the United States also are included to provide context. For convenience, the chapter is organized along environmental media—air,

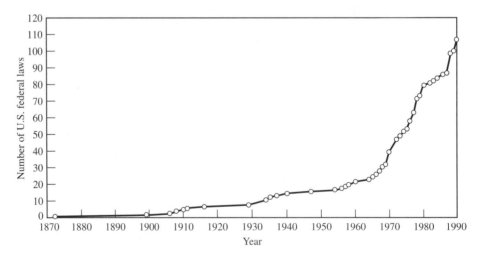

Figure 2.1   The growing number of U.S. laws on environmental protection. The increase in federal legislation that began around 1970 reflects the growing national concern about the environment. New environmental requirements have had a profound effect on the design of modern technology. (*Source:* EPRI, 1994)

water, and land. As discussed in Chapter 1, good engineering design must reflect an integrated perspective across all environmental concerns. This chapter provides the basic building blocks for that perspective.

2.2   ENVIRONMENTAL CONCERNS

Before delving into the specifics of environmental issues, let's take a moment to reflect on what we mean by an environmental "impact" or "effect." Things that affect our health tend to be highest on the list. Human health effects of greatest concern are those that can produce severe illness or death, as from drinking contaminated water. At the other end of the spectrum are less severe effects, like the reduced lung capacity or shortness of breath that can occur from exposure to high levels of ozone or carbon monoxide from auto exhaust.

In general, human health effects can be classified as *acute, chronic,* or *carcinogenic.* Acute effects occur when exposure to an environmental pollutant causes an immediate response in the human body. For example, carbon monoxide inhalation causes an acute effect by impairing the delivery of oxygen to cells, which can produce dizziness or shortness of breath. Long-term exposure to certain pollutants can result in chronic health effects; for example, exposure to high levels of particulate air pollution can lead to chronic respiratory ailments and exacerbate the symptoms of asthma. Finally, some pollutants, called carcinogens, initiate changes in cells that can lead to uncontrolled cell growth and division, known as cancer. In recent years, the health risk of exposure to chemical carcinogens has been a dominant concern in the United States and elsewhere.

Other types of environmental concerns can be classified as those related to human *welfare.* This category spans a broad range of impacts, including the effects of pollutants on plants, animals, and materials; aesthetic qualities like good visibility free from air pollution; recreational opportunities such as lakes and rivers clean enough to allow safe swimming; and effects of human activity on ecosystems, biodiversity, and natural resources. In recent years, the term *sustainable development* has emerged as a popular way of expressing concern about the long-term viability of activities that degrade the environment in order to satisfy the immediate needs of economic development but may impair the future well-being of generations to come.

In thinking about environmental impacts, keep in mind that environmental concerns reflect human values and judgments about what is important and what we care about, individually or as a society. The topics discussed in this chapter reflect some of the key issues and concerns in the United States in the late 20th century. Recognizing the breadth of environmental issues, this chapter emphasizes topics that are most closely related to the practice of engineering. More comprehensive discussions of environmental impacts may be found in the references cited at the end of this chapter.

2.3   ATMOSPHERIC EMISSIONS

Many engineered products and processes release emissions to the atmosphere that can directly or indirectly affect human health and welfare. Air quality impacts may range in scope from localized air pollution in a city or community to regional, national, and even global impacts. We begin with a discussion of the major air pollutants known to directly affect human health and welfare.

2.3.1   Criteria Air Pollutants

The first U.S. air pollution statutes were passed by the cities of Chicago and Cincinnati in 1881 to control smoke and soot from furnaces and locomotives. County governments began to pass their own pollution control laws in the early 1880s. Severe air pollution episodes began to call increased attention to this problem. In October 1948 a five-day weather stagnation confined pollutants from heavy industries in the river valley town of Donora, Pennsylvania. Twenty deaths occurred, and thousands of people experienced mild to severe respiratory distress. The event revealed a general lack of knowledge about the causes and effects of air pollution. In 1952 Oregon enacted the first statewide air pollution control legislation. Other states followed, with air pollution statutes generally aimed at smoke and particulate matter. In 1963 Congress passed the Clean Air Act (CAA), calling for studies of air pollution effects and granting responsibility to the individual states for setting and implementing air quality standards.

A landmark in modern environmental legislation was the Clean Air Act of 1970, which replaced the patchwork of different state and local regulations with a set of uniform national air quality standards that were to protect human health "with a reasonable margin of safety." These standards were developed and promulgated by the newly created U.S. Environmental Protection Agency (EPA). The original Clean Air Act of 1963 had launched studies to identify the most prevalent air pollutants and the criteria one might use to establish healthful levels of air quality. The five major pollutants identified in these studies, which became known as *criteria air pollutants,* included particulate matter (PM), sulfur dioxide (SO_2), carbon monoxide (CO), nitrogen dioxide (NO_2), and ground-level ozone (O_3). Later, in 1978, lead (Pb) was added to the list of criteria air pollutants.

Table 2.1 summarizes the national ambient air quality standards (NAAQS) for the six criteria air pollutants. These standards are expressed in terms of a concentration of pollutant per cubic meter of air. Volumetric concentrations are expressed in units of parts per million by volume (ppmv), whereas mass concentrations are expressed primarily in units of micrograms per cubic meter ($\mu g/m^3$). For most of the pollutants there are two air quality standards: a primary standard designed to protect human health, and a secondary standard related to human welfare. Standards that are specified on an hourly or daily basis are designed to protect against acute health effects, while annual average standards are related to chronic health effects. Most of the secondary standards protect against effects on materials or vegetation.

For any region of the country exceeding national standards, the 1970 CAA required each state to develop a plan to attain the standards within five years. These State Implementation Plans (SIPs) imposed emission limits on existing sources of air pollution. In addition, the federal government, through the EPA, set stringent limits on allowable emissions from certain new sources of air pollution, including new automo-

Table 2.1   National ambient air quality standards.

Pollutant	Primary (Health-Related) Standard		Secondary Standard
	Averaging Time	**Concentration[a]**	
CO	8-hour average[b]	9 ppmv (10 mg/m^3)	None
	1-hour average[b]	35 ppmv (40 mg/m^3)	None
Pb	Maximum quarterly average	1.5 μg/m^3	Same
NO$_2$	Annual arithmetic mean	0.053 ppmv (100 μg/m^3)	Same
O$_3$	Maximum daily 1-hour average[b]	0.12 ppmv (235 μg/m^3)	Same
	Maximum daily 8-hour average[c]	0.08 ppmv (157 μg/m^3)	
PM-10	Annual arithmetic mean	50 μg/m^3	Same
	24-hour average[b]	150 μg/m^3	Same
PM-2.5	Annual arithmetic mean[c]	65 μg/m^3	
	24-hour average[c]	15 μg/m^3	
SO$_2$	Annual arithmetic mean	80 μg/m^3 (0.03 ppmv)	3-hour average[b]
	24-hour average[b]	365 μg/m^3 (0.14 ppm)	1300 μg/m^3 (0.50 ppmv)

[a] Parenthetical value is an approximately equivalent concentration.

[b] Not to be exceeded more than once per year.

[c] These standards were promulgated in 1997 but rescinded by a court order in 1999. A final determination remains pending as of mid-2000.

Source: USEPA, 1998a.

biles and major industrial processes like power plants, cement plants, and petroleum refineries. Further congressional amendments to the Clean Air Act in 1977 and 1990 extended the scope and stringency of state and federal emission control requirements.

Figure 2.2 shows the total U.S. population living in areas that exceeded the NAAQS for each of the criteria pollutants in 1997. Despite significant growth in population and economic activity since passage of the 1970 Clean Air Act, nearly the entire country was in compliance with the national standards for SO$_2$ and NO$_2$, while roughly 20 percent of the population (53 million people) lived in areas that exceeded health-related standards. The greatest level of nonattainment is for ground-level ozone, with particulate matter the next most prominent pollutant. New air quality standards adopted by the EPA in 1997 would double the number of people affected.

Table 2.2 shows the magnitude and sources of criteria air pollutant emissions in the United States. The following sections briefly outline the environmental effects of each criteria pollutant, along with the progress in controlling emissions of each substance.

Particulate Matter   *Particulate matter (PM)* refers to a mixture of small solid or liquid particles suspended in air (also referred to as *total suspended particulates* or *TSP*). Some particulates can be seen as dust, smoke, or haze, while the smallest can be identified only with an electron microscope. Technology-related particulate emissions are given off by fuel combustion (such as ash or soot particles from coal and

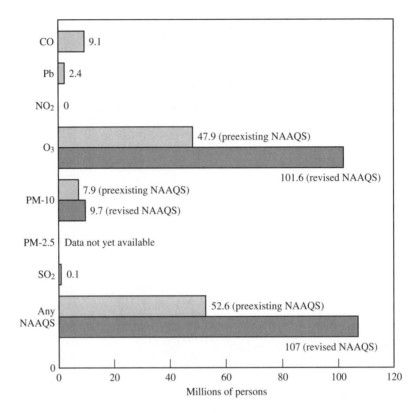

Figure 2.2   Over 50 million people in the United States lived in counties where 1997 air quality concentrations exceeded one or more of the primary (health-related) national standards shown in Table 2.1. This number would double under the new standards recently adopted by EPA. The air quality standards for ozone are the largest contributor to this total. (*Source:* USEPA, 1998a)

oil burning) and by most industrial and manufacturing processes. The health effects associated with particulate air pollution include respiratory and cardiovascular disease, damage to lung tissue, and (potentially) carcinogenesis and premature death. Particles may take on additional potency by serving as carriers that adsorb other pollutants on particle surfaces.

The size of particulate matter strongly affects the scope and severity of its impacts. Particles smaller than 10 microns (or 10 micrometers)—referred to as PM_{10}—are fine enough to penetrate deeply into the lungs, releasing pollutants on moist lung surfaces. These are the particle sizes regulated by air quality standards.[1] More recently, in 1997, new air quality standards were promulgated for particle sizes below 2.5 microns ($PM_{2.5}$). Fine particle sizes appear to be most closely linked with adverse health effects.

1 Particle size is measured in terms of aerodynamic diameter. For an irregularly shaped particle, its aerodynamic diameter is the diameter of a sphere that would fall through air at the same speed as the actual particle.

Table 2.2 Emissions of air pollutants from major U.S. source categories, 1997. (All figures in thousands of short tons)

Source Category	SO$_2$	PM$_{10}$	NO$_x$	VOCs	CO	Pb
Fuel combustion:						
Electric utilities	13,082	290	6,178	51	406	0.064
Industrial	3,365	314	3,270	217	1,110	0.017
Other sources	811	497	1,276	593	3,301	0.415
Industrial processes	1,718	1,277	917	9,836	6,052	2.897
Transportation:						
On-road vehicles	320	268	7,035	5,230	50,257	0.019
Nonroad sources	1,061	466	4,560	2,430	16,755	0.503
Miscellaneous	13	25,152[a]	346	858	9,568[b]	—
TOTAL	20,371	28,264	23,582	19,214	87,451	3.915

[a] Primarily fugitive dust from roads, plus agriculture and construction.

[b] Primarily fires.

Source: USEPA, 1998a.

In addition to health effects, particulate matter can reduce visibility (especially very fine particles of less than one micron), cause soiling or damage to materials, and pose a nuisance in the form of dust (which is composed of large particles that settle quickly out of the atmosphere and are not inhaled). The chemical nature of particulate matter also is important in determining health and environmental impacts; for example, heavy metals or pesticide residues are of greater concern than less toxic materials. In general, the complexity of chemical mixtures found in typical urban air makes it extremely difficult to sort out the effects of individual chemical species.

Figure 2.3 gives the long-term trends and shows the dramatic reduction in particulate emissions since passage of the 1970 Clean Air Act. These reductions came from steel mills, power plants, cement plants, smelters, construction sites, diesel engines, and a host of other sources. Current sources of PM$_{10}$ emissions are about evenly divided among fuel combustion, industrial processes, and transportation sources. Table 2.2 shows that on a national basis the predominant source of fine particle emissions is fugitive dust, which includes dust from paved and unpaved roads and other sources. In terms of human health effects, the chemical nature of particles emitted from industrial sources is of much greater concern than dust from unpaved roads. New control measures for fine particle emissions are anticipated over the next decade in response to the new air quality standards for PM$_{2.5}$.

Sulfur Dioxide   SO$_2$ is emitted primarily from the combustion of coal and oil, which contain sulfur as an impurity. Metal smelting and other industrial processes also emit SO$_2$. Exposure to high concentrations of SO$_2$ can lead to respiratory illnesses, alterations in the lung's defenses, and aggravation of existing cardiovascular or chronic lung disease. Asthmatics and individuals with diseases such as bronchitis or emphysema, as well as children and the elderly, are most sensitive to elevated levels of SO$_2$. Certain plants and crops are also sensitive to SO$_2$ and may be damaged.

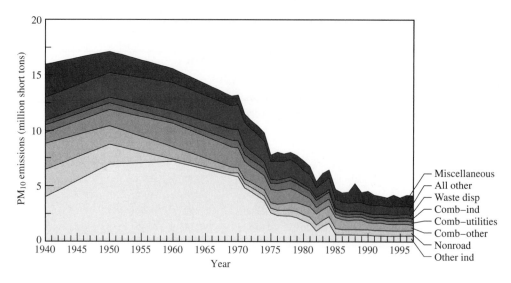

Figure 2.3   Trend in particulate matter (PM_{10}) emissions by major source category, 1940–1997. (*Source:* USEPA, 1998b)

SO_2 emissions from power plants and other sources also can undergo chemical reactions in the atmosphere to form gaseous sulfates and sulfate particles that eventually fall as acidic precipitation. Thus SO_2 emissions are implicated not only in local effects on human health and welfare, but also in widespread regional effects (see the later section on acid deposition).

Figure 2.4 shows how SO_2 emissions have fallen by 35 percent since 1970, even though coal combustion for electric power generation has doubled over that period. At present, coal-fired electric power plants continue as the dominant source of SO_2 emissions, as seen in Table 2.2. Further SO_2 reductions will occur after 2000 as a result of acid rain legislation. As seen earlier in Figure 2.2, virtually all regions of the United States are now in compliance with the ambient air quality standard for SO_2, making this one of the important success stories of the 1970 Clean Air Act.

Carbon Monoxide   Carbon monoxide is a colorless, odorless gas that is produced when fossil fuels or other carbon-containing materials are not completely combusted. When inhaled, carbon monoxide is absorbed by blood hemoglobin, which normally carries oxygen to the body. Exposure to elevated levels of CO in the atmosphere can produce a spectrum of adverse health effects such as shortness of breath and dizziness as the body's oxygen delivery system is choked off.

Figure 2.5 shows the reduction in national CO emissions since passage of the 1970 Clean Air Act. Transportation technologies, mainly automobiles, are the dominant source of CO emissions. From an air quality perspective, elevated CO levels also are attributed mainly to vehicles in urban areas. As the number of vehicles on the road has continued to grow, there has been a ratcheting down on emissions from individual vehicles and other sources in order to keep total emissions from growing.

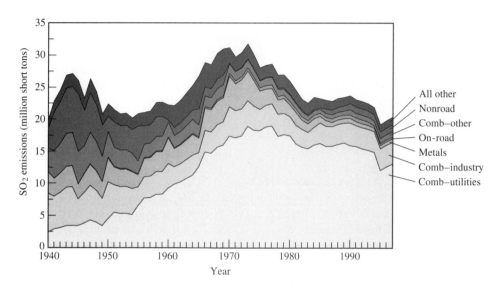

Figure 2.4 Trend in sulfur dioxide emissions by major source category, 1940–1997. (*Source:* USEPA, 1998b)

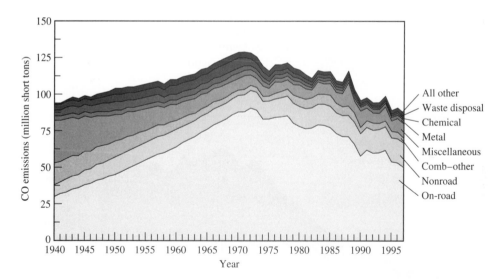

Figure 2.5 Trend in carbon monoxide emissions by major source category, 1940–1997. (*Source:* USEPA, 1998b)

Nitrogen Dioxide Nitrogen dioxide is a reddish-brown gas that is toxic in very high concentrations. At the lower concentrations typical of polluted urban air, NO_2 can irritate the respiratory system and produce respiratory illnesses such as bronchitis. Children may be particularly affected by elevated NO_2 levels.

NO_2 is primarily a result of fuel combustion. However, unlike SO_2 and CO, which are produced directly in the combustion process, only a small amount of NO_2 is emitted from the chimneys or tailpipes of fuel combustion technologies (like power plants and automobiles). Rather, NO_2 is in the family of compounds called nitrogen oxides (NO_x), which includes nitric oxide (NO). At the high temperatures found in modern combustion processes, NO is formed from the nitrogen and oxygen found in air and fuel. When the NO is released to the atmosphere, it gradually oxidizes to NO_2. Although only NO_2 (and not NO) is a criteria air pollutant, emission control requirements usually are specified on the basis of total NO_x, reported as equivalent NO_2.

Figure 2.6 shows the recent U.S. trend in NO_x emissions. In the early 1970s the transportation sector was the dominant source of NO_x, but as vehicle emissions have been reduced, stationary sources, particularly coal-fired power plants, account for a higher percentage of current emissions (Table 2.2). In terms of air quality, all of the United States is now in compliance with the national standard for NO_2. However, nitrogen oxides also are implicated in acid deposition and the formation of ground-level ozone. Furthermore, like SO_2, NO_x gases can react chemically in the atmosphere to form fine nitrate particles that contribute to $PM_{2.5}$ and its associated effects.

Ozone   The air pollutant ozone found at ground level can be thought of as "bad" ozone, in contrast to the protective layer of "good" ozone found in the stratosphere high above the earth (discussed later in Section 2.3.4). Ground-level ozone is formed by complex chemical reactions in the atmosphere involving nitrogen oxides (NO_x) and hydrocarbon gases (also known as *volatile organic compounds, VOCs, or reactive organic gases, ROGs*). These chemical reactions are triggered by summer sun-

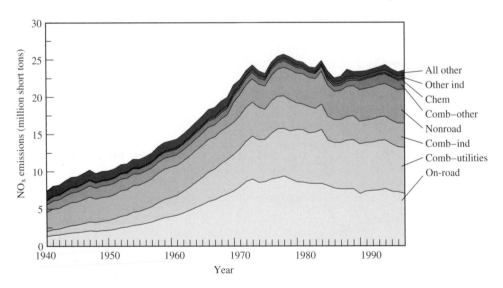

Figure 2.6   Trend in nitrogen oxide emissions by major source category, 1940–1997. (*Source:* USEPA, 1998b)

light, which provides the energy to initiate the photochemical reactions. The result is a chemical "soup" known as *photochemical smog* (or simply "smog"), produced from the NO_x and hydrocarbons put into the atmosphere by automobiles, power plants, factories, and other sources. One of the main ingredients of this soup is ozone; we refer to it as ground-level or tropospheric ozone to distinguish it from beneficial stratospheric ozone.

Ozone belongs to a class of chemicals known as *oxidants*. Because these chemicals are extremely reactive, they produce a wide range of effects. Ground-level ozone causes health problems because it attacks lung tissue, reduces lung function, and sensitizes the lungs to other irritants. Studies have shown that ambient levels of ozone affect not only people with impaired respiratory systems, such as asthmatics, but healthy people as well. Ozone causes paints and fabrics to deteriorate more rapidly and can cause the sidewalls of tires and other rubber products to become brittle and cracked. Because ground-level ozone interferes with the ability of plants to produce and store food, they are more susceptible to disease, insect attack, other environmental pollutants, and harsh weather. Ozone damage to plants and trees causes half a billion dollars per year of agricultural crop loss in the United States alone, according to EPA estimates.

Although high levels of ozone traditionally have been associated with Los Angeles (where surrounding mountains help trap air in the L.A. basin, providing a natural cauldron for the formation of photochemical smog), many U.S. regions, including much of the eastern part of the country, also have ambient ozone levels exceeding the national air quality standard. Indeed, as seen earlier in Figure 2.2, more people are exposed to levels of ozone exceeding the NAAQS than any other criteria pollutant. In 1997 the EPA tightened the ambient ozone standard, citing growing evidence of widespread health effects.

Figure 2.6 showed the major sources of nitrogen oxides that contribute to ground-level ozone formation. The other key ingredients, volatile organic compounds (VOCs), which consist mainly of hydrocarbons, come from a more diverse set of sources as shown in Figure 2.7. The transportation sector is one of the largest sources of hydrocarbon emissions. Industrial sources include petroleum refineries, chemical manufacturing plants, gasoline distribution and storage facilities, dry cleaners, and a diverse set of other processes that employ chemical solvents.

Strategies to reduce atmospheric ozone concentrations thus require reductions in NO_x and hydrocarbon emissions. Until now, the prevalent strategy has concentrated on hydrocarbon (VOC) reductions. Future strategies are likely to involve further reductions in NO_x as well as VOCs. Because the chemistry is extremely complex, and because ozone may be transported over large distances from one region to another, attainment of national air quality standards for ozone will remain one of the most challenging tasks in the years ahead.

Lead   Lead is a heavy metal that can cause neurological damage and adverse effects on organs such as the liver and kidneys. Children exposed to lead are particularly vulnerable to a range of effects that can impair normal development. Once ingested via inhalation or other means, lead tends to bioaccumulate in blood, bone, and soft tissues, so that its effects are not easily reversible.

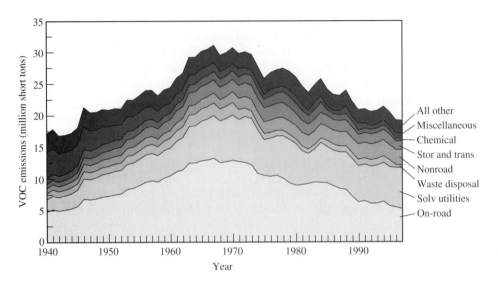

Figure 2.7   Trend in volatile organic compound emissions by major source category, 1940–1997. These VOC emissions are primarily hydrocarbon compounds. (*Source:* USEPA, 1998b)

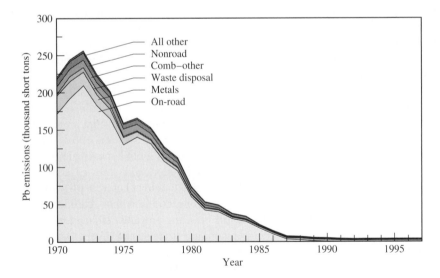

Figure 2.8   Trend in atmospheric lead emissions by major source category, 1970–1997. (*Source:* USEPA, 1998b)

Until recently, the largest source of lead emission to the atmosphere has been from automobiles using leaded gasoline. However, as seen in Figure 2.8, total lead emissions have fallen dramatically in the last two decades as newer automobiles have been designed to run on unleaded gasoline. In many other parts of the world, however, leaded gasoline use continues. The major sources of atmospheric lead in the United States now are lead smelting and manufacturing processes, as seen in Table 2.2. Domestic sources of environmental lead, such as lead erosion from water pipes and lead paint found in older buildings, do not fall under the purview of national air quality regulations; but these remain important sources of lead ingestion, especially for small children.

2.3.2   Air Toxics

In addition to regulations for the criteria air pollutants, the 1970 Clean Air Act also targeted substances called *hazardous air pollutants (HAPs)*, also known as *air toxics*. These chemicals are emitted in much smaller quantities than criteria air pollutants, but their effects can be severe, even in small doses. Carcinogenic substances like asbestos and benzene are of particular concern, as are heavy metals and other chemicals that may cause neurological, immunological, mutagenic, and other serious health effects.

The Clean Air Act procedure for establishing national emission standards for hazardous air pollutants proved to be quite cumbersome, with the result that standards for only eight HAPs were established between 1970 and 1990. In the 1990 Clean Air Act Amendments, Congress revised the procedure, specifically listing 189 chemicals as hazardous air pollutants (Table 2.3). Any industrial process emitting more than 10 tons per year (tpy) of any one of these substances, or more than 25 tpy of any combination of HAPs, is required to use "maximum available control technology" (MACT) to reduce air toxic emissions by the year 2000.[2] Additional measures must be taken if the EPA determines that an unacceptable level of risk still remains after MACT is applied.

The EPA also publishes an annual Toxics Release Inventory (TRI) that reports the annual mass emissions of toxic substances from specific facilities and industries across the United States (USEPA, 1999). The TRI lists approximately 650 chemicals and chemical categories, and reports emissions to the air, water, and land. However, it does not include any quantitative measures of the relative risk from different chemicals. Rather, individual communities must determine whether the releases from nearby industrial facilities pose any threat to people or the environment. Improved methods are being developed to help people better understand the significance of the toxic releases that are reported.

2.3.3   Acid Deposition

Acid deposition, commonly known as *acid rain*, refers to the fallout of acidic particles or any type of precipitation—rain, fog, mist, or snow—that is more acidic than

---

2 The electric utility sector was exempt from these requirements pending further study of HAPs emissions and their impacts. As of 2000, the major concern is emissions of mercury from coal-burning power plants. Like other heavy metals, human exposure to mercury can occur directly via inhalation, or indirectly via water or the food chain when atmospheric pollutants are deposited on rivers, lakes, soil, and vegetation. Ongoing studies are attempting to assess the level of risk to human health posed by direct and indirect exposure mechanisms.

Table 2.3 Hazardous air pollutants.

Specific Compounds and Chemical Categories		
Acetaldehyde	Dimethyl formamide	Pentachloronitrobenzene
Acetamide	1,1-Dimethyl hydrazine	Pentachlorophenol
Acetonitrile	Dimethyl phthalate	Phenol
Acetophenone	Dimethyl sulfate	p-Phenylenediamine
2-Acetylaminofluorene	4,6-Dinitro-o-cresol, and salts	Phosgene
Acrolein	2,4-Dinitrophenol	Phosphine
Acrylamide	2,4-Dinitrotoluene	Phosphorus
Acrylic acid	1,4-Dioxane	Phthalic anhydride
Acrylonitrile	1,2-Diphenylhydrazine	Polychlorinated biphenyls
Allyl chloride	Epichlorohydrin	1,3-Propane sultone
4-Aminobiphenyl	1,2-Epoxybutane	beta-Propiolactone
Aniline	Ethyl acrylate	Propionaldehyde
o-Anisidine	Ethyl benzene	Propoxur
Asbestos	Ethyl carbamate	Propylene dichloride
Benzene	Ethyl chloride	Propylene oxide
Benzidine	Ethylene dibromide	1,2-Propylenimine
Benzotrichloride	Ethylene dichloride	Quinoline
Benzyl chloride	Ethylene glycol	Quinone
Biphenyl	Ethylene imine	Styrene
Bis(2-ethylhexyl)phthalate	Ethylene oxide	Styrene oxide
Bis(chloromethyl)ether	Ethylene thiourea	2,3,7,8-Tetrachlorodibenzo-p-dioxin
Bromoform	Ethylidene dichloride	1,1,2,2-Tetrachloroethane
1,3-Butadiene	Formaldehyde	Tetrachloroethylene
Calcium cyanamide	Heptachlor	Titanium tetrachloride
Caprolactam	Hexachlorobenzene	Toluene
Captan	Hexachlorobutadiene	2,4-Toluene diamine
Carbaryl	Hexachlorocyclopentadiene	2,4-Toluene diisocyanate
Carbon disulfide	Hexachloroethane	o-Toluidine
Carbon tetrachloride	Hexamethylene-1,6-diisocyanate	Toxaphene
Carbonyl sulfide	Hexamethylphosphoramide	1,2,4-Trichlorobenzene
Catechol	Hexane	1,1,2-Trichloroethane
Chloramben	Hydrazine	Trichloroethylene
Chlordane	Hydrochloric acid	2,4,5-Trichlorophenol
Chlorine	Hydrogen fluoride	2,4,6-Trichlorophenol
Chloroacetic acid	Hydrogen sulfide	Triethylamine
2-Chloroacetophenone	Hydroquinone	Trifluralin
Chlorobenzene	Isophorone	2,2,4-Trimethylpentane
Chlorobenzilate	Lindane	Vinyl acetate
Chloroform	Maleic anhydride	Vinyl bromide
Chloromethyl methyl ether	Methanol	Vinyl chloride
Chloroprene	Methoxychlor	Vinylidene chloride
Cresols/Cresylic acid	Methyl bromide	Xylenes
o-Cresol	Methyl chloride	o-Xylenes
m-Cresol	Methyl chloroform	m-Xylenes
p-Cresol	Methyl ethyl ketone	p-Xylenes
Cumene	Methyl hydrazine	
2,4-D, salts and esters	Methyl iodide	*(continued)*

Table 2.3 *(continued)*

		Chemical Categories
DDE	Methyl isobutyl ketone	Antimony compounds
Diazomethane	Methyl isocyanate	Arsenic compounds
Dibenzofurans	Methyl methacrylate	Beryllium compounds
1,2-Dibromo-3-chloropropane	Methyl tert butyl ether	Cadmium compounds
Dibutylphthalate	4,4-Methylene bis	Chromium compounds
1,4-Dichlorobenzene(p)	Methylene chloride	Cobalt compounds
3,3-Dichlorobenzidene	Methylene diphenyl diisocyanate	Coke oven emissions
Dichloroethyl ether	4,4-Methylenedianiline	Cyanide compounds
1,3-Dichloropropene	Naphthalene	Glycol ethers
Dichlorvos	Nitrobenzene	Lead compounds
Diethanolamine	4-Nitrobiphenyl	Manganese compounds
N,N-Diethyl aniline	4-Nitrophenol	Mercury compounds
Diethyl sulfate	2-Nitropropane	Mineral fibers (fine)
3,3-Dimethoxybenzidine	N-Nitroso-N-methylurea	Nickel compounds
Dimethyl aminoazobenzene	N-Nitrosodimethylamine	Polycyclic organic matter
3,3-Dimethyl benzidine	N-Nitrosomorpholine	Radionuclides
Dimethyl carbamoyl chloride	Parathion	Selenium compounds

Source: CAA, 1990.

normal. Broad areas of North America, Europe, and Asia regularly experience precipitation that is substantially more acidic than natural rainfall, which has a pH of about 5.6.[3] Figure 2.9 shows the elevated level of precipitation acidity over the eastern United States in the mid-1980s.

Acid deposition gained attention in the 1960s when anglers in Sweden started noticing a decline in fish populations in high-altitude lakes. Scandinavian scientists were the first to identify the cause as increased acidity of the lake water and to link this increased acidity with precipitation having abnormally low pH. Similar effects were later observed in eastern Canada and in the northeastern United States. The principal source of the acidity was determined to be emissions of sulfur dioxide (SO_2) from power plants hundreds of miles away. In Europe, the acid rain in Scandinavia was attributed to power plants in England, Poland, and Germany. In North America, the coal-fired power plants of the midwestern United States were the primary sources. Most of this SO_2 was released high into the atmosphere through tall chimneys. Although this strategy was effective in eliminating ground-level air pollution problems in the immediate vicinity of the power plant, it gave rise to a new phenomenon: long-distance transport of air pollutants. As it was carried eastward by the prevailing winds, sulfur dioxide was gradually transformed into sulfate particles and other acidic species. Eventually, these acidic gases and particles reached the earth in wet or dry form, causing a variety of effects. Deposition of nitrate species formed from NO_x emissions also contributed to the acid loadings.

---

3 Acidity is measured in units of pH, where a pH value below 7.0 indicates an acid and above 7.0 a base. Natural rainfall is slightly acidic because of the presence of small amounts of CO_2 and other atmospheric trace gases, which form weak acids when dissolved in water. A change of one pH unit represents a 10-fold change in acidity.

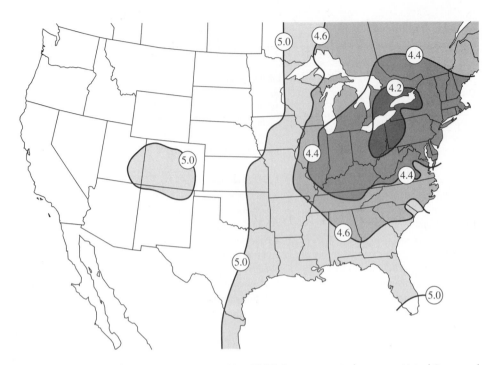

Figure 2.9   Annual average precipitation pH in 1985. Precipitation in the eastern United States and Canada was over 10 times more acidic than natural rainfall. (*Source:* NAPAP, 1985)

Acidification of freshwater lakes and streams directly affects the viability of fish and other aquatic organisms. The pH of an environment is important because it affects the functioning of virtually all enzymes, hormones, and other proteins in all living organisms. Thus acidified waters can kill sensitive fish species or prevent them from reproducing. Similarly, acid deposition contributes to the decline of some species of trees, particularly in high altitude forests with thin soils. Soil acidification also is believed to disrupt the complex soil chemistry that provides nutrients to vegetation and indirectly affects soil erosion, sedimentation of waterways, and changes in animal habitat.

Other effects of acid deposition include the deterioration of some building materials and monuments made of limestone or marble. The sulfate particles formed as a result of long-distance transport also degrade visibility and may harm human health (see the earlier section on particulate matter).

The Clean Air Act Amendments of 1990 established a program of SO_2 emission reductions to control the problems of acid deposition. For the first time in history, Congress imposed a national cap on total SO_2 emissions of approximately 9 million tons per year. This cap is to remain in perpetuity. Coal-burning power plants, the main source of SO_2 emissions, were given 10 years to reduce emissions to the required levels in two phases. Also for the first time, Congress established a national emissions trading program that allows utilities to achieve the required reduction in the most cost-effective manner (provided local air quality standards are not compro-

Acid rain damages not only lakes and forests but also a variety of building materials. The Parthenon in Rome is among the historic monuments that have been eroded by acidic air pollution.

mised). This "market mechanism" approach to regulation has been widely hailed as an improvement to the traditional "command and control" approach, in which each source is told exactly what it must do. The first phase of SO_2 emission reductions was achieved in 1995, and the second phase reductions in 2000. No further increases in total U.S. emissions of SO_2 are permitted under the current legislation.

2.3.4 Stratospheric Ozone Depletion

If you've ever suffered a bad sunburn at the beach, you know how damaging the sun's energy can be. Imagine if it were many times more potent. Under such circumstances, exposure to the sun could endanger all life on earth.

Fortunately, the earth's atmosphere protects us from the sun's most intense radiation. A thin layer of ozone (O_3) molecules in the stratosphere 10–40 km (6–25 miles) above the earth's surface absorbs high-energy solar radiation incident upon the planet. Only about 1 in 100,000 molecules in the stratosphere is an ozone molecule. But this is sufficient to prevent most of the intense radiation known as ultraviolet-B (UV-B) from reaching the earth's surface, where it would cause damaging effects by destroying protein and DNA molecules in biological tissue.

In the late 1970s and 1980s, worldwide decreases in stratospheric ozone levels began to be observed. The most severe depletion of ozone was seen during the winter months at the South Pole over Antarctica, where an "ozone hole" the size of the continental United States revealed ozone levels roughly 40 percent lower than normal. Were such a reduction in ozone levels to occur over populated areas, it could affect human health by increasing the risk of skin cancer, cataracts, and blindness from increased penetration of ultraviolet radiation. It could also reduce immune function and harm algae and phytoplankton, which form the base of the marine food chain. Less productivity of these organisms could cause a domino effect in the marine ecosystem. Although global levels of ozone depletion are not nearly as severe as those observed during the Antarctic winter, the problem remains serious and could worsen in the absence of corrective measures.

We now understand that ozone is being depleted by humanmade chemicals, most notably the family of compounds known as *chlorofluorocarbons (CFCs)*. CFCs are nonreactive, nonflammable, nontoxic, noncorrosive molecules whose properties are ideally suited for purposes such as refrigeration, air conditioning, manufacturing foam, cleaning electronics, and propelling the contents of aerosol cans. CFC production began in the 1940s, and its use rapidly grew to several million tons per year, as shown in Figure 2.10.

CFCs can enter the atmosphere directly as a gaseous emission or via evaporation from liquids such as freon refrigerants. Because they are so stable, CFC molecules remain in the air for many years, eventually finding their way up into the stratosphere. It may take decades to reach the upper levels of the atmosphere, but at altitudes above 25 km the intense UV radiation in the ozone layer can break apart a CFC molecule, releasing free chlorine atoms. Each of these atoms in turn reacts chemically to destroy large numbers of ozone molecules before being finally removed by other chemical reactions. Chapter 11 presents the details of this chemistry.

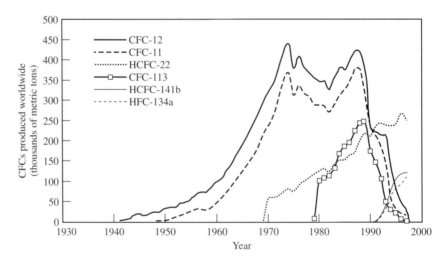

Figure 2.10   Worldwide production of CFCs and other halocarbons. After CFC production was banned in the late 1980s, the production of substitutes such as HCFCs and HFCs increased. (*Source:* Based on AFEAS, 1998)

Other ozone-depleting chemicals—such as bromine, used in fire extinguishers—also have been introduced into the stratosphere as a result of human activities.

In recognition of the global scope of the stratospheric ozone problem, in an international treaty signed in 1987 most of the industrialized world agreed to a timetable for reducing, and then phasing out, the production and use of CFCs and other ozone-depleting chemicals. Known as the Montreal Protocol on Substances That Deplete the Ozone Layer, this international accord has now been signed by 172 countries. Amendments adopted since 1990 call for the complete phaseout of CFC production by the year 2000, with a longer schedule for eliminating certain other ozone-depleting chemicals. Nonetheless, because of the long lifetimes of these chemicals, and because some developing nations are not parties to the international agreement, it will still be many decades before recovery of the ozone layer can be realized. In the meantime, continuing efforts will be necessary to find acceptable substitutes for the ozone-depleting chemicals now in wide use.

2.3.5   Greenhouse Gases

As noted earlier, the term *air pollutant* is commonly used to characterize substances like sulfur dioxide, nitrogen oxides, and particulate matter that are linked to adverse health effects and other types of environmental damage. Yet another group of compounds emitted to the atmosphere from human activity, but not currently branded as pollutants, may be responsible for environmental impacts that are even more pervasive and long-lasting than traditional air contaminants.

These gaseous emissions are called *greenhouse gases* because they trap heat in the atmosphere in much the same way that glass helps to trap solar energy in a greenhouse. The physics of the greenhouse effect is well understood and is discussed more fully in Chapter 12. Briefly, solar energy is radiated from the sun at short wavelengths (ultraviolet radiation) and is partially absorbed by the earth's surface. The warmed surface in turn radiates energy back to space, but at much longer wavelengths (infrared radiation). The balance between incoming and outgoing radiation determines the temperature of the planet. Most of the earth's atmosphere is transparent to both incoming (ultraviolet) and outgoing (infrared) radiation, but some trace gases, notably water vapor (H_2O) and carbon dioxide (CO_2), have molecular structures that absorb outgoing energy. Because less energy now escapes to outer space, these gases produce a natural warming effect. Without this greenhouse effect the earth's average temperature would be about 34°C (61°F) colder—unable to support life as we know it.

As is often the case, however, too much of a good thing can be harmful. Over the past century, the quantities of greenhouse gases (GHGs) emitted to the atmosphere from human activity have increased dramatically. The principal culprit is CO_2 emitted from combustion of the oil, gas, and coal that supply roughly 75 percent of the world's energy needs. Figure 2.11 shows the trend in CO_2 emissions from fossil fuel combustion over the past century, and Figure 2.12 shows the corresponding trend in atmospheric CO_2 concentration. Current CO_2 levels in the atmosphere are about 30 percent higher than the preindustrial levels of a hundred years ago. Current

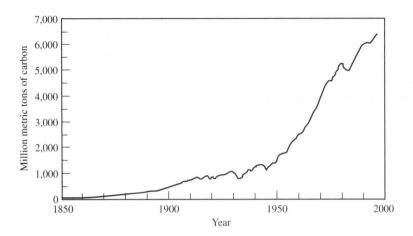

Figure 2.11   Trend in global carbon dioxide emissions from fossil fuel combustion. (*Source:* Marland et al., 1999)

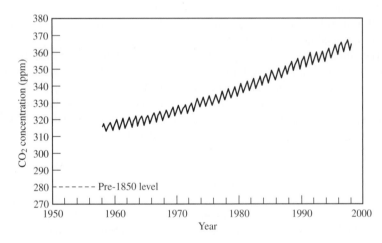

Figure 2.12   Recent trend in atmospheric CO_2 concentration. This graph shows data from the Mauna Loa observatory in Hawaii, which is the longest modern record of atmospheric CO_2 measurements. The sawtooth shape reflects the seasonal fluctuation in CO_2 uptake by biomass from summer to winter. Since preindustrial times the average CO_2 concentration has increased by 30 percent. (*Source:* Keeling and Whorf, 1998)

projections anticipate a doubling of atmospheric CO_2 concentration by the year 2100, bringing an average global temperature increase of about 2.4°C (4.3°F) (IPCC, 1996). Though seemingly small, a temperature change of this magnitude in just one century would be unprecedented in human history and would profoundly affect the earth's climate and inhabitants.

There is considerable uncertainty, however, over the future magnitude and rate of human-induced climate change and its impacts. The chief concerns include sea level rise and flooding of low-lying regions; increased precipitation and severity of storm events; increased drought; increased spread of tropical diseases; and a range of ecological effects as plants and animals attempt to cope with rapid changes. Although these concerns are based primarily on projections from mathematical models, a broad spectrum of the scientific community assessing the problem worldwide has concluded that the threat is real. In 1996 the Intergovernmental Panel on Climate Change, convened by the United Nations, concluded for the first time that "the balance of evidence suggests a discernible human influence on climate" (IPCC, 1996).

Although CO_2 is the major concern, it is not the only greenhouse gas contributing to global warming. Other such gases include methane (CH_4), nitrous oxide (N_2O), and halocarbons, which include the chlorofluorocarbons (CFCs) responsible for stratospheric ozone depletion. Other minor gases include the family of perfluorocarbons (PFCs) and sulfur hexafluoride (SF_6). Table 2.4 shows the major sources and magnitude of GHG emissions from human activity. Many of these gases are far more effective than CO_2 in trapping radiative energy, so that even small quantities in the atmosphere can have a potentially significant effect. Many halocarbon substitutes that have come into use since the ban on CFCs also are powerful greenhouse gases. An index called the Global Warming Potential (GWP) is sometimes used to estimate the CO_2 equivalence of different gases. On this basis, carbon dioxide accounts for about 85 percent of U.S. greenhouse gas emissions (USDOE, 1997a).

Greenhouse gases are distinguished from conventional air pollutants by their long lifetime in the atmosphere. Thus, unlike SO_2 or NO_x, which are chemically reactive gases that can wash out of the atmosphere in a matter of days, CO_2 is a stable, nonreactive gas that remains in the atmosphere for decades to centuries. Human-induced increases in atmospheric CO_2 levels therefore cannot be easily reversed. Indeed, because of the large quantities of CO_2 already emitted to the atmosphere, some degree of future warming is believed to be inevitable.

In recognition of the global warming problem, over 150 nations signed an accord in 1992 promising to reduce the future growth of greenhouse gas emissions. An outgrowth of this historic agreement was a formal climate change treaty signed by 84 countries in December 1998 in Kyoto, Japan. For the first time, the major industrialized nations of the world agreed to specific targets and timetables for reducing future CO_2 emissions. The United States, for example, agreed to reduce its emissions to 7 percent below 1990 levels by the year 2012 (subject to ratification by the U.S. Congress), with reductions of 6 percent and 8 percent agreed to by Japan and the European Union, respectively. Most scientists believe that much greater reductions are required to stabilize CO_2 concentrations at current levels. On the other hand, some scientists and politicians believe that no action should be taken until there is stronger evidence that human-induced warming is indeed occurring, and until countries like China and India join in international accords. Thus, while it is still too early to tell how future climate change policies may evolve, it is clear that the control of greenhouse gas emissions will be a dominant environmental issue as we begin the 21st century.

Table 2.4   Annual emissions of greenhouse gases.

Source	Annual Emissions[a] (Mt/Year)	
	World	**U.S.**
CO_2 Emissions		
Commercial energy	22,900	5,250
Cement manufacturing and gas flaring	1,000	50
Tropical deforestation[b]	5,900	—
Total CO_2	29,800	5,300
CH_4 Emissions		
Fossil fuel production related	100	
Enteric fermentation	85	
Rice paddies	60	
Landfills	40	
Animal waste	25	
Domestic sewage	25	
Total CH_4	375	31
N_2O Emissions		
Cultivated soils	3.5	
Industrial sources	1.3	
Biomass burning	0.5	
Cattle and feed lots	0.4	
Total N_2O	5.7	0.5
Halocarbon and Other Emissions		
CFC-11, -12, -113	0.7	0.1
HCFC-22	0.2	0.1
HFCs, PFCs, SF_6	n/a	0.034

[a] World energy and all U.S. data are for 1996. Other world data are from the early 1990s. There are significant uncertainties in many of these estimates. n/a = not available. Mt = million metric tons.

[b] This refers to CO_2 that would have been absorbed by the lost biomass.

Source: IPCC, 1996; Marland et al., 1999; USDOE, 1999, 1997a.

2.4   WATER POLLUTION

Contaminated water is one of the oldest environmental problems confronting people throughout the world. In this section we briefly review the principal types of waterborne contaminants and the reasons for concern. Some of the major environmental laws enacted to control U.S. water pollution also are highlighted, along with a brief status report on water quality.

2.4.1 Sources and Uses of Water

The most plentiful sources of water for human activities are *surface waters,* which include all the lakes, streams, and rivers that flow (eventually) into the oceans that cover approximately 70 percent of the earth's surface. These waters are depleted by evaporation and replenished by precipitation as part of the natural hydrological cycle. *Groundwater,* on the other hand, refers to underground water sources— aquifers that retain water percolating through the earth's surface. Groundwater is released naturally through springs, or it can be pumped to the surface via human intervention. The time required to recharge an underground aquifer is long relative to the evaporation–precipitation cycle affecting surface waters.

Chief among human needs for water are municipal water supplies, which provide water for drinking, cooking, and other domestic activities. The agricultural sector is another major consumer of water worldwide, while in industrialized countries activities such as resource extraction, manufacturing processes, and construction additionally require significant amounts of water in their operation. Surface waters also are used for transportation (barges, boats), recreation (swimming, rafting), and electric power production (via dams and waterfalls). And of course, surface waters support all aquatic life on the planet.

2.4.2 Major Water Contaminants

Surface waters and groundwater become polluted in a variety of ways. From an environmental engineering point of view, it is useful to distinguish between *point sources,* which are identifiable discharge points such as the outfall pipe from a factory or wastewater treatment plant, and *nonpoint sources,* which include runoff from agricultural lands, erosion from mining and construction activities, and fallout or deposition from the atmosphere.

Although natural sources of contamination such as volcanoes or flooding can have important impacts on water quality, the focus of this chapter is on human activities as a source of pollution. In all cases, environmental concerns reflect the quality of water needed to support various beneficial uses. There are different ways of describing the problem, but the following classes of contaminants are of principal concern.

Pathogens Pathogens are disease-causing agents such as bacteria, viruses, protozoa, and parasitic worms called helminths. These microorganisms are commonly found in the intestines of infected people or animals, and they are excreted in the feces that enter sewer systems or (as in the case of animals) fall onto the ground. Pathogens then can enter waterways from inadequately treated sewage discharges, stormwater drains, septic systems, and runoff from pastures and animal feedlots. If ingested, they can cause human illnesses ranging from life-threatening diseases such as typhoid, cholera, diarrhea, and dysentery to minor gastrointestinal, respiratory, and skin diseases. In many parts of the world, especially in nonindustrialized countries, the presence of pathogens in water supplies is still a critical public health problem that exacts a high toll on human life.

Organic Wastes   Organic wastes are the main source of *oxygen-depleting substances* in surface waters. Dissolved oxygen is the most important basic requirement for a healthy aquatic ecosystem. Most fish and insects live on the oxygen dissolved in water, and if levels fall too low the effects can range from a reduction in reproductive capacity to suffocation and death. Larvae and juvenile fish are especially sensitive and require higher levels of dissolved oxygen than more mature fish.

Although dissolved oxygen concentrations fluctuate naturally, severe oxygen depletion usually results from human activities that introduce large quantities of biodegradable organic wastes into surface waters. The most common types of wastes include human and animal excrement, food wastes, and organic residuals from industrial operations such as paper mills and food processing plants. Organic wastes find their way into water bodies via discharges from municipal and industrial wastewater treatment facilities, sewage lines, septic tanks, and urban and agricultural runoff.

Biodegradable wastes are decomposed by bacteria that use dissolved oxygen to break down waste materials. When the quantity of organic waste exceeds normal sustainable levels, bacteria growth increases and oxygen is depleted faster than it can be replenished by natural processes. The result is a water body that not only becomes unable to support aquatic life, but whose color, taste, and odor may leave it undesirable for other uses such as recreational activities or municipal water supplies.

The demand for oxygen by bacteria is called *biological oxygen demand (BOD),* which is a common measure of water quality. In addition, some oxygen-depleting substances trigger chemical reactions that place greater oxygen demand on waters, referred to as *chemical oxygen demand (COD).* High levels of BOD or COD indicate undesirable water quality.

Nutrients   Nitrogen and phosphorus are two essential nutrients needed to support vegetation and other forms of life. These chemicals are widely used in fertilizers and household detergents and are most responsible for the overenrichment of nutrients in lakes, rivers, or streams that leads to a condition called *eutrophication.* A eutrophic body of water supports an abundant growth of algae that can eventually crowd out other forms of aquatic life, leaving a water body that is unable to support fish or other life forms and is also unsuitable for human uses.

Nutrients used in fertilizers tend to bind and cling to clay and humus particles in soils and are therefore easily transported to surface water supplies through erosion and runoff. Fertilizers, especially inorganic ones, tend to leach into waterways from the soils of farmlands, lawns, and gardens. Soils with a poor capacity to hold nutrients cause even greater leaching because they require increased levels of fertilizer application. Human sewage also may contain high concentrations of nitrates, and phosphates from household detergents pass right through most municipal waste treatment systems into surface waters. Because it is less abundant in nature, phosphorus from human activity is often the limiting nutrient most responsible for accelerated freshwater eutrophication.

Nutrient enrichment also exacerbates the depletion of dissolved oxygen as decomposers, mainly bacteria, feed on increased quantities of dead and decaying algae. Decomposition under conditions of insufficient oxygen (called *anaerobic conditions*) produces odorous gases that contribute to the objectionable smell, taste, color, and aesthetics of highly eutrophic water bodies.

Whereas excessive nitrogen levels contribute to eutrophication, high nitrate levels in drinking water also can lead to a potentially fatal condition called "blue baby syndrome" (methemoglobinemia), which restricts oxygen transport in the bloodstream of newborn infants. Contaminated groundwater is the most common source of high-nitrate drinking water. Figure 2.13 shows the historical growth in U.S. nitrogen and phosphorus consumption in fertilizers over the past 50 years.

Toxic Organic Chemicals   Organic compounds built of carbon and other atoms are the stuff of all living things. But many synthetic organic chemicals, which contain additional substances like chlorine, are potentially toxic to people, plants, and animals. Some organic chemicals are known or suspected carcinogens (causing cancer) or mutagens (producing genetic mutations). Others, if ingested in sufficient quantity, can impair bodily functions and damage vital organs, with sometimes fatal effects. Oil spills are perhaps one of the better-known examples of water contamination by organic compounds, which can devastate ecosystems and marine life. In general, concern over trace toxic chemicals in water has become a major focus of water pollution control efforts in recent years.

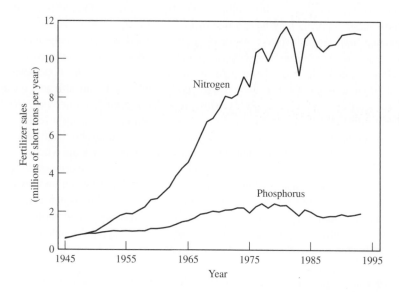

Figure 2.13   Trend in nitrogen and phosphorus use in fertilizers. (*Source:* USGS, 1998)

Industrial discharges are a major source of water pollution in many regions. However, nonpoint sources, such as agricultural runoff, are now the greatest source of impairment to U.S. waterways.

One important class of toxic organic compounds is pesticides such as the notorious DDT. Although such chemicals were devised to serve useful purposes such as protecting the food supply and controlling disease-carrying insects, chlorine-containing pesticides like DDT, DDE, aldrin, and chlordane are highly persistent in the environment. This means they do not readily break down in natural ecosystems and thus tend to accumulate in the tissue of organisms near the top of the food chain, such as birds and fish. Observed effects include a reduction in reproductive capacity, birth defects, and tumors. In some cases, humans may be exposed to increased cancer risk by eating organisms containing high levels of pesticides or by ingesting food or water contaminated with toxic organics. Other types of insecticides and herbicides that have replaced DDT and do not accumulate in body tissue (like diazinon and parathion) are nonetheless still acutely toxic to humans if ingested or absorbed through the skin in sufficient quantity.

Another class of toxic organic chemicals, volatile organic compounds (VOCs), includes substances such as vinyl chloride, carbon tetrachloride, and trichloroethylene. Chemicals in this class are often used as industrial or household solvents and as ingredients in chemical manufacturing processes. Many VOCs are known or suspected carcinogens. As with other chemicals, they may enter waterways directly in the discharge from industrial sources, through municipal stormwater and sewer systems, and by runoff into surface waters or percolation into groundwater from surface spills and accidental discharges. Because they tend to evaporate at room temperature, the concentration of toxic VOCs in surface waters is typically much lower than in groundwater, where evaporation cannot occur.

Toxic Metals   The health of humans and other living organisms requires trace levels of certain heavy metals such as chromium, cobalt, copper, iron, manganese, molybdenum, vanadium, strontium, and zinc. Excessive levels of these metals, however, can be toxic. So too with nonessential heavy metals such as mercury, lead, and arsenic. In people, ingestion of significant amounts of these metals can damage vital organs or even kill.

Some metals are found naturally in waterways, but human activities have altered the levels and distribution of toxic metals in the environment. The outfalls from metal smelting and other industrial processes, and the runoff from mining and construction activities, are among the major ways that metals enter surface waters. Metals such as lead also can originate in the pipes and plumbing of older homes and be carried into waterways by municipal discharges.

Another important route for some metals is via atmospheric deposition. Mercury, for example, is released in gaseous form from high-temperature industrial processes, including waste incineration and the combustion of coal for power generation. As it cools, the volatilized mercury can condense and settle onto nearby surface waters or be washed into waterways via runoff. Mercury is an example of a toxic metal that tends to bioaccumulate in tissues of fish and other organisms high in the food chain. Humans can then be affected through consumption of fish. Of particular concern is the compound methylmercury, which is an especially toxic form of mercury that affects the central nervous system.

Sediments and Suspended Solids Sediment consists of soil particles that enter a water body and eventually settle to the bottom. These particles range in size from fine clay and silt to large sand and gravel particles. *Siltation* refers to a suspension of small sediment particles in water. A high level of *total suspended solids (TSS)* produces a turbid water that blocks sunlight needed by aquatic vegetation. Sediments can harm aquatic organisms by clogging gills, suffocating eggs, and destroying habitat along the bottom of lakes, rivers, streams, and estuaries. Silt and sediment also interfere with the recreational uses and aesthetics of water bodies.

Land erosion from human activities such as mining, construction, logging, and farming is the major cause of sedimentation and siltation. Removal of vegetation on shorelines also can facilitate streambank erosion. As noted earlier, sediment particles carry nutrients responsible for eutrophication and depletion of dissolved oxygen.

Acidity Acidity is a key factor in water's ability to support aquatic life. Chemically, acidic water reflects a high concentration of hydrogen ions in solution. The negative logarithm of that concentration defines the quantity known as pH. Values of pH below 7 are *acidic*, whereas higher values are called *basic* or *alkaline*. Biological processes may be impaired or destroyed in waters that are too acidic. The viability of fish species such as brook trout and lake trout, for example, is noticeably diminished below a pH of 5.5, and most fish species cannot survive in waters with a pH below 5.0.

A variety of industrial processes generate acidic wastewater effluents that can contaminate surface waters. Drainage and runoff from mining operations and discarded mine wastes have long been major sources of acid loadings on streams, lakes, and rivers. Coal mining, for example, releases sulfur-bearing minerals that form dilute sulfuric acid when contacted with process water used in the mining operation, or with rainfall that leaches acidic materials from waste piles and carries them to nearby surface waters.

As discussed earlier in this chapter, acid deposition from the atmosphere directly onto surface waters or onto surrounding lands is another important source of acidic waters. Acid deposition can cause additional problems by liberating toxic metals such as aluminum that otherwise would remain chemically bound in soil. The ability of naturally occurring alkaline minerals like calcium and magnesium to neutralize (buffer) the acids in a water body also plays an important role in determining the severity of water acidification.

Salts   Although nearly 75 percent of our planet is covered with water, most of it is unfit for human use because of its high salt content. *Salts* refer to compounds of elements, including calcium, magnesium, sodium, and potassium, that produce positively charged ions in solution. These substances are referred to as cations (pronounced "cat-ions"). They combine in nature with negatively charged ions (anions) like chloride, sulfate, and bicarbonate to form the compounds we call salts. Salts tend to dissolve easily in water and are commonly measured in terms of total dissolved solids (TDS). Seawater has TDS concentrations that are roughly 60 times that of drinking water.

Salts dissolve naturally into water bodies as water flows over rocks and soils. Humanmade sources enter waterways via industrial and municipal discharges and urban runoff (such as the salt used for winter road deicing). Different organisms have different tolerances for salts, so the effects of salinity depend on the water use. The most critical demands tend to be for human drinking water and agricultural crops, which require freshwater with low TDS. Even where freshwater is used for irrigation, however, salinity problems often result from evaporation from soils and transpiration from plants, which leave an accumulation of salt deposits as water evaporates. Elevated salt concentrations can severely affect the productivity of crops and are an especially common problem in the western United States.

Heat   Thermal pollution is another way in which human activity affects water quality. The primary source is the waste heat from electric power plants. Roughly half of the fuel energy used by a power plant is dissipated as waste heat to waterways, typically to an adjacent water body. This thermal release creates a plume of warmed water that can be detrimental to fish and plant life. Species like salmon, for example, are extremely sensitive to temperature and cannot readily adjust to warmer waters. Warmer water also holds less dissolved oxygen. At the same time, the requirements for oxygen are increased because higher temperatures raise the metabolic rate of aquatic organisms. Although some fish species thrive in the warmer waters near a power plant, they too can be severely harmed by the sudden drop in temperature that occurs when a plant shuts down for scheduled maintenance or an unscheduled outage. For these reasons, most modern power plants are now required to install cooling towers that release waste heat to the atmosphere rather than to water bodies.

2.4.3   Drinking Water Quality

We turn now to a brief summary of where we stand in the United States with regard to measures and efforts to ensure that water supplies can support a variety of beneficial uses. We begin with drinking water because it is a basic need for human survival.

Residents of the United States are fortunate to be among the minority of people in the world who can reasonably assume that their community water supply is safe to drink. About half the U.S. drinking water supply comes from surface waters, which serve mainly large urban areas, and about half from groundwater sources, serving mainly rural areas. Only in 1974, however, was a uniform set of federal standards established for the approximately 200,000 public drinking water systems in the country. Under the Safe Water Drinking Act (SWDA) of 1974 (with subsequent amendments in 1986 and 1996), the U.S. Environmental Protection Agency (EPA) established maximum contaminant levels (MCLs) for a variety of organic and inorganic chemicals, along with specified treatment technologies that must be used to protect against disease-causing microorganisms.

Analogous to the Clean Air Act, the SWDA established primary standards designed to protect public health, plus secondary standards (which are nonenforceable guidelines) related to public welfare criteria such as water taste, color, odor, and aesthetics. Unlike air quality standards, however, the primary standards for drinking water quality also consider cost and technical feasibility. Tables 2.5 and 2.6 summarize the primary and secondary drinking water standards, respectively. Most are specified as MCLs given in units of milligrams per liter of water (mg/L). Because the density of pure water is 1000 g/L, an MCL of 1 mg/L is equivalent to one unit of contaminant per million units of water on a mass basis, or one part per million by weight (ppmw). For biological contaminants (pathogens), which are not easily measured in terms of an MCL, the primary standards specify a level of treatment technology consisting of filtration and disinfection, combined with testing for evidence of any fecal contamination, which is a good indicator of potential pathogens. The drinking water standards also include MCLs for several radionuclides, which are naturally occurring substances more likely to be found in groundwater than in surface waters. The measure of radioactivity is given in units of picocuries per liter of water (pCi/L).

Figure 2.14 shows one indicator used by the EPA to measure the overall success of federal, state, and local efforts to ensure safe drinking water. In 1994 nearly 20 percent of the U.S. population was served by systems that reported at least one violation of the primary drinking water standards. The EPA's goal is to reduce that figure to 5 percent by the year 2005.

As with many environmental standards, criteria for safe drinking water can be expected to become more stringent over time as treatment technology improves and as the health implications of contaminants become better understood. Toward that end, the EPA has established a number of maximum contaminant level goals (MCLGs) based solely on health considerations to indicate the direction of future standards development.

2.4.4 Surface Water Quality

National concern over the quality of surface waters in the United States dates back at least a hundred years to the federal River and Harbors Act of 1899. As with air pollution control, responsibility for water pollution control resided principally with state and local authorities until the 1970s. The Federal Water Pollution Control Act (FWPCA) of 1958 laid the groundwork for stronger federal involvement in pollution control efforts. But it

Table 2.5   Primary drinking water standards.

Primary Maximum Contaminant Levels (mg/L) for Drinking Water			
Microbiological Contaminants			
Total coliform bacteria	ML[a]	Turbidity	TT[b]
Fecal coliform and E. coli	ML[a]		
Inorganic Contaminants			
Antimony	0.006	Cyanide	0.2
Arsenic	0.05	Fluoride	4.0
Asbestos	7 MFL[c]	Lead	0.015 (AL)[d]
Barium	2.0	Mercury (inorganic)	0.002
Beryllium	0.004	Nitrate (as N)	10.0
Cadmium	0.05	Nitrite (as N)	1.0
Chromium	0.1	Selenium	0.05
Copper	1.3 (AL)[d]	Thallium	0.002
Radioactive Contaminants			
Beta and/photon emitters	4 mrem/yr	Combined radium	5 pCi/L
Alpha emitters	15 pCi/L		
Synthetic Organic Contaminants, Including Pesticides and Herbicides			
2,4-D	0.07	Endrin	0.002
2,4,5-TP (Silvex)	0.05	Epichlorohydrin	TT[b]
Acrylamide	TT[b]	Ethylene dibromide	0.00005
Alachlor	0.002	Glyphosate	0.7
Atrazine	0.003	Heptachlor	0.0004
Benzo(a)pyrene (PAH)	0.0002	Heptachlor epoxide	0.0002
Carbofuran	0.04	Hexachlorobenzene	0.001
Chlordane	0.002	Hexachloro-cyclopentadiene	0.05
Dalapon	0.2	Lindane	0.0002
Di(2-ethylhexyl)adipate	0.4	Methoxychlor	0.04
Di(2-ethylhexyl)phthalate	0.006	Oxylamyl (vydate)	0.2
Dibromochloropropane	0.0002	PCBs (polychlorinated biphenyls)	0.0005
Dinoseb	0.007	Pentachlorophenol	0.001
Dioxin (2,3,7,8-TCDD)	0.00000003	Picloram	0.5
Diquat	0.02	Simazine	0.004
Endothall	0.1	Toxaphene	0.003
Volatile Organic Contaminants			
Benzene	0.005	Ethylbenzene	0.7
Carbon tetrachloride	0.005	Styrene	0.1
Chlorobenzene	0.1	Tetrachloroethylene	0.005
o-Dichlorobenzene	0.6	1,2,4-Trichlorobenzene	0.07
p-Dichlorobenzene	0.075	1,1,1-Trichloroethane	0.2
1,2-Dichloroethane	0.005	1,1,2-Trichloroethane	0.005
1,1-Dichloroethylene	0.007	Trichloroethylene	0.005
cis-1,2-Dichloroethylene	0.07	TTHMs (total trihalomethanes)	0.10

(continued)

Table 2.5 *(continued)*

trans-1,2-Dichloroethylene	0.1	Toluene	1.0
Dichloromethane	0.005	Vinyl chloride	0.002
1,2-Dichloropropane	0.005	Xylenes	10.0

[a] Monthly limits (ML) specified for routine sampling.

[b] Treatment technique (TT) required rather than an MCL.

[c] Million fibers/liter (MFL).

[d] Action level (AL) requiring treatment.

Source: CFR, 1999.

Table 2.6   Secondary drinking water standards.

Contaminant	Maximum Contaminant Level (MCL)
Aluminum	0.05–0.2 mg/L
Chloride	250 mg/L
Color	15 color units
Copper	1 mg/L
Corrosivity	Noncorrosive
Fluoride	2.0 mg/L
Foaming agents	0.5 mg/L
Iron	0.3 mg/L
Manganese	0.05 mg/L
Odor	3 threshold odor no.
pH	6.5-8.5
Silver	0.1 mg/L
Sulfate	250 mg/L
Total dissolved solids	500 mg/L
Zinc	5 mg/L

Source: CFR, 1999.

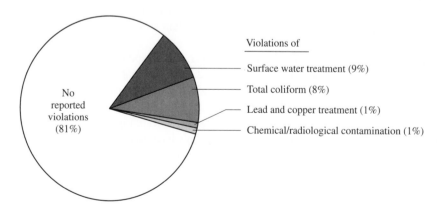

Figure 2.14   Compliance with U.S. drinking water standards, 1994. This indicator shows the percentage of U.S. population served by community drinking water systems that violated health-based requirements. (*Source:* USEPA, 1996a)

was the 1972 amendments to the FWPCA that, for the first time, established national goals for the quality of U.S. surface waters, together with a national program of control requirements and permits for wastewater discharges from industrial and municipal sources. The ambitious goal of the 1972 law was "that the discharge of pollutants into the navigable waters [of the United States] be eliminated by 1985." An interim goal was to achieve by mid-1983 a level of water quality "which provides for the protection and propagation of fish, shellfish and wildlife, and provides for recreation in and on the water." Although these goals remain to be fully met (especially the zero discharge goal), they have been the major driving force in cleaning up the nation's waterways.

The 1972 legislation and the subsequent Clean Water Act of 1977 focused strongly on the use of technology to control wastewater effluents. "Best available control technology" was required at all new facilities and at all plants discharging any of the 126 chemicals identified by the EPA as "priority toxic pollutants" (see Table 2.7). So-called "conventional" pollutants such as BOD and total suspended solids were subject to a less stringent level of control defined as "best practicable control technology currently available." These fuzzily worded requirements were turned into quantitative limits by industry and by source via federal New Source Performance Standards (NSPS) and EPA effluent guidelines. These standards and guidelines are used to set the conditions for operating permits now required for all point source discharges under the National Pollutant Discharge Elimination System (NPDES).

Complementing these regulatory requirements, a program of federal grants to state and local governments was begun in 1958 for the construction of municipal wastewater treatment systems. The national clean water program also called for ambient water quality standards to be established by each state to protect various uses of water, based on guidelines and criteria developed by the EPA. In many cases these ambient standards determine the allowable discharge limits for sources along a waterway.

State and local water authorities are required to report to the EPA biannually on progress toward attaining their water quality goals, and from these reports emerges a picture of the overall quality of U.S. surface waters. In general, these reports indicate improvements in water quality over the past several decades in the face of pressures from population growth and economic activity. However, the snapshot that emerges as of 1994 (Table 2.8) shows that more than a third of U.S. surface waters still fall short of water quality objectives.

As shown in Figure 2.15, the contaminants most responsible for water quality degradation are nutrients and bacteria. Figure 2.16 shows that the agricultural industry is the leading source of water quality degradation, impairing 60 percent of the nation's rivers. Urban runoff, including untreated discharges from storm sewers, is the second major contributor to poor water quality. Overall, Figure 2.16 suggests that after several decades of regulatory focus on point source discharges, the major problems today arise primarily from nonpoint sources of water contaminants.

2.4.5 Groundwater Quality

As with surface waters, responsibility for monitoring and reporting U.S. groundwater quality lies with individual states. The nature and extent of state programs vary widely across the country. Because of the difficulty and expense of monitoring

Table 2.7   Priority toxic pollutants for water discharges.

Chemical Name[a]		
Antimony	Trichloroethylene	Fluoranthene
Arsenic	Vinyl chloride	Fluorene
Beryllium	2-Chlorophenol	Hexachlorobenzene
Cadmium	2,4-Dichlorophenol	Hexachlorobutadiene
Chromium (III)	2,4-Dimethylphenol	Hexachlorocyclopentadiene
Chromium (VI)	2-Methyl-4-chlorophenol	Hexachloroethane
Copper	2,4-Dinitrophenol	Indeno(1,2,3-cd)pyrene
Lead	2-Nitrophenol	Isophorone
Mercury	4-Nitrophenol	Naphthalene
Nickel	3-Methyl-4-chlorophenol	Nitrobenzene
Selenium	Pentachlorophenol	N-Nitrosodimethylamine
Silver	Phenol	N-Nitrosodi-n-propylamine
Thallium	2,4,6-Trichlorophenol	N-Nitrosodiphenylamine
Zinc	Acenaphthene	Phenanthrene
Cyanide	Acenaphthylene	Pyrene
Asbestos	Anthracene	1,2,4-Trichlorobenzene
2,3,7,8-TCDD (Dioxin)	Benzidine	Aldrin
Acrolein	Benzo(a)anthracene	alpha-BHC
Acrylonitrile	Benzo(a)pyrene	beta-BHC
Benzene	Benzo(a)fluoranthene	gamma-BHC
Bromoform	Benzo(ghi)perylene	delta-BHC
Carbon tetrachloride	Benzo(k)fluoranthene	Chlordane
Chlorobenzene	bis(2-Chloroethoxy)methane	4,4'-DDT
Chlorodibromomethane	bis(2-Chloroethyl)ether	4,4'-DDE
Chloroethane	bis(2-Chloroisopropyl)ether	4,4'-DDD
2-Chloroethylvinyl ether	bis(2-Ethylhexyl)phthalate	Dieldrin
Chloroform	4-Bromophenyl phenyl ether	alpha-Endosulfan
Dichlorobromomethane	Butylbenzyl phthalate	beta-Endosulfan
1,1-Dichloroethane	2-Chloronaphthalene	Endosulfan sulfate
1,2-Dichloroethane	4-Chlorophenyl phenyl ether	Endrin
1,1-Dichloroethylene	Chrysene	Endrin aldehyde
1,2-Dichloropropane	Dibenzo(a,h)anthracene	Heptachlor
1,3-Dichloropropylene	1,2-Dichlorobenzene	Heptachlor epoxide
Ethylbenzene	1,3-Dichlorobenzene	PCB-1242
Methyl bromide	1,4-Dichlorobenzene	PCB-1254
Methyl chloride	3,3-Dichlorobenzidine	PCB-1221
Methylene chloride	Diethyl phthalate	PCB-1232
1,2,2,2-Tetrachloroethane	Dimethyl phthalate	PCB-1248
Tetrachloroethylene	Di-n-butyl phthalate	PCB-1260
Toluene	2,4-Dinitrotoluene	PCB-1016
1,2-trans-dichloroethylene	2,6-Dinitrotoluene	Toxaphene
1,1,1-trichloroethane	Di-n-octyl phthalate	
2,4-Dichlorophenol	1,2-Diphenylhydrazine	

[a] For aquatic life, there are 30 priority pollutants with some type of concentration criteria (μg/L) for acute or chronic effects in freshwater or saltwater. For human health (10^{-6} risk of carcinogenicity), there are 91 priority pollutants with consumption criteria for either "water + fish" or "fish only."
Source: CFR, 1999.

Table 2.8   Quality of U.S. surface waters, 1994.

Type of Water Body	Total U.S. Quantity	Percentage Assessed	Quality of Assessed Waters	
			Good	Impaired
Estuaries	89,100 km^2	78	63	37
Lakes	165,000 km^2	42	63	37
Rivers: Total	5.6 million km	17	64	36
Constant flow	2.1 million km	48		

Source: Based on USEPA, 1996a.

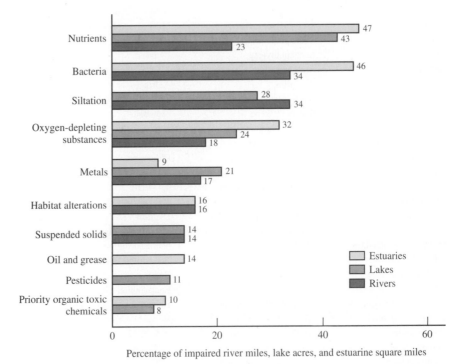

Percentage of impaired river miles, lake acres, and estuarine square miles

Figure 2.15   Leading surface water contaminants, 1994. This graph shows the percentage of impaired U.S. surface waters that are affected by a particular pollutant or stressor. (*Source:* USEPA, 1996a)

groundwater quality, most states do not yet have extensive monitoring networks; rather, sampling has focused mainly on locations where contamination problems already are known or suspected. The most prevalent information currently comes from measurements of groundwater sources of drinking water. Although these data indicate that overall groundwater quality is good, many states report a growing number of contaminants in groundwater supplies.

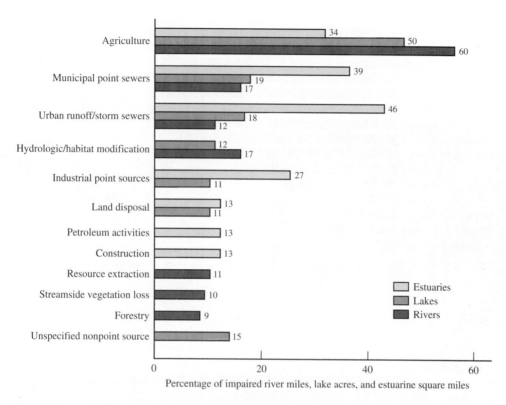

Figure 2.16 Leading sources of water quality impairment, 1994. This graph shows the human activities responsible for the impairment of U.S. surface waters. (*Source:* USEPA, 1996a)

Figure 2.17 shows the leading groundwater pollutants, and Figure 2.18 indicates the major sources of contamination. The highest-priority problem is petroleum compounds (organic chemicals) from leaking underground storage tanks (often known by their acronym, LUST). Approximately 1.2 million storage tanks are buried at over half a million sites across the country. Many of these tanks are at automobile service stations that store gasoline, diesel fuel, and other petroleum products. About 12 percent of underground storage tanks already have leaked and impacted groundwater quality (USEPA, 1995).

The second most widespread problem is nitrates, a nutrient most extensively found in agricultural fertilizers. As described earlier, high levels of nitrates in drinking water are chemically converted to nitrites that cause the "blue baby" syndrome called methemoglobinemia. Other groundwater contaminants listed in Figure 2.17 include all of the major pollutants discussed earlier. Of special note are landfill sites, including abandoned chemical waste dumps, which are a major source of the toxic metals and organic compounds that infiltrate groundwater.

An important insight from Figure 2.18 is that the principal sources of groundwater contaminants are small, dispersed facilities and nonpoint sources such as agriculture, which have been largely ignored in national environmental regulatory programs

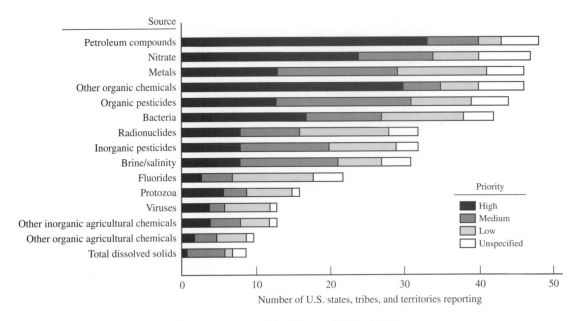

Figure 2.17   Leading groundwater contaminants, 1994. (*Source:* USEPA, 1995)

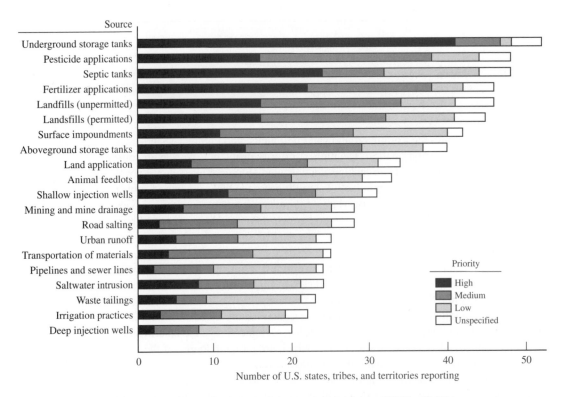

Figure 2.18   Leading sources of groundwater contamination, 1994. (*Source:* USEPA, 1995)

over the past several decades. The realization that groundwater as well as surface water has been contaminated by human activities is relatively recent, and the full extent of that contamination is still not well understood. Only recently have we also come to understand the magnitude and importance of groundwater flows into surface streams and the potential for groundwater pollutants to subsequently contaminate surface waters. Much remains to be learned about the nature of groundwater pollution, including the ability of soils and ecosystems to remove or assimilate groundwater contaminants and the effectiveness of human interventions to address the problem.

2.5 SOLID AND HAZARDOUS WASTES

Unlike contaminated air and water, which directly affect human health, pollution of the land or soil from dumping or burial of solid wastes affects most people less directly (although children and animals sometimes ingest soil directly). The primary environmental concern is that a waste material in soil may migrate into surface water or groundwater, where it can then be ingested and harm living organisms (including people). Substances such as heavy metals in soil also can enter the food chain via uptake by plants and vegetation that are subsequently consumed by animals and humans, again with potentially harmful effects. And of course, land disposal sites for human wastes historically have been smelly, unpleasant places that often serve as breeding grounds for disease-carrying insects and rodents.

This section outlines the solid waste problem and highlights some of the main regulatory developments to prevent and alleviate the environmental impacts of waste disposal on land. The discussion is organized into two parts dealing with the two categories of waste materials: hazardous and nonhazardous.

2.5.1 Hazardous Wastes

In 1976 a landmark federal law called the Resource Conservation and Recovery Act (known by its acronym RCRA, pronounced "rik-ra") created two classes of solid wastes: hazardous and nonhazardous. The definition of *hazardous* is clearly open to interpretation. In the legal language of environmental statutes, RCRA Subtitle C defines a hazardous waste as "a solid waste or combination of solid wastes which because of the quantity, concentration, or physical, chemical, or infectious characteristics may (1) cause, or significantly contribute to, an increase in mortality or an increase in serious irreversible, or incapacitating reversible illness; or (2) pose a substantial present or potential hazard to human health or the environment when improperly treated, stored, transported, or disposed of, or otherwise managed."

Thus hazardous wastes are distinguished from other pollutants by the severity of their effects and the immediacy of the danger.[4] The U.S. EPA further defines a hazardous waste as one with any of the following four characteristics:

• *Ignitability*: An ability to burn easily or cause or enhance fires.

---

4 The most damaging substances are designated as *toxics*, which are a subset of hazardous wastes. However, because these terms are often used interchangeably, we distinguish between the two only when it is important to the discussion at hand.

1991 primary hazardous waste quantity = 224 million metric tons

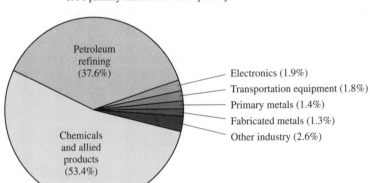

Figure 2.19   Quantities and sources of hazardous waste generation. This graph
identifies the major manufacturing industries that generate primary
hazardous wastes. (*Source:* USEPA, 1996b)

- *Corrosivity*: Strong acids and bases, or substances able to corrode metal.
- *Reactivity*: An ability to react violently or cause explosions, including reactions with water.
- *Toxicity*: An ability to threaten water supplies and health, as determined by a laboratory test of leachability.

Although hazardous wastes are defined in law as solid wastes, in fact the vast majority of these substances are liquids.[5] Among these are solvents, organic chemicals, petroleum products, paints, strong acids, and wastewater containing hazardous materials.

The EPA has developed more detailed definitions and tests for the four hazardous waste criteria just listed. For example, to determine whether a substance meets the toxicity criterion, a chemical test called the Toxicity Characteristic Leaching Procedure (TCLP) has been developed. If the metals and other chemical compounds specified in the TCLP are leached in amounts above specified thresholds, the material is considered hazardous. To further clarify the definition of a hazardous waste, the EPA also has listed many specific industrial wastes and chemicals that are declared hazardous.

Under RCRA, any waste designated as hazardous is subject to "cradle-to-grave" management, which requires tracking the waste from the time it is generated to its ultimate disposal. The treatment, storage, and disposal of hazardous wastes must be handled only by facilities that operate in accordance with strict EPA-specified procedures requiring special operating permits.

Figure 2.19 shows the quantities and sources of primary hazardous wastes regulated by the EPA under RCRA. Of the nearly 230 million metric tons generated in

---

5 For convenience, RCRA simply defines a solid waste to include solids, liquids, and gases. Such a definition could have come only from lawyers, not from engineers!

1991, more than half came from the chemical manufacturing industry. Of the total, over 97 percent was hazardous wastewater. This large quantity of liquids arises in part because one of the original motivations behind RCRA was to ensure the safe disposal of potentially hazardous residues from air pollution and water pollution control technologies mandated under the Clean Air Act and Clean Water Act. Thus, under RCRA, mixtures of hazardous residues with nonhazardous substances like water may cause the entire liquid stream to be designated as hazardous.

Accordingly, Figure 2.20 shows that the most prevalent method of disposing of hazardous wastes is via wastewater treatment systems. These technologies use physical separation, chemical reactions, and biological treatment methods to remove, neutralize, or destroy hazardous compounds. Ironically, the sludge or residue from some of these processes may generate a new hazardous waste by concentrating toxic materials from a dilute wastewater stream. This is an example of a secondary hazardous waste. In 1991 secondary wastes added 28 million metric tons to the burden of primary wastes subject to the special handling and disposal requirements of RCRA Subtitle C.

Because RCRA exempts many small waste generators and does not apply to certain specified facilities like municipal waste landfills, the EPA's figures for RCRA hazardous waste generation represent a lower bound on total hazardous waste quantities in the country. In addition to RCRA, a number of other federal laws govern specific types of hazardous wastes or substances such as pesticides and toxic chemicals. By one estimate, the total quantity of hazardous wastes from all sources would amount to over three times the RCRA quantity (Phipps, 1996).

Note in Figure 2.20 that only a very small fraction of hazardous wastes, about 1 percent, are now disposed of in surface landfills. Such was not always the case. As recently as the late 1960s, landfill disposal of what we today call hazardous wastes was common and widespread. Industrial practices ranged from burial of wastes in sealed containers to haphazard illegal dumping in the dark hours of the

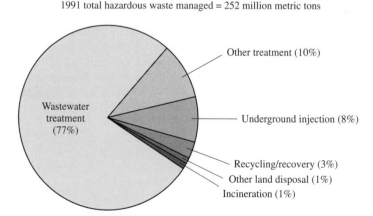

1991 total hazardous waste managed = 252 million metric tons

Figure 2.20   Hazardous waste management practices showing the quantity of total wastes managed by different treatment and disposal methods. (*Source:* USEPA, 1996b)

night. Unlike today's hazardous waste landfills, which are specially designed with impermeable barriers to prevent leaching into soils, landfills of the past took few if any precautions to prevent leaching. The legacy of such practices over many decades is now a major environmental problem. Hazardous wastes, including materials leaking from buried containers, are polluting groundwater supplies and lands across the country.

Historically, a single catastrophe often has focused public attention and galvanized political support to address an environmental problem. So it was that a 1978 incident at Love Canal near Niagara Falls, New York, highlighted the dangers of hazardous wastes. Families were relocated and a community dismantled when toxic chemicals from an abandoned industrial waste dump were found to be leaking into homes, contaminating soils, and exposing children to potential carcinogens. In 1980 Congress responded with the Comprehensive Environmental Response, Compensation, and Liability Act (CERCLA), more popularly known as Superfund.

This law and its subsequent amendments gave the federal government authority to clean up dangerous or abandoned waste sites and to force "responsible parties" to pay for cleanup costs. Special taxes imposed primarily on the chemical and petroleum industries provided most of the $1.6 billion trust fund (the source of the nickname Superfund) created to deal with emergency problems until responsible parties could be forced to pay. In 1986 this fund was increased to $8.5 billion. Nevertheless, more than a decade after CERCLA was enacted, only a handful of the 1,200 Superfund sites listed by the EPA as priority problems had been fully remedied. Part of the reason was technical: Cleanup processes are slow and costly. But the slow pace of cleanup was also due to legal issues surrounding the liability provisions of CERCLA, which held that any party involved in an EPA-listed site could be retroactively held liable for all cleanup costs. The adjudication of who is responsible for site cleanup costs, and how clean is clean, has delayed action and consumed enormous resources. Indeed, it is said that under CERCLA far more money has been spent on legal proceedings and studies than on actual site cleanup.

Widespread criticism of CERCLA has led in recent years to more flexible approaches, such as the EPA's Brownfields initiative. This program seeks to promote the redevelopment of abandoned industrial sites in ways that are environmentally sound, but without the rigid requirements imposed by CERCLA. This program is widely regarded as a step in the right direction. Nonetheless, the environmental mistakes of the past will be with us for many decades as we deal with hazardous wastes in the environment.

2.5.2   Nonhazardous Wastes

Wastes that are not designated as hazardous under Subtitle C of the Resource Conservation and Recovery Act (RCRA) are said to be *nonhazardous*. The handling and disposal of such wastes are regulated under a different section of RCRA (Subtitle D), which builds on earlier federal laws such as the Solid Waste Disposal Act of 1965 and the Resources Recovery Act of 1970.

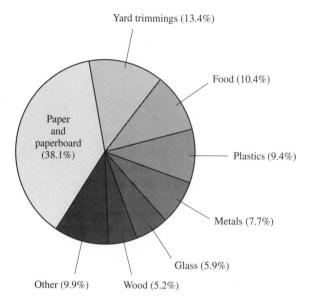

1996 total = 191 million metric tons

Figure 2.21   Composition of U.S. municipal solid waste, showing the percentage by weight of total MSW in 1996. (*Source:* USEPA, 1998c)

The most familiar type of nonhazardous waste is what we commonly call trash or garbage.[6] These are components of *municipal solid waste (MSW),* which includes all of the wastes commonly generated in residences, commercial buildings (like shopping malls, restaurants, and corporate offices), and institutional buildings (such as universities and government offices). MSW consists of such things as paper, packaging, plastics, food wastes, glass, yard wastes, wood, and discarded appliances. Similar kinds of wastes generated by industrial facilities also are part of MSW. The additional wastes generated by manufacturing processes, construction activities, mining and drilling operations, agriculture, and electric power production are distinct from MSW and are referred to as *industrial wastes.*

The total quantity of MSW generated in the United States in 1996 was 191 million metric tons, equivalent to 2.0 kg (4.4 lbs) per person per day. The average amount of waste discarded by each U.S. resident has nearly doubled since 1960. Figure 2.21 shows the composition of MSW, and Figure 2.22 shows the historical trend in municipal solid waste management practices.

6 Technically, these terms each refer to a different type of solid waste. *Garbage* refers to household food wastes, and *trash* is the combustible part of the remaining household wastes, which are defined as *rubbish.* Although these distinctions are important to environmental engineers, in colloquial use the terms are interchangeable.

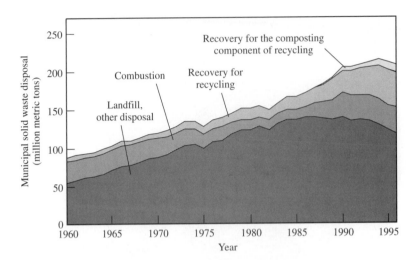

Figure 2.22   Trend in municipal solid waste management practices. In recent years the amount of MSW recovered for recycling has increased, although landfill remains the dominant waste disposal method. (*Source:* USEPA, 1998c)

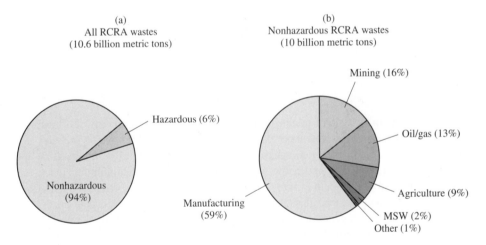

Figure 2.23   Total quantity of RCRA solid wastes. Most RCRA wastes are defined as nonhazardous. This graph displays the sources of nonhazardous wastes in 1991. (*Source:* OTA, 1992)

Big as these quantities are, the amount of MSW generated each year is dwarfed by the waste quantities generated annually by industry and agriculture. Figure 2.23 gives the breakdown by sector. The quantities now are measured in billions of tons per year, with manufacturing industries contributing the largest share of the 10 billion metric ton total. The mining industry, oil and gas industry, and agricultural sector are the next leading sources of U.S. solid waste. If one considers that these wastes are being generated to support the needs and well-being of all Americans, the average per capita waste quantity jumps to over 100 kg per person per day.

The environmental threats posed by municipal and industrial wastes are varied. Though defined as "nonhazardous" under RCRA, many of these wastes are nonetheless directly or indirectly capable of harming human health and the viability of other living species.

Consider municipal solid waste. Historically, common practice was to discard these wastes in out-of-the-way areas that were unsuited for other purposes. As noted earlier, such garbage dumps frequently served as breeding grounds for disease-carrying insects and vermin.[7] As seen in Figure 2.22, land disposal of MSW continues to be the dominant practice, accounting for about two-thirds of MSW disposal. However, modern sanitary landfills are engineered to avoid most environmental problems of the past. Nonetheless, some municipal landfills are still a potential source of groundwater and surface water contamination via runoff and leaching. Although MSW is legally classified as nonhazardous, inevitably it also contains discarded hazardous wastes like batteries, paints, solvents, and waste motor oil. Such items add heavy metals and organic compounds to the inventory of potential contaminants in soil and water.

The gradual decomposition of landfill wastes over several decades also generates new environmental problems in the form of air pollutants. Trace organic gases, including air toxics, may be emitted from landfills, along with significant amounts of methane and carbon dioxide, both of which are important greenhouse gases. In recent years some landfill operators have installed systems to collect these waste gases, especially methane, which can be burned to recover useful energy.

In many parts of the country, especially in urban areas, space for sanitary landfills is either limited or unavailable. As a practical matter, then, other methods of disposing of MSW are heavily utilized. Open burning of some types of MSW, especially garden and yard wastes, was common practice until prohibited by modern air pollution laws. Incinerators also have been widely used to burn municipal wastes in urban areas. In New York City, for example, apartment buildings often had their own small incinerator. Most of those units operated poorly and inefficiently, with signature plumes of black soot marking each source of air pollution. Today many of the chemical constituents in that incinerator smoke would be classified as hazardous air pollutants.

Large municipal incinerators also have developed a reputation as a major source of air pollution because many of these facilities were improperly designed or operated. Of particular concern are toxic organic compounds such as dioxin and furans, which are formed from chlorine-containing wastes that are not completely combusted. Even trace amounts of such compounds are carcinogenic. Incinerators that are properly designed and operated, however, can effectively reduce the volume of solid wastes without producing toxic air pollutants. Nonetheless, because of the history of environmental problems with incinerator technology, their use has declined in the face of public opposition. Even the best incinerator still leaves a solid residue of incombustible material that often is classified as hazardous waste because of high concentrations of heavy metals.

---

7 Disease carriers also are known as *vectors* (not to be confused with the vectors you use in mathematics and physics).

Disposal of municipal solid waste (MSW) is among the major environmental problems facing urban areas. The average American generates about 2 kg of MSW each day, twice the rate of forty years ago.

The problems with municipal waste also apply to industrial wastes. Because of the much larger volumes of waste materials, landfills are the preferred method of waste disposal.

Of course, the best way to avoid the environmental problems of solid waste disposal is to not generate the waste in the first place. Pollution prevention programs aimed at this objective have become widespread. Less paper and cardboard used in packaging and lighter-weight containers for soft drinks, food products, and paint cans are but a few examples. The recycling and reuse of materials are other important ways to avoid waste generation. In many industries such practices already are well established. Discarded automobiles, for example, have about 75 percent of their materials recovered for productive uses, with car batteries recycled nearly 100 percent.

At the residential level, recycling programs for newspapers, glass, and metal containers also have been established in many communities. These programs are regarded as a positive step in reducing solid wastes and environmental impacts. The economic viability of such programs, however, depends strongly on the markets for by-product materials, which can fluctuate widely. The city of Pittsburgh, for example, halted its newspaper recycling program in 1996, just four years after it began, when newsprint prices plummeted and the city was forced to pay a premium just to dispose of the wastepaper in a landfill. Several years later the program was reinstated when the demand and price for recycled newspaper returned to their previous levels.

Even from a purely environmental point of view, some municipal recycling programs have been criticized for increasing environmental emissions of air pollutants from the fuel combustion required for additional collection trucks and trips to special drop-off points. This type of "cross-media" impact—where the solution to a problem in one environmental medium (air, water, or land) creates a new problem elsewhere—

again illustrates the often complex nature of environmental problem solving. Chapter 7 explores in more detail the use of life cycle assessments for addressing such issues.

2.6 RADIOACTIVE WASTES

This category of waste material is discussed separately from other types of wastes because it is indeed a class by itself. Two key attributes distinguish radioactive waste from the pollutants and hazardous wastes discussed so far: (1) Its harmful effects on living organisms are induced by radiation rather than by chemical mechanisms, and (2) radioactive wastes remain dangerous up to hundreds of thousands of years—a time scale measured in geological terms that is arguably beyond the comprehension of most people.

Radioactive wastes are a product of the military and civilian uses of nuclear energy, which emerged in the United States in 1945 and grew rapidly through the early 1980s. Wastes are generated in all phases of the nuclear "fuel cycle" that starts with mining of uranium ore and continues through the refining and processing of that ore into nuclear fuels, used for electric power production, and other elements like plutonium, used for weapons manufacture. The fuel rod assemblies used in nuclear power plants must be replaced regularly as the fuel "burns out" and is no longer efficient at generating heat. At the end of the cycle, waste products include the spent fuel assemblies from nuclear power plants as well as wastes from nuclear weapons production. Power plant materials like concrete and other building materials may also become radioactive. In addition, smaller amounts of nuclear wastes are generated by civilian activities such as medical treatment and research. Figure 2.24 shows the quantity of nuclear waste materials awaiting disposal in the United States in 1995.

In 1954 amendments to the Atomic Energy Act empowered the Atomic Energy Commission (AEC) to oversee all civilian and military uses of nuclear energy, including waste disposal. In the 1970s the AEC's activities were taken over by the newly formed Department of Energy and Nuclear Regulatory Commission (NRC).

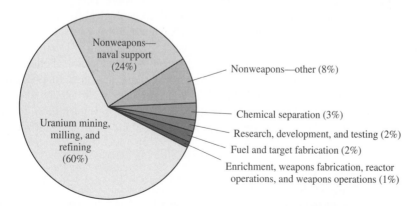

Figure 2.24   Total volume and sources of nuclear waste awaiting disposal by the U.S. Department of Energy in 1995. (*Source:* USDOE, 1997b)

These federal agencies have primary responsibility for radioactive waste disposal, with state authorities responsible for less dangerous "low-level" wastes as described in Section 2.6.3.

2.6.1   High-Level Waste

The environmental impacts of nuclear waste vary with the nature and form of the waste material. The most dangerous waste is called *high-level waste.* This includes the spent fuel from nuclear reactors, as well as the highly radioactive liquids and solids produced from any *reprocessing* of spent fuel to manufacture plutonium and uranium.[8] High-level wastes require permanent isolation from the human environment. The danger arises from the stream of subatomic particles that are released when unstable forms of uranium and its highly radioactive daughter products break apart (fission). These particles of varying energy levels (called alpha, beta, and gamma) can damage or destroy living tissue, inducing mutations and various forms of cancer. Death from exposure to intense radiation can occur over periods ranging from days to many years, depending on the intensity and duration of the exposure. Human exposure can occur via inhalation of radioactive substances and ingestion of foods or liquids containing radioactive materials.

High-level waste is characterized not only by the intensity of its radioactivity[9] but also by its very long *half-life,* which is the time required for a radioactive element to decay to half its initial mass. Table 2.9 lists the half-lives of some common radioactive elements (radionuclides). Although high-level waste includes many isotopes with relatively short half-lives (like iodine-131 with a half-life of eight days), it also contains elements like plutonium-239, whose half-life is 24,000 years. That means that 24,000 years from now, only half the plutonium currently awaiting disposal will have decayed into other, more stable materials. The remaining plutonium and other radionuclides generated in the decay chain will remain hazardous for tens to hundreds of thousands of years after that. Such is the environmental legacy of high-level radioactive waste.

As the 21st century begins, the United States has no permanent method yet in place for disposal of high-level wastes. The Nuclear Waste Policy Act of 1982 (NWPA) called for such wastes to be buried in a deep geological formation that is unlikely to be disturbed for thousands of years. Such a repository was to have been identified, built, and placed into operation by the U.S. Department of Energy (DOE) by January 1998. But strong political opposition and legal and technical challenges have thus far prevented a permanent repository from being established. In 1987 Congress amended the law, directing that a site at Yucca Mountain, Nevada, be the only location considered for high-level waste disposal. To date, several billion dollars have been spent on site evaluations and studies of safety and environmental accept-

---

8 Reprocessing is currently practiced in France and the United Kingdom. The United States, however, terminated its reprocessing activities in the 1970s, largely in the interests of nuclear nonproliferation.

9 Several different unit systems are used to measure radioactivity. One of the most common is the *curie* (Ci), named after the French scientists Marie and Pierre Curie, who studied radioactivity in the early 1900s. One curie is equivalent to 3.7×10^{10} atomic fissions per second, which is the approximate fission rate of 1 gram of radium. Because this is a very large quantity, units such as nanocuries (nCi) are often used.

Table 2.9 Half-lives of important isotopes encountered in radioactive wastes.

Element	Isotope	Half-Life (Years)	Radiation of Primary Concern
Americium	Am-241	460	Alpha particles
	Am-243	7,370	Alpha particles
Cesium	Cs-135	2,000,000	Beta and gamma particles
	Cs-137	30	Beta and gamma particles
Curium	Cm-243	32	Alpha particles
	Cm-244	18	Alpha particles
Iodine	I-229	16,000,000	Beta particles
Neodymium	Nb-94	20,000	Beta and gamma particles
Neptunium	Np-237	2,100,000	Alpha particles
Plutonium	Pu-239	24,000	Alpha particles
	Pu-240	6,580	Alpha particles
Strontium	Sr-90	29	Beta particles
Technetium	Tc-98	2,000,000	Beta particles
Thorium	Th-229	7,340	Alpha particles
	Th-230	77,000	Alpha particles

Source: Based on NRC, 1996.

ability. If the Yucca Mountain site is ultimately approved and licensed by the Nuclear Regulatory Commission (NRC), it will be 2010 at the earliest before permanent storage of high-level wastes could begin. If the site is ultimately rejected, the DOE must return to Congress for further instructions.

In the meanwhile, U.S. high-level wastes remain temporarily stored at power plants and government facilities around the country. As of 1997, about 35,000 metric tons of spent fuel were stored aboveground in large water pools at about 70 power plant sites. These pools originally were designed as temporary storage facilities to dissipate the large amount of heat and radiation generated by newly spent fuel rods removed from operating reactors. In the absence of a permanent storage facility, nuclear plants have had to increase the packing density of spent fuel assemblies and build additional pool capacity. Dry storage of older fuel rods in specially designed metal casks or concrete modules is expected to grow over the next decade, with existing pool capacity being used for the 2,000 metric tons of newly spent fuel generated each year by U.S. reactors. On a volume basis, the spent fuel assemblies currently in storage would fill a single football field about 3 meters high—a volume of roughly 14,000 m^3 (NRC, 1996).

In contrast, the volume of high-level wastes currently stored at government facilities managed by the DOE totals 380,000 m^3 (USDOE, 1997b), or nearly 30 times the volume of commercial high-level wastes. About 92 percent of this is chemical separation wastes from nuclear weapons production, with the remainder from nonweapons activities. Most of these wastes are stored underground in huge million-gallon tanks located at Hanford, Washington and Savannah River, South Carolina. The current plan is to stabilize these wastes using chemical vitrification processes that can solidify liquid and solid materials. A history of accidental releases and leakage of radioactive wastes at Hanford and other government facilities has increased the urgency of finding acceptable technical and political solutions to the high-level waste disposal problem.

2.6.2   Transuranic Wastes

This category of waste is defined by the NRC to include elements heavier than uranium (atomic numbers greater than 92) that emit radiation at the specified levels but are not as radioactive as high-level waste. Most transuranic (TRU) waste is the result of weapons production, principally the fabrication of plutonium weapons components, chemical separation of plutonium, and recycling of plutonium from production scrap, residues, or retired weapons. Because of the long half-lives of many TRU isotopes, this waste—like high-level waste—can remain radioactive for hundreds of thousands of years. Unlike high-level wastes, however, TRU waste includes a broad spectrum of waste materials, most with half-lives of tens to hundreds of years.

The U.S. DOE currently manages 220,000 m^3 of TRU wastes — about half the volume of high-level wastes. Prior to 1970, most transuranic wastes were disposed of in shallow burial trenches along with low-level wastes (discussed shortly). About two-thirds of the DOE's total TRU waste volume has been disposed of in this fashion. In 1970 the Atomic Energy Commission (predecessor to the NRC) recognized the need to isolate transuranic wastes more permanently from the environment. Thus all post-1970 wastes are currently stored in metal drums or boxes awaiting final disposal in a geological repository. A site near Carlsbad, New Mexico, known as the Waste Isolation Pilot Plant (WIPP), has been developed to store TRU wastes and began operation in 1999. There are no current plans, however, to unearth and relocate the TRU wastes disposed of before 1970.

2.6.3   Low-Level Waste

Any radioactive waste that is not officially classified as high-level waste, transuranic waste, or by-product waste from uranium mining and milling is called *low-level waste (LLW)*. About 80 percent of such wastes come from sources in the civilian sector and about 20 percent from the governmental sector. Figure 2.25 shows the recent trend in annual low-level waste volumes. The total cumulative volume of LLW managed by the DOE is about 3.3 million m^3 (USDOE, 1997b).

As its name implies, LLW consists of mainly low levels of radioactivity spread across large volumes of material. Such wastes are generated at any facility that processes, creates, or otherwise handles radioactive materials. Examples of low-level wastes include protective gloves and clothing worn by workers at nuclear power plants; bottles, test tubes, and syringes used in medical research and treatment; and contaminated water, pipes, or equipment from the chemical processing or treatment of nuclear materials. The NRC has defined three classes of low-level waste (called Class A, B, and C), ranging from the least radioactive and shortest-lived materials to

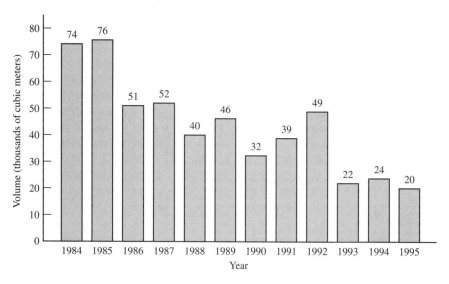

Figure 2.25   Trend in low-level waste volume received at U.S. disposal facilities. (*Source:* NRC, 1997)

Figure 2.26   Twenty-eight thousand drums of low-level radioactive waste awaiting treatment and disposal at a storage yard in Oak Ridge, Tennessee.

the most radioactive and longest-lived wastes. Although some low-level wastes can be more radioactive than some types of high-level waste, most LLW is designated as Class A (least dangerous).

Although the federal government is responsible for disposal of high-level and transuranic wastes, the disposal of low-level wastes is the responsibility of state governments, subject to federal approval. The only approved method of disposal is burial of waste containers in special landfills. Figure 2.26 shows a storage yard full of LLW drums awaiting treatment and disposal. To ensure that no single location becomes a "dumping ground" for radioactive wastes, the 1982 Nuclear Waste Disposal Act requires states to dispose of LLW within their borders or in a designated regional facility established by a compact among a group of states in a given region. Several such compacts have now been formed and are in the process of identifying suitable disposal sites. In many parts of the country, however, public opposition has slowed or prevented the siting of new LLW disposal facilities. Meanwhile, a number of older LLW disposal sites now have been closed. These sites will remain radioactive for at least several hundred years and will require active protection for that period.

2.6.4   Uranium Mill Tailings

By far the largest volume of radioactive wastes—some 32 million m^3—is the sand-like residue remaining from the processing of uranium ore. More than 2,000 kg of ore must be mined and processed to produce just 1 kg of enriched uranium for nuclear reactors. In the United States, uranium is mined primarily in the Southwest. The initial step in refining uranium ore dug from the ground is called *milling*. The residues from this process, known as *mill tailings,* still contain much of the ore's original radioactivity. Although these levels are very low compared to other types of nuclear wastes, the large volumes of material can pose a hazard, particularly from inhalation of radon gas (formed in the decay of natural radium) and from groundwater contamination from ponds created in the milling process. The Department of Energy is currently remediating many inactive milling sites using stabilization methods to prevent the windblown spread of mill tailings. Because of the long half-lives of radionuclides in mill tailings, Congress has required perpetual government custody of the tailings disposal sites.

2.7   DEPLETION OF NATURAL RESOURCES

So far, this chapter has focused on the environmental problems and impacts resulting from the gaseous, solid, and liquid wastes from human activities and technology. But other types of environmental impacts also are important. History shows that as standards of living improve, there is a concomitant increase in the use of natural resources for energy and raw materials. Figure 2.27, for example, shows the history of fuel use and materials consumption in the United States during the 20th century. This enormous growth in consumption of natural resources constitutes another way in which human activities alter the environment.

To some extent, improvements and innovations in technology have allowed human demands for goods and services to be met more efficiently than in the past, thus dampening the rate of increase in the consumption of raw materials. Improvements in energy efficiency, for example, have reduced the amount of fuels that must be extracted to provide useful energy services like lighting or transportation. Nevertheless, in the face of worldwide population growth, and with the world's population desiring an improved standard of living, the demand on the earth's natural resources undoubtedly will continue to increase. Questions then arise about the long-term

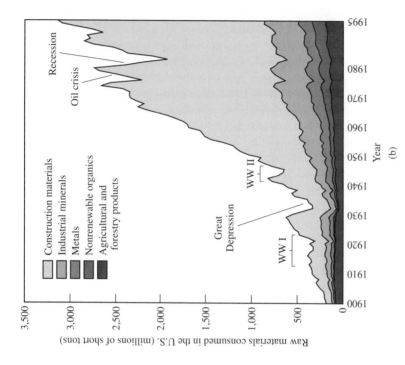

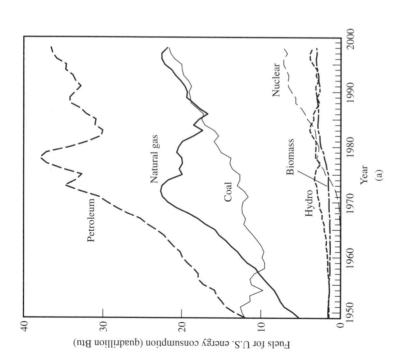

Figure 2.27 Growth of fuels and raw materials consumed in the United States. (a) Petroleum, coal, natural gas, and uranium are the natural resources that supply over 90 percent of the growing U.S. demand for energy; (b) consumption of nonfuel raw materials also has increased substantially, especially the use of construction materials such as crushed stone, sand, and gravel. (*Source:* USDOE, 1997b; USGS, 1998)

implications of natural resource consumption for people and the environment. Here we take a brief look at some of these issues.

Human use and consumption of the earth's natural resources are generally for three purposes: (1) as a source of food, (2) as a source of energy, and (3) as a source of raw materials for structures, devices, and other human endeavors. There are two general categories of natural resources: *renewable* resources, which have the capability to be replenished, and *nonrenewable* resources, which exist only in finite amounts. Examples of renewable resources include water, wind, and solar energy as well as plants, agricultural crops, and wood from forests. The latter types of resources are renewable only to the extent that they are replenished at a rate sufficient to support their level of consumption. Nonrenewable resources include all of the metals and minerals that are used to produce products and structures, as well as the coal, oil, natural gas, and uranium that are consumed to produce energy.

The consumption of natural resources is a concern for several reasons. Many people believe it is improper, selfish, and unethical for the earth's current inhabitants to deplete resources that are part of the natural environment and that may be needed or desired by future generations. Other concerns arise from the fact that the increasing consumption of nonrenewable resources is simply not sustainable over the long term. The consumption of natural resources, both renewable and nonrenewable, also is inexorably linked to the air pollution, water pollution, and solid waste impacts discussed earlier in this chapter. Metal and mineral resources, for example, must be dug out of the ground, transported, refined, and manufactured into useable products and materials.[10] Each of these steps generates air pollutants, water pollutants, and other environmental impacts. As a resource gradually becomes depleted, its quality also tends to deteriorate. For example, the concentration of pure copper metal in the ore mined in the western United States decreased from over 2 percent in 1905 to less than 0.5 percent in 1985 (Phipps, 1996). This means that more than four times more ore must be mined to obtain the same amount of metal, and more energy must be consumed to transport and refine the ore. The use of energy generates additional natural resource requirements and environmental impacts, as seen earlier and discussed more extensively in later chapters.

Renewable resources are no exception. Not only can these resources also become depleted by overconsumption—as in the case of forests lost to commercial lumber production—but their exploitation invariably affects the environment. For example, the construction of dams to utilize renewable water resources requires the flooding of large areas and the destruction of natural habitats both upstream and downstream of the dam. The full environmental consequences of dam construction are only now being recognized. Inevitably, then, any use of natural resources to satisfy human demands for food, energy, and creature comforts involves some type of environmental impact and trade-off.

How likely is it that the natural resources we depend on for materials and energy will soon be depleted? Various governmental and private organizations regularly try to estimate the quantities of commercial materials that can potentially be recovered

---

10 In the future, one can also envision extracting minerals from the ocean floor. Although the ocean bottom is known to be rich in such resources, economics and technical difficulties prevent their exploitation at this time. The future environmental consequences of ocean mining, however, have barely been thought about.

Table 2.10   Estimated reserves and remaining lifetimes of selected nonrenewable resources.

Commodity	World Reserve Base Life Index (Years Remaining)[a]
Aluminum	270
Copper	64
Iron ore	247
Lead	38
Mercury	80
Nickel	119
Tin	56
Zinc	46

[a] Estimated 1990 world reserves divided by 1992 world production.
Source: WRI, 1994.

from known reserves, along with estimates of their remaining lifetimes based on current or projected consumption rates. Table 2.10 shows an example of such estimates for common metals. Based on current consumption rates, most of these materials have remaining useful lifetimes on the order of 20 to 200 years. Such estimates, however, must be taken with a large grain of salt. Current consumption rates are seldom an accurate measure of future rates, which could either accelerate (in response to growing demand and rising living standards) or decline (as demand falls or as other materials are substituted for those currently in use). Past experience also shows that new supplies of natural resources often are discovered, substantially increasing prior estimates of known reserves. New sources of oil and natural gas, for example, continue to be found worldwide. Of course, new discoveries are not something one can count on; by definition, nonrenewable resources will one day be exhausted if they are exploited long enough. Thus a prudent approach is to utilize and manage these resources efficiently in order to sustain both economic development and environmental quality.

An emphasis on sustainable development and use of natural resources is especially important in the context of renewable resources. Viewed broadly, renewable resources include all living things on our planet. However, the technological capabilities and consumption patterns of human populations have now evolved to the point where other life forms are threatened as never before. The following section briefly addresses some of these issues and their ramifications for people and the environment.

2.8   LAND USE AND ECOLOGICAL IMPACTS

Chapter 1 of this text pointed out that environmental concerns include not only the direct impacts of human activity on our own health and welfare, but also the indirect and longer-term impacts that flow from the complex interactions among all the environmental systems that constitute our planet. The term *ecological impact* is often used to describe some of these broader implications of human activities for the environment. Ecology is the study of how organisms interact with their environment. At the highest level this includes the study of interactions among all life forms on earth. More commonly, the term *ecosystem* is used to refer to any biological community that functions as a cohesive unit within its physical environment. Forest ecosystems, aquatic ecosystems, and desert ecosystems are but a few examples of such units. An ecosystem may include organisms as small as microbes or bacteria and as large as whales and elephants. Humans, too, are creatures of the earth's ecosystems.

A basic message from the study of ecology is that everything is connected to everything else, even though these links may not be obvious or immediate. Thus, major disruptions in the earth's ecosystems caused by human activity invariably will have implications for people as well as other living things. This type of holistic perspective is extremely useful in understanding the environmental consequences of what we do in our professional lives, in our personal lives, and as a society. This section touches just briefly on a few of the ecological issues of concern today.

2.8.1   Biodiversity

One of the most important ecological concerns of our time is the issue of *biodiversity*. Around the world, the growth of human settlements and the exploitation of natural resources has caused the extinction of an increasing number of plant and animal species. The loss of biodiversity encompasses diversity at the genetic level as well as at the species level. This is a growing concern because the pressures of human population growth and economic development are likely to accelerate the loss of biodiversity in the future.

Scientists are uncertain about the total number of species on earth and the current rates of species extinction. However, estimates are that the extinction rate now is at least 50 to 100 times greater—and perhaps thousands of times greater—than the natural background rate before human civilizations entered the picture. It is very difficult to document such extinctions worldwide, but there is growing evidence of such losses in modern times. In the United States, over 500 known species are now listed as extinct, especially birds, mussels, fish, and flowering plants. Hawaii, with its diversity of native plants, and Alabama, with its species-rich waterways, are the two states with the largest number of known species extinctions (USFWS, 1999).

Loss of biological species affects the functioning and balance of an ecosystem. Historically, biological diversity has offered a means of responding to natural changes in the environment and of ensuring healthy, robust ecosystems. A loss of diversity thus endangers the long-term health of not only natural ecosystems and their inhabitants, but the human population as well. Most modern medicines, for example, are derived from the trees and plants of forest ecosystems, which are disappearing rapidly around the world.

Efforts to prevent or slow the loss of biodiversity are beginning to emerge. In the United States, the 1973 Endangered Species Act is widely credited for helping to save the American bald eagle and other threatened species from the brink of extinction.

Other federal laws have sought to protect fish and wildlife species and to preserve natural habitats. Yet today nearly 1,200 endangered species are listed by the U.S. Fish and Wildlife Service (FWS) under the Endangered Species Act. Figure 2.28 shows the number and location of listed species in the United States. Hawaii, California, and Florida head the list as of 1999, with over 100 endangered species in each state.

Enforcement of laws to protect endangered species has sometimes engendered strong controversy over the relative importance of environmental protection versus regional economic development. In one widely publicized case in the early 1990s, restrictions on logging in the northwestern United States to protect the spotted owl pitted environmental interests against local workers in the timber industry. Such conflicts are likely to occur more frequently in the future, especially in parts of the world that are undergoing rapid economic development. Internationally, the United Nations and various environmental organizations are actively engaged in programs to stem the loss of biodiversity in the face of economic pressures.

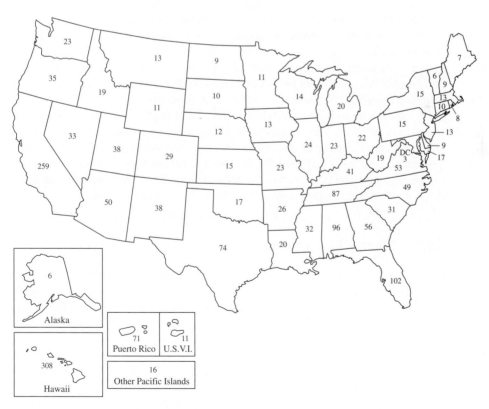

Figure 2.28   Number of endangered species in U.S. states and territories as of October 31, 1999. The total number of endangered species is 1,201, including eight whale species. The numbers in the figure are not additive because a species often occurs in more than one location. (*Source:* USFWS, 1999)

2.8.2   Loss of Habitat

One key factor contributing to the loss of biodiversity is the degradation or loss of wetlands, estuaries, coral reefs, tropical forests, and other ecologically rich habitats, both aquatic and terrestrial. The growth of modern suburbs along the coastal regions of the United States, for example, has encroached on and contaminated many of the estuaries that link inland freshwater systems with the saltwater systems of the ocean. Estuaries also serve as a vital breeding ground for aquatic and terrestrial species and as the habitat for other organisms that contribute to these ecosystems. But the encroachment of human settlements, and their increased pollutant loadings such as nutrient discharges to waterways, are contributing in a major way to the loss of vital habitat.

The United States also has suffered a significant loss of wetlands over the past century as human settlements have filled in the low-lying marshes, bogs, and other wetland areas that form breeding grounds for terrestrial and aquatic species and a buffer between land and freshwater systems. In recent years there have been regular reminders of some of the consequences of human development on floodplains and former wetlands. The severe property damage and loss of life and livelihood resulting from floods in the Midwest in 1993, California in 1995 and 1998, and North Carolina in 1999 are some of the more poignant recent examples.

Nonetheless, the pressures of population growth and economic development continue to threaten the loss of natural habitat and its ecological consequences. The fragmentation of forested areas and other habitat into smaller regions divided by roads, settlements, dams, and other human constructs has been recognized as an especially important contributor to the loss or decline of species. Figure 2.29 shows one example of how human activities have encroached on and fragmented a previously extensive growth of forest. The establishment of protected areas under state and federal laws is one important step in slowing the loss of habitat and biological diversity. More fundamentally, a better understanding is needed of the functioning of ecosystems, and the habitat requirements of diverse species, in order to better harmonize human activities with the natural environment that we draw upon for our existence.

2.8.3   Marine Ecosystems

Because most human activities occur on land, the environmental health of the world's oceans and marine ecosystems is generally less visible or prominent on most people's list of environmental concerns. Yet, as with terrestrial ecosystems, the health and resilience of the marine environment is at risk from the impacts of human activities.

Here, too, one of the principal dangers is the extinction of aquatic species as marine ecosystems are degraded from overexploitation and contamination. The world's oceans cover over 70 percent of the earth's surface and are rich in biodiversity—perhaps even more so than life on land. But the growth in toxic organic chemicals, inorganic chemicals, and radionuclides contaminating the oceans from human activity poses an increasing threat to the ocean's biological systems. Marine ecosystems are thus under continuously increasing pressure. Oil spills such as the *Exxon Valdez* incident in 1989, which dumped over 10 million gallons of crude oil into

Figure 2.29   Loss of habitat from human activities. The rapid cutting of old growth forests for lumber production (a) has destroyed the natural habitat of the spotted owl (b), one of the native species on the endangered list.

Prince William Sound, Alaska, are among the more dramatic events that gain widespread notoriety for their impact on the marine environment and the local economy. Less visible are the growing threats from routine dumping of wastes at sea and into coastal waters near populated regions.

With the majority of the world's population living near sea coasts, the adverse impacts of human activity are especially evident in coastal areas. These impacts go beyond the effects on marine life from chemicals and other wastes released into coastal waters. International shipping is transporting an increasing variety of microorganisms and larvae from one part of the globe to another in the ballast tanks of freighters. When discharged into local waters, these foreign species act as invaders that can alter or destroy local marine ecosystems. San Francisco Bay is but one U.S. port that is now ridden by exotic species of water plants that have depleted the bay's phytoplankton and changed its ecology.

Among the greatest threats to marine life is human overexploitation of marine fisheries. Modern technology has dramatically increased the productivity of commercial fishing, but with that improvement has come the more rapid depletion of species, the inadvertent killing of nontargeted species, and a change in the balance and dynamics of marine ecosystems. Commercial aquaculture operations focused on high-value products like salmon and shrimp also have altered the natural ecosystems of some coastal areas.

Because of the difficulty of exploring the depths and breadth of the marine environment, the full extent of human influence cannot be fully gauged. Scientists warn, however, of the delicate chains and interactions that can easily be disrupted by contamination and overexploitation of marine resources.

2.8.4   Land Use Practices

The environmental concerns over ecological impacts of biodiversity and loss of habitat derive from people's use of land for housing, transportation, commercial activities, and recreation. In this context, *land use* also includes the use or exploitation of waterways and oceans. The subject of land use planning and practice is a discipline of its own, and no attempt will be made here to even scratch the surface of this broad field. Rather, the point of this section is simply to emphasize to engineering students (the primary audience for this book) that no matter how well-designed or environmentally clean a "technology" (used here to mean all engineered products, processes, and structures), the act of deploying that technology means that land use patterns have been altered. When multiplied across the human population, the impacts of those land use decisions create the environmental problems described throughout this chapter.

Many countries today have some type of institutional mechanisms or laws to regulate or control land use practices. To varying degrees, such policies also may consider environmental impacts. In the United States, for example, the National Environmental Policy Act (NEPA), adopted in 1969, requires an environmental impact statement (EIS) to be prepared for any major construction project involving federal funds. In many cases, the EIS process has revealed potential problems that have led to substantial revisions or cancellations of some projects.

States and local communities in the United States also have created various agencies, commissions, and review boards that oversee land use practices. The extent to which environmental concerns are explicitly integrated into permitting and approval processes, however, is highly variable across the country. The California Coastal Commission is one example of a regional regulatory body that takes very seriously the issue of environmental protection along the California coast. Approval for new construction projects along that coast is difficult to obtain. In contrast, owners of existing homes along both coasts have regularly rebuilt beachfront houses that were damaged or destroyed by storms and hurricanes. In large part, this is because insurance companies and governmental agencies historically have subsidized the cost of reconstruction. The same has been true for homes constructed on floodplains across the United States. Thus government policies and institutions often are not of a single mind in their approach to land use practices and the environment.

Recent history, however, suggests that past practices are beginning to change. As we learn more about the complex and far-reaching relationships between human actions and the environment, there are growing calls for new land use policies and practices that are consistent with a sustainable environmental future. Finding that proper balance between environmental quality and the material needs and desires of people remains one of the biggest challenges facing humanity.

2.9 REFERENCES

AFEAS, 1998. *Alternative Fluorocarbons Environmental Acceptability Study,* Washington, DC. http://www.afeas.org/prodsales_download.html.

CAA, 1990. Clean Air Act Amendments, Public Law 101-549, Section 112, Nov. 15, 1990.

CFR, 1999. Code of Federal Regulations. *Federal Register,* 40 CFR, July 1, 1999.

EPRI, 1994. Private communication from K. Yaeger, EPRI, Palo Alto, CA.

IPCC, 1996. *Climate Change 1995, The Science of Climate Change,* J.T. Houghton, et al., Eds., Intergovernmental Panel on Climate Change, Cambridge University Press.

Keeling, C.D., and T.P. Whorf, 1998. "Atmospheric CO_2 Records from Sites in the SIO Air Sampling Network." In *Trends: A Compendium of Data on Global Change.* Carbon Dioxide Information Analysis Center, Oak Ridge National Laboratory, Oak Ridge, TN.

Marland, G.R.J., et al., 1999. "Global, Regional, and National CO_2 Emission Estimates from Fossil Fuel Burning, Cement Production, and Gas Flaring: 1751–1996 (revised March 1999)." Carbon Dioxide Information Analysis Center, Oak Ridge National Laboratory, Oak Ridge, TN. http://cdiac.esd.ornl.gov/ftp/ndp030/.

NAPAP, 1985. *Annual Report, 1985.* National Acid Precipitation Assessment Program, Washington, DC.

NRC, 1996. "Radioactive Waste." NRC Information Digest NUREG-1350, vol. 9. Nuclear Regulatory Commission, Washington, DC.

NRC, 1997. "Information Digest: 1997 Edition," NRC Report Number: NUREG-1350, vol. 9. Nuclear Regulatory Commission, Washington, DC.

OTA, 1992. *Green Products by Design: Choices for a Cleaner Environment,* Office of Technology Assessment, U.S. Congress, Washington, DC.

Phipps, E., 1996. "Overview of Environmental Problems." National Pollution Prevention Center for Higher Education, University of Michigan, Ann Arbor, MI.

USDOE, 1997a. *Emissions of Greenhouse Gases in the United States 1996.* DOE/EIA-0573(96). Energy Information Administration, U.S. Department of Energy, Washington, DC.

USDOE, 1997b. *Linking Legacies: Connecting the Cold War Nuclear Weapons Production Process to Their Environmental Consequences.* DOE/EM-0319. Office of Environmental Management, U.S. Department of Energy, Washington, DC.

USDOE, 1999. *Annual Energy Review 1998.* Energy Information Administration, U.S. Department of Energy, Washington, DC.

USEPA, 1995. *National Water Quality Inventory: 1994 Report to Congress.* EPA-841-R-95-005. Office of Water, U.S. Environmental Protection Agency, Washington, DC.

USEPA, 1996a. *Environmental Indicators of Water Quality in the U.S.* EPA-841-R-96-002. Office of Water, U.S. Environmental Protection Agency, Washington, DC.

USEPA, 1996b. *RCRA Environmental Indicators Progress Report: 1995 Update.* EPA-530-R-96-010. Office of Solid Waste and Emergency Response, U.S. Environmental Protection Agency, Washington, DC.

USEPA, 1998a. *National Air Quality and Emissions Trends Report, 1997.* EPA-454-R-98-016. Office of Air Quality Planning and Standards, U.S. Environmental Protection Agency, Washington, DC.

USEPA, 1998b. *National Air Pollutant Emission Trends Update: 1970–1997.* EPA-454-R-98-007. Office of Air Quality Planning and Standards, U.S. Environmental Protection Agency, Washington, DC. http://www.epa.gov/ttn/chief/trends97/emtrnd.html.

USEPA, 1998c. *Characterization of Municipal Solid Waste in the United States: 1997 Update.* EPA-530-R-98-007. Office of Solid Waste, U.S. Environmental Protection Agency, Washington, DC.

USEPA, 1999. *1997 Toxics Release Inventory Public Data Release Report,* U.S.
    Environmental Protection Agency, Washington, DC. http://www.epa.gov/oppintr/tri.
USFWS, 1999. U.S. Fish and Wildlife Service, Department of the Interior, Washington, DC.
    http://www.fws.gov.
USGS, 1998. *Materials Flow and Sustainability.* USGS Fact Sheet, FS-068-98. U.S.
    Geological Survey, U.S. Department of the Interior, Washington, DC.
WRI, 1994. *World Resources, 1994–95,* World Resources Institute, Washington, DC.

2.10   PROBLEMS

2.1   Figures 2.3 through 2.8 showed the national trends in U.S. emissions of criteria air pollutants. Use the Internet to find data on the associated trends in air quality of PM_{10}, SO_2, CO, NO_2, O_3, and Pb over the past few decades. Start by visiting the U.S. Environmental Protection Agency website: www.epa.gov.

2.2   Identify the principal types and sources of toxic pollutant releases in the community where you live. Use the website of the U.S. EPA (www.epa.gov/tri) or the Environmental Defense Fund (www.scorecard.org) to find the most recent information from the Toxics Release Inventory (TRI). List the five substances emitted in greatest quantity, and determine the amounts released to each environmental medium (air, water, and land).

2.3   Choose any three of the toxic releases identified in the previous problem and summarize their environmental concerns as conveyed by the information on each of the two websites noted. Which site did you find most helpful? How could the information be improved?

2.4   The data presented in Chapter 2 reflect conditions in the United States. Use the Internet to find data on the air quality concentrations of particulate matter and sulfur dioxide in any four other countries (two in the Eastern Hemisphere and two in the Western Hemisphere). Focus your search on a major city in each country you choose. Compare the air quality levels to typical values in the United States.

2.5   Locate and contact the local authority that supplies drinking water to your city or community. Request the most recent report or analysis of the chemical composition. Compare these levels to the primary U.S. drinking water standards in Table 2.5 and write a brief report summarizing your findings.

2.6   Use the Toxics Release Inventory (TRI) websites listed in Problem 2.2 to find the types, amounts, and sources of the five largest chemical releases to surface waters in your community. Write a brief report summarizing your findings.

2.7   The data on U.S. surface water quality in Table 2.8 were the most recent data available at the time this chapter was written. More recent data now are likely available. Visit the EPA website (www.epa.gov) or other Internet sites and search for data to update the information in Table 2.8. Based on your findings, summarize the trend in U.S. surface water quality since 1994.

2.8   Locate and contact the authority responsible for municipal solid waste (MSW) collection in your city or community. Determine (a) the prevailing method(s) of municipal waste disposal, (b) the location and size of the waste disposal facilities, (c) the annual quantity of MSW disposed, and (d) the annual quantity of MSW that is recycled. Also determine the population of your community, and use this information to calculate the average per capita MSW generated (kg per person per day). Compare this value to the national average reported in the text.

2.9   Write a report of approximately 1,500 words describing the history of efforts to establish a permanent U.S. repository for high-level radioactive wastes. Discuss the key technical issues that are involved, as well as the social and political factors that have influenced this effort.

2.10   Table 2.10 gave an estimate of the remaining world resources of selected nonrenewable materials. Select any one of the listed commodities with a life index of less than 100 years. Investigate the major uses of that material, and analyze the implications of its projected scarcity in the future. Include a discussion of any likely substitutes for the material or its applications.

2.11   Investigate the estimated resource base of world energy supplies of either crude oil or natural gas (choose one). One useful website is the Energy Information Administration of the U.S. Department of Energy (www.eia.doe.gov). Comment on when or whether we might be "running out" of this nonrenewable resource based on current estimates. Also discuss whether the environmental implications of future energy resource extraction might change because of the location or difficulty of exploiting the remaining reserves. Summarize your findings in a brief report.

2.12   Figure 2.28 showed the number of endangered species in each U.S. state. Choose your home state (or other state of interest) and investigate the specific species that are listed by the U.S. Fish and Wildlife Service (www.fws.gov). Choose any two species on the list and investigate the location of their natural habitat in your state, by contacting the regional FWS office (see FWS website) or a state or local agency. Prepare a map showing these locations and the nearest urban areas, highways, or other significant human encroachments. Include this information in a brief report summarizing your findings.

2

TECHNOLOGY DESIGN FOR THE ENVIRONMENT

chapter

3

Automobiles and the Environment

3.1   THE AUTOMOBILE AND SOCIETY

The automobile is arguably the most important technological development of the 20th century. In the United States, the number of cars has increased from about 8,000 in 1900 to 135 million today (see Figure 3.1). Combined with buses and trucks, the total number of vehicles owned in the United States today is nearly equal to the total U.S. population, or an average of almost one vehicle per person for every man, woman, and child in the country. Worldwide, the current vehicle population averages one for every nine people, with the United States leading the world in vehicles per capita (see Figure 3.2).

Much has been said and written about the automobile's role in our culture and society. Table 3.1, for example, summarizes some early 20th-century expectations for the automobile that envisioned many of its far-reaching effects (with some notable exceptions). Unquestionably, the social and economic impacts of the automobile are pervasive. The auto manufacturing industry directly and indirectly accounts for a major portion of the U.S. economy. Cars also are largely responsible for the growth of suburbs, which have shaped both where and how people live. The extensive network of roads and highways built to accommodate vehicle traffic also serves as a commercial lifeline, sustaining nearly every aspect of our economy. Cars provide a degree of personal freedom, convenience, speed, and ease of travel unmatched in the history of transportation. Just imagine how your own life would change if all traffic suddenly came to a halt.

Along with all the benefits that cars have brought to modern society, there are undesirable consequences as well. Traffic accidents in the United States alone kill over 40,000 people every year—the equivalent of one commercial jetliner crashing every day. And for every traffic fatality, nearly a hundred people are injured, many seriously.

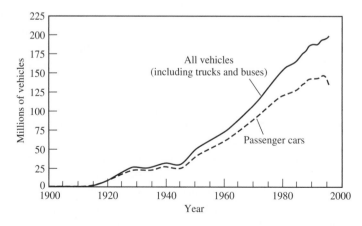

Figure 3.1   Number of U.S. registered vehicles, 1900–1995. The recent decline in passenger car vehicles reflects the growing popularity of sports utility vehicles, which are classified as light trucks. (*Source:* USDOT, 1998)

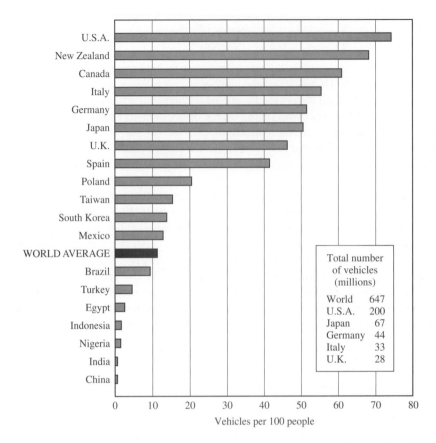

Figure 3.2   World motor vehicle registrations, 1995. (*Source:* Based on AAMA, 1997)

Cars also have been blamed for many of the adverse environmental impacts that plague modern cities throughout the world, which also affect human health and welfare. These environmental impacts are the subject of this chapter. In particular, we focus on the links between engineering design decisions and their environmental consequences.

3.2   ENVIRONMENTAL IMPACTS OF THE AUTOMOBILE

What things come to your mind when asked about the environmental impacts of the automobile? Traffic congestion? Air pollution? Smog? Junkyards? Urban sprawl? Because cars are so pervasive in our personal lives and in the U.S. economy, the environmental impacts of this technology are equally pervasive. As discussed in Chapter 1, those impacts stem both from land use decisions (related to the construction of roads, parking lots, and so on), and from engineering design decisions that determine or influence the release of pollutants to the environment (such as tailpipe emissions). Figure 3.3, for example, shows the growth in miles of surfaced and non-

Table 3.1   Early 20th-century expectations for the automobile (ca. 1900).

Expected Benefits of the Automobile

Operation of auto more economical than horse and carriage.

Offers a means of cheap mass transit.

Should provide increased storage space because auto takes up less room than horse and wagon.

Fast response/little upkeep/fast refueling/more reliable than a horse.

Should permit rapidly available medical and other emergency help.

Should enable farmers to market produce over a much wider geographical area.

Safer than horse because of better control and braking.

Offers superior maneuverability and imperviousness to weather conditions and fatigue.

Should relieve traffic congestion on city streets.

Would rid cities of unsightly and unsanitary conditions attributable to the horse.

Would improve health conditions, resulting in a reduction of infectious diseases and cases of diarrhea; would provide better ventilation for passengers.

Would restore frayed nerves and reduce nervous diseases from hectic pace of life in urban–industrial society.

Should reduce city noise.

Should extend and enhance rural life.

Would permit movement to suburbia.

Expected to halt movement of population from rural to urban areas.

Source: Burke, 1971.

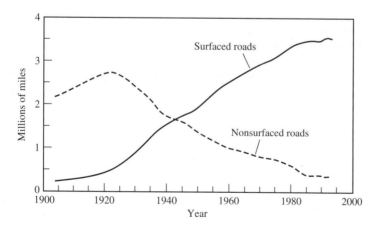

Figure 3.3   U.S. public road and street mileage, 1904–1993. During the 20th century the U.S. road system has grown by a third, from roughly 3 million to 4 million miles. Over 90 percent of public roads are now paved or surfaced. (*Source:* USDOT, 1998)

85

surfaced roads across the United States since the turn of the century. This is but the tip of the iceberg of land use impacts that are directly or indirectly attributable to automobiles. In this chapter, however, our primary focus is on the engineering aspects of automobile design and their relationship to environmental impacts. We begin with a brief overview of major environmental concerns.

3.2.1   Urban Air Pollution

As the number of vehicles on the road has increased, the concentration of cars in urban areas has made air pollution the number one environmental issue related to automobiles.

Los Angeles often is held up as the prime example of the automobile's impact on air quality and the environment. Scientists studying the L.A. smog in the late 1940s and 1950s came to understand that the thick brownish haze that regularly settled over the city—decreasing visibility and irritating eyes and throats—was caused by complex chemical reactions in the atmosphere.[1] These reactions were triggered by the same bountiful sunlight that attracted so many people to the region. The principal culprit was found to be emissions of hydrocarbons (HC, also known as volatile organic compounds, or VOCs) and nitrogen oxide (NO_x) gases emanating from Los Angeles's signature technology: the automobile. When trapped by the surrounding mountains and "cooked" in the L.A. sunlight, these gases produced ozone (O_3) and other chemical compounds that were collectively known as *photochemical smog* and were measured as equivalent ozone. A third air pollutant released from automobiles— carbon monoxide (CO)—also was found responsible for elevated atmospheric CO levels that could harm human health.

Studies showed that the main source of CO and hydrocarbons was incomplete combustion of fuel. Additional hydrocarbons were emitted in the form of gasoline vapors given off from hot engines, from the venting of the engine crankcase, and from the venting and refilling of fuel tanks. Nitrogen oxides also were formed during combustion by chemical reactions between nitrogen and oxygen in air, which occur at high temperatures.

These three air pollutants—NO_x, HC, and CO—continue to pose significant environmental problems, not only in Los Angeles but in most urban regions (see Chapter 2 for more details). As a result, federal regulations adopted since 1970 have followed the early lead of California in requiring increasingly stringent emission reductions from new cars. Table 3.2 shows the U.S. exhaust emission standards for passenger cars, which are specified in units of grams of pollutant emitted per mile of travel.[2] Meeting these requirements has posed (and continues to pose) a major challenge for engineers involved in auto design and manufacture.

---

1 The term *smog* was actually coined in England in the late 19th century to describe the mixture of smoke and fog that produced severe air pollution episodes attributed to coal burning and other industrial pollution. The term was adopted in Los Angeles half a century later to describe the air pollution phenomenon that produced similar decreases in visibility. The sources of the L.A. smog, however, are quite different from those of the London smog, as elaborated later in Chapter 8.

2 This combination of metric and English units is uniquely American, reflecting our reluctance to convert completely to the metric system.

Table 3.2   U.S. passenger car exhaust emission standards (grams per mile).

Model Year	Hydrocarbons	Carbon Monoxide	Nitrogen Oxides
Precontrol	10.6	84.0	4.1
1968–1971	4.1	34.0	–
1972–1974	3.0	28.0	3.1
1975–1976	1.5	15.0	3.1
1977–1979	1.5	15.0	2.0
1980	0.41	7.0	2.0
1981–1982	0.41	7.0	1.0
1983–1993	0.41	3.4	1.0
1994[a]	0.25	3.4	0.4
1995[b]	0.25	3.4	0.4
1996–2003[c]	0.25	3.4	0.4
2004–2006[d]	0.125	1.7	0.2

[a] Applies to 40 percent of fleet.
[b] Applies to 80 percent of fleet.
[c] Applies to 100 percent of fleet.
[d] Percentage of fleet pending EPA promulgation of final standards.
Sources: USEPA, 1998; AAMA, 1997.

Example 3.1

Exhaust emission reductions.   Calculate the percentage reduction in CO, HC, and NO_x exhaust emissions from a new passenger car in 1997 relative to an uncontrolled car in 1967.

Solution:

Use the emission standards in Table 3.2 for 1997 and pre-1968:

$$\% \text{ HC reduction} = \frac{10.6 - 0.25}{10.6} \times 100 = 98\%$$

$$\% \text{ CO reduction} = \frac{84.0 - 3.4}{84.0} \times 100 = 96\%$$

$$\% \text{ } NO_x \text{ reduction} = \frac{4.1 - 0.4}{4.1} \times 100 = 90\%$$

Despite these impressive reductions in new car emissions, the air quality impacts of automotive emissions continue to be of concern because of several offsetting factors:

Increasing vehicle population. As shown in Figure 3.1, the total number of vehicles on the road continues to grow. All else being equal, more vehicles mean more emissions and poorer air quality.

Increasing travel per vehicle. Compounding the effect of a growth in total vehicle population has been an increase in the average distance traveled per vehicle each year. From 1980 to 1995, for example, the annual miles of travel for the average U.S. motor vehicle rose from 9,500 mi (15,200 km) to 11,800 mi (18,900 km), an increase of nearly 25 percent. Multiplying miles traveled per vehicle by the total number of vehicles gives the total vehicle-miles of travel (VMT) per year, a useful measure for tracking overall vehicle usage. Figure 3.4 shows the historical trend in VMT. Over the past 15 years, VMT in the United States has grown at an average rate of 3.1 percent per year, with no slowdown in sight.

Departures from federal standards. The federal emission limits in Table 3.2 are based on a test procedure designed to mimic typical highway and city driving conditions. Actual on-road emissions, however, are generally higher than federal test cycle values, in part because many cars are driven under conditions different from the federal test procedure. Higher average speeds, for instance, cause higher NO_x emissions. Improper maintenance or tampering with emission control systems also leads to higher emissions. A clogged air filter or dirty spark plug can increase hydrocarbon and carbon monoxide emissions well above their normal design limits. Because emission control systems can degrade over time, many states require periodic inspection and measurement of tailpipe pollutants.

Greater use of light trucks. In the 1990s, there has been a sharp increase in the number of light-duty trucks as consumers shun traditional passenger cars in favor of sports utility vehicles, pickup trucks, and vans. These larger, heavier vehicles not only consume more gasoline than a passenger car, but they also emit more CO, HC, and NO_x per mile of travel. Federal emission standards

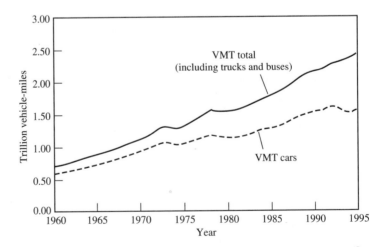

Figure 3.4   Vehicle-miles of travel, 1960–1995. The increase in vehicle-miles of travel (VMT) is due primarily to the growing number of vehicles, plus a modest increase in the average distance traveled per vehicle. (*Source:* USDOE, 1997)

prior to 1994 allowed emissions per mile to be two to three times higher for light-duty trucks than for passenger cars. Recently, however, the EPA announced plans to gradually eliminate this disparity over the coming years.

Example 3.2

Annual hydrocarbon emissions.   The average exhaust emission rate of hydrocarbons (HC) from all passenger cars on the road in 1993 has been estimated to be 2.1 g/mi. Calculate the total annual HC emissions from all passenger cars in 1993, and compare this to the emissions that would have occurred if all of these cars emitted at their precontrol (1967) level. Also compare your calculated figure to the total 1967 emissions of passenger car HC.

Solution:

From Figure 3.4, the total vehicle-miles of travel (VMT) for all passenger cars in 1993 was 1.6×10^{12} mi. The total HC emissions are then

$$1993 \text{ HC} = (1.6 \times 10^{12} \text{ mi})(2.1 \text{ g/mi})$$

$$= 3.4 \times 10^{12} \text{ g} = 3.4 \text{ million metric tons}$$

From Table 3.2, the precontrol (1967) emission rate was 10.6 g/mi. This would have resulted in emissions of

$$(1.6 \times 10^{12})(10.6) = 17 \times 10^{12} = 17 \text{ million metric tons}$$

Emission controls thus have achieved an 80 percent reduction in total annual HC emissions relative to uncontrolled levels.

A comparison to 1967 emissions is obtained using the estimated 1967 passenger car VMT from Figure 3.4 of 0.7×10^{12} mi. Then

$$1967 \text{ HC} = (0.7 \times 10^{12})(10.6) = 7 \times 10^{12} \text{ g} = 7 \text{ million metric tons}$$

Total 1993 emissions are less than half the precontrol level of 1967, even after a more than twofold increase in total VMT.

Later in this chapter we will examine in more detail the sources of automotive air pollutants and engineering solutions to reduce these emissions.

3.2.2   Greenhouse Gas Emissions

In recent years *global warming* has gained much public attention as a potential new threat to people and the environment. The major culprit is the carbon dioxide (CO_2) released to the atmosphere when fossil fuels (oil, coal, and natural gas) are burned. Unlike other automotive and industrial air pollutants that are transformed by chemical reactions or washed out of the atmosphere by rainfall and other natural processes, CO_2 is an extremely stable gas that accumulates in the atmosphere once emitted. It is called a "greenhouse" gas because it traps heat in the atmosphere in a fashion similar to the glass walls of a greenhouse. As the atmospheric CO_2 level rises, increasing amounts of heat are trapped, raising the average temperature of the earth. Chapter 2 discussed some of the concerns stemming from projected global warming trends, which include increased storm severity, sea level rise, ecological impacts, and the spread of infectious diseases. Although there are still enormous uncertainties over the magnitude and rate of future atmospheric warming and its consequent effects, more than 150 countries around the world have agreed by international treaty that action should be taken to slow the rate of greenhouse gas emissions. Chapter 12 discusses this problem in detail.

For automobiles, the engineering challenge is immense because nearly all cars now run on a single fossil fuel source—petroleum. Refined into gasoline, diesel oil, and jet fuel, petroleum supplies about 97 percent of the energy used for all forms of transportation. Figure 3.5 shows the total annual fuel consumption for all U.S. motor vehicles, as well as the average fuel consumption per vehicle. The United States currently consumes about twice as much gasoline for transportation as it did 30 years ago.

Chemically, petroleum and its derivative fuels consist mainly of carbon and hydrogen compounds. Automotive gasoline is actually a complex mixture of relatively volatile hydrocarbons blended for use in spark-ignition engines. Every time fuel is burned, the carbon is converted to CO_2, the natural end product of combustion. If we approximate the chemical formula for gasoline by the compound octane (C_8H_{18}), the stoichiometric reaction for complete combustion is

$$C_8H_{18} + 12.5O_2 \rightarrow 8CO_2 + 9H_2O \tag{3.1}$$

Thus for every mole (114 grams) of C_8H_{18} fuel that is burned, eight moles (352 g) of CO_2 are produced, along with nine moles (162 g) of water vapor. Together with other combustion products, CO_2 is released to the atmosphere through the car's exhaust pipe. The total quantity of CO_2 produced by cars throughout the world has made automobiles a major contributor to potential global warming.

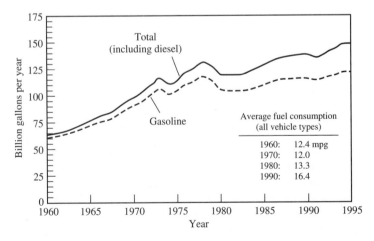

Figure 3.5   U.S. motor vehicle fuel consumption, 1960–1995. Since 1970 the average vehicle fuel economy in miles per gallon (mpg) has improved substantially, although the recent popularity of sports utility vehicles is beginning to reverse that trend. (*Source:* USDOE, 1997)

Example 3.3

Annual CO_2 emissions from cars. Use the data presented in this chapter to estimate the total annual emissions of CO_2 from gasoline-burning vehicles in the United States in 1995. The density of gasoline is 739 g/L.

Solution:

From Figure 3.5, the total gasoline consumption by motor vehicles in 1995 was 120×10^9 gal/yr = 455×10^9 liters/yr. Multiply this by the density of gasoline to find the total mass of fuel consumed:

$$m_{fuel} = (455 \times 10^9)(739) = 336 \times 10^{12} \text{ g/yr}$$

Approximate the formula for gasoline as C_8H_{18}, and assume complete combustion. From Equation (3.1) we had 352 g CO_2 per 114 g of fuel burned. So

$$CO_2 = (336 \times 10^{12})\left(\frac{352}{114}\right) = 1.04 \times 10^{15} \text{ g/yr}$$

$$= 1.04 \text{ billion metric tons/yr}$$

This estimate is within 1 percent of the U.S. Department of Energy's figure of 1.03 billion metric tons (USDOE, 1997). According to the DOE, gasoline combustion accounted for 20 percent of all U.S. emissions of CO_2 in 1995. Other petroleum fuels for transportation contributed an additional 12 percent.

Although CO_2 emissions are not currently regulated as an air pollutant, it is instructive to compare the magnitude of CO_2 released to those of the urban air pollutants CO, HC, and NO_x. This can be done using average fuel consumption data, as illustrated in the next example.

Example 3.4

CO_2 emissions per mile. Estimate the CO_2 emissions in grams per mile for a car getting 9.59 km/liter (27.6 miles per gallon) of gasoline, the U.S. average for 1995. Compare this value to the federal exhaust emission limits for CO, HC, and NO_x.

Solution:

Assume gasoline has a composition of C_8H_{18} and a density of 739 g/L, as in Example 3.3. The mass of fuel needed to travel one mile (1.602 km) is

$$m_{fuel} = \frac{1.602 \text{ km/mi}}{9.59 \text{ km/liter}} \times 739 \text{ g/L} = 123 \text{ g gasoline/mile}$$

From Equation (3.1), the rate of CO_2 emissions is 352 g per 114 g of gasoline burned, assuming complete combustion. The CO_2 emission rate per mile is then

$$CO_2 \text{ rate} = \left(\frac{352}{114}\right)(123) = 381 \text{ g } CO_2/\text{mile}$$

This is more than 100 times greater than the CO emission rate of 3.4 g/mi from new passenger cars (see Table 3.2) and roughly 1,000 times greater than the current HC and NO_x emission standards of 0.25 and 0.4 g/mi, respectively.

Note, however, that the presence of CO and HC in the exhaust gas means that not all of the carbon in fuel is completely converted to CO_2, as assumed in Equation (3.1). For a car meeting current federal standards, it can be shown that approximately 1.5 percent of the carbon is released as CO and 0.1 percent as HC (see Problem 3.5 at the end of this chapter). So reducing the CO_2 value above by 1.6 percent provides a slightly more accurate estimate. The adjusted CO_2 rate is thus 98.4 percent of the value calculated above:

$$\text{Adjusted } CO_2 \text{ rate} = (381)(0.984) = 375 \text{ g } CO_2/\text{mi}$$

The main lesson to be learned from this discussion is that any effort to reduce CO_2 emissions must involve burning less fossil fuel or using alternative energy sources that contain less (or no) carbon. Section 3.3 of this chapter explores automotive energy use in greater detail and reveals a number of ways that engineers can improve vehicle designs to reduce fuel consumption. Section 3.4 provides a brief introduction to alternative fuels and advanced vehicle concepts.

3.2.3   Materials Use and Solid Waste

Ironically, when Henry Ford started mass-producing cars in the early 1900s, they were widely hailed as the *solution* to a major environmental problem of the day. Prior to the automobile, people relied heavily on horses for personal transportation, and in urban areas the number of dead horses, often abandoned on city streets, was a growing problem. So too were the health and sanitary problems created by horses. One health official in Rochester, New York, observed in 1900 that

> The 15,000 horses [in Rochester, NY] produced enough manure in a year to make a pile covering an acre of ground 175 feet high and breeding 16 billion flies.

Today disposal of automobiles at the end of their useful life also is a significant environmental issue. The problem is especially severe in countries like Germany, which have very limited space for waste disposal. Recycling and reuse programs, which in the United States currently return about 75 percent of a car's materials (by weight) to productive uses, have significantly alleviated the solid waste problems of discarded automobiles. But the remaining 25 percent still requires disposal, typically by burial in a landfill.

Table 3.3 shows how the average weight and composition of a new automobile has changed over the past 50 years. Modern cars weigh about 25 percent less, with greater use of plastics and aluminum. Table 3.4 shows the composition of the solid waste (called automobile shredder residue or "fluff") that remains after recovery of all recyclable materials. Typically this waste is landfilled. Note that today's solid waste reflects the materials used in cars that were manufactured an average of eight to nine years ago. As the mix of materials used in auto construction changes over

Table 3.3   Weight and composition of a typical automobile.

Material	1950	1978	1985	1995
Total weight (kg)	1,901	1,623	1,449	1,458
Steel, %	67.9	59.5	55.9	55.1
Iron, %	11.6	14.3	14.7	12.4
Plastics, %	0.0	5.0	6.6	7.7
Aluminum, %	0.0	3.2	4.3	5.8
Rubber, %	4.5	4.1	4.3	4.2
Glass, %	2.8	2.4	2.7	2.9
Copper, %	1.3	1.0	1.4	1.4
Lead, %	1.2	1.1	1.0	1.0
Zinc, %	1.3	0.9	0.6	0.5
Fluids, %	5.0	5.5	5.8	5.9
Other, %	4.4	3.0	2.7	3.1

Sources: AAMA, 1995; Graedel and Allenby, 1997.

Table 3.4   Characterization of automobile shredder residue (ASR).

Material (% Total)	Constituent	Percentage
Metals (12.30%)	Ferrous	9.97
	Aluminum	0.86
	Copper	0.75
	Magnesium	0.46
	Zinc	0.26
Combustibles (49.05%)	Fiber	16.40
	Plastics	14.10
	Foam	11.75
	Paper	4.70
	Rubber	2.10
Inerts (38.65%)	Dirt	33.15
	Glass	5.50

Source: Monaco, 1993.

time, the ability to reuse or recycle those materials also may change. Table 3.5 provides historical data on vehicle production and retirement rates. The information in this table can be used together with Tables 3.3 and 3.4 to obtain first-order estimates of total raw material usage and solid wastes from automobiles, as shown in Example 3.5.

Table 3.5   Annual U.S. vehicle production and retirements (in millions), 1965–1995.

	Vehicles Produced			Vehicles Retired		
Year	Cars	Commercial	Total	Cars	Commercial	Total
1965	9.34	1.80	11.14	5.70	0.74	6.44
1967	7.41	1.61	9.02	6.98	0.95	7.93
1969	8.22	1.98	10.21	6.35	0.97	7.31
1971	8.58	2.09	10.67	6.02	1.04	7.07
1973	9.67	3.01	12.68	7.99	1.21	9.20
1975	6.72	2.27	8.99	5.67	0.91	6.58
1977	9.21	3.49	12.70	8.23	1.67	9.90
1979	8.43	3.05	11.48	9.31	1.92	11.23
1981	6.25	1.69	7.94	7.54	1.52	9.06
1983	6.78	2.44	9.22	6.24	1.50	7.73
1985	8.18	3.47	11.65	7.73	2.10	9.83
1987	7.10	3.83	10.92	8.10	2.36	10.47
1989	6.82	4.05	10.87	8.98	2.19	11.17
1991	5.44	3.37	8.81	8.57	2.28	10.85
1993	5.98	4.92	10.90	7.37	1.05	8.41
1995	6.35	5.63	11.99	7.41	2.92	10.33

Source: AAMA, 1997.

Example 3.5

Raw materials consumption for automobile production.   Calculate the total quantities of iron, steel, aluminum, plastic, rubber, and glass used in all new U.S. passenger cars produced in 1995.

Solution:

From Table 3.5, 6.35 million passenger cars were produced in 1995. From Table 3.3, the average weight was 1,458 kg (3,208 lbs), so the total fleet weight is

$$6.35 \times 10^6 \times 1,458 = 9.26 \times 10^9 \, kg = 9.26 \text{ million metric tons}$$

Table 3.3 gives the percentage of each material. Thus the total quantities used for 1995 passenger cars are as follows:

Material	Fraction of Total	Amount Used (Million Metric Tons)
Steel	0.551	5.10
Iron	0.124	1.15
Plastic	0.077	0.71
Aluminum	0.058	0.54
Rubber	0.042	0.39
Glass	0.029	0.27

As discussed in Chapter 1, these materials requirements bring with them a new set of environmental impacts beyond those from vehicle operation alone. Environmental impacts occur in the mining of raw materials; in the transport and processing of those materials into finished metals, plastics, and other goods; and in the manufacturing operations that transform and assemble these materials into the final consumer product—the automobile.

The variety of materials used in an automobile and the many components it contains (a modern automobile has over 20,000 individual parts) mean that any engineering decision that influences materials choice, shape, or quantity can have significant environmental consequences. These impacts occur both "upstream" prior to vehicle operation (in raw materials extraction, processing, and manufacturing), and "downstream" subsequent to vehicle use (in the recycling and disposal phases). The ability to reuse or recycle spent materials at the end of a car's useful life can significantly reduce the environmental impacts from materials production, transport, manufacturing, and final waste disposal.

Example 3.6

Automobile solid waste.   Estimate the total quantities of (a) recycled materials and (b) solid wastes from U.S. passenger car retirements in 1995. Based on 1995 data, assume that 95 percent of retired vehicles undergo recycling, and that 75 percent of each car's total weight is recovered. Use the average weight of 1985 vehicles to estimate the per-vehicle weight of cars retired in 1995.

Solution:

(a) From Table 3.5, the total number of U.S. passenger cars retired in 1995 was 7.41 million. Assuming the 1985 average weight per car of 1,449 kg from Table 3.3, the total quantity of materials retired was

$$(7.41 \times 10^6)(1,449) = 10.74 \times 10^9 \, \text{kg}$$

Crushed cars awaiting final disposal and recycle of materials.

Of this total, 5 percent is discarded directly, whereas 95 percent is recycled:

$$\text{Direct waste} = (10.74 \times 10^9)(0.05) = 0.54 \times 10^9 \text{ kg}$$
$$\text{Recycled material} = (10.74 \times 10^9)(0.95) = 10.2 \times 10^9 \text{ kg}$$
$$= 10.2 \text{ million metric tons}$$

(b) Seventy-five percent of the recycled material is recovered, leaving 25 percent as shredder waste:

$$\text{Recovered material} = (10.2 \times 10^9)(0.75) = 7.65 \times 10^9 \text{ kg}$$
$$\text{Shredder waste} = (10.2 \times 10^9)(0.25) = 2.55 \times 10^9 \text{ kg}$$

The total waste material disposed of in the environment is thus

$$\text{Total waste disposed} = 0.54 \times 10^9 \text{ kg} + 2.55 \times 10^9 \text{ kg}$$
$$= 3.09 \times 10^9 \text{ kg} = 3.1 \text{ million metric tons}$$

Notice that recycling and reuse of spent materials reduces the demand for virgin materials and thus reduces the associated environmental impacts of materials production. Ways of reducing environmental impacts across all phases of a product's manufacture, use, and retirement are explored more fully in Chapter 7, which addresses life cycle assessments.

3.2.4   Other Environmental Impacts

The atmospheric emissions and solid waste impacts just discussed are the most significant—but not the only—environmental impacts from automobiles. Several additional environmental issues are worth noting.

Lead Emissions   For many years, tetraethyl lead was added to gasoline to prevent engine "knock," a condition stemming from early detonation of the air–fuel mixture. Automobiles thus became a major source of lead emissions to the atmosphere. The toxic effects of lead on human health resulted in a new federal air quality standard analogous to those adopted for CO, NO_x, and ozone. Because it also poisons automotive emission control systems, lead has been gradually phased out of gasoline, and car engines have been redesigned to operate on unleaded fuel. Nonetheless, the population of older vehicles that use leaded gasoline constitutes the second largest source of U.S. lead emissions, according to the U.S. Environmental Protection Agency (EPA). Many other countries around the world, however, continue to use leaded gasoline because of its lower cost. Chapter 10 provides a more detailed look at human exposure to lead in the environment.

CFC Emissions   The chlorofluorocarbons (CFCs) used in automotive air conditioners are another source of environmental emissions associated with depletion of stratospheric ozone (producing the so-called "ozone hole"). Chapter 2 provided an overview of this issue, and Chapter 11 describes in more detail the atmospheric chemistry of ozone depletion by CFCs. In the case of automobiles, CFCs enter the atmosphere via leaks in the air conditioning system over the life of a car. To replenish the lost refrigerant, car owners typically have the air conditioner recharged with additional CFC coolant. By international agreement, however, further production of CFCs has been recently phased out in the United States and other industrialized nations. Capture and recycling of the existing stock of CFCs will be the only source of these chemicals until they are finally depleted.

The engineering challenge is to find new refrigerants that are less harmful to the environment and that function effectively. Significant progress has been made, with new coolants already in use in appliances such as refrigerators and new automotive air conditioners. Chapter 6 provides a detailed case study of engineering options for alternative coolants in refrigerators.

Waste Motor Oil   Disposal of used engine oil can pose environmental hazards, especially if waste oil finds its way into water supplies via runoff into surface waters or seepage into groundwater. The problem arises primarily from do-it-yourself oil changes, where the waste oil is either spilled directly onto the ground or discarded with household rubbish that eventually reaches a landfill. Used motor oil has been designated as a hazardous waste that requires special handling and treatment. Reputable automobile service stations collect used motor oil and return it to recycling centers, where it is processed to remove contaminants such as heavy metals. Engineering solutions to the waste oil problem lie primarily in improving the technology used to recycle this waste product and, to a more limited extent, in the chemical composition of motor oil lubricants.

Other Life Cycle Impacts   The earlier discussion of automotive materials use and solid waste introduced some of the broader life cycle environmental impacts associated with the production, manufacture, use, and disposal of materials used in automobiles. The release of particulate air pollutants from tire wear and the erosion of brake linings are additional examples of life cycle impacts. Environmental impacts also arise from the production of gasoline and other motor vehicle fuels such as diesel oil. These impacts occur across the chain of activities from the discovery and production of crude oil, its refining into gasoline and other fuels, and its storage and distribution to local service stations.

Other studies provide more comprehensive discussions and analyses of the full life cycle impacts of an automobile (MacLean and Lave, 1998). These studies quantify the additional releases of air pollutants, water pollutants, and solid wastes to the environment from activities that support automotive travel. Such activities include automotive repair operations, paint shops, parts manufacturers, and even the auto insurance industry, which requires goods and services that indirectly release pollutants to the environment. Nonetheless, the urban air pollutants and greenhouse gas emissions discussed in this chapter remain the dominant source of environmental impacts attributable to automotive technology.

3.3   FUEL AND ENERGY REQUIREMENTS

Thus far we have treated the automobile as a "black box" that consumes fuel and materials and emits gaseous, liquid, and solid wastes into the environment. To reduce these environmental emissions, it is important to look inside that black box. This section uses basic concepts from physics and chemistry to explore in greater depth the role that engineering design decisions have in reducing automotive environmental impacts. Many of these impacts are related to fuel and energy use, so we begin by examining the energy and power requirements of an automobile.

3.3.1   Power for Cruising

Let's first look at the energy and power needed to propel a car at constant speed on level ground. How much energy is needed and why? Introductory physics teaches us that the energy, E, needed to push a body along a level surface is the product of the pushing force, F, times the distance, d, through which the body moves:

$$E = Fd \tag{3.2}$$

This quantity also is known as the *work* done on the body. Next, recall that power, P, is defined as energy per unit time, while velocity, v, is distance per unit time. Thus an expression for power can be written as the product of force times velocity:

$$P = \frac{E}{t} = \frac{Fd}{t} = Fv \tag{3.3}$$

For a car moving at constant velocity (zero acceleration), Newton's second law tells us that the pushing force on the car must be equal in magnitude to the forces opposing the motion, so that the net force is equal to zero. As depicted in Figure 3.6, two major forces must be overcome to keep a car moving at a constant speed. One is the frictional force between the tires and the road. This can be expressed as the product of the rolling friction coefficient, μ, and the normal force, N, perpendicular to the ground. On level ground, N is equal to the car's weight, W. So

$$F_{friction} = \mu N = \mu W \tag{3.4}$$

The second force is the air resistance or aerodynamic drag force. From the study of fluid mechanics, we find that this force can be approximated as

$$F_{drag} = \frac{1}{2} C_d A \rho v^2 \tag{3.5}$$

where C_d is the drag coefficient, A is the cross-sectional area of the vehicle, and ρ is the density of air. The drag coefficient C_d is a measure of how aerodynamic the vehi-

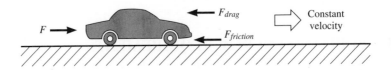

Figure 3.6   Forces acting on a car at constant velocity.

cle design is. Shapes that are blunt, such as the front of a truck or bus, have relatively high drag coefficients, whereas vehicles that are more streamlined have lower values of C_d in the range of 0.35 (for a small, sleek sports car) to 0.50 (for a typical U.S. passenger car). The total force required to propel the car is then

$$F = F_{\text{drag}} + F_{\text{friction}} \tag{3.6}$$

$$F = \frac{1}{2}C_d A\rho v^2 + \mu W \tag{3.7}$$

Substituting this expression for F into Equation (3.2) and Equation (3.3) gives the energy and power required to cruise at constant velocity on level ground:

$$E = \left(\frac{1}{2}C_d A\rho v^2 + \mu W\right)d \tag{3.8}$$

$$P = \left(\frac{1}{2}C_d A\rho v^2 + \mu W\right)v \tag{3.9}$$

Note that the quantities calculated here refer to the energy and power delivered by the wheels that propel the car. As we will see later, the energy and power supplied by the fuel must be greater than these values because of various losses in generating and transmitting energy to the wheels. In the examples below we express the car weight in units of newtons (N) to facilitate the calculations.

Example 3.7

Power for cruising at constant speed on level ground. Calculate the power needed to maintain a speed of 25 m/s (56 mph) on a level highway for a midsize car weighing 15,000 N (3,360 lbs). The vehicle has a cross-sectional area of 2.0 m^2 and a drag coefficient of 0.5. The coefficient of rolling friction is 0.02, and the density of air is 1.2 kg/m^3.

Solution:

Use Equation (3.9) to calculate the power needed:

$$P = \left(\frac{1}{2}C_d A\rho v^2 + \mu W\right)v = \frac{1}{2}C_d A\rho v^3 + \mu Wv$$

$$= \frac{1}{2}(0.5)(2.0)(1.2)(25)^3 + (0.02)(15,000)(25)$$

$$= 9,375 + 7,500$$

$$= 16,875 \text{ watts} = 16.9 \text{ kW } (22.7 \text{ hp})$$

In the United States, automotive power is often expressed in units of horsepower (1 hp = 0.746 kW). For the conditions in the above example, 56 percent of the power is needed to overcome aerodynamic drag and 44 percent to overcome rolling friction.

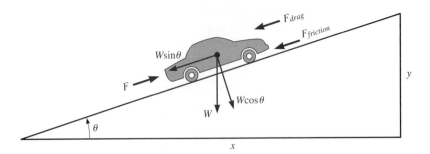

Figure 3.7   Forces acting on a car climbing a hill at constant velocity.

3.3.2   Power for Hill Climbing

The previous example assumed travel on a level road. When driving up a hill, additional work must be done against the force of gravity. The gravitational force is equal to the car's weight, W, which can be decomposed into force components parallel to and perpendicular to the road, as shown in Figure 3.7.

At constant velocity, the sum of all forces acting along the direction of motion again must equal zero. Enough "push" must now be supplied to overcome not only the forces of road friction and air resistance, but also the component of the car's weight being pulled downhill by gravity. From Figure 3.7, the magnitude of that force is $W\sin\theta$, where θ is the angle of inclination with respect to the horizontal. Similarly, the component of the car's weight perpendicular to the ground has the magnitude $W\cos\theta$. Thus the rolling friction force is now given by

$$F_{friction} = \mu N = \mu W\cos\theta$$

The total energy and power needed for hill climbing at constant speed now become

$$E = \left(\frac{1}{2}C_d A\rho v^2 + \mu W\cos\theta + W\sin\theta\right)d \qquad (3.10)$$

$$P = \left(\frac{1}{2}C_d A\rho v^2 + \mu W\cos\theta + W\sin\theta\right)v \qquad (3.11)$$

Note that when $\theta = 0$ we obtain the same equations shown earlier for level ground. Also, the steepness of a hill often is specified as the *grade,* which is defined as the ratio of vertical rise over a given horizontal distance. Thus a 5 percent grade, typical of interstate highways, corresponds to an increase in height of 5 meters for every 100 meters of horizontal distance.

Example 3.8

Power for climbing a hill.   Calculate the power needed by the car in Example 3.7 to climb a 5 percent grade at a constant velocity of 25 m/s (56 mph).

Solution:

From Figure 3.7, grade = $y/x = 0.05$. Thus, $\theta = \tan^{-1} 0.05 = 2.86°$. Now use Equation (3.11) and the results from Example 3.7 to calculate the power needed:

$$P = \left(\frac{1}{2} C_d A \rho v^2 + \mu W \cos\theta + W\sin\theta \right) v$$

$$= 9{,}375 + (7{,}500)\cos 2.86 + (15{,}000)(\sin 2.86)(25)$$

$$= 9{,}375 + 7{,}491 + 18{,}711$$

$$= 35{,}577 \text{ watts} = 35.6 \text{ kW } (47.7 \text{ hp})$$

We see that the total power needed for hill climbing in the above example is about twice that needed for cruising at the same speed on level ground. The additional power (more than half the total) is needed to lift the car against the forces of gravity.

3.3.3 Power for Acceleration

Now let's consider the energy and power needed to bring a car up to cruising speed from a complete stop. Recall that *acceleration* is defined as the change in velocity with respect to time. In general, the additional energy needed to increase a car's velocity is most easily expressed as the change in kinetic energy ($KE = mv^2/2$) after accelerating from an initial velocity v_i to a final velocity v_f. We can also express the car's mass, m, in terms of its weight, W, recalling that weight and mass are related by $W = mg$, where g is the acceleration of gravity (9.8 m/s^2). Thus the energy needed for acceleration in the absence of air resistance and road friction is

$$E_{accel} = KE_{final} - KE_{initial} = \frac{1}{2}mv_f^2 - \frac{1}{2}mv_i^2 = \frac{W}{2g}(v_f^2 - v_i^2) \qquad (3.12)$$

If the acceleration occurs over a time interval, t, then the average power supplied is simply

$$P_{accel} = \frac{E_{accel}}{t} = \frac{W}{2gt}(v_f^2 - v_i^2) \qquad (3.13)$$

This equation is useful for evaluating the power needed for passing another car on the road, as well as starting from rest where $v_i = 0$.

Example 3.9

Power to accelerate from rest. The car in Example 3.7 accelerates from rest to 25 m/s (56 mph) in 10 seconds. Calculate the average power required in the absence of air resistance and road friction.

Solution:

Use Equation (3.13). Because $v_i = 0$, we have

$$P = \frac{Wv_f^2}{2gt} = \frac{(15,000)(25)^2}{(2)(9.8)(10)}$$

$$= 47,832 \text{ watts} = 47.8 \text{ kW (64.1hp)}$$

Comparing the results of Example 3.7 and Example 3.9, we see that the power needed for acceleration is three times greater than the power needed to overcome road friction and air drag. Of course, different results would be obtained for different conditions. Equation (3.13) shows that the shorter the time for acceleration, the more power that is required. Thus high-performance cars that accelerate rapidly must have large engines to supply the needed power.

When we consider the additional energy and power to overcome road friction and aerodynamic drag during acceleration, the total energy requirement increases beyond that given by the previous equations. The additional energy to overcome road friction on level ground is again given by μWd. But the energy to overcome aerodynamic drag differs because the velocity changes as the car accelerates; thus, the aerodynamic drag force also changes with increasing speed. For a car undergoing a constant acceleration from rest, it can be shown that the energy needed to overcome drag forces during acceleration is $1/4 C_d A\rho v_f^2 d$, where d is the total distance traveled. For travel on level ground, the total energy required to accelerate from rest is thus

$$E_{accel} = \frac{W}{2g}v_f^2 + \mu Wd + \frac{1}{4}C_d A\rho v_f^2 d \tag{3.14}$$

We can also rewrite the first term of this equation using the relationship $v_f^2 = 2ad$ from elementary physics, where a is the acceleration. This gives

$$E_{accel} = \left(\frac{a}{g}W + \mu W + \frac{1}{4}C_d A\rho v_f^2\right)d \tag{3.15}$$

This equation shows more explicitly how faster acceleration requires more energy. The acceleration here is simply

$$a = \frac{v_f}{t} \tag{3.16}$$

We leave it as an exercise for students to calculate energy and power for accelerating up a hill or for passing another vehicle.

3.3.4   Energy Efficiency

The energy and power needs calculated so far represent the energy needed at the wheels to propel the car. How much fuel does it take to deliver that energy? The energy content of a liter of gasoline is about 3.5×10^7 joules. However, only a small fraction of that energy winds up as useful power to move the car. The laws of thermodynamics dictate that roughly 60 to 70 percent of the energy will be wasted as heat when the fuel is burned. Auxiliary systems connected to the engine, like power steering, radios, and air conditioners, sap away another few percent, while "standby" loss, which is the energy used to keep an engine idling and ready to power the car (while the car is stopped at a traffic light, for example), consumes up to 15 percent of the fuel's energy. Finally, an additional 5 percent or so is spent to overcome frictional forces in the transmission, axles, and gears that connect the engine to the car's drive wheels. The result is that only about 15 to 20 percent of the fuel's energy actually propels the car. In general, we can define this fraction as the overall energy efficiency, η:

$$\eta = \frac{\text{Useful energy delivered for motion}}{\text{Total fuel energy input}} = \frac{E}{E_{fuel}} \qquad (3.17)$$

where the numerator, E, refers to the energy given by Equation (3.10) or Equation (3.12). Figure 3.8 shows a simplified sketch of the energy flows and principal car components that determine overall efficiency. The car's engine converts the chemical energy in fuel into useful mechanical energy in the form of a rotating crankshaft capable of doing work. The car's drivetrain then transfers that mechanical energy to the wheels that propel the car.

Just as an energy efficiency was defined for the overall system in Figure 3.8, an expression for efficiency also can be defined for each major car component or subsystem. For simplicity, let's consider just two principal subsystems: (1) the engine, including all auxiliary systems and standby losses, and (2) the drivetrain, which includes the transmission, gears, axles, and drive wheels. We can then define the overall engine efficiency as

$$\eta_{engine} = \frac{\text{Useful energy out}}{\text{Total energy in}} = \frac{E_{shaft}}{E_{fuel}} \qquad (3.18)$$

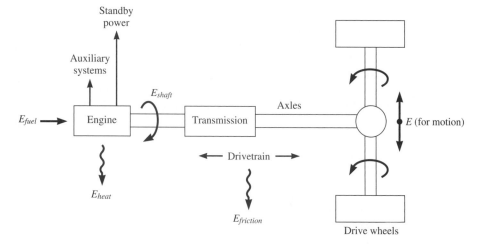

Figure 3.8   Automobile energy flows from fuel to wheels. The vehicle's engine transforms chemical energy in fuel into useful mechanical energy delivered by a rotating shaft. That energy is transmitted to the drive wheels to provide motive power for the vehicle. Most of the fuel energy, however, is lost as waste heat from the engine, plus friction in drivetrain components.

For the engine alone, *useful energy out* refers to the net shaft energy available to propel the car. Similarly, the overall drivetrain efficiency can be defined as

$$\eta_{train} = \frac{E}{E_{shaft}} \tag{3.19}$$

where E is the final energy supplied to the drive wheels. We then see that the overall efficiency can be expressed as the product of each subsystem efficiency:

$$\eta = \eta_{engine} \times \eta_{train} \tag{3.20}$$

This insight will be useful later in exploring ways that engineering design can help improve overall efficiency.

Example 3.10

Calculating engine efficiency.   During a 15-minute laboratory test at partial load conditions, a car engine consumed 1.7 liters of gasoline, and 20 kW of total power is measured at the engine drive shaft. What is engine efficiency under these conditions? How does the efficiency change if 6 kW of shaft power is used to supply energy for standby and auxiliary systems?

Solution:

To put everything on an energy basis, first convert the 20 kW of total shaft power to energy delivered during the 15-minute test:

$$E_{shaft} = (P_{shaft})(t) = (20 \times 10^3 \text{ watts})(15 \text{ min} \times 60 \text{ sec/min}) = 1.80 \times 10^7 \text{ joules}$$

The energy input is contained in the 1.7 liters of fuel consumed. Recall that 1 liter of gasoline contains 3.5×10^7 joules of energy. Thus, in the absence of standby and auxiliary power drains,

$$\eta_{engine} = \frac{E_{shaft}}{E_{fuel}} = \frac{1.80 \times 10^7}{(1.7)(3.5 \times 10^7)} = 0.30 = 30\%$$

If standby and auxiliary systems require 6 kW of power, the net shaft power available to propel the car is 14 kW. Repeating the calculation for 14 kW instead of 20 kW gives a new value for overall engine efficiency:

$$\eta_{engine} = 0.21 = 21\%$$

Recall that the engine efficiency is not the overall efficiency for the car as a whole. Other losses occur between the engine and the drive wheels.

Example 3.11

Calculating overall efficiency.   If the car engine in Example 3.10 is linked to a drivetrain with an efficiency of 80 percent, what will be the overall energy efficiency for the car at those conditions? Base your calculation on the net engine efficiency that accounts for standby and auxiliary power requirements.

Solution:

From Example 3.10, the net engine efficiency was 21 percent. Thus, from Equation (3.20),

$$\eta = \eta_{engine} \times \eta_{train}$$

$$= (0.21)(0.80)$$

$$= 0.17 = 17\%$$

In practice, the efficiency of a car engine varies with operating conditions, with peak efficiency usually occurring at a cruising speed of roughly 13 to 18 m/s (30 to 40 mph). A more detailed treatment of efficiency relationships is found in the study of thermodynamics.

3.3.5 Fuel Consumption

We saw earlier that CO_2 emissions are directly proportional to a car's fuel consumption. Rearranging the terms in Equation (3.17) relates fuel consumption to motive energy requirements and overall efficiency:

$$E_{fuel} = \frac{E}{\eta} \tag{3.21}$$

A more convenient way of expressing fuel consumption is in terms of the fuel needed to drive a given distance, d:

$$\frac{E_{fuel}}{d} = \frac{1}{\eta}\left(\frac{E}{d}\right) \tag{3.22}$$

The energy, E, required for motion depends on how the car is driven. For *constant velocity* travel, E was given by Equation (3.10). Substituting that equation into Equation (3.22) gives

$$\left.\frac{E_{fuel}}{d}\right|_{v=constant} = \frac{1}{\eta}\left(\frac{1}{2}C_d A\rho v^2 + \mu W\cos\theta + W\sin\theta\right) \tag{3.23}$$

Example 3.12

Fuel consumption at constant velocity. Calculate the volume of fuel needed by the car in Example 3.7 to drive 100 km at 25 m/s (56 mph) on a level road. Assume the overall energy efficiency is 20 percent, and the energy content of gasoline is 3.5×10^7 J/L.

Solution:

Use Equation (3.23). For a level road $\theta = 0°$; thus $\cos 0° = 1$ and $\sin 0° = 0$. This gives

$$\frac{E_{fuel}}{d} = \frac{1}{\eta}\left(\frac{1}{2}C_d A\rho v^2 + \mu W\right) = \frac{1}{0.20}\left[\left(\frac{1}{2}(0.5)(2.0)(1.2)(25)^2 + (0.02)(15{,}000)\right)\right]$$

$$= \frac{1}{0.20}(375 + 300) = 3{,}375 \text{ J/m}$$

For $d = 100$ km,

$$E_{fuel} = (3{,}375 \text{ J/m})(100 \text{ km} \times 1{,}000 \text{ m/km}) = 3.375 \times 10^8 \text{ J}$$

The quantity of fuel is then

$$(3.375 \times 10^8 \text{ J})/(3.5 \times 10^7 \text{ J/L}) = 9.64 \text{ liters}/100 \text{ km}$$

For a car undergoing *constant acceleration* from rest to a final cruising velocity, v_f, the energy required for acceleration on level ground was given earlier by Equation (3.15). The total fuel energy per unit distance traveled is then

$$\left. \frac{E_{fuel}}{d} \right|_{a=constant} = \frac{1}{\eta}\left(\frac{a}{g}W + \mu W + \frac{1}{4}C_d A\rho v_f^2 \right) \tag{3.24}$$

As before, the first term in the bracket on the right side of this equation represents the energy needed to change the car's velocity in the absence of friction, whereas the second and third terms reflect energy needed to overcome road friction and air friction (drag) during the acceleration.

Example 3.13

Fuel consumption during acceleration.   Calculate the fuel requirement per 100 km for the car in Example 3.12 accelerating from rest to a speed of 25 m/s (56 mph) in 15 sec. Include the effects of air resistance and road friction.

Solution:

From Equation (3.16), the acceleration is

$$a = \frac{v_f}{t} = \frac{25}{15} = 1.67 \text{ m/s}^2$$

Then from Equation (3.24),

$$\frac{E_{fuel}}{d} = \frac{1}{0.20}\left[\frac{1.67}{9.8}(15{,}000) + (0.02)(15{,}000) + \frac{1}{4}(0.5)(2.0)(1.2)(25)^2 \right]$$

$$= \frac{1}{0.20}(2{,}551 + 300 + 187.5)$$

$$= 15{,}193 \text{ J/m}$$

For $d = 100$ km,

$$E_{fuel} = (15{,}193)(100 \text{ km} \times 1{,}000 \text{ m/km}) = 1.52 \times 10^9 \text{ J}$$

The fuel needed is thus

$$\frac{1.52 \times 10^9 \text{ J}}{3.5 \times 10^7 \text{ J/L}} = 43.4 \text{ liters}/100 \text{ km}$$

This rate of fuel consumption during acceleration is 4.5 times greater than at the constant velocity conditions of Example 3.12! Of course, only a portion of a car's total driving time involves acceleration. Thus the overall fuel consumption depends on how a car is driven.

In Examples 3.12 and 3.13, the rate of fuel consumption was expressed in units of liters per 100 km of travel. This is the measure commonly used throughout Europe and other parts of the world. In the United States, however, we use the reciprocal measure of miles per gallon (mpg) to express fuel consumption. A convenient conversion factor is

$$\text{mpg} = \frac{235.3}{\text{liters per 100 km}} \qquad (3.25)$$

Note that miles per gallon is not a direct measure of the energy needed to travel a given distance. Rather, it gives the average distance a car can travel on a fixed amount of energy (that is, the energy in one gallon of fuel).

Example 3.14

Calculating miles per gallon. Convert the fuel economy results in Examples 3.12 and 3.13 to miles per gallon.

Solution:

Use Equation (3.25). For Example 3.12 (constant velocity),

$$\text{mpg} = \frac{235.3}{9.64} = 24.4 \text{ mpg}$$

For Example 3.13 (constant acceleration),

$$\text{mpg} = \frac{235.3}{43.4} = 5.4 \text{ mpg}$$

Notice the poor fuel economy during acceleration. This is a major reason why city driving, which requires frequent acceleration in start–stop traffic, results in lower overall mpg than highway driving at constant speed.

Table 3.6 shows the fuel economy of various cars as computed by the U.S. Environmental Protection Agency using standard laboratory driving cycles intended to simulate city and highway driving. The data in Table 3.6, together with the basic understanding of automotive fuel and energy needs developed in this section, can

Table 3.6   Characteristics of selected 1997 model year vehicles.

Vehicle Make	Model	Weight (lbs)	Height (in)	Width (in)	Power (hp)	City mpg	Hwy mpg
Acura	2.2 CL (97)	3,064	70.1	54.7	154	23	30
Volvo	S70 GLT	3,206	69.3	55.7	236	20	29
Oldsmobile	Cutlass GL	3,102	69.4	56.3	155	20	29
Ford	Escort LX Sedan	2,468	67.0	53.3	110	25	34
Honda	Accord DX Sedan	2,888	70.3	56.9	135	22	29
Mercedes	C230 Sedan	3,197	67.7	56.1	148	23	30
Plymouth	Neon Highline Coupe	2,389	67.5	54.8	132	25	33
Subaru	Outback Sport Wagon	2,835	67.1	56.3	137	23	30
BMW	540Ai	3,803	70.9	56.5	282	18	24
Eagle	Talon	2,729	68.3	51.4	140	21	31
Audi	A4 2.8 Quattro Wagon	3,428	56.7	68.2	190	17	27
Cadillac	Catera	3,770	56.3	70.3	200	18	25
Chevrolet	Cavalier Sedan	2,681	54.6	67.1	115	24	31
Chrysler	Cirrus	3,099	54.1	71.0	164	20	29
Dodge	Avenger	2,879	53.0	68.5	140	21	30
Lexus	GS 300	3,660	55.9	70.7	220	18	24
Mazda	Millenia	3,216	54.9	69.7	170	20	27
Ford	Contour LX Sedan	2,759	54.5	69.1	125	24	32
Nissan	Maxima GLE Sedan	3,095	69.7	55.7	190	21	28
Pontiac	Firebird Base Coupe	3,148	52.0	74.5	200	19	28
Toyota	Camry CE	2,998	55.7	70.1	133	23	30
Hyundai	Elantra GLS Sedan	2,734	54.9	66.9	130	23	31
Acura	Integra RS Coupe	2,643	52.6	67.3	142	24	31
Toyota	Corolla CE	2,503	54.5	66.7	120	28	35
Saturn	SW2 Wagon	2,483	54.9	66.7	124	24	34
Chrysler	Concorde LX (97)	3,552	74.4	56.3	214	18	26
Buick	Century Custom	3,371	72.7	56.5	160	20	29
Nissan	200SX Base Coupe	2,363	66.6	54.2	115	27	35
Dodge	Neon Highline Sedan	2,399	67.6	54.8	132	25	33
Kia	Sephia GS	2,608	66.7	54.7	122	23	32
Mercedes	CLK320 Coupe	3,250	53.0	67.8	215	20	27
Ford	Taurus LX Sedan	3,326	55.1	73.0	145	20	28
Lincoln	Town Car Executive	3,860	58.0	78.2	200	17	25
Mercury	Sable LS Sedan	3,388	55.4	73.0	145	20	28
Chevrolet	Camaro Convertible	3,468	52.0	74.1	200	19	29
Chrysler	LHS	3,619	55.9	74.5	214	18	26
Audi	A8 3.7	3,682	56.7	77.7	230	17	26

Source: Edmunds, 1997.

provide insights into how engineering design variables affect fuel economy. The problems at the end of this chapter explore some of these relationships in detail.

3.4 ENGINEERING CLEANER CARS

We turn now to a more detailed look at some of the ways that engineers can reduce the environmental impacts of automotive technology. We focus first on energy efficiency improvements, followed by a consideration of other design options for emissions control, including alternative fuels and alternative vehicles.

3.4.1 Designing for Energy Efficiency

Improving energy efficiency brings environmental benefits in several ways. First, CO_2 emissions that contribute to global warming are reduced in direct proportion to fuel energy saved. Second, because less fuel is required, the indirect fuel cycle impacts of petroleum production, refining, storage, and distribution also are reduced. Finally, improved energy efficiency reduces the depletion of natural resources used throughout the automotive life cycle.

In the United States, automotive energy efficiency standards in the form of a Corporate Average Fuel Economy (CAFE) requirement were first adopted in 1975. These standards currently require a fleet-averaged fuel efficiency of 27.5 miles per gallon (mpg). The original motivation for CAFE was to reduce U.S. dependence on imported oil in the wake of the 1973 Arab oil embargo. At that time the average fuel economy of U.S. automobiles was only 13 mpg. In the 1990s the problem of global warming has focused renewed attention on automotive energy efficiency as a key method of reducing fossil energy use.

Equations (3.23) and (3.24) reveal the principal engineering design parameters that affect automotive fuel consumption, as summarized in Table 3.7. Engine efficiency is perhaps the most important, as it is the primary determinant of overall efficiency (other losses being small compared to heat losses in the engine). There is only a limited ability to improve the efficiency of conventional gasoline-burning engines because of thermodynamic limitations on what is possible with combustion systems. Other types of engine designs, such as fuel cells, have a much greater potential for achieving higher efficiency. Section 3.4.5 discusses several of those options. Nevertheless, modest improvements to current engine designs are still feasible and can contribute to improved fuel economy. Technical measures that are currently available include improvements to valve control, fuel injection design, increased compression ratios, and use of smart alternators. Improvements to transmission and drivetrain components also can provide modest efficiency gains.

Vehicle weight is critical to fuel economy, as is shown by Equations (3.23) and (3.24). A car's weight can be reduced in several ways. One is to reduce overall vehicle size. A subcompact car, for example, weighs nearly half as much as a full-size vehicle. Substituting lighter materials such as aluminum and plastics for heavier materials like steel is another method. Automakers have substituted materials extensively over the

Table 3.7   Summary of key parameters affecting fuel consumption.

Parameter	Symbol
Vehicle Design Parameters	
Engine efficiency	η_{engine}
Drivetrain efficiency	η_{train}
Vehicle weight	W
Cross-sectional area	A
Drag coefficient	C_d
Rolling friction coefficient	μ
Engine size or power	P
Vehicle Operating Parameters	
Velocity	v
Acceleration	a

past two decades (review Table 3.3). For example, many cars now have an aluminum engine block and plastic rather than steel bumpers. Weight also can be reduced by design changes that eliminate auto components, such as the long driveshaft eliminated by switching from rear-wheel to front-wheel drive. Table 3.3 showed that the weight of a U.S. car has been reduced by 10 percent (165 kg or 1,620 N) between 1978 and 1995.

Reductions in cross-sectional area and drag coefficient save energy by reducing aerodynamic drag. The cross-sectional area is closely related to vehicle size (and thus indirectly to vehicle weight), whereas the drag coefficient depends on vehicle shape. Improved aerodynamic design of even minor protuberances like side-view mirrors can help reduce the overall drag coefficient. In a similar fashion, rolling friction can be reduced through improvements in tire design. For example, radial tires have a lower rolling friction coefficient than conventional four-ply tires, and this contributes to improved fuel economy.

Example 3.15

Effect of weight reduction on fuel use and emissions.   In Example 3.12, how much improvement in fuel economy would be achieved by engineering design changes that reduce the vehicle weight by 10 percent? What would be the accompanying reduction in CO_2 emissions?

Solution:

A 10 percent reduction in vehicle weight gives a new car weight of

$$W = (15,000)(0.90) = 13,500 \text{ N}$$

Repeating the calculation in Example 3.12 gives a new fuel consumption rate of 9.21 L/100 km. Compared to the earlier result of 9.64 L/100 km, gasoline consumption is reduced by 4.5 percent. Because CO_2 emissions are directly proportional to fuel use, CO_2 emissions also decrease by 4.5 percent.

Equations (3.23) and (3.24) show that a car's velocity and acceleration also are key variables determining automotive fuel consumption. These parameters depend on how a car is driven as well as on how it is designed. The engine size or power selected by the auto designer determines the maximum velocity and acceleration that can be achieved, and in general, larger engines consume more fuel. For a given vehicle, however, drivers who accelerate rapidly or cruise at high speeds will burn more gasoline than those who drive more moderately.

Example 3.16

Effect of speed on fuel use and emissions.   Compare the fuel economy of the car in Example 3.12 traveling at 25 m/s (56 mph) to the same car driving at 35 m/s (78 mph). Assume the overall efficiency (engine plus drivetrain) is 20 percent in both cases. How is the CO_2 emission rate affected?

Solution:

Repeat the calculation in Example 3.12 using $v = 35$ m/s. The result is a fuel consumption rate of 14.79 L/100 km (15.9 mpg). Compared to 9.64 L/100 km at $v = 25$ m/s (56 mph), this is an increase of 53 percent in fuel consumption! CO_2 emissions are directly proportional to fuel use, so they increase by the same percentage. Thus driving at the lower speed significantly reduces both fuel consumption and CO_2 emissions.

The astute reader may be wondering why the emission reduction benefits of improved energy efficiency did not also include reduced emissions of CO, HC, and NO_x as well as reduced CO_2. The answer is that the emission rates of these pollutants are not directly proportional to automotive fuel consumption, as is CO_2. The next section explains why this is so.

3.4.2 Understanding Pollutant Formation

Unlike CO_2, which is the natural end product of complete combustion, emissions of carbon monoxide (CO) and hydrocarbons (HC) are the result of *incomplete* combustion. Incomplete combustion occurs when there is insufficient air to combust all the fuel in an engine cylinder, or when there is insufficient time or inadequate temperature to complete the combustion. When this happens, some of the carbon in fuel is only partially oxidized to CO rather than CO_2. In a similar fashion, some of the gasoline hydrocarbons are only partially reacted and emitted in the form of unburned hydrocarbons.

Because the degree of incomplete combustion depends on the details of an engine's design, and not on its size or power, emissions of CO and HC are not directly proportional to fuel use. In fact, many smaller, more economical engines are "dirtier" than larger engines when compared on the basis of emissions per unit of fuel burned. However, passenger car emissions are regulated on the basis of emission per distance of travel (see Table 3.2), so different engine sizes are required to have the same maximum level of emissions even though their fuel consumption rates may differ dramatically.

Effect of Air–Fuel Ratio   One key parameter affecting the formation of CO and HC is the ratio of gasoline to air in the mixture that is burned. From elementary chemistry, the theoretical proportions of air and fuel needed to complete a chemical reaction are referred to as the *stoichiometric amounts*. Equation (3.1), shown earlier, is an example of a stoichiometric reaction for the chemical compound octane, which is used here to approximate the chemical formula for gasoline. That equation can be rewritten for combustion in air rather than oxygen. Taking air as a mixture of 79 percent N_2 and 21 percent O_2 by volume (a ratio of 3.76:1), the reaction is

$$C_8H_{18} + 12.5(O_2 + 3.76\,N_2) \rightarrow 8CO_2 + 9H_2O + 47\,N_2 \qquad (3.26)$$

This stoichiometric equation reveals that 12.5 moles of air are needed to react completely with each mole of fuel. If the air–fuel mole ratio were less than 12.5, it would be impossible to completely burn all of the fuel since there would not be enough air. In that case, carbon monoxide and unburned hydrocarbons would appear on the right side of Equation (3.26) as products of incomplete combustion. The greater the deficiency of air relative to the stoichiometric amount required, the more CO and HC that would appear in the combustion products, as illustrated in Figure 3.9.

Before 1970, automobile engines were commonly designed to operate with a deficiency of air relative to the stoichiometric quantity needed for complete com-

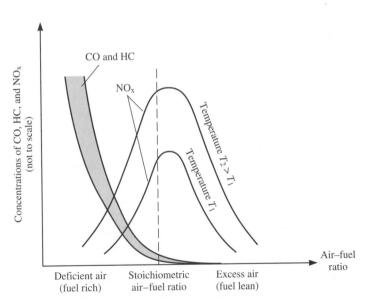

Figure 3.9   Effect of air–fuel ratio on air pollutant emissions. Emissions of CO and hydrocarbons increase dramatically as the air–fuel ratio drops below the stoichiometric value for complete combustion. Emissions of NO_x are highest at air–fuel ratios just above the stoichiometric value. Higher combustion temperatures also increase NO_x emissions.

An EPA vehicle emissions testing facility. The wheels of the truck sit on a dyna-mometer, which allows the vehicle to be "driven" under simulated road conditions. A video display outside the driver's window monitors the test conditions.

bustion (a condition known as a *fuel rich* mixture). Such designs were adopted to achieve smooth-running engines; however, they were also guaranteed to produce CO and HC in the exhaust. Controlling CO and HC emissions thus requires measures that promote complete combustion without compromising the operation of the car.

Formation of NO_x   Emissions of nitrogen oxides (NO_x) cannot be readily avoided because they are a normal product of most combustion reactions. Oxygen and nitro-gen in air, which coexist as a simple mixture at room temperatures, begin to react chemically at the very high temperatures reached during combustion. The most prevalent reaction is the one that forms nitric oxide (NO):

$$O_2 + N_2 \rightarrow 2NO \qquad\qquad (3.27)$$

Nitrogen dioxide (NO_2) also may be formed in smaller amounts by the reaction:

$$2NO + O_2 \rightarrow 2NO_2 \qquad\qquad (3.28)$$

The sum of NO and NO_2 is referred to as NO_x. In general, the production of NO_x increases exponentially with increasing temperature. Peak concentrations typically occur slightly above the stoichiometric air–fuel ratio, as shown in Figure 3.9. In the fuel rich situation, NO_x formation is inhibited because carbon and hydrogen atoms compete more favorably for oxygen atoms. NO_x production also depends strongly on the amount of time available for combustion because there is seldom enough time to reach equilibrium conditions. Controlling NO_x emissions thus requires measures that reduce peak temperatures, or alter the combustion stoichiometry, or alter the time–temperature profile of combustion events.

Overall Combustion Reaction   In general, the combustion products of a hydrocarbon fuel (C_xH_y) can be expressed as follows:

$$C_xH_y + \phi\,(O_2 + 3.76\,N_2) \rightarrow \qquad (3.29)$$

(Fuel)   (Air)

$$CO_2 + H_2O + N_2 + O_2 + CO + HC + NO + NO_2$$

(Normal end products)   (Incomplete   (Nitrogen
                             combustion) oxides)

The term ϕ on the left side represents the actual molar ratio of air to fuel. If this value exceeds the stoichiometric ratio (that is, if there is "excess air"), then some excess oxygen will appear with the combustion products. Diesel engines, for example, are designed for large amounts of excess air. Even in this case, however, some amount of CO and HC can still form because of incomplete combustion caused by uneven temperatures, poor mixing of air and fuel, and other factors.

As noted earlier, the term *volatile organic compound* (*VOC*) is often used as a substitute for HC because, strictly speaking, many of the unburned organic compounds found in automobile exhaust are not pure hydrocarbons. Formaldehyde (HCHO) is one example. Although the quantities of CO, HC, and NO_x represent very small fractions of the total combustion products, the aggregate emissions from all cars in use can give rise to significant environmental impacts, as discussed earlier.

Example 3.17

Conversion of fuel carbon to CO.   During a laboratory test, a car engine burns 123 g of octane fuel under conditions that produce 3.0 g of CO. What percentage of the carbon in the fuel is converted to CO under these conditions?

Solution:
The atomic weights of the elements are C = 12, H = 1, and O = 16. The molecular weight of octane (C_8H_{18}) is thus $(8 \times 12) + (18 \times 1) = 114$, and the molecular weight of CO is $12 + 16 = 28$. The mass of carbon in 123 g of octane fuel is thus

$$\frac{(8)(12)}{114}\,(123) = 103.6 \text{ g C}$$

Similarly, the mass of carbon in 3.0 g of CO is

$$\frac{12}{28}\,(3.0) = 1.29 \text{ g C}$$

So the percentage of fuel carbon converted to CO in this case is

$$\%\text{C as CO} = \frac{1.29}{103.6} \times 100 = 1.25\%$$

3.4.3 Designing for Low Emissions

To design engines for low emissions of CO, HC, and NO_x, it is necessary to understand how fuel is burned in an automotive engine. Figure 3.10 shows a simple diagram of this process. Fuel from the gas tank is pumped to the carburetor or fuel injector, whose function is to mix the fuel with the correct amount of air. The gaseous air–fuel mixture is then distributed via one or more manifolds to the engine's combustion chambers or cylinders. Figure 3.10 illustrates each phase of a typical four-stroke engine. On the intake stroke, the air–fuel mixture is drawn into the combustion chamber by the downward movement of the piston. Next the intake valve is closed, and the upward movement of the piston compresses the air–fuel mixture. Near the top of this compression stroke, a spark plug is fired to ignite the air–fuel mixture. The explosion and rapid expansion of combustion gases produce a downward power stroke. An exhaust valve then opens, allowing the upward piston motion of the exhaust stroke to expel the combustion gases into the exhaust manifold. The cycle then repeats itself.

During normal operation, the crankshaft of a car's engine, to which the pistons are attached, rotates at approximately 2,000 revolutions per minute (rpm). This means that each piston moves up and down approximately 33 times *each second,* leaving only 1/66 second for combustion during each downward power

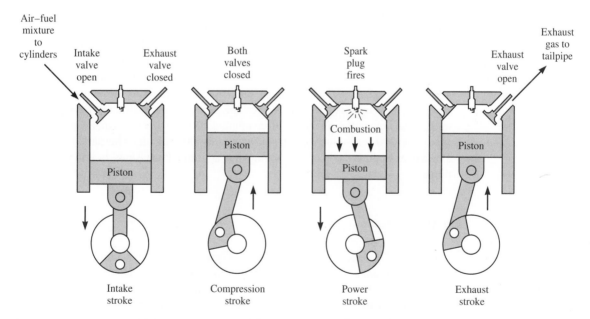

Figure 3.10   Schematic of a gasoline-powered automobile engine. The figure illustrates an engine cylinder during each stroke of the common four-stroke spark-ignition engine: (1) The intake stroke draws the air–fuel mixture into the cylinder; (2) the compression stroke compresses the mixture to a fraction of its initial volume; (3) a spark ignites the mixture, forcing the piston downward to produce power; (4) the exhaust stroke expels the combustion products from the cylinder.

stroke.[3] This extremely short time for combustion is a key reason that older car engines were designed to be fuel rich. Leaner air–fuel ratios often produced erratic combustion when fuel failed to ignite in the short time available. The difficulties of metering and controlling the ratio of air to fuel under a wide variety of operating conditions were another formidable barrier to reducing incomplete combustion.

As a result of emission control requirements, engineers have made substantial gains in reducing pollutant formation. Modern fuel injection systems now employ sophisticated sensors to accurately meter the quantities of fuel and air under changing conditions. Engine manifolds, cylinders, pistons, valves, and ignition system components all have been redesigned to improve combustion efficiency and reduce NO_x formation. Controlling CO, HC, and NO_x simultaneously is indeed a major engineering challenge because measures that improve combustion to reduce CO and HC tend to increase the formation of NO_x, as seen in Figure 3.9.

As exhaust emission requirements have grown more stringent, add-on devices called *catalytic converters* have been employed in addition to engine modifications to achieve additional emission reductions. These devices are located in the exhaust duct between the engine and the muffler. To reduce CO and HC emissions, a catalyst such as platinum oxidizes CO and HC to CO_2 and H_2O. A second catalyst, typically rhodium, converts nitrogen oxides back to molecular nitrogen and oxygen.

For catalytic converters to operate effectively, however, cars must run on unleaded gasoline because lead is a catalyst poison. Even the trace amounts of sulfur found in gasoline can reduce the effectiveness of catalysts. Furthermore, catalysts are effective only at the elevated temperatures found in engine exhaust systems. For this reason, the largest fraction of CO and HC emissions from new cars today occurs immediately upon start-up, when the engine and catalytic converter are cold. A substantial engineering challenge thus remains to achieve ultralow emissions. To improve the effectiveness of catalytic converters, the EPA proposed in 1999 to lower the average sulfur content of gasoline to 30 parts per million (ppm) by 2007, a reduction of approximately 90 percent. Alternative fuels and alternative engine designs are among the additional measures being vigorously pursued.

3.4.4   Alternative Fuels

Alternative fuels include modified forms of gasoline as well as fuels such as diesel, methanol, ethanol (produced from biomass), and compressed natural gas, all of which promise to reduce smog-forming automotive emissions. Federal requirements adopted under the 1990 Clean Air Act Amendments already require three different alternative fuel programs. One involves use of reformulated gasoline in summer months to help reduce ozone levels in the nine worst areas of the country. The refor-

---

3 To appreciate how fast that is, try wiggling your index finger up and down as fast as you can and see how many cycles you can get in one second.

mulated fuels are designed to reduce the chemical reactivity of hydrocarbon emissions that contribute most to ozone formation (see Chapter 8 for details). Limiting the benzene content of gasoline and increasing its oxygen content are among the measures being adopted.

A second type of reformulated gasoline is required during the winter months in designated areas of the country. This reformulated fuel is required to have at least 2.7 percent oxygen content by weight to reduce the formation of CO during the months when meteorological conditions tend to produce the highest levels of ambient CO concentrations. A third program required new motor vehicle fleets in about 20 ozone nonattainment areas to include some percentage of low-emission vehicles (LEVs) beginning in 1998. Such vehicles must be able to operate on alternative fuels such as compressed natural gas, liquid petroleum gas, or other clean fuels.

Figure 3.11 shows the ozone-forming potential of emissions from several alternative vehicle fuels relative to emissions from a gasoline-powered vehicle (Chang et al., 1991). The alternative fuels give modest to significant benefits, although considerable uncertainty remains about these estimates. There are also some concerns about the potential toxicity of other trace emissions from alternative fuels, such as formaldehyde emissions from methanol combustion. Water pollution problems also have been caused by storage tank leaks of methyl tertiary butyl ether (MTBE), a widely used fuel oxygenate. As a result, the EPA intends to phase out the use of MTBE in reformulated gasoline. The use, effectiveness, and indirect impacts of alternative fuels is thus the subject of ongoing investigations.

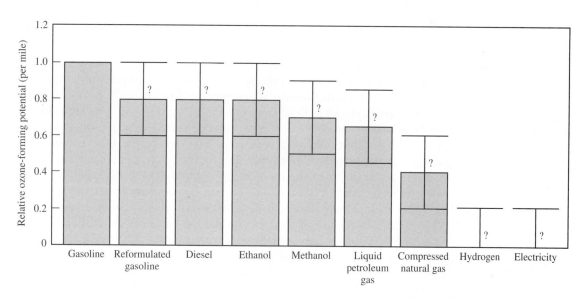

Figure 3.11   Ozone-forming potential of alternative fuels relative to emissions from gasoline-fueled vehicles (per mile of travel). The error bars and question marks indicate the range of uncertainty for alternative fuels. These estimates may change over time as details of ozone-formation chemistry become better understood. (*Source:* Reprinted with permission from *Environmental Science & Technology*, 1991, *25*, 1190–1196, ©1991 American Chemical Society)

3.4.5   Alternative Vehicles

The ideal solution to the urban air pollution problem would be vehicles that produced no emissions to the environment. Could such vehicles really exist? And would they be truly emissions-free?

In 1990 California mandated that 2 percent of the new cars sold in 1998 must be zero emission vehicles (ZEVs). That percentage was to increase to 5 percent by 2001 and to 10 percent by 2003. Subsequently, several eastern states also adopted the California standards. In March 1996, however, California relaxed its quota requirements for 1998 and 2001, insisting only that zero emission cars be *offered* for sale by 1998. However, the 10 percent ZEV requirement for 2003 remained intact. The California mandate has substantially accelerated the development of alternative vehicle concepts. In this section we examine three such concepts that are, or soon will be, available commercially: battery-powered electric vehicles (EVs), hybrid vehicles, and fuel cell–powered EVs.

Battery-Powered Electric Vehicles   The idea of a battery-powered car is not new. In fact, battery power was one of the three leading contenders (along with gasoline and steam) when automobiles were first introduced a hundred years ago. But the high cost and limited performance of batteries relative to gasoline engines were major factors preventing their widespread use. Today battery-powered EVs still have a relatively limited driving range and higher initial cost relative to conventional automotives. Their major attraction, however, is that they emit no pollutants directly and have no tailpipe. They also hold promise of consuming less energy than conventional gasoline-powered cars.

Batteries are used to power individual electric motors that are connected to the drive wheels of a car. During braking, the motors can function as generators that allow some of the car's kinetic energy to be recovered. At regular intervals, however, the depleted batteries must be recharged from an external power source. The General Motors EV1 electric vehicle, which employs 533 kg (1,170 lbs) of lead–acid batteries (see Figure 3.12), has a range of up to 145 km (90 miles) between charges.

From an environmental perspective, battery-powered vehicles fulfill their promise of zero emissions along the roads and highways where they are driven. If deployed in large numbers they could therefore help alleviate urban air pollution. However, recharging the batteries of an EV may result in air pollutants and other emissions at the power plants used to generate the electricity for recharging. Chapter 5 discusses power plant technology and its environmental impacts in greater detail. The main sources of electricity in the United States are coal and nuclear power, both of which have significant environmental impacts. Even renewable energy sources are not pollution free. Thus, battery-powered vehicles have indirect impacts because of their demand for electricity.

Other indirect environmental impacts of batteries arise from the production and recycling of battery materials such as lead. Life cycle studies indicate that lead emissions to the environment would increase substantially (in the absence of new control measures) if lead–acid batteries were widely used to power electric vehicles (Lave et al., 1996). Chapter 4 presents a more detailed discussion of battery operation, battery materials, and the environmental impacts of battery technology.

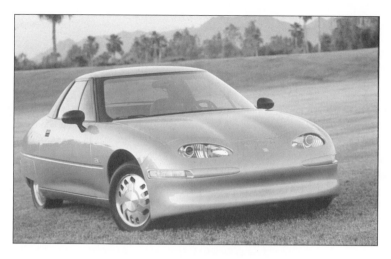

Figure 3.12   The EV1 electric vehicle from General Motors.

Advanced batteries using nickel–cadmium, nickel–metal hydride, sodium–sulfur, or other materials hold promise of substantially higher power density and vehicle driving range than are available with today's lead–acid batteries (see Table 3.8). Nevertheless, many of these batteries also employ toxic metals, and all require periodic recharging via electric power plants. Thus, although electric vehicles may indeed help solve the problem of urban air pollution, they are not the environmental panacea that is sometimes portrayed. Inevitably, a life cycle perspective reveals that there are no completely "pollution-free" energy sources for transportation.

Table 3.8   Characteristics of batteries for electric vehicles.

Battery Material	Specific Power (W/kg)	Energy Density (Wh/liter)	Specific Energy (Wh/kg)	Ultimate Cost ($/kW-hr)
Lead–acid	67–138	50–82	18–56	70–100
Nickel–iron	70–132	60–115	39–70	160–300
Nickel–cadmium	100–200	60–115	33–70	300
Nickel–metal hydride	200	152–215	54–80	200
Sodium–sulfur	90–130	76–120	80–140	100+
Sodium–nickel chloride	150	160	100	>350
Lithium polymer	100	100–120	150	50–500
Midterm goal	150	135	80	<150
Long-term goal	400	300	200	<100

Source: Industrial Ecology and the Automobile by Graedel/Allenby, © 1995. Adapted by permission of Prentice-Hall, Inc., Upper Saddle River, NJ.

Example 3.18

Emission reductions with an electric vehicle.   Assume that recharging an electric vehicle (EV) powered by lead–acid batteries requires an average of 0.18 kW-hr of electricity for each kilometer driven, and that the EV displaces a conventional automobile meeting 1997 federal exhaust emission standards. Assume too that the electricity for recharging the battery is provided by a coal-fired power plant meeting current federal emission standards equivalent to 2.6 g SO_2/kW-hr, 1.7 g NO_x/kW-hr, and 0.43 g particulates/kW-hr. What is the change in annual fuel-related air emissions using the EV, assuming both cars are driven 10,000 mi/yr?

Solution:

Because the EV emits no CO, HC, or NO_x, the annual exhaust emissions are decreased by the levels shown in Table 3.2 for a 1997 passenger car.

Decrease in vehicle emissions:

$$CO = (3.4 \text{ g/mi})(10{,}000 \text{ mi/yr}) = 34 \text{ kg/yr}$$

$$HC = (0.25 \text{ g/mi})(10{,}000 \text{ mi/yr}) = 2.5 \text{ kg/yr}$$

$$NO_x = (0.4 \text{ g/m})(10{,}000 \text{ mi/yr}) = 4.0 \text{ kg/yr}$$

Offsetting these decreases, new emissions occur at the power plant providing the electricity for battery recharging.

Increase in power plant emissions:

$$SO_2 = (2.6 \text{ g/kW-hr})(0.18 \text{ kW-hr/km})(16{,}000 \text{ km/yr}) = 7.5 \text{ kg/yr}$$

$$NO_x = (1.7 \text{ g/kW-hr})(0.18 \text{ kW-hr/km})(16{,}000 \text{ km/yr}) = 4.9 \text{ kg/yr}$$

$$\text{Particulates} = (0.43 \text{ g/kW-hr})(0.18 \text{ kW-hr/km})(16{,}000 \text{ km/yr}) = 1.2 \text{ kg/yr}$$

For the example above there is a net decrease in CO and HC emissions, but a net *increase* in NO_x, SO_2, and particulate emissions when one considers the electric power plant as well as the vehicle. Other types of electric power plants and vehicles would give different results. Other parts of the fuel cycle (like petroleum refining or coal mining) would create additional indirect emissions.

Hybrid Vehicles   A hybrid vehicle combines electric vehicle operation for city driving with a small internal combustion engine for highway driving. This combination substantially increases the vehicle driving range. From an environmental viewpoint, it also allows "zero emission" operation in urban driving situations, where it is most needed. Both Honda and Toyota now offer a hybrid car design in the United States as well as in Japan. Public infrastructure to support electric vehicle recharging is less of a problem with hybrid vehicles because the gasoline engine in current designs can be used when necessary, allowing batteries to be recharged at home. Overall, gasoline consumption and related emissions can be reduced significantly by the use of hybrid vehicle technology. However, the overall cost of a hybrid vehicle is still higher than a comparable gasoline-powered vehicle.

Advanced hybrid concepts are also being developed under a joint industry-government program called the Partnership for a New Generation of Vehicles (PNGV) coordinated by the U.S. Department of Commerce (USDOC, 1999). To date, the program has focused mainly on the combination of advanced batteries plus a small turbocharged compression-ignition, direct-injection (CIDI, or diesel) engine in the quest for high fuel economy (80 mpg, or 3 liters/100 km) with low emissions. Substantial challenges still remain to make this a viable option in the marketplace.

Fuel Cells   Electric vehicles powered by fuel cells are another promising new concept for early 21st-century automobiles. A fuel cell can be thought of as a gas-powered battery in which a continuous flow of hydrogen and oxygen gases replaces the solid electrodes of a conventional car battery. Electrical energy is generated when the hydrogen and oxygen combine to form water as a by-product. Chapter 5 discusses fuel cell technology in greater detail.

Fuel cells have been used for over three decades to provide electric power for spacecraft. Because they employ electrochemical reactions instead of combustion, they are potentially far more efficient than conventional internal combustion engines. The main technical challenges for automotive applications lie in developing improved fuel cell panels, providing a source of hydrogen fuel, and making the system compact enough to be practical. Lowering the cost of this technology is another critical need.

Substantial progress has been made in all of these areas, and all major auto companies are actively developing fuel cell–powered vehicles. The dominant fuel cell system for automotive applications employs a proton exchange membrane (PEM) to generate electricity. Production and storage of the fuel is a critical issue. Some car manufacturers are developing systems that use on-board tanks of hydrogen gas that would be periodically refilled at central "gas stations" providing hydrogen. Other designs produce hydrogen on the vehicle from liquid methanol or, as shown in Figure 3.13, from gasoline. These systems are more complex but offer the advantages of greater energy storage capacity and use of the existing infrastructure for delivering liquid fuels to vehicles. If the current cost of fuel cell systems can be reduced, and remaining technical changes overcome, this technology could soon provide an alternative to conventional gasoline-burning cars.

3.5 CONCLUSION

This chapter has merely scratched the surface in exploring the environmental consequences of engineering design in the context of automobile technology. Nonetheless, the technical fundamentals and concepts emphasized here provide a starting point for gaining insights into green design principles that minimize environmental consequences. The problems at the end of this chapter will give you additional experience in utilizing these fundamentals. A number of problems will also let you explore advanced alternative technologies more fully. Advanced engineering courses will provide deeper understanding and greater ability to pursue green design objectives related to transportation technology.

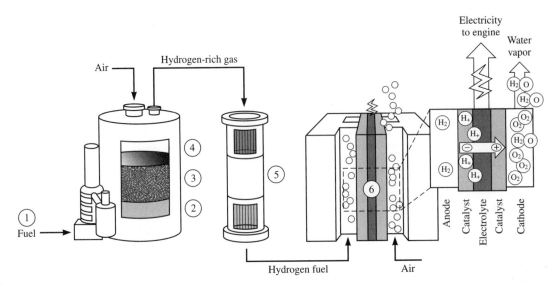

Figure 3.13   A fuel cell design powered by gasoline. In this design a fuel processor first produces a mixture of carbon monoxide (CO) and hydrogen (H_2) from a mixture of air and vaporized fuel (steps 1 and 2). Trace amounts of sulfur compounds are removed (3), then steam is added to convert most of the CO to carbon dioxide and hydrogen (4). Any remaining CO is burned (5), and the hydrogen-rich gas is then fed to the fuel cell (6), where the H_2 combines with the O_2 in air to generate electricity. Other automotive fuel cell designs employ tanks of compressed natural gas or hydrogen as alternatives to an on-board fuel processor. (*Source:* Nuvera, 1999)

3.6   REFERENCES

AAMA, 1995. *Motor Vehicle Facts and Figures.* American Automobile Manufacturers Association, Detroit, MI.

AAMA, 1997. *Motor Vehicle Facts and Figures.* American Automobile Manufacturers Association, Detroit, MI.

Burke, C.G., 1971. "The Coming of Automobile Consciousness III." *Fortune,* January, pp. 160–61.

Chang, T.Y., R.H. Hammerle, S.M. Japar, and I.T. Salmeen, 1991. "Alternative Transportation Fuels and Air Quality." *Environmental Science & Technology,* vol. 25, no. 7, p. 1190–96.

Edmunds, 1997. Edmunds.com, Santa Monica, CA. www.edmunds.com.

GMC, 1998. General Motors Corporation, Detroit, MI. www.gmev.com/gallery.

Graedel, T.E., and B.R. Allenby, 1997. *Industrial Ecology and the Automobile.* Prentice-Hall, Upper Saddle River, NJ.

Lave, L., A.G. Russell, C.T. Hendrickson, and F.C. McMichael, 1996. "Battery-Powered Vehicles: Ozone Reduction vs. Lead Discharges." *Environmental Science and Technology,* vol. 30, no. 9, 1996, pp. 402A–407A.

MacLean, H.L., and L.B. Lave, 1998. "A Life Cycle Model of an Automobile." *Environmental Science and Technology,* 32 (7), pp. 322A–330A.

Monaco, 1993. *Automobile Shredder Residue Processing with Power Production.* MS Thesis, Department of Civil and Environmental Engineering, Carnegie Mellon University, Pittsburgh, PA.

Nuvera, 1999. Private communication. Nuvera Fuel Cells, Inc., Cambridge, MA.

USDOC, 1999. Partnership for a New Generation of Vehicles, U.S. Department of Commerce, Washington, DC. www.ta.doc.gov/pngv.

USDOE, 1997. *Annual Energy Review 1996.* DOE/EIA-0384(96), Energy Information Administration, U.S. Department of Energy, Washington, DC.

USDOT, 1998. *Highway Statistics.* Federal Highway Administration, U.S. Department of Transportation, Washington, DC.

USEPA, 1998. Office of Mobile Sources, U.S. Environmental Protection Agency, Washington, DC. www.epa.gov/oms.

3.7   PROBLEMS

3.1   In 1970 there were approximately 90 million passenger cars in the United States. Assume that the life of a car is 10 years, with an equal number of cars in each age group. Thus the newest 10 percent of the 1970 fleet were manufactured in 1970, and the oldest 10 percent were built in 1961. Assume further that all vehicles meet the applicable federal exhaust emission standards for their year of manufacture (see Table 3.2) and are driven 9,500 miles/year.

(a) Based on these assumptions, what were the total hydrocarbon (HC) emissions from U.S. passenger cars in 1970? Give your answer in metric tons.

(b) Assume that each year the oldest 10 percent of the fleet retires and is replaced with an equal number of new cars meeting the applicable federal emission standards for that year. The total fleet size and annual miles driven remain constant. Under these conditions, how much total hydrocarbons would be emitted from the fleet in the year 2000?

(c) What is the percentage reduction in HC emissions from the total fleet between 1970 and 2000? Compare this to the percentage reduction in HC emissions for an individual car built in 2000 versus one built in 1970.

3.2   Repeat Problem 3.1, except now include the effect of growth in vehicle population. Assume that the number of new cars added to the fleet each year is 1.3 times the number that retire. How does this change your answers to parts (b) and (c)? What percentage of the HC reduction you initially calculated has been offset by the growth in vehicle population?

3.3   Repeat Problems 3.1 and 3.2 for carbon monoxide emissions.

3.4   Repeat Problems 3.1 and 3.2 for nitrogen oxide emissions.

3.5   To properly calculate CO_2 emissions from automobiles, one must account for the carbon in fuel that is emitted in the form of CO and unburned hydrocarbons.

(a) First calculate the grams of fuel consumed per mile of travel for a car that gets 27.6 mi/gal. The density of gasoline is 739 g/L.

(b) If the fuel composition is approximately C_8H_{18}, how many grams of carbon (total) are emitted per mile of travel?

(c) If the CO emission rate is at the 1999 federal standard, what percentage of the total carbon is emitted as CO?

(d) If the hydrocarbon emission rate is at the 1999 federal standard, what percentage of the total carbon is emitted as unburned hydrocarbons? Assume that these hydrocarbons can be represented as formaldehyde (HCHO).

(e) What percentage of the fuel carbon is emitted as CO_2?

3.6 Based on the data presented in the text, estimate the total quantities of steel, iron, aluminum, plastic, rubber, and glass used to produce all new passenger cars in 1977, assuming the same average composition as in 1978. Compare your results to those for 1995 calculated in Example 3.5. By how much has the resource requirement for each of these materials changed over this period?

3.7 Estimate the total solid waste from passenger cars retired in 1977 and compare it to that for 1995. By how much did the annual waste quantity change? State the assumptions you made to arrive at your answer. What were the key factors responsible for these changes?

3.8 Changes in land use are a source of indirect environmental impacts associated with transportation systems. Based on the total road mileage in Figure 3.3, estimate the total land area required for U.S. roads, assuming the average road accommodates three lanes of traffic. State all other assumptions made to arrive at your estimate. Express your answer both in units of land area and also in one other measure you think would be instructive (for example, a land area the size of . . . you name it).

3.9 Estimate the size of the parking lot that would be needed to hold all 200 million vehicles in the United States. State the assumptions made to arrive at your estimate. Express your answer both in units of land area and also in any other terms you think would be instructive.

3.10 The president of an auto company wants to offer a more environmentally friendly vehicle by redesigning a current car model to use a smaller engine. The car currently has a 156 kW (209 hp) engine. It weighs 15,000 N (3,370 lbs) and has a cross-sectional area of 2.5 m². It accelerates from 0 to 30 m/s (0 to 67 mph) in 10 seconds and has a top speed of 50 m/s (112 mph). A design team has come up with three alternatives for reducing the engine size: (1) Reduce the vehicle weight to 10,000 N (2,250 lbs); (2) lower the acceleration to 0 to 30 m/s in 20 seconds; (3) reduce the top speed to 40 m/s (90 mph).

(a) Which of these measures would be most effective in reducing the engine size?

(b) In what way(s) would a smaller engine size be more environmentally friendly?

3.11 A car has the following characteristics: weight = 20,000 N; cross-sectional area = 2.5 m²; drag coefficient = 0.5; rolling friction coefficient = 0.02. The energy content of gasoline is 3.5×10^7 joules/liter. The efficiency of fuel use varies with engine and drivetrain operation as described below.

(a) First consider highway driving. Assume a constant velocity of 30 m/s (67 mph) on level ground. Under these conditions the efficiency of the engine plus auxiliaries is 30 percent and the efficiency of the drivetrain is 85 percent. What is the fuel consumption in miles/gal (mpg) under these conditions? Assume the density of air = 1.2 kg/m³.

(b) Next consider city driving. As you know from your own experience, this is mostly start–stop driving. Assume that there are three start–stop cycles per kilometer of travel (five per mile). For each cycle the car starts from rest, accelerates up to 15 m/s (33 mph), then instantaneously stops. Under these conditions the net engine efficiency drops to 20 percent, and the drivetrain efficiency is 80 percent. What is the fuel consumption (mpg) for this city driving cycle?

3.12 Use the data in Table 3.6 to prepare a graph showing vehicle weight on the *x*-axis and highway miles per gallon (mpg) on the *y*-axis. Each vehicle should be represented by one data point. Then draw a smooth curve that seems to best fit the data (or use the statistical software option on your spreadsheet program to find the best-fit curve).

(a) Based on your graph, is there any apparent relationship between fuel economy (mpg) and vehicle weight? If so, how would you describe that relationship?

(b) Based on your best-fit curve, how much more gasoline would be required to drive the heaviest vehicle 100 miles compared to the lightest vehicle?

(c) On a percentage basis, how much larger would be the resulting emissions of CO_2?

(d) Would HC and NO_x emissions also be affected by vehicle weight?

3.13 Repeat Problem 3.12 plotting vehicle weight versus city mpg.

3.14 Repeat Problem 3.12 plotting vehicle horsepower versus city mpg.

3.15 Revisit the vehicle in Problem 3.11, letting the vehicle weight be a variable.

(a) Write a general equation that gives mpg as a function of the car weight, W, for (1) highway driving and (2) city driving. Assume all other parameters have the values given in Problem 3.11.

(b) Use your equations to plot a graph of mpg (*y*-axis) versus W (*x*-axis) over the range of vehicle weights from 10,000 N (2,200 lbs) to 30,000 N (5,500 lbs). Plot both highway and city fuel economy (mpg) on the same graph.

(c) Discuss the similarities and differences between your curves in part (b) and the data plotted in Problem 3.12 (highway mpg) and Problem 3.13 (city mpg). [If you did not do either of these problems, skip this part.]

3.16 Explore the relationship between energy use and technology design by examining the reciprocal of mpg, which gives the energy required for a unit distance traveled.

(a) Plot the gasoline energy requirement (in joules/meter) as a function of car weight (in newtons) for a car traveling at 30 m/s. Use the data given for the "highway driving" case in Problem 3.11.

(b) To reduce fuel consumption for all car weights, design engineers have focused on (1) improved wheel and tire design to reduce rolling friction (μ); (2) improved body design to reduce both aerodynamic drag (C_d) and cross-sectional area (A); and (3) improved engine and transmission design to improve the overall efficiency (η) of fuel use. Assume *all four* of these parameters (μ, C_d, A, η) can be reduced by 20 percent *each*. Plot a new curve, on the *same* graph as part (a), showing energy vs. weight.

(c) What is the percentage of reduction in energy use for a car weighing 20,000 N as a result of these four improvements?

(d) Is the percentage of reduction (energy savings) constant for all car weights? Explain why or why not.

3.17 In calculating the energy requirements of a car, you notice that aerodynamic drag is the only force acting on a car that does not depend explicitly on vehicle weight. However, you suspect there may be an indirect relationship because a car's cross-sectional area may vary with vehicle weight. To test this hypothesis, use the data on vehicle height and width in Table 3.6. Assume that the cross-sectional area is approximately 80 percent of the rectangular area calculated as the product of height times width. Then plot the cross-sectional area (on the y-axis) as a function of car weight (on the x-axis). Draw a best-fit line or curve through the data. Discuss the conclusions of your analysis.

3.18 A 1992 report of the National Academy of Sciences said that increasing automobile fuel economy from the current CAFE standard of 27.5 miles per gallon (mpg) to 32.5 mpg would lower CO_2 emissions at a net cost savings over the life of a car. Calculate the annual reduction in CO_2 emissions in metric tons of CO_2 per year from a new car designed to meet the proposed higher CAFE standard instead of the current one. Assume the car is driven 10,000 miles per year.

3.19 Pick any *one* of the following environmental policy proposals and discuss at least two trade-offs or consequences that would have to be evaluated before adopting the measure:

(a) A stiff tax on heavy vehicles to encourage the production and use of lighter vehicles that consume less energy and emit less CO_2.

(b) Incentives to encourage the use of advanced electric vehicles in order to improve air quality in polluted urban areas and reduce the use of imported oil for transportation.

(c) A gradually phased-in tax on gasoline that would reduce VMT and thereby reduce auto accidents, pollutant emissions, and energy use, with the tax used to improve the quality of mass transit systems.

(d) Incentives to encourage the production and use of automotive fuels derived from biomass.

3.20 An electric vehicle is powered by 500 kg of nickel–cadmium batteries whose specific power (W/kg) and specific energy (Wh/kg) are in the middle of the ranges shown in Table 3.8. The total vehicle weight is twice the battery weight. The cross-sectional area is 2.0 m², the drag coefficient is 0.3, and the rolling friction coefficient is 0.02. The overall efficiency of delivering power to the drive wheels is 90 percent.

(a) What is the top speed of this vehicle on level ground?

(b) How far could the vehicle travel at this speed on a single battery charge?

(c) How much electricity is required to recharge the battery if the efficiency of the recharging device is 80 percent?

(d) Discuss the environmental impacts you would want to analyze to determine whether it would be a good idea (environmentally) to encourage the large-scale production and use of these cars.

3.21 Diesel engines operate at much higher temperatures and air–fuel ratios compared to typical gasoline engines.

(a) Based on Figure 3.9, how would emissions of NO_x, CO, and hydrocarbons from a diesel engine likely compare to those from a gasoline-powered engine?

(b) Investigate the actual emissions from modern diesel engines and methods used to control emissions. Summarize your findings in a brief report.

3.22 Investigate the use of fuel cells for automobiles. What are its environmental advantages and disadvantages relative to battery-powered electric vehicles? What are the different methods being pursued to supply energy to the fuel cell? What are the advantages and disadvantages of each method? Summarize your findings in a brief report.

chapter

4

Batteries and the Environment*

*The principal author of this chapter was Jeffrey Rosenblum.

4.1   INTRODUCTION

Cellular phones, laptop computers, beepers, portable music players: All of these common devices require energy to operate, and this energy is provided by batteries. Battery use has grown over the century in not only the number of products and applications that utilize batteries, but also the number of consumers using these products. Currently about 4 billion household batteries, or more properly *dry cells,* are sold in the United States each year (LeCard, 1993). As shown in Table 4.1, size "AA" alkaline batteries are by far the most popular.

There are two basic categories of batteries: *nonrechargeable* (also called *primary*) and *rechargeable* (also called *secondary*) batteries. Zinc–carbon and alkaline are the most common primary battery types, whereas nickel–cadmium is the most popular secondary battery type.

From an engineering perspective, the original concept of the battery was brilliant: a portable energy source that can be manufactured in a range of sizes. Nonindustrial battery uses include recreational (electronic games, music listening devices), business convenience (cellular phones, laptop computers, beepers, calculators), safety (smoke detectors, emergency lighting, subway backup power), automotive (starter batteries), and personal (hearing aids, wristwatches, cameras, clocks) devices. Table 4.2 shows typical electrical requirements for various applications in units of milliamperes (mA) of current drain.

The battery industry has undergone tremendous growth in the past several decades. Not long ago, common uses of batteries were limited to automobiles and flashlights. Today a much wider array of battery types utilizes innovative chemical combinations. The recent push for development and production of these new technologies has been to meet the needs of new applications, ranging from electric vehicles to miniature batteries for memory devices in printed circuits, to pacemakers used in heart surgery. Each application has unique energy requirements that dictate the type of battery technology needed. These criteria include energy density, discharge characteristics, storage life, and rechargeability. Details of these characteristics are discussed later in this chapter.

Table 4.1   Annual sales of various household batteries in millions of units in 1992.

| Battery Type | Battery Size | | | | | |
	D	C	AA	AAA	9V	Total
Alkaline	228	336	1,289	181	166	2,200
Zinc–carbon	164	154	296	*	69	683
Nickel–cadmium	44	55	179	20	27	325
Total	436	545	1,764	201	271	3,208

* Data unavailable.
Source: NYSDEC, 1992.

Table 4.2   Electrical characteristics of battery-powered devices.

Application	Current Drain (mA)
LCD and quartz analog watches	0.0025
LCD calculators	0.05–0.12
Hearing aids	0.15–5
Pagers	2–20
Radios	5–190
Electronic games	20–200
Tape recorders	70–250
Portable televisions	125–500
Flashlights and lanterns	150–1,000
Cellular phones	300–700 (talk)
	40–60 (standby)
Notebook computers	200–1,500 (talk)
	50–500 (standby)
Motorized toys	400–1,500
Cameras with built-in flash	500–2,000
Movie cameras	600–1,200

Sources: Based on Crompton, 1995; Duracell, 1991; Linden, 1984.

4.1.1   Environmental Concerns

Along with the functionality and convenience that batteries offer come the environmental consequences when spent batteries are discarded with household trash or otherwise disposed of. A primary concern is the heavy metals, such as mercury, lead, and cadmium, that are an integral part of most batteries. Although batteries contribute only a small proportion of all municipal wastes by volume, they are responsible for a large share of toxic metals. In 1989 consumer batteries accounted for 88 percent of the mercury (Hg) and 54 percent of the cadmium (Cd) in the U.S. municipal solid waste stream. Although the Hg content in batteries has decreased dramatically since then, NiCad battery use is increasing, contributing almost 75 percent of the cadmium in the waste stream. Discarded cadmium in batteries and appliances nationwide was recently projected to increase from about 1,300 tons in 1990 to 2,000 tons by the year 2000 (Fishbein, 1996). Detectable levels of mercury have been traced to discarded household batteries. Although they are not a problem during use of the battery, these metals may pose environmental threats after disposal.

The two most common methods for disposing of household trash are incineration and landfilling. Each method provides pathways for toxic metals in batteries to affect human health and the environment. When buried in landfills, the battery can be crushed, or its casing can corrode. The inner materials can then leach out and feed into a surface water source or contaminate groundwater aquifers used for crop irrigation and drinking water. If household trash is incinerated, some of the heavy metals in batteries can be released into the atmosphere, although most is concentrated in the solid ash residue that typically is disposed of in a landfill. Again, if precautions are not taken, toxic metals can leach into surface water or groundwater.

When toxic metals enter the environment, human health can be damaged. For example, we can ingest trace metals directly in water and air or indirectly via the food chain (which may include trace chemicals taken up by plants, fish, and animals). Although these quantities are typically small, health problems associated with toxic levels of trace metals may affect the senses, the nervous system, the muscular system, several vital organs, and other parts of the body. The severity of effects depends on a variety of factors, such as the material's concentration in the environment and the duration of exposure. Chapters 10 and 14 discuss these factors in greater detail. Often symptoms may not be recognized until damage is already done. There is still much to learn about the effects of toxic metals on human health and ecological systems, but public concerns have prompted actions to reduce or eliminate the release of heavy metals to the environment.

Table 4.3 shows the percentage of heavy metals in four types of common household batteries, whereas Table 4.4 shows the average weight of each battery. The data in these two tables together can be used to determine the mass of heavy metals and other materials released to the environment each time one of these batteries (dry cells) is discarded.

Example 4.1

Composition of NiCad batteries. Determine the mass of nickel and cadmium discarded to the environment in a typical size "AA" NiCad battery.

Solution:

A typical size "AA" NiCad battery has a mass of 21.4 g (from Table 4.4), 28 percent of which is composed of cadmium and 15 percent of nickel (from Table 4.3). The total amount of cadmium and nickel per dry cell can be calculated as follows:

$$21.4 \text{ g} \times 0.28 = 6.0 \text{ g cadmium}$$

$$21.4 \text{ g} \times 0.15 = 3.2 \text{ g nickel}$$

In a similar fashion, the total quantity of battery chemicals can be estimated from national sales figures. The following example illustrates.

Example 4.2

Amount of cadmium in NiCad batteries sold in the United States. Determine the total amount of cadmium in all the batteries sold in the United States during 1992. Provide subtotals for each of the five battery sizes shown in Table 4.1. Which battery sizes contribute most to the total?

Solution:

The total number of size "AA" NiCad batteries sold in the United States in 1992 is 179 million (Table 4.1), and the mass of cadmium in each is 6.0 grams (Example 4.1). Thus for size "AA" batteries:

$$\frac{6.0 \text{ g cadmium}}{\text{battery}} \times \frac{179 \times 10^6 \text{ batteries}}{\text{year}} = 1.1 \times 10^9 \text{ g cadmium/yr}$$

Performing similar calculations for the four other battery sizes yields the following results:

Size D: 8.5×10^8 g cadmium/yr

Size C: 8.2×10^8 g cadmium/yr

Size AAA: 6.2×10^7 g cadmium/yr

Size 9V: 3.6×10^8 g cadmium/yr

The contribution from sizes "D," "C," and "AA" are all roughly comparable; size 9V contributes about half that amount, and size "AAA" contributes about an order of magnitude less. The total for all five battery sizes is 3.2×10^9 g/yr, or 3,200 metric tons/yr of cadmium in batteries sold in 1992.

Table 4.3   Typical composition of common household batteries.

Battery Chemicals	Percentage of Total Weight
Zinc–Carbon Batteries	
Manganese dioxide (MnO_2)	28–32
Zinc (Zn)	16–20
Steel (iron)	8–14
Carbon (C)	7–13
Zinc chloride ($ZnCl_2$)	6–10
Other	11–35
Alkaline Primary Batteries	
Manganese dioxide (MnO_2)	32–38
Steel (iron)	19–23
Zinc (Zn)	11–16
Potassium hydroxide (KOH)	5–9
Carbon (C)	3–5
Mercury (Hg)	0.001
Other	9–30
Alkaline Rechargeable Batteries	
Manganese dioxide (MnO_2)	32–38
Steel (iron)	19–23
Zinc (Zn)	11–16
Potassium hydroxide (KOH)	5–9
Carbon (C)	3–5
Silver oxide (AgO_2)	<1
Mercury (Hg)	0.001
Other	9–30
Nickel–Cadmium Rechargeable Batteries	
Cadmium (Cd)	28
Nickel (Ni)	15
Potassium hydroxide (KOH)	2.5
Cobalt (Co)	1
Other	53.5

Sources: Based on Duracell, 1995; Rayovac, 1996.

Table 4.4 Average weight of common household batteries, in grams.

Battery Size	Zinc–Carbon	Primary Alkaline	Nickel–Cadmium
AAA	9.1	11.4	11.0
AA	17.0	22.6	21.4
C	42.1	66.2	53.5
D	89.6	136.0	68.9
9V	34.9	46.0	47.5

Source: NYSDEC, 1992.

4.1.2   Recent "Green" Efforts

Efforts by the battery manufacturing industry to become more environmentally friendly have led to several developments aimed at reducing the discharge of toxic metals to the environment. For example, manufacturers of lead–acid car batteries are now required to accept and recycle spent batteries. Another important development has been the elimination of mercury from alkaline batteries. In the mid-1980s mercury accounted for 1 percent of a cell's weight. In 1982 and 1983 the major U.S. battery manufacturers (Eveready, Duracell, Panasonic/Kodak, and Rayovac) converted to a "no added mercury" design, bringing the mercury level to about 0.001 percent (owing to the natural mercury content from other materials). Mercury, though not part of the electrochemistry, had historically been added to alkaline batteries to retard corrosion of the zinc anode, which reduces battery capacity and can lead to subsequent leakage. Alternative anticorrosive techniques are currently available, such as substituting a less toxic stabilizing material in place of mercury or altering the manufacturing process to compress the powdered zinc in the anode to a greater density (CMU, 1991).

Example 4.3

Mercury content of an alkaline battery.   How much were mercury discharges reduced for a simple size "AA" alkaline battery by the design change to "no added mercury"?

Solution:

From Table 4.4, a typical size "AA" alkaline battery has a mass of 22.6 g. Prior to 1982–1983, approximately 1 percent was added mercury. Thus, the mass of mercury added to that battery was

$$\text{\textit{Before}: } 22.6 \text{ g} \times 0.01 = 0.2 \text{ g mercury}$$

With no added mercury, the amount of naturally occurring mercury is approximately 0.001 percent. Thus the amount of mercury is

$$\text{\textit{After}: } 22.6 \text{ g} \times 0.00001 = 0.0002 \text{ g mercury}$$

The rechargeable household battery was first introduced in the early 1980s using a nickel–cadmium technology. In 1990 most of the 280 million NiCad batteries purchased in the United States were sealed deep within cordless devices such as power tools, miniature vacuums, and toothbrushes (Erickson, 1991). NiCads in popular sizes (like AA and C) also are now commonplace for use in electronics such as portable radios. At first glance, this use of rechargeable batteries seems clearly beneficial to the environment because of the reduction in the number of batteries needed. Unfortunately, both nickel and cadmium are toxic materials. Eventually, when rechargeable batteries are discarded, they can release toxic metals to the environment.

As noted earlier, the pathways by which different toxic compounds may affect people and ecosystems are extremely complex. Thus it is not always a simple matter to determine which of several alternatives poses the lowest risk to human health and the environment. For heavy metals, one rough measure of relative toxicity is the *threshold limit value* (*TLV*) for occupational exposure. Table 4.5 shows such values for several common materials found in batteries. By this measure, one sees that mercury is five times more toxic than nickel, whereas nickel is four times more toxic than manganese dioxide.

In light of the potential hazards of ordinary disposal of NiCad batteries, manufacturers of products utilizing built-in NiCad batteries, such as handheld rechargeable vacuum cleaners, have modified the design of their products to allow easy removal of the battery before disposal. Improved labeling helps educate consumers about the hazards of improper disposal of the battery and informs them about proper disposal methods. Some retail stores, in conjunction with trade associations and manufacturers, provide boxes for consumers to deposit spent batteries, which are collected and sent to recyclers. Although the technology for recycling household batteries exists, only one U.S. company recycles NiCads. The only plants for recycling alkaline batteries are located in Switzerland and Japan. Nickel–metal hydride, lithium ion, and rechargeable alkaline batteries are less toxic alternatives to NiCads, although cost and performance considerations (like number of recharges and duration of charge) have maintained the popularity of NiCad technology.

Table 4.5   Threshold limit values for several common materials found in batteries.

Compound	Threshold Limit Value (TLV), mg/m^3
Cadmium	0.01
Mercury	0.01
Cobalt	0.02
Nickel	0.05
Manganese dioxide	0.2
Silver oxide	0.1
Zinc chloride	1
Zinc oxide dust	10

Source: ACGIH, 1996.

As you can see, engineers who design batteries have some control over their potential environmental impacts. Some of these decisions include the choice of materials for battery operation, design issues related to battery operating life and shelf life, and the choice of rechargeable or nonrechargeable designs. Engineers also design the electronics for recharging, and these design choices can affect the capacity of the battery and its lifespan. In addition, the design of the electric-powered devices that require the use of batteries can directly affect the type of batteries needed and the frequency with which the batteries must be changed. Thus engineers involved in a spectrum of design-related activities play a key role in determining the environmental impacts of this technology.

4.2   BATTERY BASICS

To understand in more detail how engineers can influence the environmental consequences of battery technology, this section presents some of the basic concepts underlying battery design. Factors affecting the choice and lifetime of battery materials are also discussed.

4.2.1   A Brief History

In 1800 Alessandro Volta sandwiched a strip of cardboard soaked with salt water between a silver plate and a zinc plate, as shown in Figure 4.1. Chemical reactions between the salt water and the silver plate removed electrons, which flowed

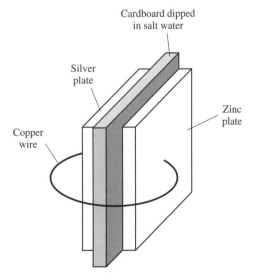

Figure 4.1   A simple electric cell similar to that made
by Volta in 1800.

through the electrically conducting salt solution to the zinc plate. When Volta connected the two plates with a copper wire, electrons then flowed from the zinc back to the silver, creating an electric current. This was the first electrochemical cell. In 1859 Gaston Plante began studies that led to the development of the first practical rechargeable battery, the lead–acid battery, which to this day is the most widely used and cost-effective rechargeable battery system (Schallenberg, 1982). French chemist Georges Leclanche developed the zinc–carbon cell in 1868. Twenty years later it was modified into a dry form (hence the term *dry cell*). This was the most common household battery type until the development of the alkaline battery in the 1960s.

Since this early work in the 19th century, many new battery systems have been discovered and developed. Figure 4.2 shows that significant improvements and advances in performance have been made over the years, both in terms of capacity and shelf life. In this figure, *energy density,* which is the amount of energy (watt-hours) that is available per mass of battery (kg), is used as a convenient measure of battery capacity. Recall that *primary batteries* are not designed to be recharged, whereas the term *secondary battery* applies to rechargeables, such as the common lead–acid car battery.

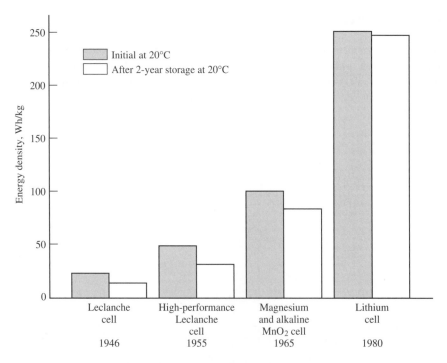

Figure 4.2   Advances in primary battery performance. (*Source:* D. Linden, *Handbook of Batteries and Fuel Cells,* © 1984, reproduced with permission of The McGraw-Hill Companies)

4.2.2   How a Battery Works

A simple chemical battery (or cell) can be obtained by immersing two dissimilar metal rods in a dilute acid solution, as shown in Figure 4.3. Most metals dissolve at least slightly in the acid.

When the metal dissolves, each atom of the metal leaves at least one electron behind on the rod. Thus the metal enters the solution as a positively charged ion, and the rod (or electrode) is left with a slightly negative charge. As the metal continues to dissolve, the electrode's negative charge becomes so strong that some of the positive ions in solution are attracted back to the negative electrode. Eventually, an equilibrium condition is reached in which the net number of ions leaving the electrode is zero. At this point, the electrode is still negatively charged relative to the solution (called the *electrolyte*) because some positive ions remain in solution. Thus an electrical *potential difference* exists between each electrode and the electrolyte.

Both electrodes are now negative, and therefore at a lower potential, with respect to the solution. However, electrodes of different materials will dissolve at different rates. Thus one of the metal rods will dissolve more rapidly than the other, causing it to be more negative relative to the solution compared to the other rod. This creates a difference in electrical potential between the two electrodes. This potential difference determines the *voltage* of the battery.

Suppose an external metal wire (or *load*) is now connected between the two electrodes A (higher potential) and B (lower potential), as shown in Figure 4.4. The potential difference forces negative charges (electrons) to flow through the wire from B to A. As soon as electrons begin to leave electrode B, the positive metal ions that were held to this electrode by the attraction of the now-missing electrons escape back into the solution (this is called a *chemical reduction reaction*). Simultaneously, as the electrons congregate on electrode A, they attract positive ions from the solution, causing them to plate out on A (called a *chemical oxidation reaction*). Therefore, as the negatively charged electrons flow through the wire from B to A, an equal

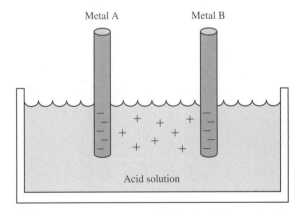

Figure 4.3   A simple chemical battery.

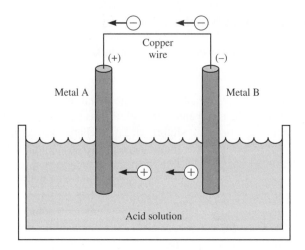

Figure 4.4   A simple battery providing current.

positive charge is carried from B to A by ions flowing through the electrolytic solution. As a result, the net charge on each electrode remains constant, thus maintaining a constant potential difference (or voltage). This process continues until electrode A becomes covered with B atoms, so that both electrodes now, in effect, are made of the same metal. At this point the battery voltage falls to zero.

Commercially practical batteries are much more complicated than the simple model just described, although they operate in basically the same way. In effect, one electrode partially dissolves, losing chemical energy in the process. Part of this energy is regained by the other electrode as plating occurs. The remainder of the energy is expended by the electrons flowing through the external wire that connects the two electrodes.

A representation of one cell of a lead–acid storage battery, such as those used in automobiles, is shown in Figure 4.5. One electrode is made of lead dioxide, PbO_2,

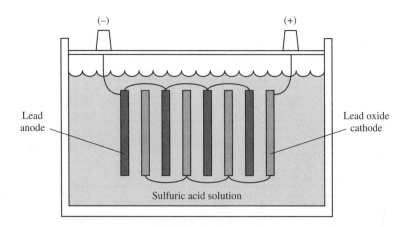

Figure 4.5   One cell of a lead–acid battery.

and the other is made of pure lead, Pb. The Pb electrode (the *anode*) has a higher potential and is therefore said to be *negative* with respect to the *positive* PbO_2 electrode (the *cathode*). The electrolyte is sulfuric acid, H_2SO_4, which in solution dissociates into H^+ and SO_4^- (sulfate) ions. When an external load is placed across the electrodes and the cell provides current, electrons leave the negative electrode and travel through the external circuit to the positive electrode. Inside the cell, lead sulfate, which is nearly insoluble, is deposited on both electrodes. When the plating of lead sulfate occurs to such an extent that the electrode materials (Pb and PbO_2) are no longer accessible to the electrolyte, the cell is "dead" and can produce no further current. To recharge the cell, an independently generated current must be passed through it in the opposite direction. This reverses the original reactions and restores the plates to their original composition.

Another common battery type is the carbon–zinc cell, typically used in flashlights and other small appliances. It consists of a zinc metal container (which serves as the anode) filled with a moist paste of ammonium chloride and zinc chloride (the electrolyte), which contains an electrode consisting of graphite and manganese dioxide (the cathode). Figure 4.6 shows the construction of a basic carbon–zinc cell. Although the electrode reactions are complex, they may be approximated by these oxidation/reduction equations:

Negative electrode: Anodic reaction (oxidation, loss of electrons)

$$Zn \rightarrow Zn^{2+} + 2e \qquad (4.1)$$

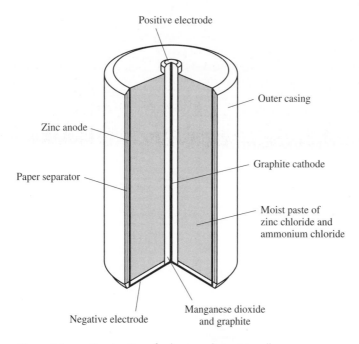

Positive electrode

Outer casing

Zinc anode

Graphite cathode

Paper separator

Moist paste of zinc chloride and ammonium chloride

Negative electrode

Manganese dioxide and graphite

Figure 4.6   Construction of a basic carbon–zinc cell.

Positive electrode: Cathodic reaction (reduction, gain of electrons)

$$2MnO_2 + 8NH_4^+ + 2e^- \rightarrow 2Mn^{3+} + 4H_2O + 8NH_3 \qquad (4.2)$$

Overall reaction: Discharge reaction

$$Zn + 2MnO_2 + 8NH_4^+ \rightarrow Zn^{2+} + 2Mn^{3+} + 8NH_3 + 4H_2O \qquad (4.3)$$

4.2.3 Theoretical Voltage

The theoretical voltage of a cell is determined by its active materials, including electrode materials and electrolyte. As we shall see later, the actual voltage of a cell is always less than the theoretical maximum voltage.

The theoretical voltage of a cell is the sum of the reduction potential at the cathode and the oxidation potential at the anode. The reduction potential at the cathode is equal to the standard reduction potential for the material used for the cathode. The oxidation potential at the anode is the *opposite* of the standard reduction potential for the material at the anode. The standard reduction potential varies if the reaction is occurring in an acidic electrolyte (as in a lead–acid cell) or an alkaline electrolyte (as in an alkaline–manganese cell). Table 4.6 lists standard reduction potentials for various materials.

Table 4.6   Standard reduction potentials for various materials.

Material	Standard Potential in Acidic Electrolyte (Volts)	Standard Potential in Alkaline Electrolyte (Volts)
Anode		
Li	*	−3.01
Zn	−0.76	−1.25
Cd	−0.4	−0.8
Pb	−0.13	*
Cathode		
Cl_2	1.36	*
MnO_2	1.23	−0.05
NiOOH	−0.25	−0.72
CuCl	0.14	*
Ag_2O	0.8	0.345
HgO	0.79	0.10
PbO_2	1.455	0.25

*Not commonly used in this type of electrolyte.
Source: Linden, 1984.

Example 4.4

Theoretical voltage of a carbon–zinc battery. Determine the theoretical voltage of a carbon–zinc dry cell.

Solution:

The overall reaction mechanism for a carbon–zinc cell can be approximated from Equation (4.3):

$$Zn + 2MnO_2 + 8NH_4^+ \rightarrow Zn^{2+} + 2Mn^{3+} + 8NH_3 + 4H_2O$$

From Table 4.6, the standard reduction potentials corresponding to the cathode and anode in an acidic electrolyte can be determined. The standard cell potential is calculated by adding the oxidation potential of the anode (the negative of the reduction potential in Table 4.6) to the reduction potential of the cathode:

Anode: $Zn \rightarrow Zn^{2+} + 2e^-$ $-(-0.76)$ V

Cathode: $2MnO_2 + 8NH_4^+ + 2e^- \rightarrow 2Mn^{3+} + 4H_2O + 8NH_3$ $+ 1.23$ V

Total theoretical voltage $= 1.99$ V

4.2.4 Theoretical Capacity

The *capacity* of a cell is expressed as the total quantity of current flow over time from an electrochemical reaction and is defined in terms of ampere-hours (Ah), which is the product of current (amperes) and time (hrs). For example, a cell that can sustain one ampere of current for two hours has the same capacity as a cell that can sustain two amperes of current for one hour: two ampere-hours (Ah). The ampere-hour capacity of a battery is directly associated with the quantity of electricity obtained from the active materials. The *theoretical specific capacity (Ah/g)* of a battery system, based only on the active materials participating in the reaction, is calculated by adding the electrochemical equivalence (the capacity per unit mass of material) of the anode and cathode. The electrochemical equivalence for several materials, in Ah/g, is presented in Table 4.7.

Example 4.5

Theoretical capacity of a carbon–zinc cell. Determine the theoretical specific capacity of a size "C" carbon–zinc cell.

Solution:

From Table 4.7, the electrochemical equivalence for the cathode and anode can be determined. The specific capacity is the sum of these two values:

$$Zn + MnO_2 \qquad = Total$$
$$0.82\ Ah/g + 0.308\ Ah/g = 1.13\ Ah/g$$

Table 4.7   Electrochemical equivalence of typical materials.

Material	Electrochemical Equivalence (Ah/g)
Anode	
Li	3.86
Zn	0.82
Cd	0.48
Pb	0.26
Cathode	
Cl_2	0.755
MnO_2	0.308
NiOOH	0.292
CuCl	0.270
Ag_2O	0.231
HgO	0.247
PbO_2	0.224

Source: Linden, 1984.

The capacity of batteries can also be expressed on an energy (watt-hour) basis by taking the voltage as well as ampere-hours into consideration:

$$\text{Energy (Wh)} = \text{Voltage (V)} \times \text{Capacity (Ah)}$$

The energy density, which is the energy density per unit of battery mass, can be calculated by multiplying the theoretical voltage (V) by the theoretical specific capacity in Ah/g:

$$\text{Energy density (Wh/g)} = \text{Voltage (V)} \times \text{Specific capacity (Ah/g)}$$

This is a useful measure for comparing different battery technologies, as seen earlier in Figure 4.2.

Example 4.6

Theoretical gravimetric energy density of a carbon–zinc cell.   Determine the theoretical energy density of a carbon–zinc cell in Wh/g.

Solution:

From Example 4.4:   Theoretical voltage = 1.99 V

From Example 4.5:   Theoretical specific capacity = 1.13 Ah/g

$$\text{Energy density (Wh/g)} = 1.99 \text{ V} \times 1.13 \text{ Ah/g}$$
$$= 2.25 \text{ Wh/g}$$

4.2.5 Actual Capacity

The actual energy density of a battery in Ah/g can be as low as 10–20 percent of the theoretical capacity because of the added mass from necessary but nonreactive materials, such as containers, separators, and electrolyte, used in battery construction.

Example 4.7

Actual energy density of a carbon–zinc cell. The actual energy density of a carbon–zinc cell has been observed to be about 30 percent of the theoretical value. Calculate the approximate capacity that can be expected from a typical size "AA" carbon–zinc battery.

Solution:

From Example 4.6, the theoretical energy density of a carbon–zinc cell is 2.25 Wh/g. The actual energy density can be estimated as follows:

$$\text{Actual energy density} \approx 2.25 \text{ Wh/g} \times 0.30$$
$$\approx 0.68 \text{ Wh/g}$$

When stating the capacity of a battery, it is common practice to state the capacity available at a particular discharge rate and battery temperature, assuming a constant discharge. The 20-hour rate at 20°C is commonly used. Thus, if a battery discharges continuously at 20°C and takes 20 hours to reach its end of life, its "nominal capacity" is referred to as C_{20}. A cell's capacity is not the same for all discharge rates because the capacity of a battery generally decreases with increasing discharge current. Table 4.8 shows that increasing the discharge rate reduces the capacity of a battery relative to the nominal capacity.

Example 4.8

Calculation of the C_{20} capacity. A battery that is discharged at a rate of 0.1 A takes 20 hours to reach the end of its life. (a) Calculate the C_{20} value. (b) If a battery with this C_{20} capacity were to be discharged at a rate of 0.2 A, how many hours would it be expected to last? In reality, would it last longer or shorter than this value?

Solution:

$$\text{(a) } C_{20} = 0.1 \text{ A} \times 20 \text{ hours}$$
$$= 2 \text{ Ah}$$

$$\text{(b) Time until full discharge} = \frac{2 \text{ Ah}}{0.2 \text{ A}} = 10 \text{ hours}$$

In reality, it would last slightly less than 10 hours because battery capacity typically decreases as the discharge current increases.

Table 4.8   Capacity of a typical lead–acid battery versus discharge rate.

Discharge Rate (Hours)	Approximate Percentage of Rated Capacity
20	100
10	97
5	88
1	62
0.5	52

Source: Crompton, 1995.

4.3   BATTERY FEATURES

Devices that use batteries have certain energy requirements that must be satisfied. Operating voltage, current drain, rate of energy usage (slow, fast, continuous, periodic), operating temperature, capacity, size, and shelf life are examples of factors that must be considered when determining which battery will suit a particular application. In turn, the choice of battery determines the types and quantities of toxic and other materials potentially entering the environment. Here we explore some of the characteristics involved in engineering an appropriate battery.

4.3.1   Voltage versus Time

Ideally, a battery should provide energy at its theoretical voltage until all of the electrode materials are consumed, at which point the voltage drops to zero. In actuality, the initial voltage is slightly lower than the theoretical voltage and drops as energy is drawn. This behavior is caused by the electrical resistance of active battery materials, polarization during discharge, and other types of energy losses associated with the chemical reaction.

Figure 4.7 illustrates the decrease in cell voltage over time for different current drain scenarios for a common size "D" carbon–zinc dry cell. Recall that current (I) and voltage (V) are related by Ohm's Law

$$V = I \times R \qquad (4.4)$$

where R is the electrical resistance in the circuit (Figure 4.8). For a constant voltage, the current draw increases as R decreases. In practice, however, as the discharge current increases, the discharge voltage actually decreases and the discharge shows a more sloping profile. The term *nominal voltage* refers to the generally accepted maximum operating voltage of a cell. The term *working voltage* refers to the actual voltage that the battery exhibits during the majority of its operation. The term *cutoff voltage* refers to the voltage at which the electrical device being powered will no longer function. Table 4.9 lists the typical number of hours of service for different carbon–zinc battery sizes and different cutoff voltages.

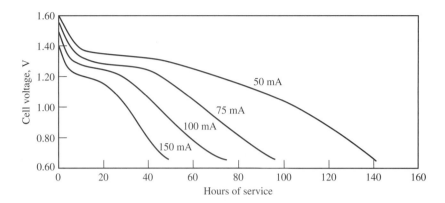

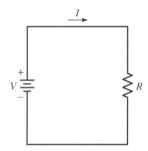

Figure 4.7 Typical discharge curves for size "D" carbon–zinc cell at 20°C discharged two hours per day. (*Source:* D. Linden, *Handbook of Batteries and Fuel Cells,* © 1984, reproduced with permission of The McGraw-Hill Companies)

Figure 4.8   A simple electrical circuit.

Example 4.9

Lifetime of a carbon–zinc cell.   Determine the expected operating life for a flashlight drawing 50 mA of current using one typical carbon–zinc size "D" battery, assuming it is used two hours per day. Typically, 0.9 V has been found to be the cutoff voltage when a cell of this type is used in a flashlight drawing 50 mA of current.

Solution:

Figure 4.7 shows a typical discharge curve for a carbon–zinc size "D" cell, discharged two hours per day at 20°C at various levels of current draw. The nominal voltage is 1.5 V.

If this flashlight is used two hours per day at 20°C, the total hours of service can be determined by reading across at the cutoff voltage of 0.9 V to the 50 mA curve, obtaining a value of about 115 hours.

The shape of the discharge curve also can vary depending on the electrochemical system. There are two general categories of discharge curves, as illustrated in Figure 4.9. A relatively flat discharge curve occurs when the working voltage stays nearly constant during battery life, whereas a sloping discharge curve occurs when

Table 4.9   Typical performance characteristics under different operating conditions for carbon–zinc batteries.

Application and Duty Cycle	Load (ohms)	Current (mA at 1.2 V)	Estimated Average Service at 70° F (Hours) Cutoff Voltage				Approx. mAh Capacity to 0.9 V
			1.2 V	1.1 V	1.0 V	0.9 V	
Size D							
Radio (4 hours/day)	39	31	145	170	184	190	6,300
Cassette (1 hour/day)	3.9	308	4.6	8.4	12.2	15.4	4,500
Flashlight (4 minutes/hour—8 hours/day)	2.2	545	1.8	4.1	6.7	8.5	4,300
Toy/game (1 hour/day)	2.2	545	1.2	2.7	4.7	6.8	3,350
Size C							
Radio (4 hours/day)	39	31	73	87	93	95	3,100
Cassette (1 hour/day)	6.8	176	6.5	10.4	13.5	15.6	2,650
Flashlight (4 minutes/hour—8 hours/day)	3.9	308	2.3	4.6	7.0	8.4	2,400
Toy/game (1 hour/day)	3.9	308	2.0	4.0	5.8	7.6	2,200
Size AA							
Radio (4 hours/day)	75	16	50	58	62	64	1,100
Cassette (1 hour/day)	10	120	3.3	4.8	6.1	6.9	800
Toy/game (1 hour/day)	3.9	256	0.3	0.7	1.4	1.7	500
Photo (15 seconds/minute—24 hours/day)	1.8	667	—	—	—	155 pulses	—
		(mA at 7.2 V)	7.8 V	6.6 V	6.0 V	5.4 V 4.2 V	Capacity to 4.2 V
Size 9 V							
Low rate (24 hours/day)	6,000	—	190	330	350	360 390	490
Radio (2 hours/day)	620	12	7.5	26.0	31.0	37.0 38.5	435
Toy/game (1 hour/day)	270	27	1.0	7.4	10.4	12.8 15.1	390
Calculator (30 minutes/day)	180	40	0.2	3.5	5.5	7.4 9.2	370
Cassette (1 hour/day)	180	40	0.2	3.0	5.1	7.1 9.0	350

Source: Based on Rayovac, 1994.

the working voltage drops continuously with operation. The voltage regulation required by the device or application is thus very important. To obtain the highest capacity of a battery, equipment should be designed to operate over a wide range of voltages. In applications where only a narrow voltage range can be tolerated, the selection of the battery system may be limited to systems having a flat discharge profile. While a disadvantage in some applications, the sloping discharge profile is quite popular in applications where sufficient warning of the end of cell life is desired—as in a flashlight, where the light gradually dims as the battery nears the end of its useful life.

When a battery stands idle after a discharge, certain chemical and physical changes take place that can cause some voltage recovery. Thus a battery voltage that has dropped during a heavy current drain will rise after a rest period, giving a sawtooth-shaped profile as illustrated in Figure 4.10. The relative improvement in operating

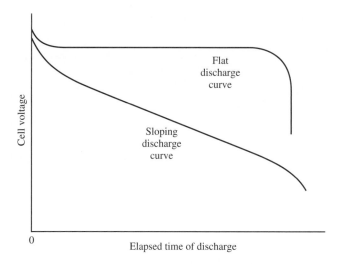

Figure 4.9   Flat versus sloping cell discharge characteristics.

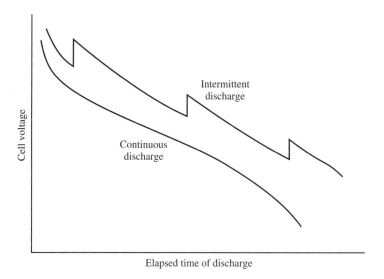

Figure 4.10   Effect of intermittent discharge on battery capacity.

life resulting from intermittent rather than continuous discharge is generally greatest at higher current drains.

A battery can be discharged under different operating modes depending on the nature and type of equipment load. Three typical modes are (1) constant resistance, (2) constant current, and (3) constant power (current load increases as the voltage drops to maintain a constant power output, $I \times V$ or V^2/R). Figure 4.11 illustrates the effects of varying discharge modes on the voltage characteristics.

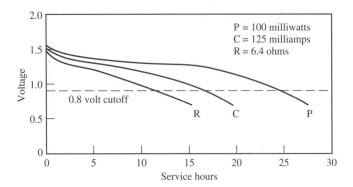

Figure 4.11   Voltage profile under different modes of discharge for a typical size "AA" alkaline MnO_2 battery. (*Source:* Duracell, 1991)

4.3.2   Effect of Operating Temperature

Operating temperature affects the performance characteristics of batteries. Certain applications require a particular operating temperature range. Table 4.10 presents typical operating temperatures for various battery systems. For example, a person using a flashlight during the winter in New England would do best using alkaline MnO_2 batteries rather than carbon–zinc batteries because carbon–zinc batteries are not recommended for temperatures below $-7°C$.

Over a range of temperatures, battery performance varies. A battery will have a lower capacity when used at a lower temperature. Figure 4.12 shows this effect for a typical size "AA" alkaline MnO_2 battery.

Example 4.10

Calculating battery lifetimes.   A 4 Ω flashlight takes one size "AA" battery. How many hours would it last on a fresh alkaline MnO_2 if operated continuously at $-10°C$? At $21°C$?

Solution:

Using the $-10°C$ line in Figure 4.12 at 4 Ω, we see that the flashlight would be expected to last approximately 1.5 hours. At $21°C$ it would be expected to last about 7 hours.

Table 4.10   Typical operating temperatures for various battery systems.

Battery Type	Operating Temperature, °C
Carbon–zinc	-7 to 55
Alkaline MnO_2	-20 to 55
Nickel-cadmium	-20 to 45
Lead-acid	-40 to 60
Button-cell silver oxide	0 to 55
Button-cell lithium–manganese	-20 to 60

Source: Based on Linden, 1984; Radio Shack, 1990.

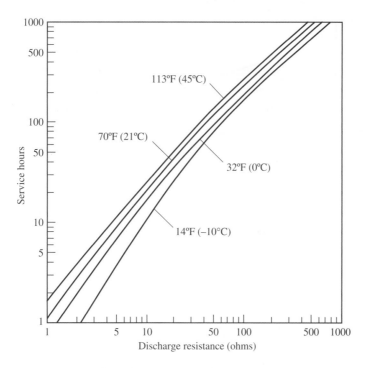

Figure 4.12   Effect of temperature and load on the performance of a typical size "AA" alkaline MnO_2 battery. (*Source:* Duracell, 1991)

4.3.3   Shelf Life

Batteries are a perishable product that deteriorates as a result of chemical action during storage. The type of cell design, choice of electrochemical system, temperature, and length of storage are factors affecting the shelf life of a battery. Figure 4.13 presents shelf life characteristics of various battery systems in terms of percentage of capacity loss as a function of storage temperature. Note that shelf life decreases with increasing temperature because most chemical reactions occur more rapidly as temperature increases.

Example 4.11

Self-discharge of a battery.   A certain alkaline MnO_2 battery has a capacity of 1.5 Ah when new. Determine its capacity after it sits on a shelf for a year at 80°F.

Solution:

From Figure 4.13, at 80°F (27°C), an alkaline MnO_2 battery loses about 1 percent of its capacity per month. It will lose about 12 times this per year, or 12 percent. Therefore, the remaining capacity after one year is

$$1.5 \text{ Ah} \times 0.88 = 1.3 \text{ Ah}$$

Note that the 12 percent annual loss is an approximation. In actuality, 1 percent of the *remaining capacity* is lost per month. So after 12 months the total loss will be slightly less than 12 percent.

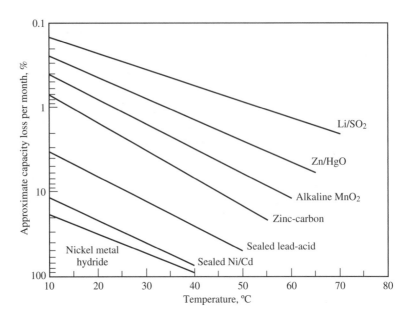

Figure 4.13   Shelf life characteristics of various battery systems. (*Sources:* Based on Linden, 1984; Duracell, 1994)

4.3.4   Lifetime of Rechargeable Batteries

Rechargeable batteries can operate through only a certain number of discharge–recharge cycles before they become unusable. One advantage of the rechargeable battery is that fewer batteries are needed to operate a device over its lifetime.

The fraction of a rechargeable battery's capacity that is discharged before recharge—referred to as the *depth of discharge*—affects the number of cycles the battery can handle. The smaller the depth of discharge, the more cycles it can sustain. Figure 4.14 shows the cycle life as a function of the depth of discharge for a typical household NiCad battery. Partial discharge followed by continued charging, though, can lead to a temporary reduction in battery capacity known as the *memory* effect. This is most often associated with laptop computer batteries. The effect is reversible by an occasional maintenance cycle consisting of a thorough discharge followed by a full recharge.

Example 4.12

Number of cycles of a NiCad battery.   A fully charged NiCad battery has an initial capacity of 350 mAh and is discharged to about 225 mAh before being recharged. How many recharging cycles will the battery take?

Solution:

The depth of discharge can be calculated as the ratio of energy discharged to the energy at full charge. On a percentage basis this is

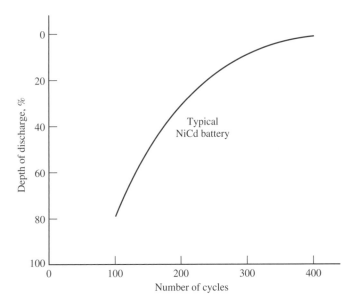

Figure 4.14 Cycle life based on the percentage depth of discharge for a NiCad battery. (*Source:* Based on Linden, 1984)

$$\text{Depth of discharge} = \frac{350 \text{ mAh} - 225 \text{ mAh}}{350 \text{ mAh}} \times 100 = 36\%$$

From Figure 4.14, a discharge depth of 36 percent allows roughly 175 cycles.

The charge that a rechargeable battery can hold is based on the number of full discharge cycles it has already gone through. Figure 4.15 shows the hours of service life per charge for a size "C" rechargeable alkaline battery as a function of the total number of recharge cycles, assuming that the battery is fully discharged after each use. The two curves refer to the use of the battery in different applications that have different cutoff voltages.

Example 4.13

Service life-hours of a reusable alkaline battery. If a device that has a cutoff voltage of 0.9 V is expected to operate for five hours on a reusable size "C" alkaline battery before being recharged, how many cycles can the battery undergo before needing to be replaced?

Solution:
From the upper curve of Figure 4.15 (0.9 V cutoff voltage), the expected service life drops to below five hours after 10 cycles.

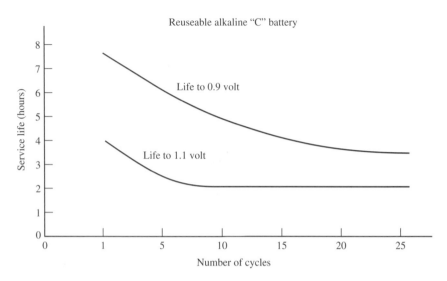

Figure 4.15   Service life based on the number of recharge cycles for a size "C" alkaline battery. (*Source:* Rayovac, 1994)

4.3.5   Battery Rechargers

Typically battery rechargers are electrical devices that plug into the wall. There are also many solar-powered rechargers on the market, as well as calculators and other devices that use built-in solar cells to keep the battery charged. Different chargers employ different designs (such as pulse versus continuous charging or different charging currents) for different battery chemistries. The power draw of rechargers is typically less than 30W.

Example 4.14

Energy required to recharge.   Assume that a size "C" rechargeable alkaline battery requires four hours to recharge, but the average person leaves the charger on for eight hours during each charge. The charger is rated at 28W. Calculate the electrical energy required to operate the charger for 10 charges. At a cost of $0.13/kWh for electricity, what is the total cost of these 10 charges?

Solution:

The charger is rated at 28W. Thus

$$\text{Energy} = \text{Power} \times \text{Time}$$

$$= 28\text{W} \times (8 \text{ hrs/charge} \times 10 \text{ charges})$$

$$= 2{,}240 \text{ Wh} = 2.24 \text{ kWh}$$

$$\text{Cost} = 2.24 \text{ kWh} \times \$0.13/\text{kWh}$$

$$= \$0.29$$

4.4 APPLICATIONS THAT USE BATTERIES

Table 4.2 showed the typical electrical current drain of common consumer products powered by batteries. The engineering design of these devices directly affects the requirements for batteries—which, in turn, affects the environmental impacts.

4.4.1 Discharge Characteristics Based on Current Draw

The actual capacity of a battery during use also depends on the amount of current drawn by a device. Figure 4.16 shows the effect of discharge rate on the capacity of two types of AA cells. Reusable alkaline cells have a greater internal resistance than NiCad batteries and therefore are not able to deliver energy as efficiently at high rates of discharge because of internal resistance losses. At low current draw, however, reusable alkalines provide significantly higher capacity than NiCads, as seen in Figure 4.16. This reflects the differences in battery characteristics discussed earlier.

The load on a battery also affects its voltage versus time characteristic. Figure 4.17 presents the typical discharge characteristics of a size "C" traditional alkaline battery in continuous use. A lower resistance (higher current draw) results in a steeper curve, which implies that fewer hours of operation can be achieved.

Figure 4.18 compares discharge profiles for primary alkaline, NiCad, and rechargeable alkaline batteries for a 100 mA discharge current. These profiles may be one of the criteria used in battery selection for a particular application.

Example 4.15

Current draw and duration of operation.   An electronic device that operates on a 1.5 V size "AA" battery has a resistance (load) of 10 Ω. What would be the initial capacity of a reusable alkaline cell for this application? How many hours would the device operate in continuous use?

Solution:

From Equation (4.4):

$$V = I \times R$$

$$I = \frac{V}{R} = \frac{1.5 \text{ V}}{10 \ \Omega} = 0.15 \text{ A} = 150 \text{ mA}$$

Note that although voltage decreases during use, it can be assumed constant when determining the equivalent current draw.

From Figure 4.16, the typical initial capacity is about 1,400 mA-hr. Therefore

$$\text{Hours} = \frac{\text{Capacity}}{\text{Current}} = \frac{1,400 \text{ mA-hr}}{150 \text{ mA}} = 9.3 \text{ hrs}$$

Example 4.16

Service hours based on load.   A device has a load of 10 Ω and uses one 1.5 V size "C" alkaline battery. The device becomes inoperable at 95 mA. Determine the number of service hours, assuming the device is used continuously at room temperature.

Solution:

First the cutoff voltage must be calculated:

$$V = I \times R$$

$$= 0.095 \text{ A} \times 10 \ \Omega = 0.95 \text{ V}$$

From Figure 4.17, service would last about 50 hours.

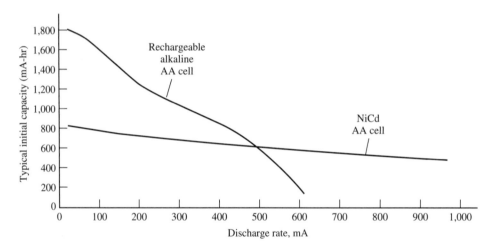

Figure 4.16   Effect of discharge rate on capacity. (*Source:* Based on Rayovac, 1996)

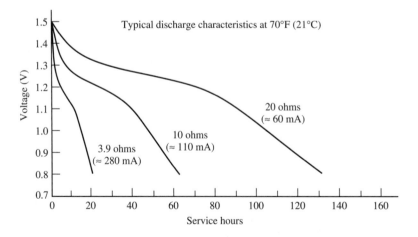

Figure 4.17   Typical discharge characteristics at various loads for a size "C" alkaline MnO_2 battery. (*Source:* Duracell, 1991)

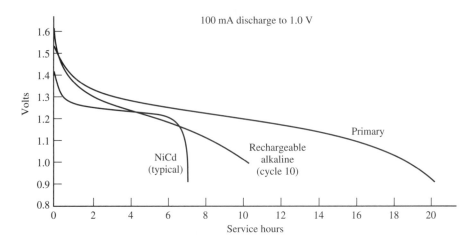

100 mA discharge to 1.0 V

Figure 4.18   Discharge profile comparison.

Example 4.17

Hours of operation based on discharge profile.   Determine the number of hours a 100 mA device with a 1.2 V cutoff voltage will operate using a NiCad battery. Calculate the capacity in Ah.

Solution:

From Figure 4.18, the battery will operate for about seven hours. Therefore

$$\text{Capacity} = \text{Current draw} \times \text{Time}$$

$$= 0.100 \text{ A} \times 7 \text{ hrs} = 0.7 \text{ Ah}$$

4.4.2   Using Multiple Batteries

Many applications take more than one battery at a time. Using multiple batteries allows an application to take advantage of an increased capacity or an increased voltage. When identical batteries are configured in series, as shown in Figure 4.19, the available voltage is the sum of all the batteries in the circuit. When they are configured in parallel, as shown in Figure 4.20, the voltage remains the same as the battery type. In either case, the available capacity is doubled.

4.5   CONCLUSION

This chapter has explored ways in which engineering design choices—in this case the design of batteries and the applications that use them—can affect environmental impacts and risks. Early in the chapter, we saw that the primary environmental impact of battery use was the solid waste generated when batteries are discarded at the end of their useful life. In particular, the toxic metals employed in the design of

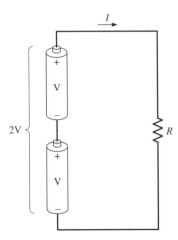

Figure 4.19   Batteries in series.

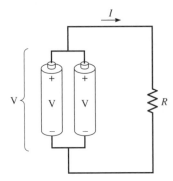

Figure 4.20   Batteries in parallel.

most common batteries are a major concern. Design choices that minimize or eliminate the release of toxic materials to the environment are thus highly desirable.

The remainder of this chapter outlined some of the engineering fundamentals involved in battery design and selection. Those basic principles can be applied to assess the environmental implications of alternative design choices for batteries and the technologies that employ them. The problems at the end of the chapter provide examples of such applications, challenging students to find acceptable engineering designs that minimize environmental impacts.

4.6   REFERENCES

ACGIH, 1996. *Guide to Occupational Exposure Values 1996.* American Conference of Governmental Industrial Hygienists, Cincinnati, OH.

CMU, 1991. "Household Batteries: Is There a Need for Change in Regulation and Disposal Procedure?" Department of Engineering and Public Policy, Carnegie Mellon University, Pittsburgh, PA.

Crompton, T.R., 1995. *Battery Reference Book.* Butterworth-Heinemann, Ltd., Oxford.

Duracell, 1991. *Technical Bulletin: Alkaline Manganese Dioxide Batteries.* Duracell, Inc., Bethel, CT.

Duracell, 1994. *Technical Bulletin: Nickel–Metal Hydride Rechargeable Batteries.* Duracell, Inc., Bethel, CT.

Duracell, 1995. *Material Safety Data Sheets.* Duracell, Inc., Bethel, CT.

Erickson, D., 1991. "Cadmium Charges: The Environmental Costs of Batteries Are Stacking Up." *Scientific American,* May.

Fishbein, B., 1996. "Industry Program to Collect and Recycle Nickel–Cadmium Batteries." Inform, Inc., New York, NY.

LeCard, M., 1993. "Recharge It: Batteries and Environmental Pollution." *Sierra,* November, pp. 42–44.

Linden, D., 1984. *Handbook of Batteries and Fuel Cells.* McGraw-Hill, New York, NY.

NYSDEC, 1992. "Report on Dry Cell Batteries in New York State." New York State Department of Environmental Conservation, Albany, NY, December.

Radio Shack, 1990. *Enercell Battery Guidebook.* Radio Shack, Fort Worth, TX.

Rayovac, 1994. *Renewal Reusable Alkaline: Application Notes & Data Sheets.* Rayovac Corp., Madison, WI.

Rayovac, 1996. *Renewal Reusable Alkaline Batteries: 1996–1997 OEM Designer's Guide & Technical Data.* Rayovac Corp., Madison, WI.

Schallenberg, R.H., 1982. *Bottled Energy: Electrical Engineering and the Evolution of Chemical Energy Storage.* American Philosophical Society, Philadelphia, PA.

4.7   PROBLEMS

4.1   Assume that 90 percent of depleted household batteries are thrown away as municipal solid waste.

(a)   Determine the total mass of cadmium disposed of each year. Present your answer in grams, pounds, and tons per year.

(b)   If battery use and the number discarded per year grow by 10 percent each year, what will be the annual mass of cadmium 10 years from now, assuming no change in the current technology?

4.2   Battery recycling has the potential to reduce the amount of cadmium in the waste stream. Construct a graph showing the amount of cadmium entering the environment as a function of the recycling rate for NiCad batteries. Discuss several obstacles that you think may limit the practical recycling rate.

4.3   Visit a local hardware store and identify as many products as you can that incorporate a rechargeable battery as part of the design. Give examples of

(a)   A design that allows for easy removal and replacement of the battery.

(b)   A product that has adequate labeling to inform the consumer about the proper disposal of the battery.

Select one design to describe in detail, and suggest modifications you would make to improve it (sketches are helpful).

4.4   What responsibility do you think each of the following organizations should have for the environmental impact of a NiCad battery after the end of its useful life, if it is sold as part of another product (such as an electric drill or cordless phone)?

(a) The battery manufacturer.

(b) The product manufacturer.

(c) The consumer.

(d) The municipality that collects the household trash.

(e) The U.S. Environmental Protection Agency.

4.5   Sketch a voltaic cell in which the reaction is $Cd + Cl_2 \rightarrow Cd^{2+} + 2Cl^-$.

On the sketch indicate

(a) The sign of each electrode.

(b) The cathode and the anode.

(c) The directions in which the ions move.

(d) The directions in which the electrons move.

(e) The cell voltage.

4.6   For each of the following battery reaction mechanisms, determine (i) the theoretical voltage for the cell (V); (ii) the theoretical specific capacity for the cell (Ah/g); and (iii) the theoretical energy density for the cell (Wh/g).

(a) Ni/Cd: Anode: $Cd(s) + 2OH^- \rightarrow Cd(OH)_2 + 2e^-$
                       Cathode: $2e^- + NiO_2(s) + 2H_2O \rightarrow Ni(OH)_2 + 2OH^-$

(b) Zinc–chloride: Anode: $Zn \rightarrow Zn^{2+} + 2e^-$
                       Cathode: $Cl_2 + 2e^- \rightarrow 2Cl^-$

(c) Lead–acid: Anode: $Pb(s) + SO_4^{2-} \rightarrow PbSO_4(s) + 2e^-$
                       Cathode: $2e^- + PbO_2(s) + SO_4^{2-} + 4H^+ \rightarrow PbSO_4(s) + 2H_2O$

(d) Mercury–oxide: Anode: $Zn + 2OH^- \rightarrow Zn(OH)_2 + 2e^-$
                       Cathode: $2e^- + HgO + H_2O \rightarrow Hg + 2OH^-$

4.7   The actual capacity of a NiCad battery is about 15 percent of the theoretical value.

(a) How big a battery (in kg) would be required if the desired capacity is 100 watt-hours?

(b) How many standard size "D" cells would it take to provide this much energy?

4.8   A battery that is discharged at a rate of 0.5 A takes 20 hours to reach the end of its life.

(a) Calculate the C_{20} value.

(b) If a battery with this C_{20} capacity were discharged at a rate of 0.4 A, how many hours would it be expected to last? In reality, would it last longer or shorter than this value?

4.9   The cutoff voltage for a particular radio application is found to be 0.75 V. If this radio is used two hours per day at 20°C, drawing a current of 75 mA from a size "D" carbon–zinc cell, determine the number of hours of service.

4.10 If a NiCad battery has a capacity of 0.9 Ah, determine the remaining capacity after it sits at room temperature for three months.

4.11 A NiCad battery has an initial capacity of 350 mAh and is usually discharged to about 200 mAh before being recharged. How many recharge cycles will the battery last?

4.12 If a device that has a cutoff voltage of 0.9 V is expected to operate for at least seven hours on a rechargeable alkaline size "C" battery before requiring recharging, how many cycles can one be expected to last before needing to be replaced? Discuss your results.

4.13 A photographer uses four size "AA" rechargeable alkaline batteries in her flash. Assume the charger power is 28 watts. If she leaves them in the charger for 12 hours, how much electricity is consumed? Compare this to the electricity drawn from the batteries during one use cycle in which the batteries are discharged to a depth of 50 percent of their capacity.

4.14 You are asked to design the power supply for an electronic device that has a load of 6 Ω and is to operate at 4.5 volts using size "AA" batteries. Sketch the battery configuration that would be required. Calculate the current draw during operation. Compare the total number of hours of device operation on one charge if you used reusable alkaline versus NiCad. Which type of battery is more desirable?

4.15 Consider a flashlight that has a load of 3.9 Ω, requires only one 1.5 V size "C" battery, and is operated continuously when used. You want to compare the cost of using a traditional nonrechargeable alkaline battery ($1 each) to the cost of using a rechargeable alkaline battery ($2 each). Assume the battery is "dead" when the voltage drops to 0.9 V.

(a) For using a rechargeable alkaline battery, calculate the current (mA) through the circuit if the battery is fully charged (1.5 V).

(b) Determine the number of hours the flashlight will operate after the first charge. Calculate the capacity of the battery (mAh) using the current found in part (a).

(c) Determine the total number of hours the flashlight will operate with 25 charges. (Hint: find the area under the curve.) Calculate the lifetime capacity of the battery using the current found in part (a).

(d) Assume that recharging requires eight hours. Calculate the energy required to operate the charger over the lifetime of the battery. At $0.13/kWh, what is the total charging cost?

(e) Calculate the total cost of purchasing and recharging the battery.

(f) Assume a traditional alkaline battery will operate the flashlight continuously for 20 hours before the voltage drops to 0.9 V. Calculate the capacity of the battery for this usage.

(g) Calculate the number of traditional alkaline batteries required to provide the same capacity as one rechargeable battery used 25 times. Determine the total cost.

(h) Compare the two batteries in terms of their advantages and disadvantages for this application.

4.16 A device has a 100 mA operational current and a cutoff voltage of 1.1 V. How many hours will a NiCad battery last? How many hours will a primary alkaline battery last? How many hours for a rechargeable alkaline?

4.17 You are an engineer trying to determine whether to incorporate a size "AAA" or size "AA" NiCad battery into the electronic device you have designed. The device draws a current of 100 mA. You are told that your device will need to last for at least five hours at a time and will be used once per day before being recharged.

Battery Type	Capacity at 100 mA Draw
NiCad AAA	350 mAh
NiCad AA	700 mAh

(a) Determine how long the device would last using each type of battery. Does each battery meet the duration specification?

(b) For each type of battery, determine the depth of discharge that occurs given (i) one half hour of use per day or (ii) three hours of use per day.

(c) For each battery type and scenario in part (b), determine the number of cycles each battery can handle (assuming a full charge at the start of each use). Determine the number of batteries needed per year and the mass of disposed cadmium per year.

(d) Discuss design alternatives in terms of recharge characteristics of NiCad batteries and environmental impacts of disposed batteries.

4.18   A store uses an electronic device for an inventory of items on its shelves. Inventory is conducted while staff members roam the store, so the device is battery operated and uses one NiCad size "AA" battery. After use, the device is connected to a charger. The device draws a current of 100 mA. The capacity of a NiCad battery at 100 mA is 700 mAh. The owners have been using the inventory device in their store for a few years now, and inventory takes about one half hour every day. They now plan to move to a bigger space and carry more items. Consequently, inventory is expected to take two hours every day.

(a) For the original store, determine the depth of discharge that occurs after each inventory (assuming a full charge at the start). From this, determine the number of cycles that the battery can handle. Finally, calculate the number of batteries needed per year.

(b) Repeat the calculation in part (a) for the bigger new store. Do battery requirements scale in direct proportion to hours of use? Explain and discuss.

(c) Discuss the design implications of your findings in terms of battery choice. Identify several ways that you might modify the design of the inventory device to minimize the number of batteries needed, and thus reduce their environmental impacts.

Electric Power Plants and the Environment

5.1   THE ROLE OF ELECTRIC POWER

If you were to draw up a top 10 list of the most important technologies of the 20th century, electric power plants would certainly be near the top. Imagine how your own life would be without electric power. No television. No stereo. No computers. And as in centuries past, you would need a burning flame to light the dark. If you have ever experienced a prolonged power outage, you've had a small taste of how life would be without electric power.

It was Thomas Edison who established the first commercial power station in 1882, located on Pearl Street in New York City. His newly invented electric lightbulbs illuminated lower Manhattan and soon replaced oil and gas lamps everywhere. Electric motors then took over the job of running machines previously powered by steam engines. Electrification of the country spread rapidly. By 1920 U.S. electric generating capacity had grown to 14 megawatts (MW), and by 1997 to 780,000 MW. Today electricity provides 38 percent of all the energy used in the U.S. residential and commercial sector, along with 13 percent of industrial energy use. Transportation is the only area of modern life that does not rely heavily on electric power as an energy source. Figure 5.1 shows the growth in U.S. electricity generation over the past 50 years.

Electricity is the fastest-growing form of energy worldwide. Historically, economic growth has been strongly correlated with the growth in electric energy use. Developing countries around the world are thus investing heavily in electric power projects to satisfy the enormous demand for electricity. In China and India, for example, electricity use doubled between 1990 and 2000. In today's age of electronics and information technology, electricity has become more critical than ever to the economic and social well-being of people everywhere.

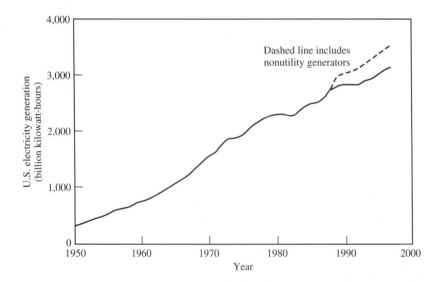

Figure 5.1   U.S. electricity generation, 1950–1998. (*Source*: USDOE, 1999)

5.2 OVERVIEW OF ENVIRONMENTAL IMPACTS

Environmentally, electricity has two distinct faces. At the point of use, electricity is pollution-free and environmentally friendly. Electric cars, for example, have no tailpipes and emit none of the pollutants of conventional automobiles (see Chapter 3). Electric appliances and lighting similarly provide important energy services while emitting no smoke, smells, or contaminants.

But to fully assess the environmental impacts of electric power, one has to ask where the electricity comes from. As we shall see in more detail later in this chapter, electricity can be produced in a variety of ways. The key point is that other forms of energy must be converted into electricity.[1] In the United States, most electricity comes from burning fossil fuels. That conversion process is never completely clean; to the contrary, it is typically a major source of environmental problems.

Figure 5.2 shows a simple schematic of the basic energy flows at a power plant. The plant converts a portion of the primary energy input, E_{in}, into useful electrical energy, E_{elec}. This conversion process is never perfect, so some of the primary energy is always released back to the environment, usually in the form of heat. In Figure 5.2 this is labeled waste energy, E_{waste}. We can then define the overall power plant efficiency, η_{plant}, as the ratio of useful electrical energy output to total energy input:

$$\eta_{plant} = \frac{\text{Electrical energy output}}{\text{Total energy input}} = \frac{E_{elec}}{E_{in}} \qquad (5.1)$$

Electrical *energy* is commonly expressed in units of kilowatt-hours (kW-hr), whereas electrical *power* (energy per unit time) is expressed in units of kilowatts (kW) or (more commonly) megawatts (MW), where 1 MW = 1,000 kW. Power can be thought of as a measure of plant size or capacity (akin to horsepower used for automotive engines), whereas electrical energy or *generation* reflects the actual usage of plant capacity.

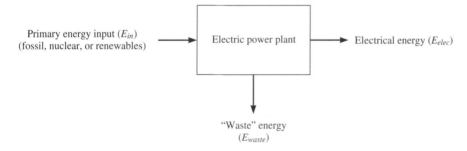

Figure 5.2   Basic power plant energy flows.

In general, the environmental impacts of electric power depend on the technology and primary energy source used to produce electricity. The most common energy source for power generation in the United States is coal, which provided 52 percent of all electricity in 1997 (USDOE, 1998a). The next most important energy source is uranium used by nuclear power plants, providing 18 percent of U.S. electricity. The remaining 30 percent is shared by natural gas (14 percent), hydroelectric power (10 percent), oil (3 percent), and miscellaneous renewable sources (3 percent) consisting of geothermal, biomass, municipal solid waste (including landfill gas), wind, and solar energy. Figure 5.3 shows the historical trend in the U.S. energy mix for electric power production. For the past century, coal has been the mainstay of electric power generation, not only in the United States but worldwide. Globally, coal and other fossil fuels account for nearly two-thirds of all electricity generation, with hydroelectric and nuclear power providing most of the remaining energy supply (USDOE, 1999).

We begin this chapter with an overview of how environmental problems arise from the three major types of energy sources used for electric power production: fossil fuels, nuclear energy, and renewables. For this discussion the electric power plant is treated as a "black box." Later in the chapter we look in more detail at different technologies for power generation.

5.2.1   Environmental Impacts of Fossil Fuels

Fossil fuels consist of coal, oil, and natural gas formed from the decomposition of plant and animal matter buried in geological formations. This organic matter consists mainly of carbon and hydrogen, which release chemical energy when burned. In addition,

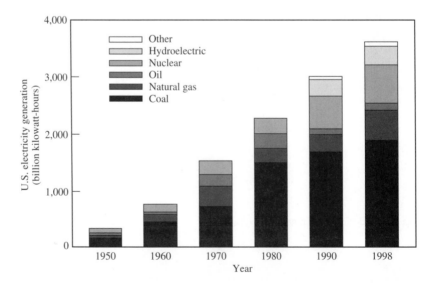

Figure 5.3   Energy sources for U.S. electricity generation, 1950–1998. (*Source:* USDOE, 1999)

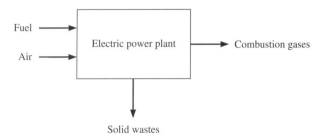

Figure 5.4   Basic mass flows in a fossil fuel power plant.

there are impurities such as sulfur, nitrogen, oxygen, and ash, which vary from fuel to fuel. As we discuss later, these impurities are a major source of environmental problems, especially air pollution and solid wastes, shown schematically in Figure 5.4.

Table 5.1 shows the typical composition and energy content of fossil fuels used for electric power generation. The energy per unit mass of fuel is known as the *heating value,* commonly specified in units of kilojoules per kilogram of fuel (kJ/kg).[2] This energy comes mainly from the carbon and hydrogen in the fuel.

Coal, oil, and natural gas differ in the amounts of hydrogen and carbon they contain. This difference also accounts for the differences in fuel energy content and

---

2 There are (unfortunately) two different definitions of heating value in use today. These are known as the *higher heating value (HHV)* and the *lower heating value (LHV).* All values in this text refer to the higher heating value, which is the prevailing U.S. standard. Chapter 12 contains additional information on this distinction.

Table 5.1   Typical composition of fossil fuels used for power generation.

Fuel Property	Coal[a]		Oil[b]		Natural Gas[c]
	Bituminous	**Subbituminous**	**Distillate**	**Residual**	
Heating value (kJ/kg)[d]	28,400	19,400	45,200	42,500	54,400
Composition (wt%)					
Carbon	67.0	48.2	87.2	85.6	74.1
Hydrogen	5.0	3.3	12.5	9.7	23.9
Sulfur	1.5[e]	0.4	0.3	2.3	0.0
Nitrogen	1.5	0.7	0.02	1.2	1.7
Oxygen	8.7	11.9		0.8	0.3
Ash	9.8	5.3		0.1	
Moisture	6.7	30.2		0.3	

[a] Properties of specific coals vary within roughly ±15% of these values, except for sulfur, as noted below.

[b] Distillate refers to No. 2 fuel oil, residual to No. 6 fuel oil. Sulfur content may vary by up to a factor of 3.

[c] Composition based on 85.3% methane, 12.6% ethane, and 2.1% other gases by volume.

[d] All heating values refer to the higher heating value (HHV) convention.

[e] Average 1995 sulfur content of utility fuels. Range is approximately 0.5–4.0%.

Source: Based on EPRI, 1997a; USDOE, 1997b.

physical state (solid, liquid, and gas). In coal, the ratio of hydrogen atoms to carbon atoms is approximately 1:1, while in oil the ratio is approximately 2:1, and in natural gas 4:1. Natural gas consists mostly of methane (CH_4) mixed with small quantities of ethane, propane, and butane. In contrast, coal and oil (or petroleum) are not well-defined chemical substances, but rather are mixtures of complex organic molecules. Oils that are more highly refined are called *distillate oils,* as opposed to the less highly refined *residual oils,* which are thicker and contain higher levels of impurities. Coals are also categorized according to their properties and composition. The most common types of coal are called *bituminous coal* (also known as soft coal) and *subbituminous coal,* a category that is lower in energy content. A third category, called *anthracite* (or hard coal), is far less common and not typically used for power generation. *Lignite* is the lowest rank of coal, often referred to as brown coal.

When a fossil fuel is burned, the atomic bonds between carbon and hydrogen atoms are broken, releasing chemical energy. This energy provides heat, which is subsequently converted to electrical energy in the power plant. The quantity of fuel burned largely determines the amounts of environmental contaminants that are released. The following example illustrates how fuel quantities are determined.

Example 5.1

Power plant fuel consumption.   A 500 MW power plant burns bituminous coal having the properties listed in Table 5.1. The overall plant efficiency is 36 percent. The plant operates at an annual capacity factor of 65 percent, meaning that the electrical energy generated in a year is 65 percent of the maximum possible. (This is a typical value for U.S. coal plants, which sometimes operate at less than full capacity and are occasionally shut down for regular maintenance.) Calculate the annual quantity (mass) of coal burned at this power plant. Also calculate the annual quantities of ash and sulfur in the coal *entering* the plant.

Solution:

The quantity of coal burned is determined by the fuel energy input needed to generate a given amount of electricity. The maximum electrical output from this plant during a year is

$$\text{Maximum energy output} = \left(500 \text{ MW} \times 1{,}000 \frac{kW}{MW} \right) \left(24 \frac{hrs}{da} \times 365 \frac{da}{yr} \right)$$

$$= (5 \times 10^5 \text{ kW}) \, (8{,}760 \text{ hrs/yr})$$

$$= 4.380 \times 10^9 \text{ kW-hr/yr}$$

This is the output of the plant if it operated 24 hours a day, 365 days per year. But the capacity factor is 65 percent, so the actual electrical output is

$$\text{Annual energy output} = (0.65) \, (4.380 \times 10^9)$$

$$= 2.847 \times 10^9 \text{ kW-hr/yr}$$

From Equation (5.1), the overall plant efficiency is

$$\eta = \frac{E_{elect}}{E_{in}}$$

Substituting the known values for η and E_{elec} gives

$$0.36 = \frac{2.847 \times 10^9 \text{ kW-hr}}{E_{in}}$$

$$E_{in} = 7.908 \times 10^9 \text{ kW-hr/yr}$$

Because fuel energy input is normally expressed in thermal units of kilojoules, it is convenient to first convert to those units. Recalling that 1 watt = 1 joule/sec (or 1 kW = 1 kJ/s), the conversion factor is

$$E_{in} = 7.908 \times 10^9 \frac{\text{kW-hr}}{\text{yr}} = 7.908 \times 10^9 \frac{\text{kJ-hr}}{\text{sec-yr}} \times 3{,}600 \frac{\text{sec}}{\text{hr}}$$

$$= 2.847 \times 10^{13} \text{ kJ/yr}$$

This amount of energy input can be represented as the product of the mass of coal burned in a year (kg/yr) times the coal heating value (kJ/kg):

$$E_{in} = (m_{coal})(HV_{coal})$$

From Table 5.1, the coal heating value is 28,400 kJ/kg. Thus

$$m_{coal} = \frac{E_{in}}{HV_{coal}} = \frac{2.847 \times 10^{13} \text{ kJ/yr}}{28{,}400 \text{ kJ/kg}}$$

$$= 1.002 \times 10^9 \text{ kg/yr}$$

This is the amount of fuel burned at the plant during one year. The quantities of coal ash and sulfur entering the plant are then determined by the coal composition data in Table 5.1, which shows the ash weight percentage to be 9.8% and sulfur 1.5%. Thus

$$m_{ash} = (0.098)(1.002 \times 10^9) = 9.82 \times 10^7 \text{ kg/yr}$$

$$m_{sulfur} = (0.015)(1.002 \times 10^9) = 1.50 \times 10^7 \text{ kg/yr}$$

These quantities will be used in later examples to calculate environmental impacts.

Major Environmental Concerns   The environmental concerns about fossil fuel power plants today center primarily around the atmospheric emissions of carbon dioxide (CO_2), sulfur dioxide (SO_2), nitrogen oxides (NO_x), and particulate matter (PM). Table 5.2 shows the quantities of these emissions released per kilowatt-hour of electricity generated in the United States by different types of power plants. Note that these emission rates reflect the current differences in power generation efficiency using different types of fuels and technologies.

    Because of the potential impacts of these pollutants on human health and the environment, federal and state regulations limit the amounts that can be released from any particular facility. In general, federal emission limits apply to newly constructed plants, whereas state and local regulations cover existing power plants. An exception is the set of federal rules for acid rain control, which applies to all facilities. Examples of federal emission limits are given in Table 5.3. The New Source

Table 5.2   Average pollutant emission rates from U.S. power plants, 1996 (grams per kilowatt-hour of electricity generated).

Plant Type	CO_2	SO_2	NO_x	PM[a]
Coal-fired plants	989	6.38	3.69	0.35
Oil-fired plants	1,020	8.96	2.01	0.15
Natural gas-fired plants	803	0.00	2.87	0.005
All fossil fuel plants[b]	1,030	5.32	3.51	0.29
All plant types[c]	701	3.62	2.39	0.20

[a] Total particulate matter (PM), of which approximately 37% is PM-10 (particles less than 10 microns).

[b] Also includes plants burning blast furnace gas, wood, and refuse.

[c] Includes nuclear and renewables.

Source: Based on USDOE, 1998a; USEPA, 1998a; USEPA, 1998b.

Table 5.3   Federal emission standards for fossil fuel power plants.

Plant Type and Vintage	Maximum Allowable Emissions[a]		
	SO_2	NO_x	PM
New plants built after Aug. 17, 1971[b]:			
Coal-fired	520 ng/J	300 ng/J	43 ng/J
Oil-fired	340 ng/J	129 ng/J	—
Gas-fired	—	86 ng/J	—
New coal plants built after Sept. 18, 1978[b]:	70–90% reduction[c]	260 ng/J (bitum.)	13 ng/J
		210 ng/J (subbit.)	
All new plants built after July 9, 1997[b]:	(no change)	0.72 g/kW-hr[d] or 65 ng/J	(no change)
All existing plants:	40% below 1990[e]	173–373 ng/J[f]	

[a] Units of ng/J means nanograms of pollutant per joule of fuel heat input to the boiler.

[b] Applies to steam electric power plants with heat inputs greater than 73 MW.

[c] Required reductions in SO_2 depend on the coal sulfur content. Equivalent emission rates are approximately 260 ng/J for plants burning bituminous coal and 100 ng/J for subbituminous coal.

[d] This NO_x limit applies to all fuel types and is based on electrical output rather than heat input.

[e] Average reduction imposed by the 1990 acid rain control requirements, which establishes a national emissions cap of approximately 9 million tons SO_2/year from all U.S. power plants as of January 1, 2000. Other federal, state, and local standards still apply to all pollutants. Average emission levels are approximately 520 ng/J based on 1985–1987 average fuel consumption.

[f] Limits vary according to the type of boiler. The average value is 217 ng/J.

Source: CFR, 1998; CFR, 1999; Siegel, 1997.

Performance Standards (NSPS) established by the U.S. Environmental Protection Agency (EPA) are intended to require the use of "best available control technology" to minimize pollutant emissions from new plants. These standards usually are expressed as a maximum allowable mass emission per unit of fuel energy into the power plant. More recently, some limits have been expressed as a percentage reduction or as an emission rate per unit of electrical output.

The air pollutants noted here are not the only environmental concerns. The following sections review the types and sources of all major pollutants from fossil fuel power plants.

Formation of Carbon Dioxide When completely combusted, the carbon in fossil fuels is converted to carbon dioxide (CO_2), the main greenhouse gas of concern for global warming (see Chapter 12). Unlike other pollutants that arise from fuel impurities, carbon dioxide is the natural end product of fossil fuel combustion. The overall reaction is the oxidation of carbon to CO_2, where air is the usual source of oxygen:

$$C + O_2 \rightarrow CO_2 \tag{5.2}$$

Similarly, the hydrogen in fuel is oxidized to H_2O and released as water vapor. Thus the combustion of a fossil fuel such as natural gas, idealized as pure methane, can be written as follows:

$$CH_4 + 2O_2 \rightarrow CO_2 + 2H_2O \tag{5.3}$$

Similar equations can be written for oil and coal combustion, given the chemical composition of the fuel. Although the chemistry of CO_2 formation is similar in each case, an important difference between coal, oil, and natural gas in terms of CO_2 emissions is the quantity of CO_2 formed *per unit of energy released*. This difference is illustrated by the following example.

Example 5.2

CO_2 emissions from coal and natural gas combustion. Calculate the CO_2 emissions from bituminous coal and natural gas combustion on the basis of emissions per unit of energy released. Based on Table 5.1, assume that coal contains 67.0 percent C by weight with an energy content of 28,400 kJ/kg, while natural gas contains 74.1 percent C by weight and releases 54,400 kJ/kg.

Solution:

For both fuels the relevant reaction is

$$C + O_2 \rightarrow CO_2$$

The atomic weights of carbon and oxygen are 12 and 16, respectively (see Table A.2 in the Appendix), so the molecular weight of CO_2 is

$$12 + 2(16) = 44$$

Thus 12 kg of C produces 44 kg of CO_2.

Because coal is 67.0 percent C by weight, in 1 kg of coal there is 0.67 kg of C, which produces

$$(CO_2)_{coal} = (0.67 \text{ kg C}) \left(\frac{44 \text{ kg } CO_2}{12 \text{ kg C}} \right)$$

$$= 2.46 \text{ kg } CO_2/\text{kg coal burned}$$

Each kg of coal burned releases 28,400 kJ of energy, so the CO_2 per unit of fuel energy is

$$(CO_2)_{coal} = \frac{2.46 \text{ kg } CO_2/\text{kg coal}}{28,400 \text{ kJ/kg coal}}$$

$$= 86.5 \times 10^{-6} \text{ kg } CO_2/\text{kJ of fuel energy}$$

Repeating the same calculation for natural gas with a carbon content of 74.1 percent and energy of 54,000 kJ/kg gives

$$(CO_2)_{gas} = 49.9 \times 10^{-6} \text{ kg CO}_2/\text{kJ of fuel energy}$$

The ratio of CO_2 emission from natural gas to that from coal is thus

$$\frac{(CO_2)_{gas}}{(CO_2)_{coal}} = \frac{49.9 \times 10^{-6}}{86.5 \times 10^{-6}} = 0.58$$

Note that the solution to this problem required information only about the fuel properties.

Example 5.2 shows that natural gas releases 42 percent less CO_2 than coal while supplying the same amount of energy. Different fuel compositions will give slightly different results, but in all cases natural gas emits substantially less CO_2 than coal per unit of energy supplied, provided that the conversion efficiencies are the same. This environmental benefit is one of the reasons that natural gas has become increasingly attractive as a fuel for electric power generation.

Formation of Sulfur Dioxide One of the most important impurities in fossil fuels is sulfur. Sulfur occurs in the mineral form as pyrite (a compound of iron and sulfur, also known as "fool's gold" because of its bright yellow color) and also in the organic form within the molecular structure of coal. Oil and natural gas also contain sulfur compounds when extracted from the ground. However, most of these impurities are removed in oil refineries and gas treatment plants, which clean and refine the raw fuel before it is distributed for use. Coal, on the other hand, undergoes little or no processing for sulfur removal before it is burned. Thus the highest levels of sulfur impurities usually are found in coal.

When coal is burned, a small amount of the sulfur, typically 2 to 5 percent of the total (depending on coal type), is retained in the solid ash particles. Most of the sulfur (95 percent or more) is oxidized to gaseous sulfur dioxide (SO_2):

$$S + O_2 \rightarrow SO_2 \tag{5.4}$$

This SO_2 is a major contributor to acid rain and a source of adverse health effects (see Chapter 2). A very small amount (less than 1 percent) of the SO_2 formed during combustion may be further oxidized to SO_3, which reacts with water vapor in the combustion gas to form gaseous sulfuric acid (H_2SO_4). When released to the atmosphere, this gas condenses into fine liquid droplets that remain suspended in the air. This acid aerosol is another source of the environmental impacts associated with sulfur impurities in fossil fuels.

Example 5.3

Sulfur dioxide emissions from a coal-fired power plant.   Calculate the annual mass emission of SO_2 from the power plant in Example 5.1, assuming that 97 percent of the sulfur entering the plant is converted to SO_2. Also express the result as an emission rate per unit of fuel energy input.

Tall stacks, such as the 800-ft chimneys shown here, were built to disperse pollutants and avoid high groundlevel concentrations. However, they also contribute to long-range pollutant transport that results in acid deposition far from the plant.

Compare the value to the federal acid rain control limits of 1.076 g/MJ (2.5 lbs/MBtu) for Phase I (compliance by 1995) and 0.516 g/MJ (1.2 lbs/MBtu) for Phase II (compliance by 2000).

Solution:

From Example 5.1, the total mass of sulfur entering the plant is 1.50×10^7 kg/yr. Of this, 97 percent is converted to SO_2. The relevant chemical reaction is

$$S + O_2 \rightarrow SO_2$$

The atomic weight of sulfur is 32, and the molecular weight of O_2 is also 32. Thus the molecular weight of SO_2 is $32 + 32 = 64$. The mass ratio of SO_2 to S is therefore

$$\frac{\text{Mass } SO_2}{\text{Mass S}} = \frac{64}{32} = 2.0$$

The total annual mass of SO_2 is therefore

$$m_{SO_2} = (0.97)\left(1.50 \times 10^7 \frac{\text{kg S}}{\text{yr}}\right)\left(2.0\frac{\text{kg } SO_2}{\text{kg S}}\right)$$

$$= 2.91 \times 10^7 \text{ kg } SO_2/\text{yr}$$

Expressing this result in terms of emissions per unit of fuel energy input gives

$$\frac{m_{SO_2}}{E_{in}} = \frac{2.91 \times 10^7 \text{ kg } SO_2/\text{yr}}{2.85 \times 10^{13} \text{ kJ/yr}} = 1.02 \times 10^{-6} \text{ kg } SO_2/\text{kJ}$$

$$= 1.02 \text{ g } SO_2/\text{MJ}$$

This emission rate would meet the Phase I acid rain emission limit of 1.076 g SO_2/MJ, but not the Phase II limit of 0.516 g/MJ.

Formation of Particulate Matter   Another major impurity in fossil fuels is mineral matter, commonly called *ash*. This solid incombustible material includes compounds of iron, silicon, and other common elements of the earth's crust. Indeed, most natural elements—including toxic heavy metals—may be found in the ash of fossil fuels, especially coal. Because oil and natural gas are refined before use, they contain little or no ash when combusted (see Table 5.1). Coal, on the other hand, is relatively high in ash, including rock and mineral matter torn from the ground in the mining process. Some of this material is separated from coal prior to combustion. Still, Table 5.1 shows that roughly 10 percent of the coal burned in power plants is incombustible material (akin to the gray ash residue left in your charcoal barbecue).

Unless preventive measures are taken, coal combustion releases into the atmosphere large amounts of this ash (also known as *flyash* because it "flies" out of the chimney). All modern power plants, however, have environmental control systems that remove nearly all the ash particles entrained in the combustion gas stream. Solids that escape collection may pose an environmental concern. Such particulate matter emissions are among the criteria air pollutants discussed in Chapter 2.

Example 5.4

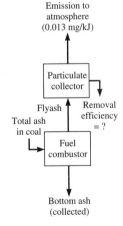

Emission to atmosphere (0.013 mg/kJ)

Particulate collector

Flyash

Total ash in coal

Removal efficiency = ?

Fuel combustor

Bottom ash (collected)

Particulate emissions from a coal-fired power plant.   Current federal regulations limit the atmospheric emissions of particulate matter from new coal-fired power plants to a rate of 0.013 g/MJ (0.03 lbs/MBtu) based on fuel energy input. A collection device is installed in the exhaust gas stream to control atmospheric emissions. For the power plant in Example 5.1, what removal efficiency is needed for the particulate collector to meet the federal standard? Assume that 80 percent of the ash in coal is entrained in the combustion exhaust gas, while the remaining 20 percent (called *bottom ash*) is collected as a solid waste at the combustor.

Solution:

A simple sketch may be helpful to visualize the problem. Because we already calculated the total ash and energy inputs in Example 5.1, we can use those results in this example. Flyash is 80 percent of the total ash, so this gives

$$\frac{\text{Flyash mass}}{\text{Fuel energy input}} = \frac{0.80 \, m_{ash}}{E_{in}} = \frac{(0.80)(9.82 \times 10^7 \text{ kg ash/yr})}{2.847 \times 10^{13} \text{ kJ fuel/yr}}$$

$$= 2.76 \times 10^{-6} \text{ kg ash/kJ fuel} = 2.76 \text{ g/MJ}$$

This quantity represents the potential emission rate of particulate matter. The allowable emission rate, however, is only 0.013 g/MJ. Thus

$$\frac{\text{Allowable emissions}}{\text{Potential emissions}} = \frac{0.013}{2.76} = 0.0047, \text{ or } 0.47\%$$

Only 0.47 percent of the flyash can be emitted, so the particulate collection device must remove 99.53 percent of the flyash in order to achieve the particulate emission standard. Later in this chapter we will look at the design of technology to achieve this objective.

Note that we could have obtained the same result without having to first calculate the total annual quantity of ash and energy input. Because the emission limit is specified in terms of mass per unit of fuel energy, the only parameters really needed to work this problem are the fuel composition and heating value. See the problems at the end of this chapter for additional examples of this type.

Formation of Nitrogen Oxides   Another major class of air pollutants from fossil fuel combustion is oxides of nitrogen (NO_x). As noted in Chapter 2, NO_x consists mainly of nitric oxide (NO) and nitrogen dioxide (NO_2) formed from high-temperature reactions between oxygen and nitrogen. NO_x can be formed solely from the molecular nitrogen and oxygen in air, as well as from the atomic nitrogen and oxygen impurities in fuel. Because its formation during the combustion process is very complex, the amount of NO_x usually must be determined experimentally. For oil and coal, NO_x emissions per unit of fuel energy are usually higher than for natural gas combustion, as reflected in Table 5.3. The higher NO_x levels are due mainly to the presence of fuel-bound nitrogen, which is readily converted to NO when the fuel is combusted. The amount of "fuel-bound NO_x" is typically two to three times greater than the "thermal NO_x" produced only from the reactions between N_2 and O_2 in air. Though NO_x constitutes only a small fraction of the total combustion products, its environmental impacts are nonetheless significant. Example 5.5 illustrates how NO_x emissions can be calculated.

Example 5.5

Nitrogen oxide emissions from a coal-fired power plant.   Assume the power plant in Example 5.1 emits NO_x at a rate equal to the 1979 Federal New Source Performance Standard (NSPS) for coal-fired power plants, which establishes a maximum limit on NO_x emissions from newly constructed units. Estimate the total mass of NO_x emitted per year by this plant.

Solution:

From Table 5.3, the nitrogen oxide NSPS for coal-burning plants built after 1978 is 260 ng/J, or 0.260 g NO_2/MJ (0.6 lb/MBtu) of fuel energy input to the power plant. In Example 5.1 we found the annual energy input to be

$$E_{in} = 2.847 \times 10^{13} \text{ kJ/yr}$$

Thus the annual mass of NO_x emitted is

$$m_{NO_x} = \left(0.260 \frac{\text{g } NO_2}{\text{MJ}} \right) \left(2.847 \times 10^{13} \frac{\text{kJ}}{\text{yr}} \right) \left(10^{-3} \frac{\text{kg}}{\text{g}} \right) \left(10^{-3} \frac{\text{MJ}}{\text{kJ}} \right)$$

$$= 7.40 \times 10^6 \text{ kg } NO_2/\text{yr}$$

Note that even though NO_x consists of a mixture of NO and NO_2, the total mass is expressed as equivalent NO_2. This is because NO_2 is the criteria air pollutant associated with adverse health effects (see Chapter 2). Furthermore, even though NO is the principal compound formed during combustion, once emitted to the atmosphere it quickly oxidizes to the pollutant NO_2. The mass equivalence between NO and NO_2 is found using the ratio of molecular weights:

$$\frac{m_{NO_2}}{m_{NO}} = \frac{\text{Mol. wt. of } NO_2}{\text{Mol. wt. of NO}} = \frac{46}{30} = 1.533$$

Thus

$$m_{NO_2} = 1.533 \, m_{NO}$$

Solid and Liquid Wastes   Collected ash from coal-burning plants creates a new environmental problem of solid waste disposal. Water pollution problems also may be created because water is often used to transport collected ash within the power plant. When plant water streams are discharged to the environment, they may carry ash compounds in the form of dissolved or suspended solids. Additional water pollution problems may arise from the discharge of waste heat to a river, lake, or stream. Chapter 2 discussed the environmental effect of such discharges on plant and fish life. Modern power plants thus must control not only atmospheric emissions, but water pollutants and solid wastes as well. Later in this chapter, Section 5.5.1 discusses the engineering approaches to environmental control.

Other Life Cycle Impacts   Environmental impacts occur not only at the electric power plant, but everywhere along the chain of processes that includes the extraction, refining, transport, and storage of fuels, as well as the transmission and distribution of electricity. Here we highlight just a few of these life cycle impacts.

Because coal supplies most of the electricity in the United States (and the world), the environmental impacts of coal mining are of particular concern. Coal is extracted both from underground mines deep in the earth and from surface mines (known also as *strip mines*), where large mechanical shovels dig coal directly out of the ground. Like other types of mining operations, coal mining has a checkered environmental past. Streams and rivers have been polluted, regions of the country have been scarred and littered with refuse piles, and homes and other structures have been damaged when the land above underground mines caves in or settles over time (a phenomenon known as *subsidence*). Modern mining methods and land reclamation practices have, to a large extent, eliminated the major environmental problems of the past. Nonetheless, coal mining remains a source of land use impacts and disturbances, which inevitably arise from any resource extraction process (including drilling for oil and natural gas).

Land use impacts occur throughout the fuel cycle. Transportation of natural gas, for example, requires large pipelines to be constructed over distances of hundreds or thousands of kilometers. The expanded use of natural gas for power generation could thus exacerbate land use disturbances. Because methane is also a powerful greenhouse gas, leakage of gas from pipelines is another environmental concern. Past leakage losses have been estimated to be approximately 1 to 2 percent of natural gas consumption (USEPA, 1996). Improved monitoring technology is expected to substantially reduce or eliminate such leakage losses, but any direct emission of methane to the atmosphere offsets some of the benefits of lower CO_2 production found in Example 5.2.

Transmission and distribution lines that carry electricity from power plants to customers are another source of environmental impacts. Although not unique to fossil fuels, this aspect of electricity use often arouses public concern, especially near heavily populated areas.

Various studies have addressed the full life cycle environmental impacts of electric power generation using fossil fuels (ORNL, 1992). The impacts of mining, drilling, transport, and distribution often can be important at the local level, but the dominant environmental impacts identified in life cycle assessments arise from air

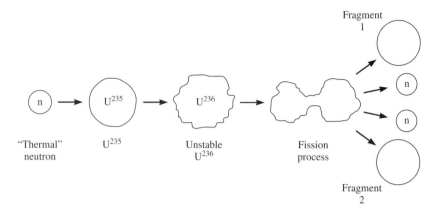

Figure 5.5   Fission of uranium in a nuclear reactor. Each fission process produces two different radioactive elements (labeled fragments 1 and 2) plus two or three neutrons.

pollutants released at the power plant. Later in this chapter we take an in-depth look at the engineering measures available to alleviate these environmental problems.

5.2.2   Environmental Impacts of Nuclear Power

Nuclear fuels play a role similar to that of fossil fuels in power generation by supplying a source of heat that is ultimately converted to electricity. In nuclear fission reactions, the nucleus of an atom is broken apart, releasing large amounts of energy; whereas in fossil fuel combustion the chemical bonds between carbon and hydrogen atoms are severed, yielding relatively smaller energy releases. The main attraction of nuclear energy, therefore, is that each kilogram of processed nuclear fuel releases approximately 80,000 times more energy than an equal mass of fossil fuel. This enormous difference in energy density led early nuclear enthusiasts of the 1950s to declare that nuclear power would simply be "too cheap to meter."

The nuclear fuel cycle begins with the mining and processing of uranium ore, whose environmental impacts were discussed in Chapter 2. As in digging for gold, a large quantity of material must be processed to extract the valuable uranium oxide (U_3O_8), which constitutes only about 0.25 percent of the ore by weight. Furthermore, only about 0.7 percent of the uranium is in the isotope form needed for nuclear power plants. Elaborate chemical processing is thus required to extract, separate, and concentrate the uranium isotope $_{92}U^{235}$ from the more abundant $_{92}U^{238}$ isotope of uranium.[3] The U^{235} is what powers nuclear reactors.

This form of uranium can readily absorb slow-moving neutrons, producing an unstable atom of U^{236} as sketched in Figure 5.5. The U^{236} atom then breaks apart

---

3 Recall that the number in the lower left is the atomic number of an element, equal to the number of protons in the nucleus. The number in the upper right is the atomic mass number, equal to the number of protons plus neutrons. Isotopes of an element are distinguished by different atomic mass numbers.

(fissions) into two radioactive elements (called fission fragments) plus two or three neutrons. Energy is also released, mainly in the form of kinetic energy of the fission fragments. There are many different ways in which the unstable U^{236} can decompose, so a variety of fission fragments can be produced. For example, two possible reactions are

$$_{92}U^{235} + {_0}n^1 \rightarrow {_{92}}U^{236} \rightarrow {_{56}}Ba^{137} + {_{36}}Kr^{97} + 2{_0}n^1 + \text{Energy} \tag{5.5}$$

$$_{92}U^{235} + {_0}n^1 \rightarrow {_{92}}U^{236} \rightarrow {_{54}}Xe^{140} + {_{38}}Sr^{94} + 2{_0}n^1 + \text{Energy} \tag{5.6}$$

A key feature of U^{235} fission reactions is that they produce at least two neutrons for every one absorbed. After these fast-moving neutrons have been slowed down by collisions with other atoms or reactor "moderator" materials, they are able to initiate new fission reactions, which in turn release even more neutrons. The result is a chain reaction that produces new radioactive substances while releasing substantial amounts of energy. Typically only about 3 or 4 percent of the nuclear fuel is fissile U^{235}. Plants are carefully designed to control the chain reaction so it does not get out of hand.

As the fission products of U^{235} decay into other radioactive elements, different forms of radiation are released. In addition, the more abundant (but far more stable) U^{238} also is partially transformed through a series of reactions into long-lived radioactive elements such as plutonium ($_{94}Pu^{239}$).

As discussed in Chapter 2, the major environmental issues surrounding nuclear power center around the effects of radiation on living organisms. This radiation originates in three forms. One is the release of *alpha a particles,* which are pieces of atomic nuclei each consisting of two protons (positively charged particles) plus two neutrons (particles with weight similar to protons but with no electrical charge). For example, U^{238} releases an alpha particle as it decays into thorium:

$$_{92}U^{238} \rightarrow {_{90}}Th^{234} + \alpha \tag{5.7}$$

A second form of radiation is a negatively charged free electron, known as a *beta (β) particle.* These are produced in reactions such as the decay of xenon into cesium:

$$_{54}Xe^{140} \rightarrow {_{55}}Cs^{140} + \beta \tag{5.8}$$

Finally, *gamma (γ) radiation* refers to radiative energy released in the form of electromagnetic waves. Such waves also exhibit the behavior of particles, referred to as *photons.* The energy releases shown in Equation (5.5) and Equation (5.6) are examples of gamma radiation.

The biological effects induced by exposure to α, β, or γ particles are a function of the energy level of each particle type and the overall dose that is received. Radioactive materials are produced at each step in the nuclear fuel cycle (mining, processing, enrichment, power generation, and disposal), and some releases of radiation inevitably occur at each stage, either accidentally or by design within allowable limits. The highest levels of radioactivity are found in the spent fuel that is periodically removed from nuclear reactors when it is no longer efficient at generating heat. Spent fuel is far more radioactive than the original uranium in fuel rods because it contains a variety of radioactive decay products (radionuclides), such as plutonium, that have extremely long half-lives. This means that such material will remain dangerous for tens of thousands of years. As discussed in Chapter 2, the safe and permanent disposal of high-level radioactive waste represents one of the greatest environmental challenges facing modern society.

Example 5.6

Production of high-level radioactive waste.   A 1,000 MW nuclear power plant operates at an annual capacity factor of 74 percent. Its overall thermal efficiency is 32 percent. The energy supplied by the nuclear fission reaction of U^{235} shown in Equation (5.5) is 7.5×10^{10} kJ/kg U^{235}. The concentration of U^{235} in the reactor fuel is 3.0 percent. Estimate the annual quantity of high-level waste (spent fuel) produced by this power plant.

Solution:

First we must find the annual quantity of U^{235} required to operate the plant. This calculation follows the same procedure used in Example 5.1 to find the annual quantity of coal burned:

$$\text{Annual energy input} = \frac{\text{Annual energy output}}{\text{Plant efficiency}}$$

$$= \frac{(1{,}000 \text{ MW} \times 1{,}000 \text{ kW/MW})(0.74 \times 8{,}760 \text{ hrs/yr})(3{,}600 \text{ kJ/kW-hr})}{0.32}$$

$$= 7.293 \times 10^{13} \text{ kJ/yr}$$

From this, the required mass of U^{235} is

$$m_{U^{235}} = \frac{7.293 \times 10^{13} \text{ kJ/yr}}{7.5 \times 10^{10} \text{ kJ/kg}} = 972 \text{ kg } U^{235}/\text{yr}$$

Because only 3 percent of the fuel is U^{235}, the annual quantity of fuel is

$$m_{fuel} = \frac{972 \text{ kg } U^{235}/\text{yr}}{0.03 \text{ kg } U^{235}/\text{kg fuel}} = 32{,}400 \text{ kg fuel/yr}$$

This amount of fuel must be supplied each year, so the same amount must be removed each year as spent fuel. The spent fuel is highly radioactive and is classified as high-level waste.

In contrast to the "too cheap to meter" optimism of the 1950s, environmental and safety concerns have driven up the cost of nuclear power generation and delayed or halted most new plant construction. In the United States, no new nuclear power plants have been ordered since 1979, when an accident at the Three Mile Island power station in central Pennsylvania provoked widespread public concern about nuclear safety and led to a series of costly new measures to improve plant operations. The subsequent 1986 Russian accident at the Chernobyl power plant further increased public concern about nuclear energy. Although U.S. plant designs cannot suffer the type of disaster that occurred at the type of plant used at Chernobyl, opposition to the construction of new nuclear plants nonetheless has remained strong. In the past 20 years more than 60 nuclear projects have been canceled in the United States.

At the same time, over 100 nuclear power plants continue to operate in the United States today. According to Department of Energy projections (USDOE, 1998b), many of these plants will retire over the next 10 to 20 years as they reach their licensed operating lifetime or because they are no longer economical. U.S. nuclear power production is thus expected to decline over the next two decades. Elsewhere in the world the outlook for nuclear power is mixed. A number of industrialized countries, especially in Europe, are pursuing paths similar to the United States by continuing to operate existing nuclear facilities but building no new ones. In developing countries throughout Asia, on the other hand, the construction and planning of nuclear power plants is proceeding aggressively. Thus the extent to which nuclear power will ultimately fulfill the positive expectations of its supporters, or the dire consequences of its detractors, will not be known with certainty until sometime in the next 10,000 years.

5.2.3   Environmental Impacts of Renewable Energy

In light of the environmental concerns about fossil fuels and nuclear energy, the use of renewable energy is widely viewed as a more attractive alternative for future electric power production. Renewables include hydroelectric, wind, biomass, refuse, geothermal, and solar energy. A popular belief is that such renewables represent sources of clean, pollution-free energy. Indeed, in many respects renewable energy offers substantial environmental benefits over conventional fossil and nuclear fuels. However, as Chapter 2 pointed out, the technical reality is that renewable energy sources also produce adverse environmental impacts, which vary with the particular energy source.

Table 5.4 summarizes the major environmental issues related to renewable energy sources. One common element is the need for relatively large amounts of land to generate electricity because renewables represent a much more diffuse energy source than fossil or nuclear fuels. In some cases, however, the land occupied by renewable energy sources can be used for dual purposes. For instance, the land occupied by modern wind farms often supports cattle grazing.

Table 5.4   Environmental impacts of renewable energy sources.

Energy Source	Impact Issues
Geothermal	Hydrogen sulfide releases; local seismic effects; noise; land use
Biomass	Land requirements; ecological impacts of harvesting and transportation; loss of species diversity; use of fertilizers; atmospheric emissions during harvesting and conversion
Solar	Land requirements for large-scale implementation; life cycle manufacturing and disposal issues for PV solar cells
Wind	Land use requirements; visual impact; electromagnetic interference; birds
Ocean-based	Visual impacts; conflicts with shipping; water quality and sedimentation pattern changes; ecosystem disruptions

Hydroelectric power is today the most widely used form of renewable energy. However, hydroelectric plants are now under scrutiny for their detrimental long-term impacts on land use and ecosystems downstream of hydroelectric dams. Because of such concerns, renewal of long-term operating permits for hydroelectric facilities in the United States is no longer assured. In Arizona local environmental organizations have called for the removal of the Glen Canyon dam to restore lands that were flooded to create Lake Powell. In Maine the Edwards Dam has become the first in the country to be denied a renewal of its operating permit. Constructed in 1837, the dam recently has been demolished to restore the Kennebec River to its natural habitat. Elsewhere in the world, environmental assessments of the Three Gorges dam under construction in China—to be the largest hydroelectric facility in the world—point to major adverse impacts. The planned damming of the Yangtzee River will displace some 1.2 million people and flood lands to create a water reservoir more than 600 km long. Both the World Bank and the U.S. Export–Import Bank have refused to help finance this project, mainly because of its adverse environmental effects (USDOE, 1998b).

The need for some type of energy storage system is another feature required for primary reliance on intermittent energy sources like wind and sunlight. The environmental impacts of energy storage, in turn, depend on the system or technology employed. Storage batteries, for example, are currently constructed of heavy metals such as lead, nickel, and cadmium—which are considered toxic substances of environmental concern (see Chapters 2 and 4). More benign methods of energy storage are thus needed to avoid creating environmental problems.

As these examples illustrate, environmental impacts, whether direct or indirect, occur with each form of renewable energy. In many cases, improved engineering design can ameliorate or eliminate some of the impacts, as we will see later in this chapter. In other cases, impacts may be far less severe and much more acceptable than those of nonrenewable energy sources. The key point is that there is no such thing as a completely pollution-free source of electricity.

The remainder of this chapter takes a more detailed look at the technology of electric power generation and how engineering fundamentals can be applied to help minimize or avoid environmental impacts. Because the subject of electric power is so broad, the discussions in this chapter are necessarily limited. The emphasis on fundamentals, however, should provide students with the building blocks to independently pursue specific topics of interest.

5.3 ELECTRIC POWER FUNDAMENTALS

In 1831 the English scientist Michael Faraday made an interesting discovery. He found that moving a coil of wire through a magnetic field caused an electrical current to flow in the wire. With that experiment the modern electric generator was born. It would take another 50 years before Thomas Edison developed a practical way to use Faraday's discovery for commercial power generation. The rest, as they say, is history.

5.3.1   Current, Voltage, and Power

Electrical *current* is the flow of charged particles through a conducting medium such as a copper wire. The basic unit of current is the *ampere* (*amp*), defined as a flow of one coulomb per second, where a *coulomb, C,* is the basic unit of electrical charge. Current can arise from the flow of either positively or negatively charged ions, but in metal wires only the electrons move. The electrical charge on each electron is 1.6×10^{-19} C. The work required to move a unit of charge along a conductor is known as the *voltage*. For common metallic conductors, current, I, and voltage, V, are related according to Ohm's Law:

$$V = I \times R \tag{5.9}$$

where R is the value of an electrical *resistance* (measured in *ohms*) in a simple circuit. Because voltage is a measure of work per unit charge, and current is defined as charge per unit time, the product of the two gives the electrical power, P, measured in *watts:*

$$P = V \times I \tag{5.10}$$

Finally, because power is defined as energy per unit time (recall that a watt is one joule per second), the electrical *energy, E,* supplied during a time interval, Δt, is

$$E = P\Delta t \tag{5.11}$$

As noted earlier, electrical energy is commonly measured in units of kilowatt-hours (kW-hr).

In addition to metallic conductors like copper wires, the class of materials known as *semiconductors* also is important in some electric power applications. Such materials allow current to flow easily in one direction but not the other, and thus they do not obey a linear voltage–current relationship as given by Ohm's Law. Note too that electrical conductors do not necessarily have to be solid materials. Electrically conducting liquids (called *electrolytes*) and electrically conducting gases (called *plasmas*) also are important in some power generation applications.

5.3.2   Energy, Heat, and Work

To understand the basics of an electric power plant, the reader must clearly understand the technical meanings of three commonly used terms: energy, heat, and work. The study of thermodynamics draws important distinctions between these terms. *Heat* and *work* are both forms of *energy. Heat* is energy transferred by virtue of a temperature difference, whereas *work* is manifested by the ability to lift a weight or move a mass against a resisting force. Electricity represents energy in the form of work. The distinction is important because work can be fully transformed into heat, but heat cannot be completely converted to work in any real device (this principle is known as the *Second Law of Thermodynamics*). The practical implications of this important difference will become clearer as we consider different ways of generating electricity.

5.3.3 Electromechanical Generators

From an engineering perspective, the key question affecting power plant technology is how one supplies the voltage or force needed to generate an electrical current. Surprisingly, there are many more ways than the electromechanical generator discovered by Michael Faraday (see Table 5.5). Later in the chapter, we discuss some of these other types of generators. First we examine the dominant technology for generating electricity.

The term *electromechanical generator* implies the use of some form of mechanical energy to move an electrical conductor through a magnetic field. Most electrical generators in the world today employ this technique. Figure 5.6 shows the basic configuration of such a generator. The electrical conductor is a coil of copper wire attached to a shaft that is mechanically rotated in a magnetic field. The laws of physics tell us that electrons in the wire will be driven in a direction perpendicular to both the magnetic field and the wire velocity. Electric current will thus flow along the length of the wire to the commutator rings in Figure 5.6, where a connection can be made to an external circuit. Because the velocity of the wire conductor changes direction as it rotates, the voltage generated also changes direction, producing the sinusoidal shape sketched in Figure 5.6. This, in turn, produces the sinusoidal variation in current flow known as *alternating current* or *AC* for short. The frequency with which the current changes direction depends on the speed of the rotating shaft. In the Americas, the standard frequency is 60 cycles per second, or 60 hertz (Hz). In Europe and other parts of the world, the standard is 50 Hz.

A key feature of this generator design is that it can produce the high voltages and currents required for large-scale use of electricity. Rather than the single wire shown in Figure 5.6, commercial generators are wound with large numbers of coils to increase current flow. Some of this electricity is used to power large electromagnets that create the magnetic field. Power levels in the hundreds of megawatts can be produced from a single generator. Transmission lines carry current at high voltage to local distribution systems, where transformers reduce the voltage to either 110 v or 220 v typical of household power supplies.

Table 5.5 Getting electrons to move: A variety of fundamental mechanisms are capable of generating a voltage that can drive an electrical current.

Generator Type	Principle of Operation
Electromechanical	Movement of a conductor in a direction perpendicular to a magnetic field
Electrochemical	Two dissimilar conductors immersed in an electrolyte
Photovoltaic	Excitation of electrons in semiconductor materials by photons from the sun (or another light source)
Thermoelectric	Temperature difference between the junction and free ends of two dissimilar metals
Thermionic	Heat supplied to high-temperature conductor
Piezoelectric	Pressure applied to certain crystalline materials

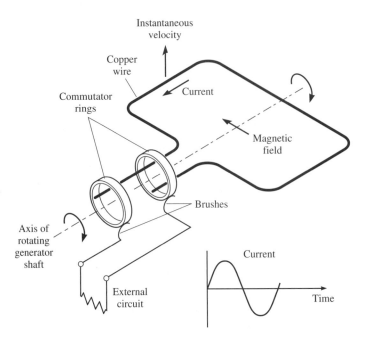

Figure 5.6   A simple electromechanical generator. Rotation of a metallic conductor in a magnetic field produces an alternating current when connected to an external load. Multiple wire windings are used for commercial generators.

The laws of physics offer no free lunch, so the generation of electricity requires some external source of energy to rotate the generator shaft. Indeed, physics (Lenz's Law) teaches us that as the generator shaft in Figure 5.6 begins to rotate, forces are created that oppose the direction of motion of the wire coil. The faster the rotation, the greater the opposing force. Thus work must be supplied to overcome the resisting force in order to keep the generator shaft turning. (You may have experienced this effect by turning the crank of a small hand generator or pedaling an exercise bike connected to an electrical generator.) In most commercial power plants, the work needed to turn the generator is supplied by a device called a *turbine*.

5.3.4   Turbines and Energy Sources

As illustrated in Figure 5.7, a simple turbine resembles a wheel whose circumference is fitted with curved blades that are impacted by a moving fluid. That fluid could be air or water, either at ambient temperature or as a hot gas at high temperature. The energy of the impinging fluid causes the turbine blades to rotate, thus turning the generator shaft to produce electricity.

The design of a turbine varies according to the type of working fluid or energy source that is used. Most power plants today use steam or hot combustion gas to

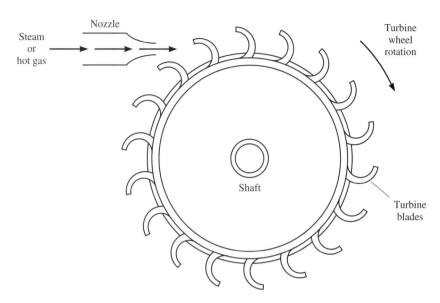

Figure 5.7   Schematic of a simple impulse turbine. A variety of turbine designs are employed for different applications.

drive the turbine. Long before these methods were devised, turbine wheels were powered by two of nature's oldest energy sources, wind and water. Early turbine designs used for grinding grain and other mechanical tasks later evolved into turbines for generating electricity. In modern times, fossil fuels supply most of the energy used to drive turbines for electric power generation.

5.4   PERFORMANCE OF FOSSIL FUEL POWER PLANTS

At the beginning of this chapter we examined the major environmental impacts of fossil fuel power plants by treating the power generation technology as a "black box." Now we peer inside that box to gain a deeper understanding of how engineering fundamentals and design choices affect technology performance and consequent environmental impacts. In particular, this section emphasizes the importance of thermodynamic efficiency as a critical parameter in reducing environmental impacts. Efficiency is defined and examined for individual power plant components as well as for the overall system. At all of these levels, engineering design plays a major role in the environmental outcome.

5.4.1   Steam Electric Plants

Steam turbines are the most prevalent method used worldwide for spinning the shaft of an electromechanical generator. By combining two of nature's most basic elements—

fire and water—high-temperature steam can be generated to produce a hundredfold increase in the electrical output of a turbine generator compared to liquid water alone. The term *fire* refers to any method of generating heat to produce steam. Because steam can be produced using a wide variety of fuels and energy sources, it offers the flexibility to generate electricity reliably and continuously at any location and at any time of day. Steam turbines today provide approximately two-thirds of all the electric power generated in the United States.

Energy Content of Steam   The energy available in hot steam depends on the steam temperature and pressure entering the turbine. In thermodynamics, the term *enthalpy* is used to quantify this thermal energy content. Students unfamiliar with this term may more readily recognize the property called *specific heat, c_p,* which is used in introductory chemistry and physics to calculate the amount of thermal energy needed to raise the temperature of a unit mass of liquid or gas from an initial value T_1 to a higher value T_2 at constant pressure (hence the subscript p). Total enthalpy, H, can be thought of as the product of specific heat multiplied by mass and temperature.[4] Thus, when a mass m undergoes a temperature change from T_1 to T_2, the corresponding enthalpy change is given by

$$H_2 - H_1 = mc_p\,(T_2 - T_1) \qquad (5.12)$$

In this equation the value of specific heat is assumed to be constant. For most substances this is a good assumption, provided that the temperature difference is not too large. In general, however, the value of specific heat varies with the temperature and pressure of a substance. This is especially important for gases such as steam, which undergo very large changes in temperature and pressure in an electric power plant. In such cases enthalpy gives a more robust measure of changes in energy.

The values of enthalpy for a given substance are tabulated as a function of temperature and pressure in terms of *specific enthalpy, h,* which is the enthalpy per unit mass of a substance. Thus

$$h = \frac{H}{m} \qquad (5.13)$$

The specific enthalpy is considered a property of the substance.

From Equation (5.12) we also see that

$$h_2 - h_1 = c_p\,(T_2 - T_1)$$

Thus the change in specific enthalpy of a substance (in our case, steam) is approximately proportional to the change in temperature, recognizing that the value of specific heat is not truly constant.

Table 5.6 shows the enthalpy values of water and steam at selected temperatures and pressures. At any given pressure, there is a large increase in specific enthalpy as

4 Students familiar with thermodynamics know that *specific heat* is actually defined as the rate of change of enthalpy with temperature at either constant pressure (c_p) or constant volume (c_v). For present purposes, the distinction between c_p and c_v is not important.

Table 5.6   Specific enthalpy of water and steam (kilojoules per kilogram).

Pressure kPa	T_{sat} °C	Saturated Liquid h_f	Saturated Vapor h_g	Superheated Steam Temperature, °C 100	200	300	400	500	600	700	800
1	7.0	29.3	2,514.4	2,688.6	2,880.1	3,076.8	3,279.7	3,489.2	3,705.6	3,928.9	4,158.7
5	32.9	137.8	2,561.6	2,688.1	2,879.9	3,076.7	3,279.7	3,489.2	3,705.6	3,928.8	4,158.7
10	45.8	191.8	2,584.8	2,687.5	2,879.6	3,076.6	3,279.6	3,489.1	3,705.5	3,928.8	4,158.7
50	81.3	340.6	2,646.0	2,682.6	2,877.7	3,075.7	3,279.0	3,488.7	3,705.2	3,928.6	4,158.5
100	99.6	417.5	2,675.4	2,676.2	2,875.4	3,074.5	3,278.2	3,488.1	3,704.8	3,928.2	4,158.3
200	120.2	504.7	2,706.3		2,870.5	3,072.1	3,276.7	3,487.0	3,704.0	3,927.6	4,157.8
400	143.6	604.7	2,737.6		2,860.4	3,067.2	3,273.6	3,484.9	3,702.3	3,926.4	4,156.9
600	158.8	670.4	2,755.5		2,849.7	3,062.3	3,270.6	3,482.7	3,700.7	3,925.1	4,155.9
800	170.4	720.9	2,767.5		2,838.6	3,057.3	3,267.5	3,480.5	3,699.1	3,923.9	4,155.0
1,000	179.9	762.6	2,776.2		2,826.8	3,052.1	3,264.4	3,478.3	3,697.4	3,922.7	4,154.1
2,000	212.4	908.6	2,797.2			3,025.0	3,248.7	3,467.3	3,689.2	3,916.5	4,149.4
5,000	263.9	1,154.4	2,794.2			2,925.5	3,198.3	3,433.7	3,664.5	3,897.9	4,135.3
10,000	311.0	1,408.0	2,727.7				3,099.9	3,374.6	3,622.7	3,866.8	4,112.0
15,000	342.1	1,611.0	2,615.0				2,979.1	3,310.6	3,579.8	3,835.4	4,088.6
20,000	365.7	1,826.5	2,418.4				2,820.5	3,241.1	3,535.5	3,803.8	4,065.3
30,000							2,161.8	3,085.0	3,443.0	3,739.7	4,018.5

Source: Based on ASME, 1977.

water changes from a liquid state to a vapor.[5] Students may recognize this increase in enthalpy as the *heat of vaporization,* which is the energy required to boil and completely evaporate a unit mass of water. During this change of state from liquid to vapor, the temperature of the water remains constant at the boiling point (100°C for water at atmospheric pressure). After the water is completely evaporated, its temperature again begins to rise as more heat is added. The enthalpy of this "superheated" steam increases accordingly, as illustrated in Example 5.7. In thermodynamics, the term *saturated liquid* refers to the state at which liquid water just begins to evaporate. The enthalpy at this state is designated as h_f. The term *saturated vapor* refers to the point at which evaporation is completed (during heating) or at which condensation to a liquid just begins (during cooling). The enthalpy of saturated vapor is designated as h_g.

Example 5.7

The enthalpy of steam.   Compare the enthalpy of saturated liquid water at a pressure of 100 kPa (approximately one atmosphere) to steam at the following conditions: (a) saturated vapor at 100 kPa and (b) superheated vapor at 15 MPa and 560°C.

5 Note that the term *vapor* means a gas that condenses into a liquid at normal atmospheric conditions. The words *steam* and *water vapor* often are used interchangeably. *Steam* also can mean the mixture of vapor and liquid that occurs when water is evaporating or condensing (commonly referred to as *wet steam*).

Solution:

From Table 5.6, the enthalpy of saturated liquid (h_f) at 100 kPa is 417.5 kJ/kg.

(a) From Table 5.6, the enthalpy of saturated vapor (h_g) at $P = 100$ kPa is 2,675.4 kJ/kg. This represents 6.4 times more energy per unit mass relative to saturated liquid.

(b) To obtain the enthalpy of steam at 15 MPa and 560°C, interpolate between the values in Table 5.6 for temperatures of 500°C and 600°C. Thus, at $P = 15,000$ kPa,

$$h_{560} = h_{500} + (60/100)(h_{600} - h_{500})$$

$$= 3,310.6 + 0.6 (3,579.8 - 3,310.6)$$

$$= 3,472.1 \text{ kJ/kg}$$

This represents 8.3 times more energy per unit mass relative to the enthalpy of saturated liquid at P = 100 kPa.

From this example one can readily see the advantage of using steam instead of falling water to drive a turbine. At the steam conditions shown, one kilogram of water vapor has roughly 7,000 times more energy than a kilogram of liquid water used to generate electricity at a typical hydroelectric facility (discussed later in Section 5.6.4). Even after accounting for differences in the energy conversion efficiency, the thermodynamic advantages of steam are substantial.

Extracting Work from Steam   As noted earlier, the conversion of thermal energy to mechanical energy (work) cannot be achieved with 100 percent efficiency by any real device. The Second Law of Thermodynamics, which addresses the concept of *entropy,* requires that some of the thermal energy in hot steam be dumped to the environment in order to transform heat into work continuously.

The job of the steam turbine is to extract as much useful work as possible, consistent with the laws of thermodynamics. The work extracted by the turbine is then used to turn the generator that produces electricity. In general, the way to maximize the turbine work output is to maximize the inlet steam enthalpy, h_{in} by increasing the steam temperature and pressure), and to minimize the outlet enthalpy by allowing the steam to expand and cool as much as possible before leaving the turbine.

As a practical matter, inlet steam conditions are limited by materials of construction for the turbine blades, whereas the outlet conditions are limited by the amount of condensation the turbine blades can tolerate. Any significant quantity of water droplets in condensing steam can cause rapid erosion and failure of the turbine blades. For this reason, turbine exit conditions are set by a requirement to have no more than about 10 percent liquid in the steam exiting the turbine.

One can imagine a simple open-cycle system in which steam exiting the turbine is simply discarded to the environment at atmospheric pressure. This limits the exit temperature to 100°C if condensation is to be avoided in order to prevent turbine blade erosion. As illustrated in the next example, the amount of useful work extracted is also fairly low.

Example 5.8

A hypothetical open-cycle power plant.   Steam enters a turbine at 560°C and 15 MPa and leaves as saturated vapor at atmospheric pressure (approximated as 100 kPa). What fraction of the inlet steam energy is extracted by the turbine?

Solution:

From Example 5.7, the inlet steam enthalpy is 3,472 kJ/kg, and the outlet enthalpy is 2,675 kJ/kg. The useful energy (work) extracted in the turbine per kg of steam is therefore

$$E_{turbine} = (h_{in} - h_{out})_{turbine}$$
$$= 3,472 - 2,675 = 797 \text{ kJ/kg}$$

As a fraction of the inlet steam enthalpy this is

$$\frac{E_{turbine}}{h_{in}} = \frac{797}{3,472} = 0.230, \text{ or } 23.0\%$$

To take advantage of the energy-producing capacity of steam, engineers have devised more complex cycles, as discussed next. These improvements allow more useful work to be extracted in the steam cycle. This reduces the fuel requirements of the power plant, thus also reducing environmental impacts.

A Basic Steam Electric Plant   In 1765 James Watt discovered that the efficiency of a steam engine could be substantially improved by using a closed-cycle arrangement with a *condenser* to cool the waste steam. This same technique applied to a steam electric plant allows more useful energy to be extracted in the turbine, thereby raising its efficiency.

Figure 5.8 shows a schematic of this basic steam electric plant. The closed-cycle arrangement permits the turbine exit pressure to fall below atmospheric pressure (because a vacuum forms in the condenser as vapor turns to liquid). At the lower pressure, condensation begins at temperatures below 100°C. More useful energy can thus be extracted in the turbine without encountering the problems of water condensation.

Besides the turbine and low-pressure condenser, two additional pieces of equipment are shown in Figure 5.8. One is the *boiler,* which generates steam. In Figure 5.8 the symbol E_{in} is the energy provided to the boiler by the primary energy source (coal, oil, nuclear, or other fuels), and Q is the amount of this energy transferred to water as heat for generating steam. These quantities are often specified on a rate basis, denoted by a dot over the symbol (that is, $\dot{E}_{in}$ and $\dot{Q}_{in}$).

Another piece of equipment is a *pump* (called the *feedwater pump*), which pumps the fully condensed steam (liquid water) exiting the condenser to the boiler at high pressure. The pump requires energy (work), but for steam cycles the amount of work required is small relative to the energy extracted in the turbine.

The overall steam cycle depicted in Figure 5.8 is known as the *Rankine cycle,* named after the Scottish engineer William Rankine, who was a pioneer in the field of thermodynamics in the late 19th century. More complex forms of this basic cycle

A modern steam turbine design for a combined cycle system. Steam flows from right to left through multiple stages of turbine blades of increasing length. The diameter of the last (low-pressure) stage is approximately two meters. When encased in a pressure vessel, this unit will power a 100 MW electrical generator.

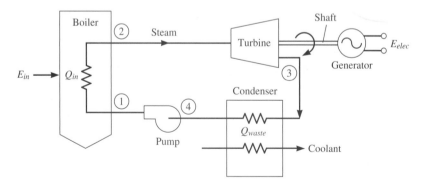

Figure 5.8   Components of a basic steam electric generator.

have since been developed to increase the overall cycle efficiency. Most commonly, steam is extracted from the middle stages of the turbine and reheated to improve the turbine performance.

An overall efficiency for the steam cycle can be defined based on the measured electrical energy output from the turbine generator after subtracting the energy needed to operate the feed water pump, plus any frictional losses in the turbine and generator. This quantity is called the *gross electrical output*, $E_{elec,gross}$. A *steam cycle efficiency*, η_{cycle}, is typically defined relative to the thermal energy *added* to the steam in the boiler:

$$\eta_{cycle} = \frac{\text{Gross electrical energy output}}{\text{Heat added to steam}}$$ (5.14)

Typically thermodynamic designs used at modern steam electric power plants achieve steam cycle efficiencies on the order of 40 to 45 percent. However, this is not the overall efficiency of the power plant. Two other factors also need to be considered.

Generating the Steam   To determine the *overall* efficiency of the power plant, one also has to consider the method and efficiency of generating steam. The most common method of steam generation is to burn a fossil fuel like pulverized coal or natural gas. Fuel and air are introduced through burners protruding from the walls or corners of the boiler. The boiler itself resembles a large box, which can be up to several stories high. The energy released from fuel combustion inside the chamber heats the boiler feed water, which flows through pipes embedded in the boiler walls. As the water turns to steam, it is collected in a manifold and routed to the steam turbine.

To maximize the boiler efficiency, the hot (typically 425°C) combustion gases exiting the combustion chamber are passed through a heat exchanger (called the *air preheater*) to warm the incoming combustion air. This reduces the amount of fuel needed to heat the air inside the furnace. The *boiler efficiency* can then be expressed as

$$\eta_{boiler} = \frac{\text{Rate of heat added to steam}}{\text{Rate of fuel energy input}} = \frac{\dot{Q}}{\dot{E}_{in}}$$ (5.15)

where $\dot{Q}$ and $\dot{E}_{in}$ are as depicted in Figure 5.8. The heat addition, $\dot{Q}$, increases the enthalpy of the boiler feed water. Thus,

$$\dot{Q}_{steam} = \dot{m}_{steam} \left(h_{out} - h_{in} \right)_{boiler}$$ (5.16)

where $\dot{m}_{steam}$ is the mass flow rate of steam, and h_{in} and h_{out} correspond to the specific enthalpy of steam entering and exiting the boiler at points 1 and 2 in the flow diagram of Figure 5.8.

The rate of fuel energy input to the boiler can be expressed as the product of the fuel mass flow rate, $\dot{m}_{fuel}$, times the fuel heating value, *HV*, which is the energy released from burning a unit mass of fuel (see Table 5.1 for typical numerical values):

$$\dot{E}_{in} = \dot{m}_{fuel} (HV)$$ (5.17)

Using these relationships, the boiler efficiency in Equation (5.15) can now be rewritten as

$$\eta_{boiler} = \frac{\dot{m}_{steam} \left(h_{out} - h_{in} \right)_{boiler}}{\dot{m}_{fuel} (HV)}$$ (5.18)

A typical thermal efficiency for modern boilers is about 88 percent. Most of the remaining 12 percent of the fuel energy is lost with the hot combustion gases that are released to the atmosphere.

Gross Plant Efficiency　　Two different efficiencies are used to describe the overall power plant performance: *gross efficiency* and *net efficiency*. The gross power plant efficiency, η_{gross}, refers to the combined efficiency of the boiler and steam cycle:

$$\eta_{gross} = \frac{\text{Gross electrical output}}{\text{Fuel energy input}} = \frac{E_{elec,\,gross}}{E_{in}} \tag{5.19}$$

This can also be expressed as the product of the boiler efficiency times the steam cycle efficiency:

$$\eta_{gross} = \left(\frac{\text{Heat added to steam}}{\text{Fuel energy input}} \right)\left(\frac{\text{Gross electrical output}}{\text{Heat added to steam}} \right)$$

$$= \eta_{boiler} \times \eta_{cycle} \tag{5.20}$$

This value is known as the *gross* plant efficiency because it accounts only for energy losses associated with the use of steam for power generation, but not for the electricity used within the power plant itself to operate pollution control systems, fans for combustion air intake, and other plant equipment. Example 5.9 illustrates how the gross efficiency is calculated.

Example 5.9

Gross efficiency of a steam electric plant.　　The gross electrical output of a steam electric power plant is 500 MW. The steam flow rate through the boiler and turbine is 642 kg/s. Steam leaves the boiler as superheated vapor at 15 MPa, 560°C, after entering the boiler as saturated liquid at a pressure of 15 MPa. The boiler fuel is bituminous coal with the properties listed in Table 5.1. Coal is burned at a rate of 47.5 kg/s. Calculate (a) the boiler efficiency, (b) the steam cycle efficiency, and (c) the gross plant efficiency.

Solution:

Express all energy flows on a rate basis:

(a)　The rate of fuel energy input to the boiler is given by Equation (5.17):

$$\dot{E}_{in} = \dot{m}_{fuel}\,(HV)$$

$$= (47.5 \text{ kg/s})(28,400 \text{ kJ/kg})$$

$$= 1.349 \times 10^6 \text{ kJ/s}$$

The energy transferred to steam in the boiler is given by Equation (5.16):

$$\dot{Q} = \dot{m}_{steam}\,(h_{out} - h_{in})_{boiler}$$

The inlet enthalpy is obtained from Table 5.6 for saturated liquid at 15 MPa, while the outlet enthalpy at 560°C was calculated in Example 5.7. Thus

$$\dot{Q} = (642)(3{,}472.1 - 1{,}611.0)$$

$$= 1.195 \times 10^6 \text{ kJ/s}$$

The boiler efficiency from Equation (5.15) is then

$$\eta_{boiler} = \frac{\dot{Q}}{\dot{E}_{in}} = \frac{1.195 \times 10^6}{1.349 \times 10^6}$$

$$= 0.8858, \text{ or } 88.6\%$$

(b) The overall steam cycle efficiency is given by Equation (5.14), where the gross electrical output of 500 MW can be expressed as

$$\dot{E}_{elec,\,gross} = 500 \text{ MW} = 500{,}000 \text{ kW} = 500{,}000 \text{ kJ/s}$$

$$\eta_{cycle} = \frac{\text{Gross electrical output}}{\text{Heat added to steam}}$$

Thus

$$= \frac{500{,}000 \text{ kJ/s}}{1.195 \times 10^6 \text{ kJ/s}}$$

$$= 0.4184, \text{ or } 41.8\%$$

(c) The gross plant efficiency is then found from Equation (5.19), expressed in terms of power (energy per unit time) rather than energy:

$$\eta_{gross} = \frac{\text{Gross electrical output rate}}{\text{Fuel energy input rate}} = \frac{\dot{E}_{elec,gross}}{\dot{E}_{in}}$$

$$= \frac{5 \times 10^5 \text{ kJ/s}}{1.349 \times 10^6 \text{ kJ/s}}$$

$$= 0.371, \text{ or } 37.1\%$$

Alternatively, this same result can be obtained from Equation (5.20) as the product of the boiler and steam cycle efficiencies:

$$\eta_{gross} = \eta_{boiler} \times \eta_{cycle}$$

$$= (0.8858)(0.4184)$$

$$= 0.371, \text{ or } 37.1\%$$

Net Plant Efficiency   Real power plants require electrical energy to operate fans, motors, and pumps and to run equipment other than the basic steam cycle. Let us define this total auxiliary energy requirement as E_{aux}. Then the *net* electrical energy leaving the power plant is

$$E_{elec,net} = E_{elec,gross} - E_{aux} \qquad (5.21)$$

Analogous to Equation (5.19), we can now define the *net plant efficiency* as

$$\eta_{net} = \frac{\text{Net electrical output}}{\text{Fuel energy input}} = \frac{E_{elec,net}}{E_{in}} \qquad (5.22)$$

This value corresponds to the plant efficiency defined earlier in Equation (5.1), in which the overall power plant was treated simply as a "black box."

Example 5.10

Net efficiency of a steam electric plant.   The auxiliary systems of the power plant in Example 5.9 require 5.6% of the gross electrical output to operate environmental control equipment, fans, conveyors, and other equipment besides the steam cycle. (a) How much power is available for transmission to customers outside the power plant? (b) What is the net plant efficiency?

Solution:

Because the gross plant output is 500 MW, the auxiliary power requirements amount to

$$\dot{E}_{aux} = (0.056)(500) = 28 \text{ MW}$$

Thus the net power available for transmission to customers is

$$\dot{E}_{elec,net} = \dot{E}_{elec,gross} - \dot{E}_{aux}$$
$$= 500 - 28$$
$$= 472 \text{ MW}$$

Using the fuel energy input value from Example 5.9, the net plant efficiency is found to be

$$\eta_{net} = \frac{\dot{E}_{elec,net}}{\dot{m}_{fuel}(HV)}$$
$$= \frac{472{,}000 \text{ kJ/s}}{1.349 \times 10^6 \text{ kJ/s}}$$
$$= 0.350, \text{ or } 35.0\%$$

As you can see, the net plant efficiency is built up from a variety of component efficiencies, each of which has its own engineering design parameters. Improvements in any of these components can thus improve the overall plant efficiency. Environmental control systems are one of the auxiliary systems of a power plant. We see here that the energy required for environmental controls unavoidably decreases the net electrical power available to customers outside the power plant. Thus engineering design measures that can prevent pollutants from forming in the first place can also yield efficiency benefits by reducing the need for additional environmental control technology.

5.4.2   Gas Turbine Plants

Just as the combination of water and fire to produce steam greatly increases the ability to generate electricity, so too does the combination of air and fire offer substantial advantages over air (wind) alone. Again, *fire* refers to any means of heating the working fluid. Modern gas turbines are driven by a flow of high-temperature, high-pressure air heated by the combustion of a clean fuel such as natural gas or distillate oil. Dirtier fuels such as coal cannot be used directly to power a gas turbine because the impurities and ash particles in the combustion gas would quickly destroy the fast-moving turbine blades. Thus a very clean fuel is required for this type of system.

The study of thermodynamics introduces methods for analyzing turbine performance in more detail. Such methods provide important insights into ways of engineering more efficient designs that ultimately reduce environmental impacts.

A Basic Gas Turbine Power Plant   When hot gas rather than steam is used to drive a turbine, the energy conversion system is simpler because the working fluid is heated directly without the need for a large boiler. As depicted in Figure 5.9, air is first compressed to increase the mass flow rate, then heated in a combustion chamber fired by a clean fuel such as natural gas. Nearly all of the energy released during combustion serves to heat the working fluid, which is a mixture of air plus combustion products (primarily CO_2, H_2O, and NO_x). The thermodynamic properties of this mixture are similar to those of air.

As with a steam turbine, the hot gas (at about 1,200–1,300°C) is directed through a nozzle onto the gas turbine blades, causing rotation of the shaft that turns a generator. At the turbine exit, the gas has cooled to a temperature of roughly 550°C—still quite hot, but not hot enough to efficiently extract additional work in the turbine. At most gas turbine power plants this exhaust gas is vented directly to the atmosphere. This simple design is known as the *Brayton cycle,* named after the American inventor George Brayton.

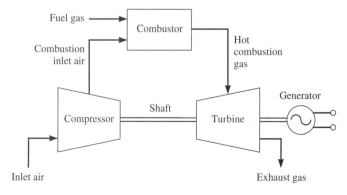

Figure 5.9   Components of a simple gas turbine power cycle.

Net Plant Efficiency   The net plant efficiency of a gas turbine plant can be expressed in the same way as for a steam plant:

$$\eta_{net} = \frac{E_{elec,net}}{m_{fuel}\,(HV)} \qquad (5.23)$$

Historically, most gas turbine power plants operate in a simple open-cycle mode, typically to supply electricity during periods of peak demand. Such plants have relatively low net efficiencies of about 25 to 30 percent. A substantial part of the turbine output is required to compress the air prior to combustion. Other auxiliary energy requirements are typically negligible. The design of more efficient air compressors is thus one of the ways that engineers can help improve the overall performance of gas turbine power plants.

5.4.3   Combined-Cycle Plants

Recent years have seen rapid growth in *combined-cycle* plants, which offer the advantage of higher overall efficiency compared to the simple cycles just described. A combined-cycle plant fired by natural gas is shown schematically in Figure 5.10. This design combines the gas turbine cycle with a steam turbine cycle. Rather than venting the hot turbine exhaust gas to the atmosphere as in the once-through Bray-

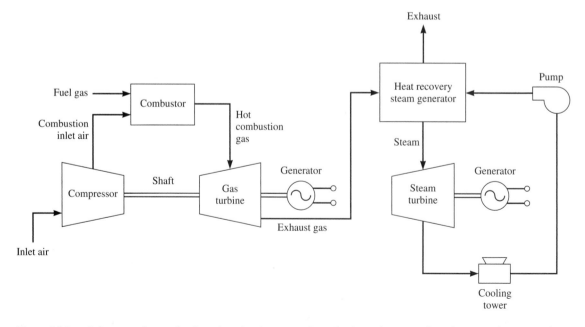

Figure 5.10   Schematic of a gas-fired combined-cycle power plant. The hot exhaust gas from the gas turbine is used to produce steam, which drives a steam turbine to generate additional electricity.

ton cycle, the hot gas is used to generate steam in a type of boiler called a heat recovery steam generator (HRSG). Although the steam temperatures and pressures are not as high as in a conventional steam electric power plant, the conditions are adequate to generate enough additional electricity to warrant the cost of adding the HRSG. Modern combined-cycle power plants of this type currently achieve efficiencies of 50 percent or more.[6]

Economically, a combined-cycle plant is attractive only if the savings in fuel cost resulting from higher plant efficiency are sufficient to offset the higher capital and maintenance costs of a more complex system. In the 1990s several factors came together to make gas-fired combined-cycle plants economically attractive. First, natural gas prices in the United States and elsewhere fell to levels well below those of the early 1980s. Second, the efficiency of gas turbines improved as new technology (derived from the aerospace industry) increased turbine operating temperatures. Finally, the capital cost of combined-cycle systems fell dramatically in response to global competition. The result was a dramatic decrease in the cost of electricity produced by a natural gas-fired combined-cycle plant compared to a traditional simple-cycle plant. At current fuel and equipment prices, gas-fired combined-cycle plants are usually the most economical way of generating electricity for a new facility. Use of this technology is thus expected to grow substantially over the next few decades as a source of continuous "base load" power, not just peaking power.

Example 5.11

Gas turbine fuel use. A conventional simple-cycle 250 MW gas turbine power plant operates at full capacity for 4,000 hrs/yr with a net efficiency of 28 percent. A combined-cycle plant offers an efficiency of 48 percent. The plant burns natural gas costing $2.60 per million kJ. What is the annual gas cost savings using the combined-cycle plant assuming no change in operating hours?

Solution:

The annual electrical energy produced is

$$E_{elec} = (\dot{E}_{elec})(time) = (250,000 \text{ kW})(4,000 \text{ hrs/yr})$$

$$= 1.00 \times 10^9 \text{ kW-hr/yr}$$

With an efficiency of 28 percent, the fuel energy required for the simple-cycle plant is

$$E_{fuel} = \frac{E_{elec, net}}{\eta_{net}} = \frac{(1.00 \times 10^9 \text{ kW-hr/yr})(3,600 \text{ kJ/kW-hr})}{0.28}$$

$$= 1.286 \times 10^{13} \text{ kJ/yr}$$

For the combined-cycle plant at 48 percent efficiency, the fuel requirement is

$$E_{fuel} = \frac{(1.00 \times 10^9)(3,600)}{0.48} = 7.500 \times 10^{12} \text{ kJ/yr}$$

---

6 Manufacturers of gas turbines prefer to use the convention of *lower heating value (LHV)* to quantify fuel energy content. This results in efficiency values that are about 10 percent higher than those reported here (that is, about 55 percent or more for current combined cycles). Chapter 12 discusses the difference between higher and lower heating values in greater detail.

The fuel energy savings are thus

$$(1.286 \times 10^{13}) - (7.500 \times 10^{12}) = 5.357 \times 10^{12} \text{ kJ/yr}$$

The fuel cost savings are therefore

$$(5.357 \times 10^{12} \text{ kJ/yr})\left(\frac{\$2.60}{10^6 \text{ kJ}}\right) = \$1.393 \times 10^7/\text{yr}$$

$$= \$13.9 \text{ million/yr}$$

A more complete economic analysis of these systems would also consider differences in their capital costs, maintenance costs, and other factors. Chapter 13 discusses economic analysis methods in detail.

5.5   REDUCING ENVIRONMENTAL IMPACTS

Three basic approaches are available to reduce the environmental impacts of power plants. The first involves technological measures to control or remove a pollutant before it is released to the environment. This approach is sometimes referred to as "end of pipe" treatment. Although this is less desirable than pollution prevention approaches, which avoid creating the problem in the first place, the use of environmental control technology is nonetheless a vital and widely used method for substantially reducing environmental impacts at both new and existing facilities.

The second approach is the application of "green design" principles to increase the efficiency of electric power generation. Higher efficiency means less primary energy is needed to generate a desired amount of electricity; environmental impacts are reduced proportionally. This approach is most fruitful when designing new facilities, although modest efficiency gains often can be found at many existing plants. Green design also includes pollution prevention approaches such as innovations that produce useful by-products rather than solid wastes.

The third approach to pollution abatement involves selecting and utilizing cleaner energy sources and alternative technologies with lower environmental impacts. Similar in concept to green design, this option applies to situations where more than one technology can do a given job—in this case, producing a given amount of electricity. This approach to environmental design may sound straightforward, but we shall see that in practice it may be difficult to implement. Nor is it always clear which option is most environmentally preferable.

To illustrate modern methods of reducing environmental impacts, this section focuses on a case study of coal-fired steam electric plants, which supply most of our electricity today. As a result of environmental regulations, the design of such plants has been dramatically affected by the need to generate electricity more cleanly.

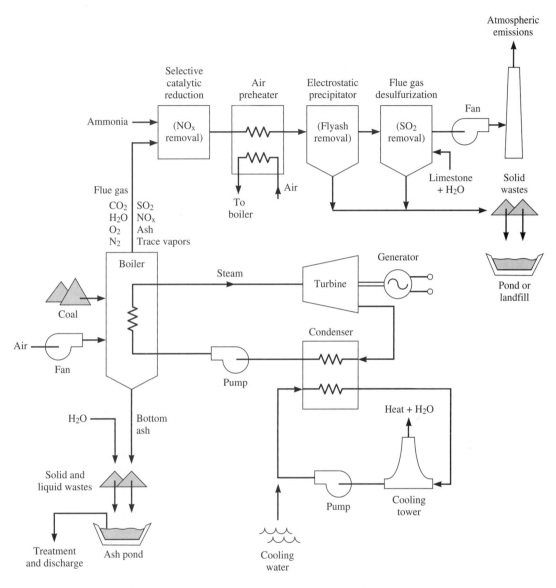

Figure 5.11   Schematic of a coal-fired steam electric plant with environmental controls for major air pollutants, water pollutants, and solid wastes.

5.5.1   Environmental Control Technology

In response to environmental regulation, modern fossil fuel power plants are equipped with a variety of technologies to reduce or eliminate pollutant emissions. Figure 5.11 shows a schematic of a coal-fired plant equipped with emission control systems for particulate matter, SO_2, NO_x, heat, water pollutants, and solid wastes. At the core of the diagram is the steam cycle shown earlier in Figure 5.8. One sees that environmental control technologies add substantial complexity to the overall system. The major technologies shown in Figure 5.11 are described next to highlight the key engineering principles and design parameters related to environmental control.

Particulate Emissions Control   The emission of flyash or soot particles is the oldest and most noticeable form of air pollution from coal combustion. Methods to control these emissions were the first to be developed. Most modern coal-burning plants employ a device called an *electrostatic precipitator* (*ESP*) to capture flyash emissions before they are released to the atmosphere. The basic principle behind an ESP is illustrated in Figure 5.12. A DC electric field is imposed between a wire and a pair of flat plates suspended in the gas flow path. As the flue gas flows between the plates, ash particles in the gas are bombarded with negative ions, taking on a negative electrical charge. Electrostatic attraction then pulls the charged particles toward the positively charged plates, where they are collected. At intervals the electric field is momentarily relaxed, allowing the particles to fall into a collection hopper. Commercial ESPs contain many plates to ensure that all the gas is treated.

The most common measure of ESP performance is the overall particulate removal efficiency, defined as the mass ratio of particulate matter (PM) flowing out of the ESP (not captured) to the total particulate mass flow entering the device:

$$\eta_{ESP} = 1 - \frac{\dot{m}_{PM,\,out}}{\dot{m}_{PM,\,in}} \tag{5.24}$$

ESPs, first developed in the early 1900s, are today one of the most efficient methods of controlling dust particle emissions, able to achieve removal efficiencies of 99.9 percent or more. The overall removal efficiency is a function of the total plate collector area, A (m^2), the flue volumetric gas flow rate, G (m^3/s), and the effective particle "drift" velocity, w (m/s), perpendicular to the gas flow. These parameters are related by an equation called the modified Deutsch–Anderson equation:

$$\eta_{ESP} = 1 - e^{-\left(\frac{A}{G}w\right)^n} \tag{5.25}$$

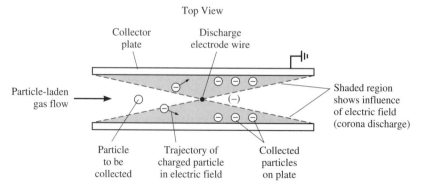

Top View

Collector plate   Discharge electrode wire

Particle-laden gas flow

Shaded region shows influence of electric field (corona discharge)

Particle to be collected   Trajectory of charged particle in electric field   Collected particles on plate

Figure 5.12   Schematic of an electrostatic precipitator.

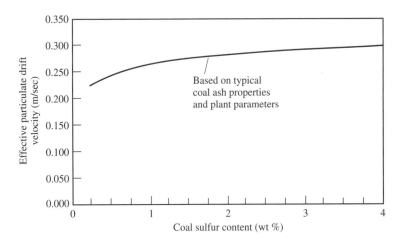

Figure 5.13   Particle drift velocity for electrostatic precipitator design.

The general form of this equation is derived from basic physics by considering the motion of a charged spherical particle in an electric field. The equation is modified by adding the exponent n to account for the fact that real dust particles are neither spherical nor uniform in size. A value of $n = 0.6$ best fits the data for most situations. In addition, the effective drift velocity parameter, w, must be determined empirically. This parameter is a function of the electrical resistivity of the ash particles, which in turn depends on the chemical constituents in ash as well as on the flue gas temperature, humidity, and sulfur content. Computer models have been developed to predict these relationships for different types of coal (Rubin et al., 1997). Figure 5.13 displays typical values of the drift velocity parameter as a function of sulfur content, with all other parameters held fixed at typical values.

The terms in Equation (5.25) also can be rearranged to yield the following expression, assuming a value of $n = 0.6$:

$$\frac{A}{G} = \frac{1}{w}\left[\ln\left(\frac{1}{1 - \eta_{ESP}}\right)\right]^{1.667} \tag{5.26}$$

This form of the equation is useful for calculating the total ESP plate collector area needed to achieve a given overall removal efficiency. The exponent 1.667 is the value of $1/n$, where $n = 0.6$.

Example 5.12

Design of an electrostatic precipitator.   The power plant described earlier in Example 5.4 required a particulate removal efficiency of 99.53 percent to comply with federal emission regulations. Estimate the ESP plate collector area needed for this 500 MW facility, whose design flue gas flow rate is 750 m^3/s.

Solution:

From Example 5.4, the bituminous coal burned at this plant has a sulfur content of 1.5 percent. The effective drift velocity, from Figure 5.13, is therefore estimated at

$$w = 0.28 \text{ m/s}$$

Inserting this value and the desired removal efficiency in Equation (5.26) gives

$$\frac{A}{G} = \frac{1}{0.28}\left[\ln\left(\frac{1}{1 - 0.9953}\right)\right]^{1.667}$$

$$= (58.67 \text{ m/s})^{-1}$$

This ratio is known as the specific collection area (SCA). Multiplying it by the gas flow rate gives the total plate collector area required:

$$A = (A/G) \times G$$

$$= (58.67 \text{ s/m})(750 \text{ m}^3/\text{s})$$

$$= 44{,}000 \text{ m}^2$$

In practice, some degree of overdesign is often incorporated to account for future changes in fuel quality or plant operation.

Notice that the total collection area required is quite large. In the example above, the total plate area corresponds to a square whose sides are each roughly the length of two American football fields! To achieve this large surface area, ESPs use a large number of individual collector plates spaced about 30 cm apart, with both sides of each plate serving as collectors. The final ESP housing can be as large as a two- or three-story building.

Modern ESPs can be designed to remove nearly all of the flyash particles, as can other types of efficient particulate collectors such as fabric filters (also known as a *baghouse,* named for the structure that houses a set of filter bags, and that operates much like a large vacuum cleaner). Particles that escape collection, however, are typically very fine particles in the submicron size ranges, which are of increasing environmental concern (see Chapter 2). Thus there is continuing interest and ongoing study of the particulate control technology needs for electric power plants.

Sulfur Dioxide Control   We saw earlier that coal combustion is the main source of SO_2 emissions. Many power plants reduce SO_2 emissions by switching to low-sulfur coals. Given the wide ranges in fuel sulfur content (see Table 5.1), this is often the most economical solution when modest reductions in SO_2 are required. Other plants find it more economical to burn cheaper high-sulfur coals and install an SO_2 removal system like the one illustrated in Figure 5.11. Such *flue gas desulfurization (FGD)* systems have been required on all new coal-fired plants constructed in the United States since 1978 (see Table 5.3). FGD technology achieves the highest levels of SO_2 removal.

The most widely used FGD technology employs a slurry of pulverized limestone mixed with water to remove SO_2 via chemical reactions that take place in a vessel commonly known as a *scrubber.* Limestone is calcium carbonate, $CaCO_3$, the same chemical used in antacids to neutralize an upset stomach. Gardeners also use limestone to neutralize acidic soils. In a scrubber, a mixture of limestone and water is sprayed into the SO_2-laden combustion gas. SO_2 is removed according to the following reaction:

$$SO_2 + (CaCO_3 + 2H_2O) + \tfrac{1}{2}O_2 \rightarrow CO_2 + (CaSO_4 \cdot 2H_2O) \qquad (5.27)$$
$$\text{Limestone slurry} \qquad\qquad\qquad\qquad \text{Gypsum}$$

A small amount of oxygen is introduced in this reaction so that the product formed is hydrated calcium sulfate, otherwise known as gypsum. Equation (5.27) shows that theoretically a stoichiometric ratio of one mole of calcium is required to remove one mole of sulfur. In practice, however, there is not enough time available to complete this equilibrium reaction, so the actual SO_2 removal efficiency is always less than 100 percent. Some amount of excess reagent typically is needed to obtain a high level of SO_2 removal.

We can define the overall SO_2 removal efficiency in the scrubber as follows:

$$\eta_{FGD} = \frac{\text{Mass of } SO_2 \text{ removed in the scrubber}}{\text{Mass of } SO_2 \text{ into the scrubber}} \qquad (5.28)$$

The overall FGD removal efficiency depends on a number of process parameters that influence the mass transfer rate at which SO_2 can be removed from the gas stream. Improvements in the design and understanding of FGD chemistry over the past two decades have yielded systems that can achieve over 95 percent SO_2 removal using limestone and 98 to 99 percent removal using lime (CaO), which is more chemically reactive (but also more expensive) than limestone. The following example illustrates the use of an FGD system for SO_2 removal.

Example 5.13

SO_2 removal with an FGD system. To comply with national acid rain control requirements, the power plant described in Example 5.3 decides to install a wet limestone FGD system to reduce its current SO_2 emissions by 95 percent. The FGD system requires a reagent stoichiometry of 1.03 (that is, 3 percent more limestone than theoretically needed to remove a mole of SO_2). Calculate for this plant (a) its new annual SO_2 emissions and (b) the annual quantity of limestone required.

Solution:

(a) From Example 5.3, the current SO_2 emissions are 2.910×10^7 kg/yr. The FGD system will reduce emissions by 95 percent. Thus the new emissions will be 5 percent of the current value:

$$SO_2 \text{ removed} = (0.95)(2.910 \times 10^7) = 2.765 \times 10^7 \text{ kg/yr}$$

$$SO_2 \text{ emissions with FGD} = (0.05)(2.910 \times 10^7) = 1.46 \times 10^6 \text{ kg/yr}$$

(b) Equation (5.27) shows that the theoretical reagent requirement is 1.00 mol Ca/mol SO_2 removed, versus an actual requirement of 1.03 mol Ca/mol SO_2. The molecular weights are

$$SO_2 = 32 + 2(16) = 64$$

$$CaCO_3 = 40 + 12 + 3(16) = 100$$

Thus a stoichiometric ratio of 1.03 corresponds to a mass ratio of

$$\frac{m_{CaCO_3}}{m_{SO_2}} = \left(1.03 \frac{\text{mol Ca}}{\text{mol } SO_2}\right)\left(\frac{100 \text{ kg } CaCO_3/\text{mol Ca}}{64 \text{ kg } SO_2/\text{mol } SO_2}\right)$$

$$= 1.6094 \frac{\text{kg } CaCO_3}{\text{kg } SO_2 \text{ removed}}$$

The annual mass of limestone required is thus

$$m_{CaCO_3} = (1.6094)(2.765 \times 10^7)$$

$$= 4.45 \times 10^7 \text{ kg CaCO}_3/\text{yr}$$

As of 1997, approximately 27 percent of U.S. coal-burning plants (80,400 MW) were equipped with FGD systems. The relatively high cost of FGD technology has been one of the major factors limiting its deployment for SO_2 removal. Environmentally, a major drawback of FGD is the solid waste produced by most current systems. Solid waste management strategies are discussed a little later in this section.

Nitrogen Oxide Control   Tables 5.2 and 5.3 showed typical levels of NO_x emissions for different types of boilers and fuels used for electric power generation, along with current federal emission limits. To reduce the level of NO_x emissions, two general methods are used. The first method, known as *combustion modification,* alters the design of the burners and/or combustion chamber to affect the temperature, time, and other parameters that control NO_x formation. Reengineered burner designs are generically known as "low-NO_x burners." Typically, combustion modification methods reduce NO_x up to about 50 percent.

The second method of NO_x control employs a chemical treatment system to remove NO_x from the flue gas. This type of system is analogous to the FGD system described earlier for SO_2 removal. Worldwide, the most prevalent postcombustion NO_x control technology is *selective catalytic reduction (SCR).* As the name implies, a chemical catalyst is used to help selectively reduce nitrogen oxides in the flue gas back to molecular nitrogen. Catalytic converters used on automobiles reduce NO_x emissions in a similar fashion (see Chapter 3). In power plants, however, ammonia (NH_3) is injected into the flue gas stream to achieve higher levels of NO_x reduction via the following chemical reactions:

$$NO + NH_3 + \tfrac{1}{4}O_2 \rightarrow N_2 + \tfrac{3}{2}H_2O \tag{5.29}$$

$$NO_2 + 2NH_3 + \tfrac{1}{2}O_2 \rightarrow \tfrac{3}{2}N_2 + 3H_2O \tag{5.30}$$

SCR operation requires temperatures of approximately 400°C, so the SCR reactor is usually located at the boiler exit as shown in Figure 5.11. In practice, some excess ammonia is required to obtain desired levels of NO_x removal. Current SCR system designs typically achieve 70 to 90 percent NO_x removal. Typically, SCR is used in conjunction with low-NO_x burners or other combustion modification methods so as to minimize the overall cost of emissions control.

A technology known as *selective non catalytic reduction (SNCR)* employs the same basic chemistry as SCR but without the aid of a catalyst. These systems are lower in cost, but also less efficient in NO_x removal (typically about 50 percent). For optimal performance, ammonia is injected directly into the upper portion of the furnace.

Example 5.14

NO$_x$ control technology requirements.   The U.S. Environmental Protection Agency recently revised the New Source Performance Standard (NSPS) for NO$_x$ emissions from all future *new* power plants. For coal-fired plants built after 1997 the new NSPS corresponds to 65 ng/J (0.15 lb/MBtu). What percentage reduction in NO$_x$ emissions would be required for bituminous coal-fired boilers relative to the previous NSPS for units built after 1978? What current technology would be able to meet the new 1997 standard?

Solution:

From Table 5.3, the NSPS for bituminous-fired units built after 1978 is a NO$_x$ limit of 260 ng/J (0.6 lb/MBtu), and the new standard is equivalent to 65 ng/J. This would require a further NO$_x$ emissions reduction of

$$\text{NO}_x \text{ reduction} = \frac{260 - 65}{260} = 0.75, \text{ or a 75\% reduction}$$

Based on the discussion in the text, this level of reduction could not be met with combustion modification methods or SNCR alone, which reduces emissions by up to about 50 percent. Instead, postcombustion removal employing selective catalytic reduction (SCR) technology would be required. In many cases a combination of combustion modifications and SCR could be the least costly way of achieving the required emission reduction.

Prior to 1998, environmental regulations in the United States did not require the relatively high levels of NO$_x$ reduction afforded by SCR technology. Although a handful of SCR units have been installed on coal-fired plants in the United States in response to state and local regulations, the most widespread applications worldwide have been in Germany and Japan, where NO$_x$ reduction requirements are far more stringent. U.S. regulations, however, are in the process of changing. To achieve ambient air quality standards for ozone, the U.S. Environmental Protection Agency (EPA) estimates that NO$_x$ emissions from existing power plants must be reduced by up to 85 percent below current levels. Thus many U.S. utilities are now planning to employ SCR on existing power plants, and the use of this technology will grow substantially in the years ahead.

Water Pollution Controls   The major water pollutant traditionally associated with steam electric power plants is thermal pollution. Waste heat from the power plant condenser is typically transferred to cooling water drawn from a nearby stream, lake, or river. In passing through the condenser, the cooling water temperature increases by about 10°C. In a simple once-through cooling system, this warmed water is discharged back to the stream, lake, or river, producing the environmental effects described in Chapter 2.

The heat released by the condensing steam is the difference in total steam enthalpy entering and exiting the condenser. If the mass flow of steam is $\dot{m}_{steam}$, the rate of heat transferred to the cooling water is

$$\text{Waste heat released} = \dot{m}_{steam} (\text{h}_{in} - \text{h}_{out})_{condenser} \qquad (5.31)$$

This is the rate of waste heat transferred from the steam to the cooling water and subsequently released to the environment.

When water pollution regulations prohibit the dumping of waste heat to water bodies, the heat is instead transferred to the atmosphere using a *cooling tower*. A common type of power plant cooling tower is a large parabolic structure like the one shown in Figure 5.14. Warm cooling water exiting the condenser is pumped into the tower at a height of about 10 to 15 m, where it is sprayed into the interior of the tower. As the water falls to the base of the tower, it releases heat by evaporation in much the same way that you might cool a cup of hot coffee by pouring it from one cup to another. Heat given up to the surrounding air causes the warmed air to rise due to natural buoyancy. The parabolic shape of the tower provides a natural draft chimney that allows cooler ambient air to enter the base of the tower and flow upward, carrying away the waste heat. At the bottom of the tower the cooled water is collected and pumped back to the condenser in a closed cycle.

Although parabolic cooling towers are often associated with nuclear power plants, all types of steam-generating power plants employ this technology. Many power plants use smaller forced draft cooling towers with fans to blow cooling air through the structure. The release of this waste heat to the atmosphere does not pose any major environmental concern, but the release of moisture entrained in the warm air exiting a cooling tower often produces a visible plume of white clouds as the moisture condenses. In some cases, especially on cold days, the vapor plume from a cooling tower can produce a low-lying fog near the power plant. Thus the cooling towers must be located away from nearby roadways, where visibility could be affected.

In addition to thermal discharges, power plants also are a potential source of other water pollutants such as suspended solids, acidity, and heavy metals. These pollutants originate mainly from the ash collected at coal-burning plants. In most plants, ash is continuously collected and mixed with water to transport the ash to a

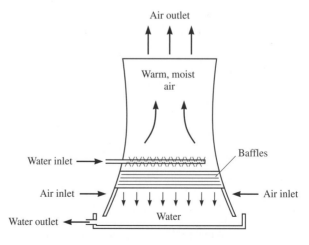

Figure 5.14   Schematic of a power plant cooling tower employing a hyperbolic natural draft design.

disposal site, typically a large pond. The water discharged from that pond carries with it traces of chemical constituents found in the ash. As discussed in Chapter 2, the National Pollutant Discharge Elimination System (NPDES) restricts the release of such pollutants to the environment. Most power plants achieve NPDES limits by allowing ash particles to settle out in an ash pond, leaving an effluent water stream that is relatively free of contaminants. Some plants, however, require chemical treatment systems to achieve NPDES permit conditions. Most commonly, chemicals such as lime are added to neutralize acidic waters prior to discharge.

The leaching of metals from the bottom of an ash pond also is a potential source of groundwater contamination. To control such problems, ash ponds are constructed with an impermeable layer of natural or synthetic materials such as clay, compacted soil, or rubber linings to prevent the infiltration of contaminants into the ground.

Solid Waste Management   The large quantities of ash collected at coal-burning power plants represent a major solid waste management problem that is being addressed via several methods. For plants with large ash ponds, the final ash disposal site is often the pond itself. In other cases, the pond is periodically dredged, and the ash is trucked to an engineered landfill site. In recent years many utilities have begun to utilize dry flyash collection systems to avoid the use of water altogether. The collected ash is disposed of directly in an engineered landfill.

At plants equipped with flue gas desulfurization (FGD) systems, the most common disposal practice is to mix the FGD waste with collected ash. The ash properties yield a cementlike mixture that is less susceptible to leaching and is disposed of in an engineered landfill.

Some power plants are able to recycle or reuse ash and FGD wastes to minimize or avoid waste disposal problems. Environmentally this is clearly the most attractive approach. About 30 percent of the roughly 100 million tons of ash collected at U.S. power plants each year is sold and used as aggregate for road construction, additives in cement manufacturing, or other useful purposes. Research programs have demonstrated other potential uses for waste materials, but in most instances it is still cheaper to use conventional raw materials rather than power plant by-products. Thus, there are major challenges (opportunities) for technology innovations to more fully utilize power plant by-products.

A similar situation applies to FGD wastes. With some additional processing to remove impurities, the gypsum produced in modern FGD systems can be sold as a commercial by-product to produce wallboard or other building materials; this is common practice in Japan and Germany. In the United States, however, supplies of natural gypsum are relatively plentiful, as is land for waste disposal. Thus for most U.S. power plants it is usually cheaper to dispose of the FGD gypsum as a solid waste in a landfill rather than process and ship it to a manufacturing plant. However, growing environmental pressures and the increasing cost of waste disposal are beginning to provide stronger incentives for productive use of solid waste from power plants. Recently, for example, new wallboard manufacturing plants have been built adjacent to several U.S. power plants in order to utilize the gypsum from existing FGD systems.

Example 5.15

Solid waste from an FGD system.   Calculate the mass of solid waste produced each year by the flue gas desulfurization (FGD) system described in Example 5.13. Assume that the FGD waste consists of gypsum produced by the chemical reaction in Equation (5.27) plus unreacted limestone. Further assume that the final waste product contains 15 percent moisture (water) by weight, typical of the dewatering technology used to separate FGD solids from the wet scrubber slurry.

Solution:

From Example 5.13, the mass of SO_2 removed by the FGD system is 2.765×10^7 kg/yr. From Equation (5.27), each mole of SO_2 removed results in one mole of gypsum ($CaSO_4 \div 2H_2O$). (Note that the H_2O in this expression is part of the crystalline structure of the gypsum and is referred to as *water of hydration*). Based on the atomic weights of its constituents (Ca = 40, S = 32, O = 16, H = 1), the molecular weight of gypsum is

$$40 + 32 + 4(16) + 2(2 + 16) = 172$$

Because one mole of SO_2 removed produces one mole of gypsum, the mass ratio of gypsum to SO_2 removed is proportional to the molecular weights. Thus

$$\frac{m_{gypsum}}{m_{SO_2}} = \frac{172}{64} = 2.6875 \frac{\text{kg gypsum}}{\text{kg } SO_2 \text{ removed}}$$

We must also account for unreacted limestone. From Equation (5.27) the SO_2 removal reaction requires one mole of $CaCO_3$ per mole of SO_2 removed. On a mass basis this theoretical amount is $100/64 = 1.5625$ kg $CaCO_3$/kg SO_2 removed.

From Example 5.13, the actual limestone requirement is 1.6094 kg/kg SO_2 removed, or 3 percent more than the amount theoretically needed. The excess (unreacted) reagent is therefore

$$\text{Unreacted } CaCO_3 = 1.6094 - 1.5625 = 0.0469 \frac{\text{kg } CaCO_3}{\text{kg } SO_2 \text{ removed}}$$

The total dry waste produced is the sum of gypsum plus unreacted limestone:

$$\text{Total dry waste} = 2.6875 + 0.0469 = 2.7344 \frac{\text{kg dry waste}}{\text{kg } SO_2 \text{ removed}}$$

The final wet waste for disposal contains 15 percent moisture (also known as occluded water), which is mixed with the dry solids. This means that there is one kg of wet waste for every 0.85 kg of dry waste. Thus

$$\text{Total wet waste} = \frac{2.7344}{0.85} = 3.2169 \frac{\text{kg wet waste}}{\text{kg } SO_2 \text{ removed}}$$

Note that up to this point we have expressed all waste quantities in terms of "per kg SO_2 removed." This provides a generalized result that can be applied to different SO_2 removal rates. Multiplying the above result by the mass of SO_2 removed gives the total annual mass of FGD waste disposed:

$$\dot{m}_{waste} = (3.2169)(2.765 \times 10^7)$$

$$= 8.89 \times 10^7 \, \text{kg FGD waste/yr}$$

5.5.2 Improving Energy Efficiency

A key means of reducing environmental impacts is to improve the efficiency of the energy conversion process. If less primary energy is needed to produce electricity, the magnitude of environmental impacts also will be reduced. We examine here some technical limitations and opportunities for improving the overall energy efficiency of fossil fuel plants.

Increased Operating Temperatures   Large improvements in the efficiency of electric power generation are not easy to obtain. The laws of thermodynamics pose a substantial barrier for power plants using conventional Rankine or Brayton cycles, guaranteeing an efficiency of less than 100 percent. How much less depends on the *absolute temperature* (K) of the working fluid and the environmental surroundings. It can be shown that the maximum theoretical efficiency is given by this relationship:

$$\eta_{max} = 1 - \frac{T_o}{T} \qquad (5.32)$$

This is the efficiency of an ideal cycle (called a *Carnot cycle* after the French scientist Sadi Carnot) operating between the maximum fluid temperature, T, and the environmental surrounding temperature, T_o. These temperatures are often referred to as the *hot* and *cold sinks* or *reservoirs*.

A Carnot cycle efficiency.   Calculate the maximum theoretical efficiency of a power plant using steam at a peak temperature of 560°C, assuming an ambient temperature of 20°C.

Solution:

First convert the given temperatures to absolute temperature on the kelvin scale:

$$\text{Steam temperature} = 560 + 273 = 823 \text{ K}$$

$$\text{Surroundings temperature} = 20 + 273 = 293 \text{ K}$$

From Equation (5.32),

$$\eta_{max} = 1 - \frac{T_o}{T}$$

$$= 1 - \frac{293}{833}$$

$$= 0.648, \text{ or } 64.8\%$$

Real power plants operate well below the theoretical maximum because of inefficiencies and irreversibilities that are unavoidable in real devices. Nonetheless, a major insight from Equation (5.32) is that the overall efficiency may be improved by increasing the maximum temperature of the working fluid. Over the past several decades, engineering improvements in boiler and turbine designs have increased efficiency via this method. For example, gas turbines today operate at inlet temperatures of approximately 1,300°C, compared to about 1,100°C a decade ago. Each 55°C increment adds approximately one percentage point to the overall efficiency of a simple cycle power plant. For steam electric plants burning coal, the use of so-called supercritical boilers currently can achieve net plant efficiencies of about 42–43 percent (including full environmental controls) by increasing the steam pressure and temperature to 31 MPa and 590°C, compared to typical conditions of 17 MPa and 540°C for conventional boilers with net efficiencies of about 35 percent. Advanced boilers of this type are being used in Japan and Europe (where coal costs are high), but are not found in the United States because of their higher capital cost.

In all cases, the maximum temperature of the power generation cycle is limited primarily by the construction materials available for boilers and turbine blades. Continued advances in materials science and engineering are thus critical to achieving significant advances in power generation efficiency.

Cogeneration   A question that naturally arises in the study of steam cycles is whether it is really necessary to dump all the condenser heat into a nearby water body. Can't that heat be recovered for some useful purpose?

The answer to this question is yes. For instance, the heat released by the condensing steam could provide hot water to nearby residences or industries, saving on their fuel bills. If higher temperatures are needed, steam can be extracted from the turbine at higher temperatures and pressures. This type of combined heat and power (CHP) system is known in the United States as *cogeneration*. It raises the overall efficiency of fuel use significantly by utilizing heat that would otherwise be dumped directly to the environment.[7]

An economic price must be paid, however, for the additional piping and other equipment needed to supply the steam or hot water to customers outside the power plant. In most situations, this cost far exceeds the value of the heat supplied because most potential customers are remote from the power plant. Thus cogeneration systems have found only limited applications in the United States, typically at large industrial complexes that require both heat and electrical power continuously. New opportunities continue to arise, however, especially with the increased use of smaller gas-fired plants located closer to potential customers.

Elsewhere in the world, as in some of the more densely populated regions of Europe, cogeneration systems are more widely used to provide more efficient energy services. Commercial technology includes the use of reciprocating (diesel) engines rather than steam turbines for stand-alone applications such as office buildings or shopping centers. Cogeneration systems typically increase the efficiency of fuel utilization to 70 or 80 percent. However, such systems are effective only when there are simultaneous demands for both heat and power in the proportions delivered by the system. In particular, a year-round demand for steam is often a limiting factor in the viability of cogeneration technology.

---

7 This does not violate the laws of thermodynamics because the heat eventually winds up in the environment. First, however, it may warm a home or office building or heat water used for washing. In this way an energy service is provided before the heat is released to the environment more diffusely and at a lower temperature.

Advanced Cycles   The use of advanced cycles is another important way to increase the efficiency of electric power generation. A number of such cycles are under active development by the U.S. Department of Energy and other organizations (NRC, 1995). Two advanced systems are highlighted here.

Integrated Gasification Combined Cycle (IGCC)   One promising method to make coal-based power systems more efficient is a combined-cycle system integrated with *coal gasification*. In this system, depicted in Figure 5.15, coal reacts chemically with steam and oxygen at high temperature and pressure to form a *fuel gas* mixture consisting mainly of hydrogen (H_2), carbon monoxide (CO), and methane (CH_4). Oxygen levels in the gasifier are kept low to create a chemically reducing environment that forms the fuel gas, rather than the H_2O and CO_2 that are produced when coal is burned (oxidized). Impurities such as sulfur and ash are removed by efficient environmental control systems, leaving a clean fuel gas whose energy content is about 15 to 20 percent that of natural gas.

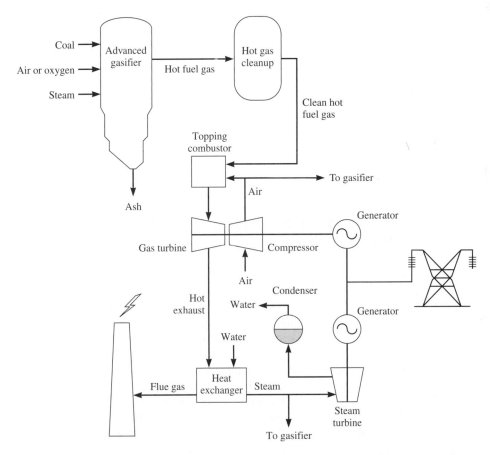

Figure 5.15   Schematic of an integrated coal gasification combined-cycle power plant. (*Source:* USDOE, 1996)

This clean fuel gas is then used to generate electricity in a combined-cycle system similar to the one shown earlier in Figure 5.10. Several recently constructed IGCC plants have achieved overall thermal efficiencies of approximately 40 percent, and further process improvements, coupled with advanced (higher-temperature) gas turbines, are expected to raise this to around 50 percent over the next decade (NRC, 1995). A major attraction of IGCC systems is that the fuel gas produced in this process also can be used to synthesize liquid fuels and chemicals as well as produce electricity. Advanced IGCC systems can thus serve as a "coal refinery" producing a variety of products. Environmentally, IGCC systems also reduce solid wastes by converting sulfur impurities into sulfuric acid or elemental sulfur byproducts. At present, the principal drawback of an IGCC system is its high capital cost.

Integrated Gas Turbine Fuel Cell   The highest-efficiency plants under development combine advanced gas turbines with advanced fuel cells. This concept is still in the early developmental stages, but efficiencies of more than 80 percent are projected using natural gas as an energy source. A high-temperature fuel cell, of the type described later in this chapter, would first generate electricity electrochemically. The exhaust gas from the fuel cell would then drive a gas turbine, generating additional power. An alternative design using coal would first employ a coal gasifier to generate a clean fuel gas for the fuel cell. Another advantage of gasification is that if offers an efficient route for separating and capturing CO_2 as a potential carbon management strategy to address the problem of global warming (see Chapter 12). Although such systems are at least a decade or two away from commercialization, they promise to nearly double the efficiency of electric power generation and substantially reduce environmental impacts. Indeed, the Department of Energy's Vision 21 program has the goal of developing such a system with zero (or near-zero) emissions by 2015 (NRC, 2000).

5.6   ALTERNATIVE ENERGY SOURCES AND TECHNOLOGY

A general approach to reduce the environmental impacts of electric power plants is to change the energy source or technology used to generate electricity. Indeed, as one delves deeper into power generation cycles, energy sources, and environmental control technologies, the number of options for electric power production expands considerably beyond the fossil fuel systems considered so far. Table 5.7 provides a more detailed list of candidate power generation options.

In this section we briefly discuss a number of these alternative technologies for generating electricity. Because the subject is so broad, no attempt is made to cover all options. Rather, the objective is to highlight fundamental principles and distinguish the key features of alternative energy technologies. From that point of view, the various options described fall into two general categories: (1) alternative energy sources and working fluids for driving turbine-based electromechanical generators and (2) direct energy conversion methods that produce electricity without the intermediate step of a turbine generator.

Table 5.7 Candidate electric power generation technologies.

Combustion-Based Power Generation and Environmental Control	Gasification-Based Power Generation and Environmental Control	Renewable Electric Power Generation and Environmental Control

Fuels: Coal, natural gas, oil, refuse-derived fuels, municipal solid waste, biomass, shale, etc.
 	
Types of Technologies:
 	
Gas turbine-based systems
  Cycle
    Simple cycle
    Combined cycle
    Steam-injected
    Other
  Type of gas turbine
    Aero-derivative
    Heavy-duty
Mass-burn combustors
Pulverized coal combustors
Slagging combustors
Fluidized bed combustors
  Atmospheric
  Pressurized
  Circulating
Internal combustion engines
Fuel cells
  Phosphoric acid
  Sodium carbonate
  Solid oxide
 	
Environmental Controls for Combustion-Based Systems
 	
Gas turbine-based systems
  Wet injection
  Selective catalytic reduction
  CO catalyst
  Premixed lean-burn combustor
  Staged rich/lean combustor
  Catalytic combustion
Coal preparation
  Physical coal cleaning
  Chemical coal cleaning
  Combustion-based approaches
Sorbent injection
Low-NO$_x$ burners
	Fuels: Coal, biomass, petroleum coke, sewage sludge, etc.
	
	Types of Technologies:
	  Gasification technologies are modular, and are defined by the design and configuration of key components, including the type of cycle, gasifier, oxidant, gas cleanup system, and by-product options.
	
	Cycles
	  Integrated gasification
	  Combined cycle
	  Integrated gasification steam-injected gas turbine
	  Integrated gasification humid air turbine
	  Integrated gasification compressed air storage and humidification
	  Integrated gasification fuel cells
	Types of gasifiers
	  Fixed bed
	  Fluidized bed
	  Entrained-flow
	  Multiple stages/hybrids
	Oxidant
	  Oxygen
	  Air
	  Partial air separation
	  Highly integrated air separation
	Environmental
	  Cold gas cleanup
	    Gas cooling
	    Particulate scrubbing
	    Solvent-based acid gas removal
	    Gasifier in-bed sulfur capture
	    Elemental sulfur recovery
		Solar
		  Photovoltaics
		    Crystalline silicon cells
		    Amorphous silicon compound semiconductors
		  Solar steam (thermal)
		    Parabolic dishes/central receivers
		  Hybrid solar/natural gas systems
		  Battery and other energy storage systems
		  On- vs. off-grid
		  Life cycle: manufacture, disposal
		Geothermal
		  Direct steam
		  Single vs. dual flash
		  Binary
		  Hydrogen sulfide control
		  Water treatment/handling
		Wind turbines
		  Decentralized vs. wind farms
		  On- vs. off-grid
		  Backup and storage systems
		Hydroelectric
		  Large scale
		  Micro-hydro (rural)
		Ocean-based
		  Ocean thermal energy
		  Conversion
		  Tidal systems
		  Wave-driven systems
		  Current-driven systems
		Biomass (see combustion and gasification-based systems)

(continued)

Table 5.7 (continued)

Combustion-Based Power Generation and Environmental Control	Gasification-Based Power Generation and Environmental Control	Nuclear-Based Electric Power Generation and Environmental Control
Other low NO_X (e.g., overfire air)	Hot gas cleanup	*Fission-based energy conversion systems*
Postcombustion approaches	Barrier filtration	Boiler water reactors
Particulate control	Dry, regenerable acid gas removal	Pressurized water reactors
Electrostatic precipitator	Gasifier in-bed sulfur capture	Heavy water reactors
Fabric filters	By-product vs. throwaway	Gas-cooled reactors
Flue gas desulfurization	Coproduction	Breeder reactors
Throwaway	Methanol	*Fission-based safety/ environmental*
Regenerable	Hydrogen	Passive safety design
Dry regenerable combined SO_2/NO_x	Carbon dioxide	Design standardization
Highly integrated combined removal systems (e.g., $SO_2/NO_x/PM$)	Ammonia	Uranium mining
	Synthetic natural gas	Spent fuel disposal/recycling
Solid waste		Plant decommissioning
Disposal options		
By-product recovery		
Miscellaneous		
CO_2 scrubbers		
Retrofit vs. new		

Source: Frey et al., 1995.

5.6.1   Nuclear Energy

Nuclear power plants are steam electric generators that use uranium rather than fossil fuels to produce the steam that drives a turbine generator. Figure 5.16 shows a simple schematic of a common nuclear-powered steam electric plant. The right portion of the diagram is identical to the Rankine cycle discussed earlier. The main difference is that in a nuclear power plant the steam is at much lower pressure and temperature compared to fossil fuel plants (typically 7 MPa and about 400–600°C). This means that nuclear plants are less efficient, with net thermal efficiencies typically in the range of 30 to 32 percent.

In the pressurized water design shown in Figure 5.16, energy released from nuclear fission reactions heats a circulating stream of water known as the *primary coolant.* This water is kept at a pressure of about 15 MPa to keep it from boiling (hence the term *pressurized water reactor* or *PWR*). The pressurized hot water flows through a heat exchanger to generate steam for the power cycle in a secondary loop. Other nuclear plant designs operate the primary water coolant loop at a lower pressure (about 7 MPa), allowing the water in the reactor core to boil (this is a *boiling water reactor* or *BWR*). Saturated steam from the reactor is then fed directly to a steam turbine, elimi-

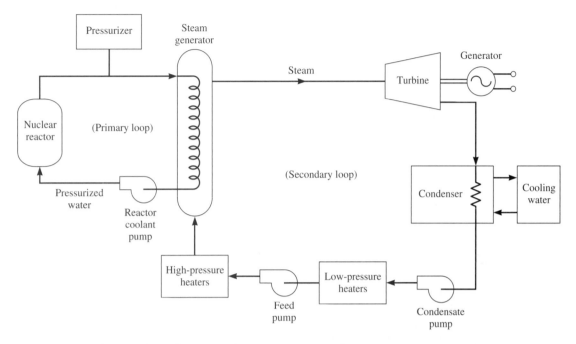

Figure 5.16   Schematic of a nuclear power plant using a pressurized water reactor.

nating the steam generator and secondary loop shown in Figure 5.16. Other nuclear plant designs use a gas such as helium to cool the reactor and supply heat for the steam cycle. Nearly all U.S. nuclear power plants, however, employ a PWR or BWR design.

From an environmental point of view, the key focus of nuclear plant designs is to keep radioactive materials from being intentionally or accidentally released to the environment. The reactor core is the main source of highly radioactive materials.

The core consists of assemblies of hundreds of nuclear fuel rods, each of which resembles a long, thin pipe into which cylindrical pellets of uranium oxide fuel are inserted. A key safety feature is the use of control rods of neutron-absorbing materials that can shut down the nuclear chain reaction described earlier in Section 5.2.2. Various types of shielding are employed, together with safety systems designed to ensure against overheating of the reactor core. Readers interested in more details about nuclear plant designs and safety issues can find a great deal of literature on the subject. As discussed in Chapter 2, a critical environmental issue still unresolved is how to dispose of the high-level radioactive wastes (spent fuel) now being temporarily stored aboveground at power plants across the country.

5.6.2   Biomass and Refuse Energy

Long before fossil fuels were discovered or electric power plants invented, biomass was the principal energy source for heating, cooling, and early industrial processes. Even today a large fraction of the world's population relies on biomass for their primary energy needs. Biomass fuels include wood, grain, and herbaceous crops, as well as animal and crop residues. Many types of urban or municipal solid wastes (MSW) and their by-products also may be classified as biomass, especially wastepaper, cardboard, and other materials produced from crops or wood products.

For electric power generation, biomass and waste fuels represent another potential source of energy that can generate steam for a conventional steam electric power plant. U.S. capacity of such plants totaled 16,700 MW in 1997 (USDOE, 1998a). Biomass also can be processed in a gasifier to produce a fuel gas capable of powering a gas turbine. The capture and use of landfill gas to power small gas turbines also has emerged in recent years as another way of utilizing biomass-based fuels.

Most biomass fuels, including MSW, have an energy content roughly half that of coal, in the range of approximately 10,000–15,000 kJ/kg. The use of biomass as a boiler fuel thus requires a much larger throughput of material to provide a given amount of energy. Operationally, the use of biomass fuels, especially municipal wastes, can also introduce technical problems related to the high moisture content and chemical impurities in wastes. Some power plants therefore burn biomass or refuse-derived fuel (RDF) only in small amounts (for example, no more than 10 to 15 percent of total energy input) as a supplemental fuel mixed with coal.

For the most part, however, biomass and waste fuels are not widely used for electric power generation at this time. Nonetheless, concern over global warming has stimulated interest in biomass for power generation and conversion to other energy forms such as liquid fuels. During their growth phase, trees and other forms of biomass absorb CO_2 from the atmosphere and store it as carbon in plant tissue. Even if this carbon is eventually released as CO_2 when the biomass is used as a fuel, there has been no *net* release of CO_2 to the atmosphere. Thus a managed biomass plantation that can provide useful energy with no net CO_2 emission is one of the concepts being examined, and progress is being made in reducing the acreage required to support a sustainable biomass fuel source as a potential future energy option.

5.6.3   Geothermal Energy

Another energy source for steam electric generators is geothermal energy. This renewable energy source is derived from the molten core of the earth, which transfers heat toward the surface, producing underground deposits of magma, steam, hot water, and hot dry rocks. Such deposits can be found throughout the globe. When they are sufficiently close to the surface (within a few kilometers) and sufficiently hot (above 150°C), it can be economical to drill a well and extract this thermal energy for use in a steam turbine generator.

In the United States, geothermal plants are a commercial source of renewable energy for power generation, with 3,200 MW of installed capacity in 1997. The largest and oldest concentration of geothermal plants is in the Geysers region of northern California. Steam extracted from the wellhead is first treated to remove liquids and other debris that may impair turbine performance. Typical turbine steam inlet conditions are inferior to those of fossil fuel plants, with temperatures of roughly 200°C at pressures of 400 to 800 kPa. Thus geothermal power plants are much less efficient than conventional fossil fuel plants.

Schemes also have been developed or proposed to utilize geopressurized hot water as an energy source for power generation and other energy needs. Although the natural resource base of geothermal energy is potentially very large, its utilization remains limited by the technical difficulties and associated high costs of energy extraction and processing. The environmental impacts of geothermal energy production also are potentially significant, as indicated earlier in Table 5.4. Recent experience in California also shows that long-term exploitation of geothermal wells can deplete or diminish their energy-producing potential. As noted in Chapter 2, many renewable resources are renewable only if utilized at a sustainable level.

5.6.4 Hydroelectric Energy

All of the power generation technologies examined thus far have employed a mechanical turbine generator powered by steam or hot gas. Now we consider a turbine-based system driven by water rather than steam. Hydraulic turbines are the modern incarnation of one of the oldest of energy conversion technologies, the water wheel. Modern hydraulic turbines are designed to more efficiently capture the energy of flowing water at hydroelectric dams and natural waterfalls like Niagara Falls.

Figure 5.17 shows a simple sketch of a hydroelectric facility. Water collected behind a dam is carried through large ducts to the inlet of one or more hydraulic turbines at the base of the dam. The energy available for conversion to electricity is the

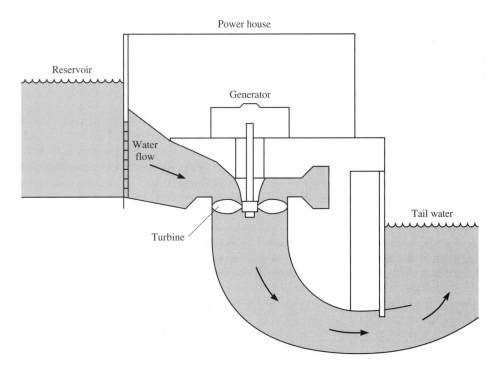

Figure 5.17 Schematic of a hydroelectric power plant.

potential energy of the impounded water (whose initial kinetic energy is usually negligible). Thus

$$\text{Potential energy of water} = m_w g (\Delta z) \tag{5.33}$$

where m_w is the mass of water, g is the acceleration of gravity (equal to 9.807 m/s^2), and Δz is the change in elevation from the surface of the impounded water to the axis of the hydraulic turbine or water outflow surface (depending on the turbine design). The height, Δz, is known as the operating *head*. This value affects not only the theoretical power available but also the turbine design. For example, low-head facilities (Δz less than about 20 m) often use a turbine resembling the propeller of a ship, while other designs are commonly installed at high-head power plants (Δz usually over 100 m) to reduce frictional losses and maximize efficiency.

As stored water falls through the height Δz, potential energy is converted to kinetic energy. The turbine design attempts to efficiently convert this energy flow into rotational energy to spin the generator. Some useful energy inevitably is lost to frictional heating between the moving fluid and the turbine surfaces. Additional friction occurs in the bearings that support the turbine and generator shafts and in the water intake and outlet structures of the power plant. The kinetic energy of water exiting the turbine also represents unrecovered energy. For engineering design, it is important to identify each of these sources of energy loss individually so as to better understand how each can be reduced to improve overall efficiency.

For our purposes, however, we will define and quantify only the overall hydroelectric conversion efficiency, η_{hydro}, which is the ratio of electrical energy output to potential energy input:

$$\eta_{hydro} = \frac{E_{elec}}{m_w g (\Delta z)} \tag{5.34}$$

Often it is more convenient to express this efficiency as a ratio of power rather than energy:

$$\eta_{hydro} = \frac{P_{elec}}{\dot{m}_w g (\Delta z)} \tag{5.35}$$

where P_{elec} is the electrical power output and $\dot{m}_w$ is the mass flow *rate* of water through the turbines. The efficiency of a hydroelectric plant is calculated in the next example.

Example 5.17

Efficiency of a hydroelectric power plant.   A hydroelectric plant has an electrical capacity of 150 MW. The operating head is 35 m, and the water flow rate into the turbines is 500 m^3/s. Determine the overall efficiency of this plant under these conditions.

Solution:

The plant capacity represents electrical power, so the overall efficiency is given by Equation (5.35). First, however, we must determine the mass flow rate of water, given the volumetric flow rate. The density of water is 1 g/cm^3 or 1,000 kg/m^3. Thus

$$\dot{m}_w = (500 \text{ m}^3/\text{s})(1,000 \text{ kg/m}^3) = 500,000 \text{ kg/s}$$

Then

$$\eta_{hydro} = \frac{P_{elec}}{\dot{m}_w\, g(\Delta z)}$$

$$= \frac{150 \times 10^6 \text{ W}}{(500{,}000 \text{ kg/s})(9.807 \text{ m/s}^2)(35 \text{ m})} = \frac{1.50 \times 10^8 \text{ W}}{1.72 \times 10^8 \text{ W}}$$

$$= 0.87, \text{ or } 87\%$$

(As an additional exercise, students may wish to confirm that the denominator in this calculation indeed has the unit of watts.)

A major attraction of hydropower is that water is a renewable energy resource, replenished by the natural cycle of evaporation and precipitation. At the same time, however, there are only a limited number of natural waterfalls and river flows capable of generating significant amounts of electricity. In the United States, the most attractive sites for hydroelectric power plants already have been exploited to provide the 76,700 MW of generating capacity installed as of 1997. Locations still remain where smaller, low-head hydro facilities could be constructed, but such sites are typically in remote locations where the difficulty and expense of transmitting the electricity diminishes the attractiveness of such projects, even in the absence of environmental concerns. However, elsewhere in the world, especially in the developing countries of Asia, substantial growth in hydroelectric power is planned over the next few decades (USDOE, 1998b). The environmental impacts of such developments were discussed earlier in Section 5.2.3.

5.6.5 Wind Energy

Just as early civilizations used the energy of falling water to extract useful work, they also captured the energy of wind using rudimentary turbine blades to turn a shaft and perform useful work. Devices such as the anemones used in ancient Persia for grinding grain were one of the earliest technologies, later surpassed by the more familiar windmills found in various parts of Europe.[8] Today's wind turbines are far more efficient at converting the kinetic energy of air into the mechanical energy of rotation needed to drive an electric generator. Wind turbine blades are designed in the shape of an airfoil, similar to the wings of an airplane, so as to be able to take maximum advantage of the aerodynamic forces that contribute to the rotating turbine blade. Modern wind turbines resemble a propeller with either two or three blades.

The amount of wind energy impinging on the turbine is the kinetic energy of the moving air:

$$\text{Wind energy} = \tfrac{1}{2} m_{air} v^2 \qquad (5.36)$$

---

[8] Many people think of the windmill as unique to Holland. However, their use in Europe was much more widespread. For example, in the story of Don Quixote de la Mancha, the classic novel by Miguel de Cervantes, Don Quixote does battle with a windmill on the plains of 16th-century Spain, believing the windmill to be a dragon.

Here m_{air} and v refer to the mass and velocity, respectively, of the moving air. The mass of air can be expressed as the product of the air density, ρ, times the volume, V, of air striking the wind turbine during a given period. Because the air is constantly in motion, the air volume can be expressed as the product of the wind speed, v, times the time interval, Δt, times the cross-sectional area, A, enveloped by the turbine blades. Substituting these terms into Equation (5.36) gives this relationship:

$$\text{Wind energy} = \tfrac{1}{2}(\rho V)v^2 = \tfrac{1}{2}(\rho A v \Delta t)v^2 = \tfrac{1}{2}\rho A v^3 \Delta t \qquad (5.37)$$

Power, P, is defined as energy per unit time; and for a conventional wind turbine rotating on a horizontal axis, the swept area, A, is given simply by $A = \pi r^2$, where r is the radius of the rotor. Therefore

$$\text{Wind power} = P_{wind} = \tfrac{1}{2}A(\rho v^3)_{air} = \tfrac{1}{2}\pi r^2 (\rho v^3)_{air} \qquad (5.38)$$

Equation (5.38) shows that one way to increase the maximum amount of power obtainable from a wind turbine is to increase the size of the turbine blade. Doubling the length of the blade, for instance, increases the maximum power by a factor of four. As a practical matter, however, the size of a rotor is limited both by cost and by engineering considerations such as stress and vibrations. Thus, rather than making individual turbine blades extremely large, the preferred solution to increase overall power and energy levels is to deploy multiple wind turbines. Such arrays are commonly called *wind farms*.

In terms of energy efficiency, wind turbines have certain limitations. Because the air impinging on a wind turbine does not halt as it crosses the device, only a fraction of the kinetic energy in wind is recoverable for conversion to electricity. It can be shown from a theoretical analysis for a steady flow, constant velocity air stream that the maximum useful power available from the wind is

$$P_{max} = \frac{8}{27}\pi r^2 (\rho v^3)_{air} \qquad (5.39)$$

This quantity is 59.3 percent of the total wind power given by Equation (5.38). This represents an upper limit on the overall efficiency of a wind turbine generator (known as the *Betz limit*). In general, the overall efficiency of the combined turbine generator system can be expressed as the ratio of electrical power output to the wind power input:

$$\eta_{wind} = \frac{P_{elec}}{P_{wind}} = \frac{P_{elec}}{\tfrac{1}{2}\pi(\rho\, r^2 v^3)} \qquad (5.40)$$

The most efficient wind machines today have overall efficiencies in the range of approximately 40 percent.

Example 5.18

Wind turbine power.   A wind turbine with a rotor diameter of 50 m operates on a hilltop where the wind is blowing at 9.0 m/s. The air temperature is 10°C, and the corresponding air density is 1.25 kg/m³. Determine (a) the equivalent electrical power of the wind under these conditions, (b) the maximum theoretical power for the wind turbine, and (c) the actual power generated if the overall turbine generator efficiency is 34 percent.

Solution:

(a) The wind power is given by Equation (5.38):

$$P_{wind} = \tfrac{1}{2} \pi r^2 \left(\rho v^3 \right)_{air}$$

$$= \tfrac{1}{2} (3.1416)(\tfrac{50}{2})^2 (1.25)(9.0)^3$$

$$= 8.945 \times 10^5 \text{ W} = 895 \text{ kW}$$

(b) The maximum theoretical power is given by Equation (5.39):

$$P_{max} = \frac{8}{27} \pi r^2 \left(\rho v^3 \right)_{air} = 0.593 \, P_{wind}$$

$$= (0.593)(895)$$

$$= 531 \text{ kW}$$

(c) The actual power generated is found from Equation (5.40):

$$\eta_{wind} = \frac{P_{elec}}{P_{wind}} = 0.34$$

$$P_{elec} = (0.34)(895)$$

$$= 304 \text{ kW}$$

For simplicity, Example 5.18 assumed a constant wind speed. In practice, of course, the wind speed at any location is highly variable. Furthermore, since wind speed is a critical determinant of the maximum available power, geographic locations with high average wind speeds are the most desirable sites for wind turbines. In the United States such locations are found predominantly in the upper Midwest, as well as in several coastal locations.

Commercial wind turbines in use today have electrical power ratings ranging from about 250 kW to 750 kW. Typically, a number of such turbines are connected together to produce a wind farm capable of generating larger amounts of power. A typical 25 MW wind farm might include 50 turbines, each rated at 500 kW. The total land area required for such a facility would be roughly 20–40 hectares/MW, depending on the terrain and turbine configuration. The turbine structures themselves, however, occupy only about 5–10 percent of the total land area. Figure 5.18 shows one view of a modern wind farm.

In most U.S. regions the cost of wind power today is still not competitive with conventional power generation sources. Nonetheless, costs have declined substantially over the past two decades, and this trend is expected to continue as turbines become larger and more efficient (NREL, 1998). The use of wind turbines is thus expected to grow in the years ahead from the 1997 capacity of 1,800 MW (0.2 percent of total U.S. generating capacity). As discussed later in this chapter, the use of wind energy could accelerate even more rapidly if measures are adopted to control

Figure 5.18   A view of a modern wind farm in northern California. (*Source:* NREL, 1998)

global warming, or if state agencies continue to provide mandates or incentives for the use of renewable energy, as many are now doing (see Section 5.8).

5.6.6   Electrochemical Generators

All of the electrical generating technologies discussed up to now have been based on the electromechanical generator principles discovered by Michael Faraday. However, more than 30 years before Faraday produced electricity by moving a wire in a magnetic field, Alessandro Volta in 1800 discovered a different way of generating electricity. Volta found that two dissimilar metals separated by an electrolyte produced a small voltage that could drive a current through an external circuit connecting the two metal electrodes. Thus was invented the electrochemical cell, predecessor to the modern battery. Chapter 4 discussed this technology in detail, including the underlying chemistry and physics. Different combinations of electrode materials produce different voltages, typically in the range of 1–2 v. Higher voltages are achieved by stacking several electrochemical cells in series to form a battery. The voltage of an electrochemical system has a constant polarity, so, current always flows in a single direction. This type of current flow is called a *direct current* (*DC*), in contrast to the alternating current produced by electromechanical generators.

The electrochemical cells in common use today (more commonly called batteries) deliver relatively low levels of energy and power suitable for small electronic appliances. Larger batteries, such as those used in automobiles, provide greater amounts of energy (with currents in excess of 100 amps) for short periods. Eventually, however, all batteries become depleted and must be replaced, including rechargeable batteries. Chapter 4 addressed the environmental impacts of battery use, including issues related to disposal and recycling.

Fuel cells are another form of electrochemical generator whose operation is similar to that of a conventional battery, except that the electrode reactants are delivered continuously in the form of a gas. Hence a fuel cell can operate continuously for long periods. Figure 5.19 shows the operation of a typical fuel cell powered by hydrogen. Small fuel cells running on hydrogen and oxygen, which combine to form water in the process of generating electricity, have been used for many years to provide electrical power for spacecraft. Several hundred commercial fuel cells using a phosphoric acid electrolyte are in operation around the world in niche applications.

Table 5.8 lists the characteristics of several fuel cell designs currently under development. Although most fuel cells are powered by hydrogen, intensive research efforts are under way to develop cells that can operate on more readily available fuels such as methanol or methane. This usually involves an additional chemical processing step, known as *reforming,* that converts the fuel feedstock into hydrogen, which is then fed to the fuel cell. Chapter 3 (Figure 3.13) showed an example of such a system being developed for automotive applications. For large-scale power generation, the most promising fuel cell technologies are the high-temperature systems such as the solid oxide fuel cell, which is the most efficient. Recent demonstration projects using high-temperature fuel cells have produced power levels ranging from 200 kW to 2 MW (NRC, 1995). At present, however, the cost of fuel cell technology is prohibitive for most applications. But if continued research succeeds in producing a reliable and economical device, the use of fuel cells for large-scale as well as small-scale power generation (including use in automobiles and buildings) could emerge as an important new method of generating electricity. This technology could thus lead the way to the *distributed generation* of electricity, as opposed to the large *central station generation* systems in place today.

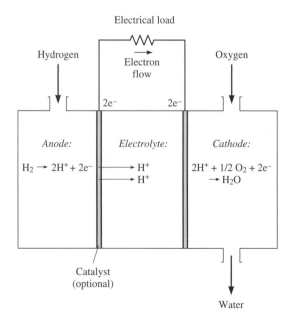

Figure 5.19   Schematic of a typical fuel cell.

Table 5.8   Characteristics of fuel cells under development.

Fuel Cell Characteristic	Type of Fuel Cell			
	Proton Exchange Membrane	Phosphoric Acid	Molten Carbonate	Solid Oxide
Electrolyte	Polymer	Phosphoric acid	Potassium carbonate, lithium	Zirconium oxide + yttrium oxide carbonate
Conductor	H^+	H^+	CO_3^{2-}	O^{2-}
Catalyst	Pt	Pt	(None)	(None)
Operating temperature	~80°C	180–250°C	500–750°C	800–1,000°C
Start-up time	Several seconds	10–20 minutes	Several hours	10 or more hours
Fuel	Hydrogen	Hydrogen	Hydrogen, carbon monoxide	Hydrogen, carbon monoxide
Fuel source[a]	Natural gas, methanol	Natural gas, methanol	Natural gas, methanol, petroleum, coal	Natural gas, methanol, petroleum, coal
Commercial uses	Electric vehicles, spacecraft	Distributed power plants, on-site generators, electric vehicles	Central power plants	Central power plants

[a] Systems using natural gas or methanol require a reformer to produce hydrogen. Systems using petroleum or coal also typically require a gasifier to produce fuel gas.
Source: Toyota, 1998.

Many of the current environmental problems of power generation technology also may be alleviated with fuel cell technology, although further study is required to understand its full environmental implications.

5.6.7   Photovoltaic Generators

Solar energy originates in the nuclear fusion reactions of the sun. Part of this energy reaches earth in the form of electromagnetic waves (radiation), which also exhibit the behavior of particles, called *photons*. The energy of these photons can directly mobilize electrons in certain materials, causing a current to flow. This phenomenon is known as the *photoelectric effect*, first observed in 1839 by the French scientist Edmond Becquerel. It was not until 1954, however, that a workable solar cell was introduced by researchers at Bell Laboratories in New Jersey. Those early designs have since evolved into more efficient cells made of semiconductor materials.

Semiconductors are materials that allow electrons to flow in only one direction, similar to a diode. They are typically made of silicon, which has a latticed crystalline structure consisting of silicon atoms bound to each other by four valence electrons. Adding small amounts of impurities to this lattice structure (a process called *doping*) creates the electrical properties desired. Thus replacing a silicon atom with an atom having five valence electrons, such as arsenic or phosphorus, leaves one electron free to conduct current. This electron-rich material is called an *n*-type semiconductor. Similarly, doping silicon with atoms having only three valence electrons, such as boron or gallium, produces a *p*-type semiconductor with a deficiency of electrons. The missing electrons act like "holes" that can be filled by excess electrons from the *n*-type materials when the two materials are bound together to form a junction.

Before electricity can flow, a potential energy barrier at the junction (termed the *band-gap*) must be overcome. In photovoltaic applications, the photon energy of sunlight is sufficient to excite free electrons and allow them to cross the junction. Wires connected to each side of the photovoltaic cell then supply current to an external circuit, as shown schematically in Figure 5.20.

Most commercial photovoltaic cells consist of wafers of crystalline silicon about 0.5 mm thick, composed of two layers of oppositely doped semiconductors. A cell 10 cm in diameter can produce about one watt of power. Modules of dozens of cells can be further grouped into panels and then arrays, which may produce several kilowatts of power. Commercial photovoltaic cells today convert about 12–13 percent of the energy in incident sunlight into usable electricity. Experimental cells have achieved twice that efficiency, but only under carefully controlled laboratory conditions.

The cost and efficiency of photovoltaic (PV) systems have improved markedly over the last few decades. New developments such as thin-film photovoltaics have substantially lowered the cost of PV technology and hold substantial promise for future commercial applications. Nonetheless, this technology is not yet competitive with conventional power generation systems based on fossil fuels. At present, its principal markets lie in niche applications where PV power can replace or supplement conventional electric power at the customer end of the system. The need for an energy storage system sufficient to cover periods of little or no sunlight is one of the major impediments to widespread reliance on solar energy. To contribute significantly to U.S. power generation, photovoltaic systems will have to interconnect with the electrical grid and be competitive with other power generation sources.

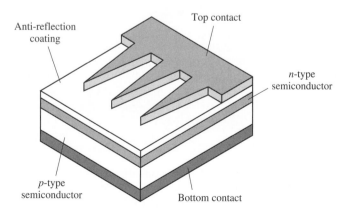

Figure 5.20   Schematic of a simple photovoltaic cell.

Table 5.9 Projected applications of photovoltaic systems.

Parameter	Residential System[a]		Utility System[b]	
	Base Case 1997	**Projected 2005**	**Base Case 1997**	**Projected 2005**
Plant size:				
DC rating (peak kW)	2.8	3.4	20	10,000
AC rating (peak kW)	2.3	2.8	16	8,000
Module area (m²)	20	20	333	91,000
PV module efficiency:				
Laboratory cell (best) (%)			18	20
Power module (best) (%)			10	15
Commercial module (%)	14	17	6	11
System efficiency (%)	11.3	14.1	4.8	8.8
PV module peak output (W/m²)			60	110
System energy produced:				
Average insolation (MW-hr/yr)	4.08	5.07	29	15,000
High insolation (MW-hr/yr)	5.22	6.48	37	18,600
Total capital cost (1997$/$W_{peak}$)	6.46	3.90	9.3	2.9
Total O&M costs (¢/kW-hr)	1.0	0.8	2.3	0.3

[a] Based on expected development of crystalline silicon modules for an individual home.
[b] Based on expected development of thin-film technology for a fixed flat-plate system.
Source: EPRI, 1997b.

View of a house employing building integrated photovoltaics (BIPV) on the sloped roof to supply electricity. BIPV is an advanced concept that uses photovoltaic panels as a construction material as well as a power generator. This reduces the cost of the PV system relative to traditional installations that are mounted on an existing roof or wall.

Further technological improvements, however, are expected to increase the markets and use of photovoltaic electricity in the years ahead. Table 5.9 shows one recent projection of future photovoltaic systems for both residential and utility-scale power generation. In the near term, the most promising applications are likely to be for residential power, where PV systems can provide electricity during peak demand, when electricity is most expensive. The U.S. Department of Energy is promoting an initiative to install 1 million photovoltaic systems on the rooftops of residential and commercial buildings by the year 2010 (EPRI, 1997b). Subsequently, it is expected that larger-scale utility systems may become competitive based on continued improvements in thin-film PV technology. A sustained research and development program is essential to the long-term success of photovoltaics in the marketplace.

5.7 COMPARING ENVIRONMENTAL IMPACTS

Now that we have considered a variety of power generation methods and energy sources, the major question regarding the environmental impacts of electric power generation is how to compare alternative technologies to determine which is environmentally preferable.

In some cases the answer may be fairly straightforward. For example, replacing a conventional coal-burning plant with one burning natural gas using an efficient combined-cycle system provides clear environmental benefits in terms of reduced air pollution, reduced greenhouse gas emissions, and reduced solid wastes. But what about replacing a gas-fired combined-cycle plant with a nuclear plant or with a renewable energy technology such as wind turbines or a hydroelectric facility? In these cases comparisons are not at all straightforward. For example, use of the alternative technologies could eliminate the atmospheric emissions of CO_2 and NO_x from natural gas combustion, but would exacerbate the land use and ecological impacts of the renewables and the radioactive waste problems of nuclear plants.

These types of "apples versus oranges" comparisons of environmental impacts across different pollutants and different environmental media do not readily lend themselves to easy answers or generalized results regarding environmental preferability. Rather, assessments of environmental impacts in the context of specific projects or locations often are necessary to explicate and weigh the environmental trade-offs that are involved. Many factors in addition to environmental impacts become important in deciding on the relative merits of alternative power generation options.

5.8 LOOKING AHEAD

The future development of electric power systems in the United States will be strongly influenced by technology innovation and environmental concerns. Economics, politics, and institutional factors also will play critical roles. Some of the factors affecting future power generation technology and its environmental impacts are briefly highlighted here.

5.8.1   Environmental Outlook

The following are among the key environmental issues that will affect electric power plants over the next decade and beyond.

Ozone Attainment   To achieve national ambient air quality standards for ozone, substantial reductions in NO_x emissions from electric power plants will be required, as noted earlier. The technological response to future requirements is likely to involve a combination of combustion modification methods and more costly flue gas treatment processes such as selective catalytic reduction. The added cost of NO_x controls will adversely affect the economics of fossil fuel power plants—especially coal-fired plants—relative to other alternatives. However, fuel costs and other factors are likely to be much more significant in the overall economic equation.

Fine Particulates   The new ambient air quality standards for fine particulate matter less than 2.5 microns in size ($PM_{2.5}$)—announced by the EPA in 1997—could require additional reductions in SO_2 and NO_x emissions because these gases react chemically in the atmosphere to form fine sulfate and nitrate particles. There is still substantial uncertainty, however, about the relationships between gaseous pollutant emissions and the subsequent formation of fine particles. Legal challenges to the new air quality standards add further uncertainty. Thus the implications for new emission control requirements for fossil fuel plants cannot yet be determined with any confidence. Nonetheless, power plants are a major source of the precursor SO_2 and NO_x emissions, so there remains a strong potential for control requirements beyond those already mandated. Attainment of $PM_{2.5}$ standards also could require further reductions in primary particulate emissions from some power plants.

Air Toxics   The electric utility industry was exempted by Congress from the 1990 Clean Air Act requirements to control emissions of air toxics, pending further study of potential health risks (see Chapter 2). The results of these studies found emissions of mercury to be the main concern. Mercury is one of the many trace elements found in coal. Although it is emitted in only very small quantities, total power plant mercury emissions account for approximately a third of the total U.S. inventory of mercury entering the environment. The persistence of mercury in the environment and the fact that it can bioaccumulate in fish and other organisms make it an environmental concern. There is no proven technology currently available to capture mercury emissions from power plants, but several approaches are being investigated. Thus some form of mercury control requirements could be imposed on coal-burning plants in the near future.

Global Climate Change   Of all the environmental issues looming on the horizon, none is more far-reaching in its consequences than potential requirements to reduce carbon dioxide emissions to control global warming. Chapter 12 is devoted entirely to this subject. The consequences of sizeable CO_2 reductions from electric power plants would be profound and would be felt most sharply at coal-burning facilities, which emit 90 percent of the CO_2 from the utility sector (USDOE, 1998a). The technological response to any future regulations will depend strongly on the magnitude and timetable of any CO_2 reduction requirements. The reduction level agreed to by the United States under the 1997 Kyoto Protocol is a 7 percent decrease in CO_2 below 1990 levels by 2008–2012. However, Congress has been reluctant to ratify that agreement, so there remains considerable doubt as to whether, or when, a requirement to reduce emissions would actually take effect. Nonetheless, some see the Kyoto agreement as merely the first step of a longer-term effort to reduce the growth of CO_2 emissions worldwide.

The technical challenges in meeting a significant CO_2 reduction requirement are widespread. They include the development and deployment of more efficient power generating systems; efficiency improvements in lighting, appliances, motors, and all other electric equipment so as to reduce the demand for energy; increased emphasis on zero- and low-carbon energy sources such as natural gas and renewables; and the development of new CO_2 sequestration technologies as an option for carbon management. Chapter 12 explores these strategies in greater detail.

5.8.2   Technology Outlook

We have seen that electricity can be produced from a wide variety of technologies and energy sources. In considering the environmental impacts of electric power plants, a key uncertainty is what types of power plants will be built and used in the future. Depending on the time frame of interest, one can envision futures that look very different from the present. For example, one long-term change could be the replacement of our current system of large central-station power plants, which ship electricity to many customers over long distances, with small distributed generation systems that supply individual households, neighborhoods, or buildings. Small gas turbines or fuel cells powered by natural gas, or low-cost photovoltaics coupled with advanced energy storage systems, could be the technologies of choice for such a future.

In general, two major factors will affect the rate of new technology deployment. One is the need for new generating capacity to accommodate the growth in demand for electricity. The other is the need for new power plants to replace existing ones as they retire. Forecasting these needs, however, is not an easy task (see Chapter 15). One well-known annual forecast comes from the Energy Information Administration of the U.S. Department of Energy (DOE/EIA), which uses an elaborate set of computer models to project the U.S. energy outlook over the next two decades based on current environmental requirements (USDOE, 1999).

EIA also modeled five scenarios for achieving the Kyoto Protocol reductions (USDOE, 1998c). The most stringent case required the United States to reduce its own CO_2 emissions by 7 percent below the 1990 levels by 2010 and hold to that level thereafter. Other scenarios allowed U.S. CO_2 emissions to *increase* by up to 24 percent above the 1990 levels. This is possible because the Kyoto agreement allows a country

Table 5.10 Projected effect of the Kyoto Protocol on U.S. electricity generation in 2020.

Energy Source	2020 Reference Case (10^9 kW-hr)	Percent Increase for Scenario[a]	
		1990 + 24%	1990 − 7%
Coal	2,186	−41	−99
Natural gas	1,362	36	57
Nuclear	356	33	95
Renewables:	383	14	123
Hydropower	313	0	33
Municipal waste	30	0	0
Geothermal	20	26	170
Wind	8.7	400	1,500
Biomass	8.7	160	3,400
Solar thermal	1.5	0	0
Photovoltaics	1.4	0	64

[a] The two cases shown bracket the five scenarios modeled in this study.
Source: Based on USDOE, 1998c.

to offset its own emissions by achieving CO_2 reductions in other countries where it is more economical. Reductions in greenhouse gases besides CO_2 also contribute to the overall Kyoto targets.

Table 5.10 summarizes the range of EIA scenarios in terms of the percentage change in electricity generated from different energy sources in 2020 compared to the reference case without the Kyoto Protocol. Coal use falls dramatically and is all but eliminated in the "1990−7 percent" scenario. These losses are taken up by substantial increases in the use of natural gas, nuclear energy, and renewables, especially wind and biomass. A decrease of 6–14 percent in the total demand for electricity also contributes to overall CO_2 reductions in these EIA scenarios. This decline in electricity use stems from the much higher electricity prices and other economic consequences induced by the changes in energy mix.

These EIA scenarios were among the first attempts to analyze the technological implications of the proposed Kyoto Protocol agreements. The EIA results have been controversial because of the many assumptions made in the analysis (see Chapter 15 for a more complete discussion of environmental forecasting). Nevertheless, it is clear from this and other studies that global efforts to significantly reduce CO_2 emissions will have a dramatic impact on future power generation technology.

5.9 CONCLUSION

This chapter has presented an overview of the environmental issues surrounding electric power generation. It also has explored a variety of ways in which electricity can be generated, highlighting the fundamental principles of engineering design that affect environmental impacts. The future design of electric power systems will continue to be strongly influenced by environmental concerns. The engineering challenge lies in developing cleaner, more efficient technologies at affordable costs. Students interested in specific technologies will find a wealth of information in the references listed at the end of this chapter as well as other sources. The problems at the end of the chapter also will help you explore the environmental impacts and design of current and advanced technologies for electric power generation.

5.10 REFERENCES

ASME, 1977. "ASME Steam Tables in SI (Metric) Units." American Society of Mechanical Engineers, New York, NY.

CFR, 1998. Code of Federal Regulations. *Federal Register,* September 16, 1998, vol. 63, no. 179, p. 49, 442.

CFR, 1999. Code of Federal Regulations. *Federal Register,* July 1, 1999, 40CFR, ch. I.

EPRI, 1997a. *PISCES: Power Plant Chemical Assessment Model Version 2.0.* CM-107036-V1, Electric Power Research Institute, Palo Alto, CA, July.

EPRI, 1997b. *Renewable Energy Technology Characteristics.* EPRI TR-109496, Electric Power Research Institute, Palo Alto, CA.

Frey, H.C., et al., 1995. *A Method for Federal Energy Research Planning: Integrated Consideration of Technologies, Markets and Uncertainties.* Report to Lawrence Livermore National Laboratory, Livermore, CA, April.

NRC, 1995. *Coal: Energy for the Future.* National Research Council, National Academy Press, Washington, DC.

NRC, 2000. *Vision 21: Fossil Fuel Options for the Future,* National Research Council, National Academy Press, Washington, DC.

NREL, 1998. http://www.nrel.gov, National Renewable Energy Laboratory, Golden, CO.

ORNL, 1992. *External Costs and Benefits of Fuel Cycles: A Study by the U.S. Department of Energy and the Commission of the European Communities,* Report Nos. 1–5, Oak Ridge National Laboratory, Oak Ridge, TN.

Rubin, E.S., et al., 1997. "Integrated Environmental Control Modeling of Coal-Fired Power Systems." *Journal of Air & Waste Management Association,* vol. 47, pp. 1,180–88.

Siegel, S.A., 1997. *Evaluating the Cost Effectiveness of the Title IV 1990 CAAA.* PhD Thesis, Department of Engineering and Public Policy, Carnegie Mellon University, Pittsburgh, PA.

Toyota, 1998. Y. Nonobe and Y. Kimurh, "Development of Electric Vehicles Powered by Fuel Cell." *Toyota Technical Review,* vol. 47, no. 2, April, pp. 67–72, Toyota Motor Corporation.

USDOE, 1996. *Clean Coal Technologies Research, Development, and Demonstration Plan.* DOE/FE-0284, U.S. Department of Energy, Washington, DC.

USDOE, 1997a. www.eia.doe.gov/fuelcoal.html. Energy Information Administration, U.S. Department of Energy, Washington, DC.

USDOE, 1997b. *Annual Energy Outlook 1998.* DOE/EIA-0383(98), Energy Information Administration, U.S. Department of Energy, Washington, DC.

USDOE, 1998a. *Annual Energy Review 1997.* DOE/EIA-0384(97), Energy Information Administration, U.S. Department of Energy, Washington, DC.

USDOE, 1998b. *International Annual Energy Outlook 1998.* DOE/EIA-0484(98), Energy Information Administration, U.S. Department of Energy, Washington, DC.

USDOE, 1998c. *Impacts of the Kyoto Protocol on U.S. Energy Markets and Economic Activity.* SR/OIAF/98-03, Energy Information Administration, U.S. Department of Energy, Washington, DC.

USDOE, 1999. *Annual Energy Review 1998.* DOE/EIA-0384(98), Energy Information Administration, U.S. Department of Energy, Washington, DC.

USEPA, 1996. *Methane Emissions from the Natural Gas Industry.* EPA-600/R-96-080, U.S. Environmental Protection Agency, Research Triangle Park, NC, June.

USEPA, 1998a. *National Air Quality and Emissions Trends Report, 1997.* December 1998, EPA 454/R-98-016, Office of Air Quality Planning and Standards, U.S. Environmental Protection Agency, Washington, DC.

USEPA, 1998b. AIRS database, www.epa.gov/airsweb/sources.htm. U.S. Environmental Protection Agency, Washington, DC.

5.11   PROBLEMS

5.1   A coal-fired power plant produces a net electrical output of 500 MW. The composition of the coal burned is C = 72%, H = 5%, S = 2%, O = 8%, N = 1%, and ash = 12% by weight. Its higher heating value (HV) is 30,940 kJ/kg. The net thermal efficiency is 33 percent.

   (a)   How much coal (in metric tons) does this plant consume in one day running at full capacity?

   (b)   Assume that 75 percent of the ash in coal is entrained in the flue gas stream as fly-ash and the remainder is collected in the boiler as bottom ash. How many tons per day of flyash would be emitted from this plant in the absence of any particulate controls?

   (c)   How many tons/day of SO_2 are emitted if 97 percent of the sulfur in coal is converted to SO_2?

   (d)   How many tons/day of NO_x are emitted if the NO_x emission rate is 300 ng/J (0.7 lbs/10^6 Btu)?

5.2   The 1971 federal New Source Performance Standard (NSPS) for power plant SO_2 emissions specified a maximum allowable rate of 520 ng/J (1.2 lbs SO_2/10^6 Btu) for coal-fired plants. This is also the target rate for acid rain control requirements at existing plants. Based on typical coal heating values, what coal sulfur content (%S) would satisfy this standard without requiring a scrubber or other type of SO_2 removal system for (a) bituminous coal and (b) subbituminous coal?

5.3   For the power plant and coal in Problem 5.1, assume that an electrostatic precipitator (ESP) is sized to just meet the current federal NSPS for particulate matter (PM) of 13 ng/J (0.03 lbs PM/10^6 Btu).

   (a)   How many metric tons/day of flyash now would be emitted when the plant runs at full capacity?

   (b)   What particulate removal efficiency is needed?

   (c)   What specific collection area (SCA) is needed?

   (d)   To comply with acid rain control requirements, the plant is considering switching to a low-sulfur western coal with 0.6 percent sulfur, 14.0 percent ash, and a heating value of 26,500 kJ/kg. However, the plant operator is uncertain whether the existing ESP will be adequate to meet the particulate emission standard using the new coal. Give your expert analysis. Discuss any assumptions needed to reach or support your conclusion.

5.4 Plot the required specific collection area (A/G) of an electrostatic precipitator as a function of the design particulate removal efficiency (η_{ESP}) for coal sulfur contents of 0.5 percent, 1.5 percent, and 3.0 percent. Restrict your plot to values of η_{ESP} greater than 0.98.

(a) What do you conclude about the effects of sulfur content and particulate removal requirements on the size of an ESP?

(b) For the 1.5 percent S coal and a removal efficiency of 99.5 percent, what percentage increase in plate collector area is needed for an additional 50 percent reduction in emissions?

(c) Repeat part (b) for 0.5 percent coal.

5.5 Carbon dioxide (CO_2) is not presently a regulated air pollutant, although it is of growing concern because of its role as a greenhouse gas. For the power plant in Problem 5.1, calculate

(a) The annual mass of CO_2 emissions in metric tons/yr.

(b) The CO_2 emission rate in ng/kJ of energy input to the plant.

(c) The CO_2 emission rate in g/kW-hr of electricity generated.

Which of these measures seems most appropriate to you as a basis for comparing CO_2 emissions from different fuels or technologies used for power generation? Explain your reasoning.

5.6 Consider further the 500 MW coal-fired power plant described in Problem 5.1. Assume the plant uses low-NO_x burners and selective catalytic reduction (SCR) for NO_x control, an electrostatic precipitator (ESP) for particulate control, and a wet limestone ($CaCO_3$) flue gas desulfurization (FGD) system with forced oxidation for SO_2 removal.

(a) Draw a schematic diagram of the plant, identifying the location of the major environmental control systems. On the sketch show the major stream flows into and out of the plant (coal, air, flue gas, chemical reagents, and solid wastes).

(b) Perform a mass balance for the plant, assuming that it meets the 1997 federal New Source Performance Standards shown in Table 5.3. Normalize all flows on one kilowatt-hour (kW-hr) of net electricity produced for sale. Assume for this analysis that collected flyash is dry. Assume also that the FGD system requires 1.05 moles Ca/mol S removed and produces by-product gypsum ($CaSO_4 \cdot 2H_2O$), which is dewatered to 90 percent solids by weight (the remaining 10 percent is water). On a sketch of the plant show the following mass flow rates per kW-hr: (i) coal input; (ii) FGD reagent input; (iii) NO_x, SO_2, and PM emitted to the atmosphere; and (iv) all solid wastes. Be sure to show your supporting calculations.

5.7 Improved engineering design and higher steam temperatures raised the steam cycle efficiency of a fossil fuel power plant from 42 percent to 44 percent. Additional design changes increased the boiler efficiency from 87 percent to 89 percent. What was the resulting improvement in overall plant efficiency? Express your answer two ways:

(a) As an absolute value (in percentage points), referring to the difference between final and initial efficiency.

(b) As a relative value referring to the percentage improvement over the original efficiency.

5.8 (a) Explain the difference between "higher heating value" and "lower heating value" for a fuel.

(b) Modern gas turbine combined-cycle systems sold by U.S. manufacturers for electric power generation have an overall thermal efficiency of about 45 percent based on the higher heating value of 54,400 kJ/kg for natural gas. What would be the efficiency based on the lower heating value of 48,700 kJ/kg (which is the basis used by many European manufacturers)?

(c) If you were a turbine manufacturer, which heating value (higher or lower) would you prefer to use as the basis for efficiency claims about your product? Why?

(d) Does the choice of a heating value convention affect the magnitude of air pollutant emissions from a particular system? Explain.

5.9 Consider a 250 MW power plant burning natural gas (idealized as pure methane, CH_4) with air at the stoichiometric quantity. The net power plant efficiency is 33 percent, and the fuel heating value is 55,800 kJ/kg. The plant was designed to meet a NO_x emission limit of 86 ng/J (0.2 lbs NO_x 10^6 Btu) expressed as equivalent NO_2.

(a) How much fuel energy must be supplied to this plant when it is running at full capacity?

(b) Based on the designed NO_x emission rate, how much NO_x is emitted per day if the plant runs at full capacity?

5.10 (a) For the plant in Problem 5.9, write the chemical reaction equation for stoichiometric fuel combustion.

(b) How many moles of flue gas are produced per mole of fuel burned?

(c) How many moles of NO_x are produced per mole of fuel burned, based on the designed emission rate?

(d) Estimate the concentration (in parts per million by volume, ppmv) of NO in flue gas.

(e) Determine the percentage reduction in NO_x emissions needed to meet a new local regulation requiring the plant to emit no more than 75 ppm NO_x by volume.

5.11 Natural uranium is composed of two isotopes: fissionable U^{238} (99.3 percent by weight) and fissile U^{235} (0.711 percent by weight). Typical assays of ores from U.S. mines show about 0.25 percent uranium metal by weight, primarily in the form of U_3O_8.

(a) Using the assumptions and results from Example 5.6, how much ore must be mined each year to provide the fuel for a 1,000 MW nuclear power plant operating at a 74 percent annual average capacity factor? Express your answer in terms of both total mass per year and mass per kilowatt-hour of electricity generated.

(b) How much of the total ore mass is used for power production, and how much is left as solid waste (mill tailings)?

5.12 A wind turbine generator with a rotor diameter of 50 m is designed for maximum power output at a wind speed of 50 km/h. The overall efficiency of the turbine generator system is 42 percent.

(a) For an air temperature of 20°C, what is the maximum power that can be delivered?

(b) How much electricity is generated in a year if the annual average capacity factor based on this design is 0.40?

(c) Based on the emission factors in Table 5.2, how much air pollution would be avoided if this electricity could displace power generated by an average coal-fired plant?

5.13 (a) How many turbine generators of the type described in Problem 5.12 would be needed to supply the annual electricity demand of a residential community of 30,000 people whose per capita electricity consumption is 6,000 kW-hr/yr?

(b) How would your answer change if the rotor diameter of the wind turbines could be increased to 75 m?

5.14 Suppose all the electricity supplied by coal-fired power plants in 1998 were instead provided by wind turbines.

(a) How many megawatts of installed wind turbine capacity would be needed based on an annual capacity factor (or load factor) of 40 percent and a reserve margin of 15 percent excess capacity?

(b) How much total land area would be required based on an average land area of 30 hectares/MW? Express your answer not only in units of area, but also in terms of two additional measures you think would be informative (for instance, a land area the size of . . . you name it).

5.15 Consider the 150 MW hydroelectric plant in Example 5.17.

(a) How much electricity is generated during a year if the annual average capacity factor is 90 percent?

(b) What is the total volume of water that passes through the turbines during six months?

(c) If all of this water were stored in a reservoir, how much land surface area would be needed if the reservoir were 35 m deep? Express your answer not only in conventional units of area, but also in terms of two additional measures you think would be informative (like a land area the size of . . .).

5.16 A hydroelectric plant has been proposed to meet future electricity demand in a newly industrialized country in lieu of a fossil fuel plant. The projected facility size is 800 MW. The overall efficiency of the plant is estimated to be 87 percent, and the operating head is 70 m.

(a) What is the water flow rate required for this facility?

(b) If the reservoir serving this facility were designed to hold a three-month supply of water based on full plant operation, how much surface area would be needed based on a reservoir depth of 50 m?

5.17 Choose one of the following renewable energy sources and investigate its potential for providing a significant share of U.S. electric power production. In particular, what are the barriers to widespread commercialization? What are the key technical issues that still must be addressed? How much progress has been made in improving the efficiency and reducing the cost of this technology? What environmental impacts, if any, must be addressed? Prepare a brief report summarizing your findings, including a list of references used.

(a) Wind

(b) Solar photovoltaic

(c) Biomass

(d) Geothermal

5.18 Repeat Problem 5.17 for a selection of alternative power generation technologies (see Table 5.7) which were not explicitly discussed in this chapter, but which have been widely studied. Again summarize the results of your investigation in a brief report.

   (a) Pressurized fluidized bed combustion

   (b) Breeder nuclear reactors

   (c) Solar thermal power plants

   (d) Ocean thermal power plants

   (e) Magnetohydronamic generators

chapter

6

Refrigeration and the Environment

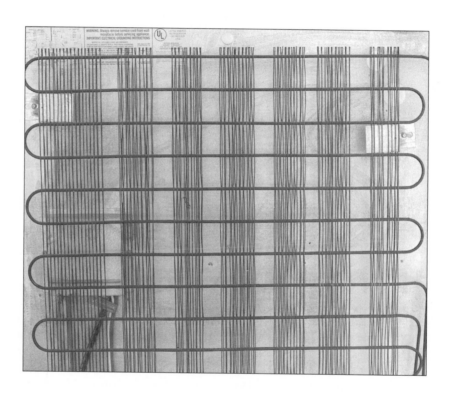

6.1   INTRODUCTION

The refrigerator in your kitchen is something you probably take for granted. But consider how your life would change without it. No cold drinks. The milk and juice would spoil quickly, as would the vegetables, fish, and fruit. Frozen foods would be a thing of the past, and shopping for food could become a daily chore. Your grocery store or supermarket would also look quite different without refrigeration. And because many food items are shipped via refrigerated trains and trucks, some things you normally buy simply would not be available.

Even as far back as 2000 B.C., snow, ice, and cold water were used to cool drinks and preserve food. The first known artificial refrigeration system was demonstrated in 1748 by William Cullen at the University of Glasgow. He turned water into ice by reducing the pressure in a beaker of ethyl ether placed inside a beaker of water. The low pressure caused the ethyl ether to evaporate, absorbing heat from the water and causing it to freeze. There were several subsequent attempts to use this concept for artificial refrigeration, but it was not until a century later that refrigeration systems began to find commercial use.

In 1844 the American physician John Gorrie designed a refrigeration machine based on compressed air, which was the first system built for commercial refrigeration and air conditioning. In 1851 the Australian James Harrison introduced refrigeration to the brewing and meat-packing industries. His system used a sulfuric ether as a refrigerant. In 1859 the Frenchman Ferdinand Carré found that ammonia was able to absorb larger quantities of heat, and he introduced this as the working fluid. The use of ammonia became very popular despite the fact that if spilled in significant quantities it was lethal as well as flammable. Other 19th-century refrigerants included carbon dioxide, methyl chloride, and sulfur dioxide.

A good refrigerant has the property of being a liquid at room temperature under high pressure while vaporizing at lower temperature and pressure. A practical device for utilizing these properties was the vapor compression cycle, first patented in England in 1834 by the American engineer Jacob Perkins. The energy for compression was provided by fuel combustion until electricity came on the scene in the early 20th century. In 1918 the first electrically powered household refrigerator (called the Kelvinator) was introduced on the American market, using sulfur dioxide as a refrigerant.

Engineers continued to search for better alternatives. The seemingly ideal solution appeared in the late 1920s when Thomas Midgley Jr. and his team of researchers created a family of substances consisting of hydrocarbon compounds with chlorine and fluorine atoms attached. These substances, called *chlorofluorocarbons* (*CFCs*), were thermodynamically superior to other refrigerants. They were stable and nontoxic and appeared to be environmentally benign. CFCs soon replaced all other refrigerants in commercial applications, eliminating the safety hazards of the more reactive and potentially dangerous chemicals previously used. As the affluence of Americans grew during the 20th century, the number of households with refrigerators and freezers also grew dramatically, and with it the production and use of CFCs (see Figure 6.1). Elsewhere in the world similar trends evolved. Refrigeration systems based on CFCs became an integral part of the modern technological age.

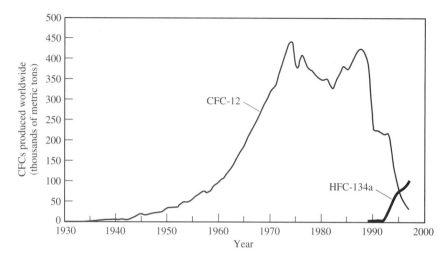

Figure 6.1   Annual worldwide production of CFC-12 (freon), the refrigerant commonly used in household refrigerators until the 1990s. The recent ban on CFC production has led to increased production of HFC-134a as a substitute refrigerant. (*Source*: AFEAS, 1998)

6.2   ENVIRONMENTAL OVERVIEW

Not until the 1980s did the world realize that this ideal refrigerant was not so environmentally benign after all. Rather, CFCs were found to be the cause of an environmental problem, the likes of which were unprecedented in human history. As outlined in Chapter 2, CFCs are responsible for destroying the protective layer of ozone that blankets the earth high in the atmosphere. Never before had humans created an environmental problem on a global scale.

Chapter 11 describes the mechanisms of stratospheric ozone depletion in more detail. The process begins with the release of CFCs into the air. For refrigerators, this occurs via leaks in equipment as refrigerators age and via accidental or intentional spills when refrigerators are discarded. Releases also occur during CFC production. Because CFCs are used for a variety of purposes besides refrigeration (see Chapter 11), there are multiple paths by which these compounds are released to the environment.

Once in the atmosphere, CFCs contribute not only to depletion of the ozone layer, but also to global warming, as discussed in Chapter 12. Because CFCs are extremely stable compounds, their contribution to both ozone depletion and global warming persists for many decades. Although the production of CFCs is now banned under U.S. and international laws, the existing stock of chemicals continues in use and will eventually enter the atmosphere.

Environmental concerns also arise from the use of energy. Modern refrigerators require electricity for their operation. The generation of electricity results in regional air pollution, CO_2 emissions that contribute to global warming, dangerous radioactive wastes, and a variety of other environmental impacts discussed in Chapter 5. The aggregate impact from the millions of refrigerators in operation is quite significant.

Refrigeration systems in general—including air conditioning and other applications—account for over 20 percent of U.S. electricity use, and they are one of the fastest-growing segments of energy use technologies. Thus engineering design measures that can reduce the energy requirements for refrigeration will have important environmental benefits.

These two issues—CFCs and energy use—are the major topics of this chapter. Additional environmental concerns arise throughout the life cycle of a refrigerator, including its manufacture and ultimate disposal. Although the lifetime of a refrigerator is relatively long, life cycle issues are nonetheless important because of the large number of devices in use. A general discussion of life cycle assessments comes later in Chapter 7. First, the examples here provide a perspective on both the direct and indirect environmental emissions associated with household refrigerators.

Example 6.1

CFC production for refrigerators.   The U.S. Energy Information Administration reports that there were a total of 101 million U.S. households in 1997, of which 85 percent had one refrigerator, while 15 percent had two or more. If each refrigerator contains 0.25 kg of CFC refrigerant, what is the total amount of CFC that can eventually get into the atmosphere when these refrigerators are discarded?

Solution:

First estimate the total number of household refrigerators, assuming that the two or more category is simply 2.0:

$$\text{Total number of refrigerators} = (10 \times 10^6)(0.85) + (101 \times 10^6)(0.15)(2)$$

$$= 116.2 \times 10^6 \text{ units}$$

To account for households with more than two refrigerators, let us simply round this estimate up to two significant figures, giving us 120 million units. If an average unit contains 0.25 kg of CFC,

$$\text{Total mass of CFCs} = (120 \times 10^6)(0.25)$$

$$= 30 \times 10^6 \text{ kg}$$

$$= 30{,}000 \text{ metric tons}$$

In Example 6.1 the *rate* at which CFCs enter the atmosphere will depend on the rates of leakage and refrigerator disposal and on whether efforts are made to collect CFCs from discarded units. As a result of environmental regulations to protect the ozone layer, new refrigerants are now being introduced in place of CFCs, and stricter efforts are being made to collect and recycle the CFCs from older refrigerators as they retire.

Example 6.2

Energy-related emissions from refrigerators.   If the average household refrigerator–freezer uses 950 kW-hr of electricity per year, estimate the total U.S. electricity generation required for all of these appliances. Also use the data in Chapter 5, Table 5.2, to estimate the associated annual environmental emissions of CO_2, SO_2, and NO_x.

Solution:

From Example 6.1, the total number of household refrigerators is estimated to be 120 million. Thus

$$\text{Total annual electricity use} = (120 \times 10^6)(950)$$

$$= 11.4 \times 10^{10} \text{ kW-hr/yr}$$

This amounts to 9 percent of the total U.S. residential electricity use. From Chapter 5, Table 5.2, the average emission factors for air emissions from U.S. power plants in 1996 were

$$CO_2 = 701 \text{ g/kW-hr}$$

$$SO_2 = 3.62 \text{ g/kW-hr}$$

$$NO_x = 2.39 \text{ g/kW-hr}$$

Multiplying these values by the total annual electricity use of 11.4×10^{10} kW-hr/yr gives the following annual emissions (1 metric ton = 1,000 kg = 10^6 g):

$$CO_2 = 80 \text{ million metric tons/yr}$$

$$SO_2 = 413{,}000 \text{ metric tons/yr}$$

$$NO_x = 272{,}000 \text{ metric tons/yr}$$

Example 6.2 gives a simple snapshot of the indirect air pollution consequences from electricity used to operate household refrigerators. The U.S. average emission factors for air pollutants conceal large variations from one region of the country to another because of differences in regional energy mix for electric power generation (such as heavy use of coal and nuclear power in the Midwest versus a larger proportion of hydroelectric power in the Northwest). These regional differences are important in analyzing the local and regional impacts of air pollutant emissions like SO_2 and NO_x. For CO_2 associated with global warming, however, the aggregate emission level is the major concern.

Environmental impacts also may change over time as the mix of energy sources and technologies for electric power generation changes in response to environmental regulations and other factors. Chapter 5 discussed some of the projections. The most important message, however, is that the indirect impacts of energy use represent a significant environmental concern in the design of refrigeration technology in addition to the direct effects of CFCs or other refrigerants.

6.3   ALTERNATIVE REFRIGERANTS

The overall goal of this chapter is to identify engineering design options that can eliminate or reduce the environmental impacts of refrigerators. To begin that study we need first to identify the available alternatives to CFCs and assess their relative

Table 6.1   Environmental attributes of alternative refrigerants.

Refrigerant	Chemical Formula	Atmospheric Lifetime (Years)[a]	Ozone Depletion Potential (ODP)[b]	Global Warming Potential (GWP)[c]
Chlorofluorocarbons (CFCs)				
CFC-11	$CFCl_3$	50	1.00	4,000 (1.00)
CFC-12	CF_2Cl_2	102	0.82	8,500 (2.13)
CFC-113	$C_2F_3Cl_3$	85	0.90	5,000 (1.25)
CFC-114	$C_2F_4Cl_2$	300	0.85	9,300 (2.33)
CFC-115	C_2F_5Cl	1700	0.40	9,300 (2.33)
Hydrochlorofluorocarbons (HCFCs)				
HCFC-22	CF_2HCl	13.3	0.034	1,700 (0.43)
HCFC-123	$C_2F_3HCl_2$	1.4	0.012	93 (0.023)
HCFC-141b	$C_2FH_3Cl_2$	9.4	0.086	630 (0.16)
HCFC-142b	$C_2F_2H_3Cl$	19.5	0.043	2,000 (0.50)
Hydrofluorocarbons (HFCs)				
HFC-23	CHF_3	264	$<4 \times 10^{-4}$	11,700 (2.93)
HFC-125	C_2HF_5	32.6	$<3 \times 10^{-5}$	2,800 (0.70)
HFC-134a	$C_2H_2F_4$	14.6	$<1.5 \times 10^{-5}$	1,300 (0.33)
HFC-152a	$C_2H_4F_2$	1.5	0	140 (0.035)
Other Refrigerants				
Propane	C_3H_8	<0.1	0	0
Ammonia	NH_3	<0.1	0	0

[a] This is the time for an initial mass to decay exponentially to $1/e = 0.368$ of its initial value. After three "lifetimes" only 5 percent of the initial amount remains.

[b] All values are shown relative to CFC-11 based on current models.

[c] Based on a 100-year averaging time, relative to a unit mass of CO_2. Values in parentheses are normalized relative to CFC-11, as with ODP values.

Source: Based on IPCC, 1996; WMO, 1999.

merits from an environmental viewpoint. Then we can assess the technical merits of environmentally acceptable alternatives.

Table 6.1 lists a number of chemical compounds that have been studied, used, or proposed as refrigerants. The first class of compounds listed are the CFCs. The numerical designation of each CFC is related to its molecular structure, as explained in Chapter 11. Of greater importance are the two environmental index values shown for each compound. The ozone depletion potential (ODP) gives a relative measure of the potency of each substance in destroying ozone molecules in the stratosphere. CFC-11 is the reference substance with an ODP value of 1.00. CFC-12 (also known as R-12 or freon-12) is the refrigerant found in household refrigerators. It also has a high ODP value. In both cases, the chlorine atoms in CFCs are responsible for ozone depletion.

The global warming potential (GWP) values in Table 6.1 provide a relative measure of the ability of different substances to trap heat in the atmosphere (see Chapter 12 for details). Here the reference substance is CO_2, whose GWP value is 1.00. As Table 6.1 shows, CFC molecules are thousands of times more potent than a comparable mass of CO_2 in terms of global warming. As it turns out, the loss of stratospheric ozone caused by CFCs offsets much of the warming effect of CFC molecules because ozone too is a heat-trapping gas. This indirect effect is extremely complex and is not reflected in the GWP index in Table 6.1.

Adding hydrogen to CFCs produces a class of chemicals called hydrochlorofluorocarbons (HCFCs), which also are suitable refrigerants in applications such as residential air conditioning. Environmentally, HCFCs are more benign than CFCs because they are less stable and decompose more readily in the atmosphere. Accordingly, the ODP values are relatively small. Nonetheless, HCFCs still play a role in both ozone depletion and greenhouse warming. Thus the international treaties that have banned further production of CFCs also call for a phaseout of HCFCs by the year 2030.

Removing the chlorine from HCFCs yields the class of compounds called hydrofluorocarbons (HFCs). As seen in Table 6.1, these refrigerants have extremely small (nearly zero) ozone depletion potential and so are among the prime candidates to replace CFCs in refrigeration applications. Indeed, new refrigerators and automotive air conditioners are now being built using HCF-134a as a refrigerant. Figure 6.1 shows the increased production of this chemical beginning in the 1990s. One environmental drawback, however, is that HFCs also are greenhouse gases that contribute to global warming. However, there are still no environmental regulations governing the release of greenhouse gases to the atmosphere.

Stripping away the fluorine atoms from HFCs leaves the chemical class of pure hydrocarbons (HCs), some of which also are suitable refrigerants, notably propane. Hydrocarbons are not commonly thought of as a refrigerant, but rather as a fuel or as an air pollutant that contributes to urban smog (see Chapters 2 and 8). As we shall see later in this chapter, however, the thermodynamic properties of propane as a refrigerant are actually comparable or superior to the CFCs now in use. Environmentally, propane has the desirable attributes of zero ODP and zero GWP. The flammability of propane remains the major barrier to its adoption as a refrigerant, despite the safety claims of technology developers.

Other possible refrigerants include common inorganic compounds such as ammonia, which is widely used as a refrigerant for industrial and commercial applications. Environmentally, ammonia is attractive because it too has zero ODP and GWP. But the pungent odor of ammonia and its flammability and toxicity make it less desirable for use in household refrigerators.

The basic conclusion from this discussion is that there is no perfect solution to the problem of finding a CFC substitute for household refrigerators that is environmentally benign, safe to use, and thermodynamically suitable. Rather, as is often the case, some trade-offs are necessary to achieve a solution that is technically, environmentally, and socially acceptable. The balance of this chapter focuses on the technical understanding needed to evaluate such trade-offs.

6.4   FUNDAMENTALS OF REFRIGERATION

We turn now to a more detailed look at how a refrigerator works. The objective is to gain a fundamental understanding in order to analyze engineering options for reducing the direct and indirect environmental impacts described earlier. Although the focus of this chapter is on the household refrigerator, the same principles apply to room air conditioners and other types of refrigeration devices.

6.4.1   Primary Energy Flows

Figure 6.2 shows a simple schematic of the primary energy flows for a refrigerator. The function of a refrigerator is to cool the interior space by transferring heat from inside the refrigerator to the outside space, which in thermodynamics is called the *surroundings*. If we denote the temperature of the surroundings as T_o and the colder temperature inside the refrigerator as T_c, we see that the refrigerator must move heat "uphill" from a low temperature to a higher temperature.

This, of course, is exactly opposite to the natural flow of heat. When two bodies of different temperatures are brought together, heat flows from the warmer body to the cooler body until the two bodies reach the same equilibrium temperature. For a refrigerator to function, therefore, an additional source of energy is required to help pump heat from the cooler region to the warmer region. In modern refrigerators that energy source is electricity.

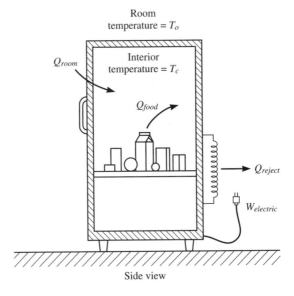

Figure 6.2   Schematic of energy flows for a household refrigerator. Heat flow into the refrigerator from the surrounding room, plus heat from recently stored foods, is rejected back to the surroundings with the aid of electrical energy (work) supplied to the refrigeration system.

Before looking at how a refrigerator works, let us first examine the primary sources and flows of unwanted heat in a refrigerator. The first and most obvious is the heat that must be extracted from the food items to be refrigerated. If these items are initially at the room temperature, T_o, the heat, Q_{food}, that must be extracted to cool these items down to the refrigerator temperature, T_c, is the product of the mass, m, the specific heat, c_p, of each item, and the temperature difference between the initial and final states. Summing over all items gives

$$Q_{food} = \sum_{all\ items} (mc_p)_{food}(T_o - T_c) \tag{6.1}$$

Once the food items have cooled to the desired temperature, T_c, there is no further heat flow from this source. However, because the temperature inside the refrigerator is cooler than the surrounding room temperature, some additional heat invariably leaks into the refrigerator from the room. This occurs via the processes of natural convection and conduction through the refrigerator walls, or when the refrigerator door is opened. Only if the walls were made of perfect thermal insulation could this leakage heat transfer be avoided, provided the refrigerator door was never opened! From the study of heat transfer, the rate of room heat conducted through the walls of a refrigerator can be expressed as

$$\dot{Q}_{room} = U_o A_o (T_o - T_c) \tag{6.2}$$

where A_o is the external surface area of the refrigerator and U_o is the *overall heat transfer coefficient*, typically specified in units of W/m²-K. The value of U_o is determined by the thickness and thermal properties of the refrigerator walls and by the nature of air circulation patterns inside and outside the refrigerator. Recall that the "dot" over the Q means "per unit time," which gives us a heat flow *rate* (like joules/second or watts).

The total heat to be removed from the refrigerator compartment is thus the sum of the heat leakage from the room plus the heat extracted from the food as it cools. On a *rate* basis this is

$$\dot{Q}_{frig} = \dot{Q}_{room} + \dot{Q}_{food} \tag{6.3}$$

Determining how quickly food cools when placed inside a refrigerator is a problem in transient heat transfer that is beyond the scope of this text. However, the following example illustrates the magnitudes of the heat flows involved.

Example 6.3

Heat flows in a refrigerator.   A refrigerator with an interior temperature of 4°C and an external surface area of 5.6 m² stands in a room at 22°C. The overall heat transfer coefficient between the refrigerator and the room is 0.46 W/m²-K. A two-liter bottle of mineral water ($\rho = 1.0$ kg/L, $c_p = 4.19$ kJ/kg-K) initially at room temperature is placed inside the refrigerator and takes 3.0 hours to cool to 4°C. Calculate and compare the heat flow rates from food cooling and room leakage. Also calculate the total heat flow that must be removed from the refrigerator while the food is cooling and after it has cooled.

Solution:

The total heat removed from the two-liter bottle of water is given by Equation (6.1). Because the density of water is 1.0 kg/L, the total mass is 2.0 kg. Thus

$$Q_{food} = mc_p(T_o - T_c)$$

$$= (2.0)(4.19)(22 - 4)$$

$$= 150.8 \text{ kJ}$$

This is the total amount of heat extracted in 3.0 hours. Thus the average heat flow *rate* is

$$\dot{Q}_{food} = \frac{Q_{food}}{\Delta t}$$

$$= \frac{150.8}{3.0}$$

$$= 50.3 \text{ kJ/hr}$$

Next find the heat leakage from the room with the refrigerator door closed. The heat transfer rate from the room is given by Equation (6.2):

$$\dot{Q}_{room} = U_o A_o(T_o - T_c)$$

$$= (0.46)(5.6)(22 - 4)$$

$$= 46.37 \text{ W}$$

Converting to units of kJ/hr gives

$$\dot{Q}_{room} = 46.37 \text{ J/s} \times 3{,}600 \text{ s/hr} \times 0.001 \text{ kJ/J}$$

$$= 166.9 \text{ kJ/hr}$$

The rate of heat flow leakage from the room is thus three times greater than the average heat flow from the bottle of water.

While the food is cooling, the total heat flow that must be transferred out of the refrigerator to the surroundings is given by Equation (6.3):

$$\dot{Q}_{frig} = \dot{Q}_{room} + \dot{Q}_{food}$$

$$= 166.9 + 50.3$$

$$= 217 \text{ kJ/hr}$$

Once the food reaches the inside temperature of the refrigerator, only the room leakage heat must be removed:

$$\dot{Q}_{frig} = \dot{Q}_{room} = 167 \text{ kJ/hr}$$

Of course, opening the refrigerator door also adds to the burden of heat that must be transferred. Although the processes of heat transfer and air exchange with an open door are complex, a simple way to think about them is that warm air entering the refrigerator is just another type of mass that must be cooled. The mass, temperature, and specific heat of air are the appropriate quantities to include when applying Equation (6.1).

The electrical energy, W_{elec}, supplied to the refrigerator provides the work needed to remove heat from the interior and dump it to the surroundings. A refrigerator thus serves as a *heat pump* in which energy is input in the form of work (electricity) and the useful output in the flow of heat from the cooled space to the warmer surroundings. The electrical energy input provides the needed assist to pump this unwanted heat uphill from a low temperature to a higher temperature. The smaller the assist that is required, the more efficient the device.

An overall energy balance for the refrigerator in Figure 6.2 also reveals that the total rate of heat rejected from the heat pump cycle to the surrounding room, $\dot{Q}_{reject}$, is

$$\dot{Q}_{reject} = \dot{Q}_{room} + \dot{Q}_{food} + \dot{W}_{elec} \qquad (6.4)$$

This equation tells us that the electrical energy supplied to the refrigerator is ultimately converted to heat rejected to the surroundings. In most refrigerators heat rejection occurs from external coils either at the back of the refrigerator or at the bottom front of the refrigerator, where a small fan helps convect heat into the room. If you hold your hand near one of these locations while a refrigerator is running, you will notice the warm air resulting from this heat rejection.

6.4.2   The Refrigeration Cycle

To understand how a refrigerator works, we begin with a qualitative look at the major components of a typical refrigeration system. Figure 6.3 shows these components. The cycle begins at point 1, which is the inlet to the *evaporator*. This is where heat from inside the refrigerator is transferred to a refrigerant such as a CFC, which flows through tubes in the interior of the refrigerator. *In order for heat to flow out of the refrigerator compartment, the refrigerant temperature at point 1 must be lower than the temperature,* T_c, *inside the refrigerator.* Heat then flows naturally to the refrigerant, which enters the evaporator as a liquid mixed with some vapor. It boils (vaporizes) as it picks up heat from inside the refrigerator. As it exits the evaporator at point 2, the refrigerant is in the form of a gas. The evaporator function is thus identical to that of the boiler described in Chapter 5 for power plants.[1]

Most of the heat absorbed by the refrigerant is due to a phase change from liquid to gas, as shown schematically in Figure 6.4. In order to now "dump" this heat to the surroundings, the refrigerant must condense back into a liquid. But if the pressure remains constant, condensation occurs at the same temperature as evaporation (for example, water at atmospheric pressure boils and condenses at 100°C). Increasing the pressure of the refrigerant after it vaporizes will allow it to condense at a higher temperature, in much the same way that a pressure cooker causes water to

---

1 Students may wonder why different words are used for components that perform the identical function in different devices. Some might say it is simply to keep students confused! A more benign answer is that terms like *boiler* and *evaporator* were originally coined by different groups of people working in different industries. That usage persists in our modern vocabulary.

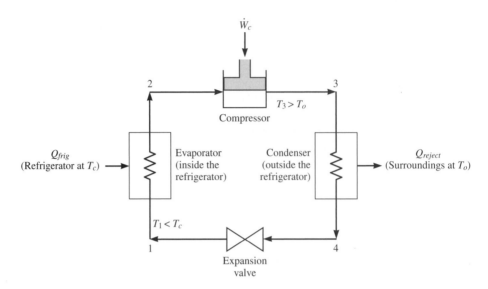

Figure 6.3   Components of a refrigeration system. The refrigerant picks up heat as it passes through the evaporator and compressor. This heat is rejected to the surroundings by the condenser. The refrigerant then passes through an expansion valve to reduce the pressure before returning to the evaporator.

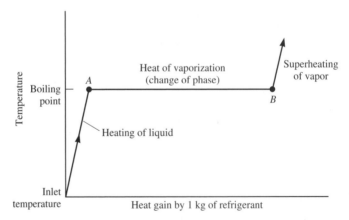

Figure 6.4   Heat gain by a liquid refrigerant increases its temperature. At some point (which depends on the pressure) the liquid begins to evaporate. During this phase change the temperature remains constant. Further heat addition then increases the temperature of the vapor.

boil or condense at temperatures above 100°C. *The temperature of the pressurized refrigerant must be greater than the surrounding room temperature if heat is to flow out of the refrigerant to the surroundings.* A *compressor* is used to raise the pressure of the refrigerant vapor to the required level. Work is required for that purpose, as seen in Figure 6.3. That work is typically supplied in the form of electricity.

The high-pressure, high-temperature vapor at point 3 now enters another heat exchanger, called the *condenser*. For many refrigerators, this can be seen as the coiled tubing protruding from the back of the refrigerator. If you carefully (and very quickly) touch this coil while the compressor is running, you will find that it is quite hot. In the condenser, heat flows from the hot refrigerant vapor to the surrounding room air. As it cools, the refrigerant condenses into a liquid at high pressure.

The final step in the refrigerator cycle is to lower the pressure and temperature of the refrigerant back to the evaporator inlet conditions. This is accomplished by passing the coolant through a capillary tube or valve. The flow restriction creates a lower pressure at the valve exit. As the refrigerant expands into this lower-pressure region it also cools, and a small amount of evaporation typically occurs. The cold, lower-pressure liquid–vapor mixture is now ready to repeat the refrigeration cycle.

6.4.3   Some Basic Questions

The key questions of environmental concern are

- How much CFC is needed to operate a typical refrigerator?

- What substitute refrigerants can be used in place of CFCs?

- How much electricity is needed to operate a refrigerator using CFCs?

- How much is needed using substitute refrigerants?

- What engineering measures are available to reduce energy use and to minimize the need for chemical refrigerants?

In order to answer these questions, a more basic, quantitative look at the underlying thermodynamics of the refrigeration cycle is necessary.

6.4.4   Thermodynamic Relationships

The quantity of refrigerant needed to absorb a given amount of heat in the evaporator depends on the physical and chemical properties of the refrigerant. Commonly, the property known as *specific heat* is defined as the quantity of heat needed to raise the temperature of a unit mass of a substance by one degree. Similarly, the *heat of vaporization* of a substance quantifies the heat absorbed by a unit mass as it boils and changes state from a liquid to a gas at constant temperature. Chapter 5 discussed these properties in the context of power generation cycles.

A more rigorous and generalized framework for quantifying the heat absorption capabilities of a substance involves the thermodynamic property called *enthalpy*. This too was introduced in Chapter 5 but is defined here in more detail because it plays a critical role in refrigerator design. (If you have already studied thermodynamics, all of this will look quite familiar. Still, a brief review may be helpful.)

Consider a mass m of a substance with an initial state defined by pressure P_1, temperature T_1, and volume V_1, as depicted in Figure 6.5. An amount of heat, Q, is then added, and an amount of work, W, is extracted. Consider the case where this

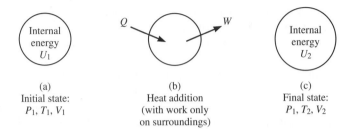

(a)   (b)   (c)
Initial state: Heat addition   Final state:
P_1, T_1, V_1 (with work only P_1, T_2, V_2
                            on surroundings)

Figure 6.5   A thermodynamic process at constant pressure. The addition of heat changes
the internal energy of the system as well as the temperature and volume. Work
is done on the surroundings as the volume changes.

process occurs at constant pressure: The final state of the substance is defined by P_1, T_2, and V_2. The First Law of Thermodynamics prescribes certain relationships between the initial and final states. This law, known as the *conservation of energy,* was introduced in Chapter 1. It states that energy cannot be created or destroyed, but only changed from one form to another. The initial energy of the mass m includes its kinetic energy, potential energy, and *internal energy.* The latter depends only on the temperature of a substance and is commonly denoted by the symbol U [not to be confused with the U_o defined earlier as the overall heat transfer coefficient in Equation (6.2)]. If the kinetic and potential energy of the mass are zero or do not change during the process, then the First Law can be expressed simply as

Energy input = Energy output + Change in internal energy

Referring to Figure 6.5, this equation becomes

$$Q = W + (U_2 - U_1) \tag{6.5}$$

where $(U_2 - U_1)$ is the difference in internal energy at temperatures T_2 and T_1. In this case, the work W also can be expressed in terms of thermodynamic properties. As heat is added, the volume of the substance usually must increase if the pressure is to remain constant. Work is thus required to "push" the boundaries of the system to its final state, as depicted in Figure 6.5. Recalling that pressure is defined as force per unit area, this work can be expressed as

$$W = P(V_2 - V_1) \qquad (P = \text{constant}) \tag{6.6}$$

In this case, no other work is done by the system. Substituting this expression in Equation (6.5) gives

$$Q = P(V_2 - V_1) + (U_2 - U_1)$$
$$= (U_2 + PV_2) - (U_1 + PV_1) \tag{6.7}$$

For convenience, the total internal energy and total volume can be divided by the total mass to define a *specific internal energy, u,* and *specific volume, v*:

$$u = \frac{U}{m} \text{ (Specific internal energy)} \tag{6.8}$$

$$v = \frac{V}{m} \quad \text{(Specific volume)} \qquad (6.9)$$

Then

$$Q = m(u_2 + Pv_2) - m(u_1 + Pv_1) \qquad (6.10)$$

The combination $(u + Pv)$ shows up often in thermodynamics. So instead of always computing these properties separately, they are lumped together to define a new property called the *specific enthalpy, h*:

$$h = u + Pv \quad \text{(by definition)} \qquad (6.11)$$

Because the specific enthalpy of a substance depends only on its temperature, pressure, and specific volume (these three properties are themselves interrelated, so that only two of them can be specified independently), enthalpy values for different substances can easily be tabulated to facilitate calculations. Although it was derived here for the case of constant pressure, the concept of enthalpy applies generally to any changes in pressure.

Substituting the definition for enthalpy into Equation (6.10) gives

$$Q = m(h_2 - h_1) \qquad (6.12)$$

This relationship is used extensively in thermodynamics. It is a more general form of the commonly used expression involving specific heat:

$$Q = mc_p(T_2 - T_1) \qquad (6.13)$$

The above expression is useful for engineering analysis when the temperature difference is not too large, or when no phase change occurs. Over large temperature ranges the specific heat of many substances is not constant, especially for gases. Furthermore, during a phase change, as when a liquid boils and becomes a gas, the specific heat is not defined. Rather, we must use the heat of vaporization.

Enthalpy, however, is defined over all ranges of temperature and pressure, including phase changes. It is actually the property used to define the specific heat and the heat of vaporization. As noted earlier, the latter refers to the thermal energy required to evaporate a unit mass of a substance. The specific enthalpy of a liquid as it just begins to boil (corresponding to point A in Figure 6.4) is designated as h_f and is known as the enthalpy of *saturated liquid*. The enthalpy at the point where the liquid is completely vaporized is designated as h_g and is known as the enthalpy of *saturated vapor*. The points h_f and h_g have the same temperature and pressure, but the difference in their values reflects the heat absorbed during evaporation (or released during condensation). This quantity defines the heat of vaporization commonly used in chemistry and physics textbooks. In thermodynamics this quantity is denoted by the symbol h_{fg}:

$$h_{fg} = h_g - h_f \quad \text{(Heat of vaporization)} \qquad (6.14)$$

Between the saturated liquid and saturated vapor states, there is a mixture of liquid and vapor. The enthalpy of this mixture is based on the relative amounts of liquid and vapor. A fraction called the *quality* of the mixture is defined as the ratio of vapor mass to total mass. The quality is usually denoted by the letter x:

$$x = \frac{\text{Mass of vapor}}{\text{Total mass of vapor } + \text{ liquid}} \qquad (6.15)$$

A quality of zero refers to saturated liquid, where $h = h_f$. At $x = 1$, we have saturated vapor, where $h = h_g$. For liquid–vapor mixtures, the enthalpy increases in proportion to the quality fraction:

$$h_{mix} = h_f + xh_{fg} \qquad (6.16)$$

Example 6.4

Enthalpy of a liquid–vapor mixture.   The specific enthalpy of ammonia (an early refrigerant) is 200.0 kJ/kg in its saturated liquid state at 0°C and 429.4 kPa. The enthalpy of saturated vapor is 1,461.8 kJ/kg. For these conditions, determine (a) the heat of vaporization and (b) the specific enthalpy of a mixture of 70% vapor, 30% liquid ammonia by mass.

Solution:

(a)   Equation (6.14) defines the heat of vaporization:

$$h_{fg} = h_g - h_f$$
$$= 1{,}461.8 - 200.0$$
$$= 1{,}261.8 \text{ kJ/kg}$$

This is the amount of heat required to evaporate 1 kg of ammonia at 0°C, 429.5 kPa.

(b)   The quality of the liquid–vapor mixture is given by Equation (6.15):

$$x = \frac{\text{Vapor mass}}{\text{Total mass}} = \frac{0.70}{1.00} = 0.70$$

From Equation (6.16), the specific enthalpy of the mixture is then

$$h_{mix} = h_f + xh_{fg}$$
$$= 200.0 + 0.70\,(1{,}261.8)$$
$$= 1{,}083.3 \text{ kJ/kg}$$

6.4.5   Refrigerant Properties

Table 6.2 shows the tabulated values of enthalpy for CFC-12, the common refrigerant known as *freon*. For comparing the properties of different refrigerants, it is often helpful to display the enthalpy graphically as a function of pressure. Such a diagram is illustrated in Figure 6.6.

The dome-shaped line in Figure 6.6 represents the saturated liquid and saturated vapor states of the refrigerant. Points 1 through 4 on this figure correspond to the numbered points of Figure 6.3. The cycle 1–2–3–4 represents a basic refrigeration cycle. In real refrigerators, the gaseous refrigerant leaving the evaporator is often

Table 6.2 Thermodynamic properties of CFC-12.

Properties of Saturated Liquid and Saturated Vapor							
Temperature, °C	Pressure, MPa	Density, kg/m³ Liquid	Volume, m³/kg Vapor	Enthalpy, kJ/kg Liquid	Vapor	Entropy, kJ/kg-K Liquid	Vapor
−30.00	0.10044	1,487.2	0.16029	172.81	338.81	0.8951	1.5779
−20.00	0.15088	1,457.6	0.10965	181.72	343.53	0.9309	1.5701
−10.00	0.21893	1,427.1	0.07731	190.78	348.17	0.9658	1.5639
0.00	0.30827	1,395.6	0.05593	200.00	352.68	1.0000	1.5590
10.00	0.42276	1,362.8	0.04134	209.41	357.05	1.0335	1.5550
20.00	0.56651	1,328.6	0.03111	219.03	361.23	1.0666	1.5516
30.00	0.74379	1,292.5	0.02376	228.89	365.16	1.0992	1.5487
40.00	0.95909	1,254.2	0.01837	239.03	368.81	1.1315	1.5459
50.00	1.2171	1,213.0	0.01432	249.51	372.07	1.1638	1.5431
60.00	1.5227	1,168.2	0.01123	260.37	374.86	1.1961	1.5398
70.00	1.8814	1,118.5	0.00882	271.73	377.01	1.2288	1.5356

Selected Properties of Superheated Vapor							
Temperature K	°C	P = 0.15 MPa v m³/kg	h kJ/kg	s kJ/kg-K	P = 1.00 MPa v m³/kg	h kJ/kg	s kJ/kg-K
260.0	−13	0.11374	347.14	1.5849			
270.0	−3	0.11879	353.05	1.6072			
280.0	7	0.12378	359.04	1.6290			
290.0	17	0.12873	365.11	1.6503			
300.0	27	0.13363	371.27	1.6711			
310.0	37	0.13849	377.50	1.6916			
320.0	47	0.14333	383.81	1.7116	0.018122	372.67	1.5559
330.0	57	0.14814	390.20	1.7313	0.019158	380.11	1.5788
340.0	67	0.15293	396.67	1.7506	0.020141	387.45	1.6007

Source: Based on ASHRAE, 1993.

superheated to a temperature several degrees above the boiling point, as is indicated by the enthalpy increase from point 2 (saturated vapor) to point 2′. The superheating ensures that no liquid enters the compressor because a liquid–vapor mixture cannot be easily compressed. Similarly, the liquid refrigerant exiting the condenser may be subcooled to a temperature slightly below the boiling point, as reflected by the point 4′. Subcooling brings the refrigerant slightly closer to the saturated liquid state after expansion from point 4′ to point 1′. The cycle 1′–2′–3′–4′ has some practical advantages over the basic cycle 1–2–3–4 but requires more complex design calculations. In

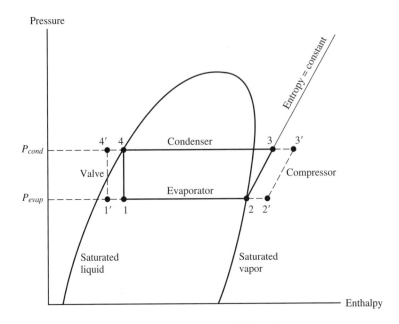

Figure 6.6   Pressure–enthalpy diagram for a typical refrigerant. The points on refrigerant cycles (1–2–3–4 or 1′–2′–3′–4′) correspond to the points on the diagram of Figure 6.3. The dashed lines show the cycle with superheating and subcooling of the refrigerant.

this chapter we will analyze only the basic cycle because it illustrates all of the key points with less computational complexity.

Another thermodynamic property called *entropy* also plays an important role in the refrigeration cycle. This is seen in Figure 6.6 where the enthalpy increase during compression (point 2 to point 3) occurs along a line of constant entropy. This represents an ideal process known as *isentropic* (constant entropy) compression. A detailed discussion of entropy and its role in engineering analysis is left for a first course in thermodynamics. Suffice it to say that entropy affects the amount of work required to compress a refrigerant. In real processes, the entropy tends to increase, requiring more work than the ideal isentropic process. Later in the chapter we will introduce a simple method of accounting for this inefficiency.

6.5   DESIGNING A CFC-FREE REFRIGERATOR

We are now ready to examine in greater detail some of the engineering approaches available to reduce the direct and indirect environmental impacts of refrigerators. We begin by examining the amount of refrigerant that is needed and the options for CFC-free refrigerants that do not deplete the ozone layer.

6.5.1 Refrigerant Mass Flow

We saw in Example 6.1 that the environmental impact of CFCs depends on the total amount of CFC in use. The first questions one might ask, therefore, are how much refrigerant is required, and whether that amount can be reduced through improved engineering design. As we shall see, the answers to these questions come partly from engineering calculations and partly from trial-and-error experiments.

First, the required mass flow *rate* of refrigerant can be determined directly from Equation (6.12), which relates the refrigerator heat load to the change in specific enthalpy of the refrigerant used:

$$\dot{m} = \frac{\dot{Q}_{frig}}{h_2 - h_1} \tag{6.17}$$

where h_2 and h_1 are the specific enthalpies leaving and entering the evaporator, as depicted in Figure 6.6. Note that the refrigerant mass flow is expressed here on a rate basis because the refrigerator heat load, $\dot{Q}_{frig}$, is typically specified as a rate of heat removal, as seen earlier in Equation (6.3).

Since the mass flow rate of refrigerant is proportional to the heat load $\dot{Q}_{frig}$, an engineer must choose the maximum or design value of $\dot{Q}_{frig}$, which we will call $\dot{Q}_{design}$. This value, in turn, will determine the maximum refrigerant flow rate that is required.

The concept of a *pull-down time*, Δt_{pd}, is commonly used to determine the design heat load of a refrigerator. This is defined as the time required to cool the air inside an empty refrigerator from room temperature down to the refrigerator and freezer operating temperatures. The next example illustrates this calculation.

Example 6.5

Calculation of the design heat load. A design pull-down time of two minutes is specified for a 0.5 m³ (18 ft³) household refrigerator-freezer. The design operating temperatures are 4°C for the refrigerator fresh food section (volume = 0.36 m³) and −15°C for the freezer section (volume = 0.14 m³). The design ambient room temperature is 32°C. The average density of air in the refrigerator is 1.2 kg/m³, and its specific heat is 1.0 kJ/kg-K. Calculate the design heat load.

Solution:

Equation (6.1) is used to determine the total heat that must be removed from the air inside the refrigerator–freezer. The air is initially at 32°C (a worst-case value). The total mass of air is equal to the product of density times volume. Separate calculations are needed for the refrigerator and freezer compartments because they have different volumes and design temperatures:

$$Q_{frig} = m_{air}c_p(T_o - T_c)_{frig} = (\rho_{air} V_{frig})c_p(T_o - T_c)_{frig}$$
$$= (1.2)(0.36)(1.0)(32 - 4)$$
$$= 12.096 \text{ kJ}$$
$$Q_{freezer} = (\rho_{air} V_{freezer})c_p(T_o - T_c)_{freezer}$$
$$= (1.2)(0.14)(1.0)[32 - (-15)]$$
$$= 7.896 \text{ kJ}$$

Thus

$$Q_{total} = Q_{frig} + Q_{freezer}$$
$$= 12.096 + 7.896$$
$$= 19.99 \text{ kJ}$$

Dividing this total by the design pull-down time of two minutes gives the design heat load:

$$\dot{Q}_{design} = \frac{Q_{total}}{\Delta t_{pd}}$$

$$= \frac{(19.99 \text{ kJ})(60 \text{ min/hr})}{2 \text{ min}}$$

$$= 600 \text{ kJ/hr}$$

Recall that this design is based on a high year-round room temperature of 32°C (90°F).

After the design heat load is established, the enthalpy change in Equation (6.17) also must be specified in order to calculate the refrigerant mass flow rate. This enthalpy difference is known as the *refrigeration effect, RE*. This is the quantity of heat removed in the evaporator by a unit mass (1 kg) of the selected refrigerant:

$$RE = h_2 - h_1 \text{ (by definition)} \tag{6.18}$$

Once engineers specify the design conditions for the refrigerator, the refrigeration effect can be easily determined for any selected refrigerant. The key design parameters that must be specified are the refrigerant temperatures entering the evaporator and the condenser. Any superheating at the evaporator outlet also must be specified. As depicted earlier in Figure 6.3, the design evaporator temperature, T_1, must be lower than the design refrigerator or freezer temperature, T_c, in order for heat to flow to the refrigerant. Similarly, the design condenser temperature, T_3, must be higher than the design ambient temperature, T_o, in order for heat to flow to the surroundings. The choice of these temperature differences is based in part on the need to keep the heat exchanger sizes reasonable. Because the heat transfer to or from the refrigerant is based on equations just like Equation (6.2) (that is, $Q = U_o A_o \Delta T$), if ΔT is small, the heat exchanger surface area, A_o, must be big. The following example illustrates how the refrigeration effect is determined for a specified design.

Example 6.6

Calculation of refrigeration effect.   An evaporator design temperature of -20°C is selected for the refrigerator–freezer in Example 6.5. This is 5°C below the design freezer temperature of -15°C. For the condenser, a design temperature of 40°C is selected, 8°C above the design room temperature. For simplicity, assume the evaporator outlet state is saturated vapor (no superheating) and the condenser outlet is saturated liquid (no subcooling). The refrigerant is CFC-12. Sketch the cycle on an enthalpy–pressure diagram, and determine the value of the refrigeration effect for this design.

Solution:

The cycle diagram for this design corresponds to points 1–2–3–4 in Figure 6.6. The evaporator and condenser pressures and exit enthalpies are found from Table 6.2 for saturated conditions at the design temperatures:

Evaporator $(T = -20°C)$: $P_{evap} = 0.15$ MPa $= P_1 = P_2$

Thus

$$h_2 = h_g = 343.53 \text{ kJ/kg}$$

Condenser $(T = 40°C)$: $P_{cond} = 0.96$ MPa $= P_3 = P_4$

Thus

$$h_4 = h_f = 239.03 \text{ kJ/kg}$$

We also know that $h_1 = h_4$ for the constant enthalpy expansion. Thus from Equation (6.18) the refrigeration effect is

$$RE = h_2 - h_1$$

$$= 343.53 - 239.03 = 104.5 \text{ kJ/kg}$$

The value of refrigeration effect is an important parameter in comparing different refrigerants because it gives a direct measure of the amount of heat removed by a unit mass of refrigerant. The greater the refrigeration effect, the smaller the mass of chemical refrigerant needed to meet a given design requirement.

We can now calculate the mass flow rate of refrigerant needed for a specified design. The next example illustrates the calculation.

Example 6.7

Calculation of refrigerant mass flow rate. Based on the design assumptions in Examples 6.5 and 6.6, determine the mass flow rate of CFC-12 that is required for the refrigerator–freezer in those examples.

Solution:

From Example 6.5 the design heat load is 600 kJ/hr, and from Example 6.6 the refrigeration effect is 104.5 kJ/kg. Equation (6.17) then gives the required mass flow rate:

$$\dot{m}_{CFC} = \frac{\dot{Q}_{design}}{h_2 - h_1} = \frac{600}{104.5}$$

$$= 5.74 \text{ kg/hr}$$

Table 6.3 summarizes the key design parameters that went into determining the required mass flow rate of refrigerant. Notice that all of these specifications reflect a considerable amount of engineering judgment as to what constitutes an acceptable refrigerator design.

Table 6.3   Engineering design parameters affecting refrigerant mass flow rate.

Design Parameter	Symbol
Refrigerated volume	V
Interior temperature (minimum)	T_c
Outside temperature (maximum)	T_o
Pull-down time	Δt_{pd}
Evaporator temperature	T_1
Condenser temperature	T_4

The mass flow rate of refrigerant together with refrigerant transport properties such as viscosity, thermal conductivity, density, and specific heat are subsequently used to calculate the surface area and length of tubing needed in the evaporator and condenser to transfer heat at the design rates. These detailed heat transfer calculations are beyond the scope of this text.

6.5.2   Refrigerant Charge

The total mass, m, of refrigerant flowing through the system is also known as the refrigerant *charge*. Environmentally, this important quantity represents the total amount of refrigerant that might ultimately enter the atmosphere. For a substance flowing through a pipe or tube, we can relate the total mass to the mass flow rate by knowing the pipe length (L) and the velocity (v) of refrigerant through the pipe:

$$m = \dot{m}\frac{L}{v} \tag{6.19}$$

Example 6.8

Calculation of refrigerant charge.   For the refrigerator design in Example 6.7, detailed heat transfer calculations reveal that the total heat exchange area required for the evaporator and condenser is 0.90 m². That surface area is provided by metal tubing with a total length of 30 m. The velocity of circulating refrigerant is 0.25 m/s. Calculate the mass of refrigerant charge.

Solution:

From Example 6.7, the mass flow rate of CFC refrigerant is 5.74 kg/hr. Use this value in Equation (6.19) to find the total mass for the given length and velocity:

$$m_{CFC} = \dot{m}_{CFC}\frac{L}{v} = \frac{(5.74 \text{ kg/hr})(30 \text{ m})}{(0.25 \text{ m/s})(3{,}600 \text{ s/hr})}$$

$$= 0.19 \text{ kg}$$

In practice, the total refrigerant charge is determined partly from calculations and partly from additional considerations related to the design of an actual device. Trial-and-error experiments traditionally have been used to account for physical and thermal processes within an actual refrigeration cycle that cause degradation or loss of refrigerant. Consequently, the total mass of refrigerant is usually somewhat greater than theoretical calculations based only on heat transfer considerations. A typical value for a household refrigerator like the one in Example 6.7 is about 0.25 kg, 30 percent greater than the calculated value in Example 6.8.

As we shall see, however, only the refrigerant mass flow rate—and not the total refrigerant mass—is required for engineering design calculations related to refrigerant effectiveness and energy consumption. These are the two major topics of concern in this chapter.

6.5.3 Refrigeration Cycle Efficiency

Another important design parameter in selecting a refrigerant is the thermal efficiency of the refrigeration cycle. The *thermal efficiency* of a system is usually defined as the ratio of the useful energy output to the total energy input (sometimes referred to as the energy that "costs"). For technologies such as automobiles (Chapter 3) or electric power plants (Chapter 5), the energy input is the chemical energy in the fuel, and the useful energy output is the work done by a rotating shaft. These two technologies are examples of *heat engines*. In contrast, the useful energy derived from a refrigeration cycle is the unwanted heat removed by the evaporator. Refrigerators are an example of a *heat pump*. The energy input that costs is the work needed to compress the refrigerant. The term *coefficient of performance (COP)*, rather than efficiency, is used to define the ratio of useful energy delivered to the energy input that costs for a refrigeration cycle:

$$\text{Coefficient of performance (COP)} = \frac{\text{Heat removed in evaporator}}{\text{Work input for compression}}$$

$$= \frac{Q_{frig}}{W_c} \tag{6.20}$$

For a unit mass of refrigerant, the numerator is simply equal to the refrigeration effect, $RE = h_2 - h_1$. Referring to Figure 6.7, a simple energy balance shows that the compressor work per unit mass, w_c, is equal to the specific enthalpy difference $h_3 - h_2$. Thus another way of expressing the COP is

$$\text{COP} = \frac{RE}{w_c} = \frac{h_2 - h_1}{h_3 - h_2} \tag{6.21}$$

The following example illustrates the calculation of the COP.

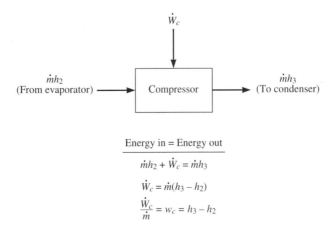

Figure 6.7   Energy flows for an ideal compressor.

Example 6.9

Coefficient of performance for a refrigerator cycle.   Calculate the COP for the refrigeration cycle in Example 6.6 assuming isentropic compression.

Solution:

From Example 6.6 we have

$$RE = h_2 - h_1 = 104.5 \text{ kJ/kg}$$

This is the heat absorbed in the evaporator by a unit mass of refrigerant. Next we must calculate the compressor work per unit mass, given by the specific enthalpy change $h_3 - h_2$ in the cycle diagram from Example 6.6. From that example, we know the properties of saturated vapor at point 2 exiting the evaporator, including the specific entropy, s_2, obtained from the tables of thermodynamic properties for CFC-12:

$$P_2 = 0.15 \text{ MPa}, h_2 = 343.53 \text{ kg/kJ}, s_2 = 1.5701 \text{ kJ/kg–K}$$

We also know from Example 6.6 that $P_3 = 0.96$ MPa, and because the compression is isentropic (constant entropy), we have $s_3 = s_2 = 1.5701$ kJ/kg-K. Knowing both P_3 and s_3 allows us to find all other conditions at point 3 from the tables of thermodynamic properties (Table 6.2). The value sought is the enthalpy, h_3. Using tabulated values for $P = 1.0$ MPa, and interpolating to find the conditions corresponding to $s_3 = 1.5701$ kJ/kg-K, the result for this case is approximately

$$h_3 = 377.3 \text{ kJ/kg} \ (T_3 = 53°\text{C})$$

The compressor work per unit of mass is thus

$$w_c = h_3 - h_2$$
$$= 377.3 - 343.53 = 33.8 \text{ kJ/kg}$$

Equation (6.21) then gives us the value of COP:

$$\text{COP} = \frac{h_2 - h_1}{h_3 - h_2} = \frac{104.5}{33.8} = 3.09$$

Notice that the COP has a value greater than 1.0, in contrast to the thermal efficiency values for automobile engines, power plants, and other types of heat engines, which are all much less than 1.0. This is because a heat pump merely assists the flow of thermal energy from a low temperature to a higher temperature, rather than converting thermal energy to mechanical energy (work), as in a heat engine. Still, it can be said that *the higher the COP, the more efficient the device,* because a higher COP means that less external work is needed to remove the desired heat. As we shall see shortly, different refrigerants have different COPs for a given refrigerator design. For this reason, COP is an important criterion in selecting a refrigerant.

6.5.4 Comparison of Alternative Refrigerants

As discussed earlier, the major environmental challenges for refrigerator design are to eliminate the CFCs that contribute to stratospheric ozone depletion and global warming, and to improve energy efficiency so as to reduce the adverse impacts of electric power production. Accordingly, two desirable attributes of a non-CFC refrigerant include a high refrigeration effect (*RE* value) so as to minimize the mass of refrigerant needed to meet a design heat load, plus a high COP so as to maximize energy efficiency. The question then is whether such refrigerants exist. If so, do any other factors militate against their use? If there are no perfect substitutes for CFCs, what trade-offs or penalties are involved in the use of non-CFC refrigerants?

To answer such questions, one must begin by repeating the thermodynamic analyses of Examples 6.6 and 6.9 for other candidate refrigerants. Figure 6.8 and Table 6.4 show the results of such an analysis for four alternative refrigerants, including CFC-12 as a reference case (Naser et al., 1994). The alternatives include one

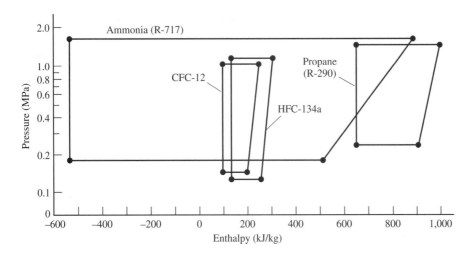

Figure 6.8 Comparison of refrigeration cycles for alternative refrigerants. Note that the reference state for zero enthalpy differs from Table 6.2. (*Source:* Based on Naser et al., 1994)

Table 6.4   Performance characteristics of alternative refrigerants.

Chemical Class	Refrigerant Name(s)	Relative Mass[a]	Relative COP[b]	Relative Volume[c]	T_3 (°C)[d]	P_3 (MPa)[e]	Pressure Ratio[f]
CFC	CFC-12 (R-12)	1.00	1.00	1.00	66	1.11	7.3
HFC	HFC-134a (R-134a)	0.82	0.93	1.24	60	1.19	9.1
HC	Propane (R-290)	0.42	1.01	0.69	60	1.59	7.2
Inorganic	Ammonia (R-717)	0.10	0.92	0.57	171	1.83	9.9

[a] Refrigerant mass flow rate relative to CFC-12. The reciprocal of this value is the relative refrigeration effect, *RE* (kJ/kg).
[b] Coefficient of performance relative to CFC-12.
[c] Effective compressor volume (displacement) relative to CFC-12.
[d] Compressor discharge temperature for the design refrigeration cycle.
[e] Compressor discharge pressure to the condenser.
[f] Ratio of compressor discharge pressure to inlet pressure.
Source: Based on Naser et al., 1994.

hydrofluorocarbon (HFC-134a), one pure hydrocarbon (propane), and one inorganic compound (ammonia). The cycle analyzed in this case is slightly different from the one used in our previous examples; it includes a small amount of superheating and subcooling corresponding to the cycle 1′–2′–3′–4′ shown earlier in Figure 6.6.

Figure 6.8 shows graphically that the four different refrigerants require different operating pressures to achieve the specified evaporator and condenser design temperatures. The magnitude of the refrigeration effect—indicated by the length of the lower horizontal line segment of each cycle diagram—also varies greatly among the four candidate refrigerants.

To compare these options, Table 6.4 shows the value of several key performance parameters. Several of these parameters are normalized on the value for CFC-12. Thus the column labeled "Relative Mass" shows the value of refrigerant mass flow rate relative to CFC-12.

Mass and Energy Requirements   The relative mass value of 0.10 for ammonia means that only one-tenth as much ammonia is needed compared to CFC-12. This reflects the much larger refrigeration effect seen for ammonia in Figure 6.8. On the other hand, the relative COP value for ammonia is 8 percent lower than CFC-12, which means that 8 percent more energy is required to operate the ammonia refrigeration cycle. Relative to CFC-12, the only refrigerant in Table 6.4 with both a lower mass flow requirement and a higher COP is propane.

Compressor Requirements   The additional performance parameters in Table 6.4 relate to the compressor requirements. The relative compressor volume reflects the size of compressor needed to operate the refrigeration cycle. The compressor volume primarily reflects the density of refrigerant in the compressor, though an allowance also is made for "dead space" and flow leakage in typical compressor designs (as discussed later). One sees that ammonia requires a compressor size only 57 percent that of CFC-12. This is beneficial because a smaller compressor takes up less space and reduces the cost of the refrigerator.

System Temperature and Pressure   On the other hand, Table 6.4 shows that ammonia also has a very high compressor discharge temperature, which could aggravate operating problems and increase the need for maintenance. Ammonia also has the highest operating pressure and pressure ratio of the four systems. These too are disadvantages because high pressure requires heavier and more complex construction to prevent leakage, while high pressure ratios tend to reduce the compressor efficiency by causing greater departures from ideal isentropic conditions.

In contrast, propane (also known as R-290) has a slightly lower pressure ratio than CFC-12, with a compressor size 31 percent smaller. The compressor discharge temperature also is slightly lower than for CFC-12, although the maximum pressure is higher. Overall, the data in Table 6.4 indicate that thermodynamically propane is indeed an attractive alternative to CFC-12.

Safety Considerations   The principal disadvantage of propane is its flammability. Because of this factor, the use of propane in sealed systems is subject to stringent safety and regulatory requirements that inhibit its use as a household refrigerant. Technology developers, however, argue that current regulatory restrictions are inappropriate given the very small quantities of propane that are required and the ability to design systems that are virtually leakproof and incapable of exploding. The viability of propane as a refrigerant is thus tied primarily to policies regarding safety.

Overall System Design   The difficulty in finding alternative refrigerants that meet all desired design criteria underscores the historical choice of CFCs from a purely engineering and safety standpoint, before their environmental impacts were fully understood. Table 6.4 shows that one of the next best alternatives that avoids the ozone-depleting potential of CFCs is HFC-134a. This refrigerant has a slightly higher refrigeration effect than CFC-12, but a lower COP and a larger compressor volume and pressure ratio. Like CFCs, however, it is nonflammable, nontoxic, and extremely stable. Its principal technical drawback is that it is incompatible with the mineral oils that are typically mixed with refrigerants to provide internal lubrication for the compressor. However, new classes of lubricants recently have been developed that have the proper solubility and other required characteristics. At this time HFC-134a has emerged as the preferred substitute for CFC-12 for new refrigerators and automotive air conditioners, although it cannot replace CFCs in existing devices because of the differences in compressor requirements just noted.

This discussion illustrates the often complex nature of trade-offs involved in finding environmentally friendly solutions to engineering design problems. We continue that discussion with a look at factors affecting the energy requirements of a refrigerator.

6.6   REDUCING ENERGY CONSUMPTION

Figure 6.2 earlier showed the refrigerator as a simple "black box" requiring electricity for its operation. We are now in position to analyze the amount of electricity required and ways of reducing the total annual energy consumption. This will involve a detailed look at the energy-using components of a refrigerator and the engineering design choices that affect overall energy consumption.

6.6.1   Compressor Energy Requirements

The compressor is the principal consumer of energy in a refrigerator. From Equation (6.20), the electrical energy needed for an ideal (isentropic) compressor is

$$W_c = \frac{Q_{frig}}{COP} \text{ (Ideal process)} \tag{6.22}$$

In practice, however, compressors require more energy than this because of three types of inefficiencies. One is the fact that compression does not actually take place under ideal isentropic conditions. In real devices there is usually an increase in entropy that requires additional work for compression. An *isentropic efficiency,* η_s, can be defined to account for this effect:

$$\eta_s = \frac{\text{Work required for isentropic compression}}{\text{Work required for actual compression}} \tag{6.23}$$

The value of isentropic efficiency must be determined experimentally for each refrigerant, but typical values are in the range of 0.65–0.75.

Another source of inefficiency, mentioned in the discussion of Table 6.4, is the fact that the compressors used in refrigerators are designed with a small clearance volume between the end of the piston stroke and the cylinder wall. This clearance volume is typically 5 to 10 percent of the total volumetric displacement. Thus not all of the refrigerant is expelled on each stroke of the piston; some remains in the clearance volume. In addition, small amounts of refrigerant vapor leak past the piston, and some refrigerant is absorbed in the compressor lubricating oil. Miscellaneous heat losses can further reduce the effective volume of the compressor. To account for all these factors, an overall *volumetric efficiency,* η_v, can be defined as

$$\eta_v = \frac{\text{Actual gas volume exiting compressor}}{\text{Total compressor displacement volume}} \tag{6.24}$$

All volume losses translate directly into losses in the total refrigerant energy delivered by the compressor to the condenser. Because less heat is being pumped than the design value for an ideal cycle, the overall efficiency of the system is diminished.

An equation can be derived to express the volumetric efficiency in terms of the clearance volume fraction, f_c, the specific volume, v, of the refrigerant entering and leaving the compressor, and a fraction, f_v, that accounts for volumetric losses besides those due to the clearance volume. This equation is

$$\eta_v = f_v\left[1 - f_c\left(\frac{v_2}{v_3} - 1\right)\right]$$

(6.25)

The values v_2 and v_3 are the specific volumes (m³/kg) of refrigerant entering and leaving the compressor, corresponding to points 2 and 3 on the cycle diagram in Figure 6.6. Numerical values of v for any refrigerant can be found in tables of thermodynamic properties. Data from refrigerator compressors shows that typical values of f_v are around 0.9, indicating a 10 percent effective volume loss from sources other than the clearance volume.

A third type of inefficiency associated with refrigerant compression is in the electric motor used to drive the compressor (see Figure 6.9). Losses in the motor arise from friction and heating in the bearings and coils of the motor. The *motor efficiency*, η_m, is defined as

$$\eta_m = \frac{\text{Motor energy output to compressor}}{\text{Total electrical input to motor}}$$

(6.26)

Based on the three types of losses just described, an *overall compressor efficiency*, η_c, can be expressed as the product of the isentropic efficiency, volumetric efficiency, and motor efficiency:

$$\eta_c = \eta_s \times \eta_v \times \eta_m$$

(6.27)

This overall efficiency relates the theoretical compressor work given by Equation (6.22) to the actual electrical energy consumption:

$$\eta_c = \frac{\text{Theoretical compressor work}}{\text{Actual energy input for compression}} = \frac{W_c}{W_{c,act}}$$

(6.28)

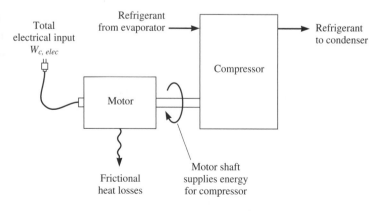

Figure 6.9   Schematic of energy flows for refrigerant compression.

Substituting this expression into Equation (6.22) gives the actual compressor energy requirement as a function of the refrigerant heat load, coefficient of performance, and overall compressor efficiency:

$$W_{c,act} = \frac{Q_{frig}}{(COP)\eta_c}$$ (6.29)

If energy flows are expressed on a rate basis (energy per unit time), this equation also can be used to find the electrical power needed to operate the compressor, as illustrated in the following example.

Example 6.10

Compressor efficiency and power consumption.   Assume that the CFC-12 refrigeration cycle in Example 6.9 has an isentropic efficiency of 70 percent. The compressor clearance volume is 5 percent, and the miscellaneous loss factor is 0.9. The motor efficiency is 80 percent. Find the overall compressor efficiency and the compressor power needed to meet the design refrigeration load.

Solution:

The overall compressor efficiency is given by Equation (6.27):

$$\eta_c = \eta_s \times \eta_v \times \eta_m$$

The isentropic efficiency is given as 0.70, and the motor efficiency is given as 0.80. To calculate the compressor volumetric efficiency, use Equation (6.25). The specific volume of CFC-12 entering and leaving the compressor is found from tables of thermodynamic properties for the design pressure and temperature at the compressor inlet and outlet:

Inlet = Saturated vapor at 0.15 MPa, $-20°C$: $v_2 = 0.1097$ m³/kg
Outlet = Superheated vapor at 0.96 MPa, 53°C: $v_3 = 0.0188$ m³/kg

Substituting these values in Equation (6.25) gives

$$\eta_v = f_v\left[1 - f_c\left(\frac{v_2}{v_3} - 1\right)\right]$$

$$= 0.9\left[1 - (0.05)\left(\frac{0.1097}{0.0188} - 1\right)\right] = 0.68$$

Then the overall compressor efficiency from Equation (6.27) is

$$\eta_c = (0.70)(0.68)(0.80)$$

$$= 0.381, \text{ or } 38.1\%$$

Using this result in Equation (6.29), along with the COP value of 3.09 from Example 6.9, gives the electrical power (in watts) required at the design heat load:

$$\dot{W}_{c,act} = \frac{\dot{Q}_{design}}{(COP)(\eta_c)}$$

$$= \frac{(600 \text{ kJ/hr})(1,000 \text{ J/kJ})}{(3.09)(0.381)(3,600 \text{ s/hr})}$$

$$= 142 \text{ watts}$$

Note the relatively low value of overall compressor efficiency of 38.1 percent in this example. Because this efficiency directly affects the electrical energy requirements of a refrigerator, improvements in compressor design and motor design can dramatically reduce overall energy use.

Of course, as you probably know from your own experience, a refrigerator compressor runs only intermittently, not continuously. This occurs because the actual heat load at any moment is often less than the design load. Thus the total compressor energy consumption over a year is given by

$$W_{c,elec} = \dot{W}_{c,act}(8{,}760 \times f_{run}) \tag{6.30}$$

where $W_{c,elec}$ is the annual compressor energy use, $\dot{W}_{c,act}$ is the electrical power for the design heat load, 8,760 is the total number of hours in a year, and f_{run} is the fraction of hours in a year the compressor is running. This fraction will depend on the actual operating conditions and use of the refrigerator, as illustrated shortly.

6.6.2 Auxiliary Energy Requirements

Besides the compressor, household refrigerators require electricity to run two small fans, typically one for the evaporator and one for the condenser. Blowing air across these heat exchangers improves the rates of heat transfer to and from the circulating refrigerant. These two fans operate whenever the compressor is running. In many refrigerators electrical energy also may be used for a small electrical heater strip that prevents condensation around the refrigerator doors in humid weather. This heater also operates only a fraction of the time, similar to the small fans. We can then define the total auxiliary power requirement, $\dot{W}_{aux}$, as

$$\dot{W}_{aux} = \dot{W}_{fans} + \dot{W}_{heater} \tag{6.31}$$

For a standard household refrigerator–freezer such as used in the illustrative examples, a typical value for auxiliary power is about 43 watts, consisting of 10 watts for the evaporator fan, 14 watts for the condenser fan, and 19 watts for the anticondensation heater (Naser et al., 1994). The annual auxiliary energy requirement is then

$$W_{aux} = \dot{W}_{aux}(8{,}760 \times f_{run}) \tag{6.32}$$

Note that this equation neglects the small amount of energy used by the interior lightbulbs that shine when the refrigerator or freezer door is opened.

6.6.3 Total Energy Consumption

Combining Equation (6.30) and Equation (6.31) gives the total electrical energy used by the refrigerator in a year:

$$W_{elec} = (\dot{W}_{c,act} + \dot{W}_{aux})(8{,}760 \times f_{run}) \tag{6.33}$$

Example 6.11

Refrigerator annual energy consumption.   Calculate the total power requirement and annual electrical energy consumption (in kilowatt-hours) for the refrigeration in Example 6.10, assuming that the auxiliary power is 43 watts and the compressor runs 43 percent of the time (as derived later in Example 6.12).

Solution:

From Example 6.10, the compressor power requirement is 142 W. The total power requirement is thus $142 + 43 = 185$ W. The total energy consumption from Equation (6.33) is

$$W_{elec} = (185 \text{ W})(0.43 \times 8{,}760 \text{ hrs/yr})\left(\frac{1 \text{ kW}}{1{,}000 \text{ W}}\right)$$

$$= 697 \text{ kW-hr/yr}$$

We see that a key factor affecting the annual energy consumption is the fraction of time the compressor is running. This fraction is determined by the actual refrigerator heat gain, $\dot{Q}_{frig}$, defined earlier by Equation (6.3). From an engineering perspective we are most interested in how the heat gain is affected by engineering design (as opposed to the additional heat gain from storing different amounts of food and warm air each time the door is opened). We therefore now take a closer look at how the thermal design of a refrigerator affects the actual heat gain.

6.6.4   Effect of Thermal Insulation Design

As illustrated earlier in Example 6.3, once the food items in a refrigerator have cooled to the desired temperature, the actual heat load on the refrigeration cycle is determined by the heat leakage from the surrounding room, given earlier by Equation (6.2):

$$\dot{Q}_{room} = U_o A_o (T_o - T_c)$$

The Overall Heat Transfer Coefficient   The overall heat transfer coefficient, U_o, is an important engineering design variable. For a given refrigerator size (as reflected by the external surface area, A_o), the overall heat transfer coefficient indicates how well the refrigerator walls are insulated so as to minimize the unwanted heat flow $\dot{Q}_{room}$. The lower the value of U_o, the lower the heat flow into the refrigerator.

In its most basic form, the overall heat transfer coefficient is a composite of three distinct heat transfer processes: (1) convection of heat from the room air to the exterior surface of the refrigerator; (2) conduction of heat through the refrigerator walls; and (3) convection of heat from the interior surface to the air inside the refrigerator.

Figure 6.10 shows a sketch of the temperature profile resulting from these three processes for a wall of thickness x. The rate of heat transfer through the wall material itself depends on the wall thickness, x, and the *thermal conductivity*, k, of the material (typically given in units of W/m-K). The lower the value of k and the thicker the wall, the lower the heat transfer rate. At the exterior and interior wall surfaces,

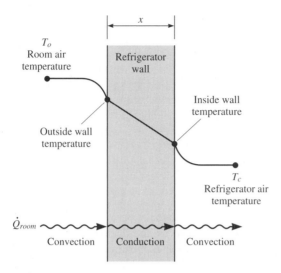

Figure 6.10 Heat transfer through a refrigerator wall. The process of convection transfers heat between air and the wall surfaces. The process of conduction transfers heat across the wall material.

however, the heat transfer rate to or from the surrounding air is governed by a parameter called the *convective conductance,* denoted by the symbol h_c. The value of h_c (typically in units of W/m²-K) depends mainly on whether the air adjacent to the wall is moving or stagnant. Around a refrigerator the air is usually still, so the convective conductance is relatively small, which helps to lower the overall heat transfer rate. A more detailed heat transfer analysis for the simple (one-dimensional) wall in Figure 6.10 yields the expression

$$\dot{Q}_{room} = \frac{A_o(T_o - T_c)}{\dfrac{1}{h_{c,o}} + \dfrac{x}{k} + \dfrac{1}{h_{c,i}}} \tag{6.34}$$

where $h_{c,o}$ and $h_{c,i}$ are the convective conductance values at the outer and inner wall surfaces, respectively. This equation actually defines the value of U_o in Equation (6.2):

$$U_o = \left(\frac{1}{h_{c,o}} + \frac{x}{k} + \frac{1}{h_{c,i}} \right)^{-1} \tag{6.35}$$

Properties of Insulating Materials   Table 6.5 shows the value of thermal conductivity for several materials that are (or could be) used to insulate refrigerator walls. Common insulators often are denoted by an R-value, which is a measure of thermal resistance (the reciprocal of thermal conductivity). The numerical values of R ratings used in the United States correspond to the reciprocal of k for a one-inch thickness of material, expressed in English units. Thus an insulation R-value of 4 means the thermal conductivity is 1/4 Btu-inch/hr-ft²-°F. The most common insulating material currently

Table 6.5   Thermal conductivity of selected materials.

Material	R-Value	K(W/m-K)
Fiberglass	4	0.036
Urethane foam	7	0.021
Vacuum powder panels (current)	25–30	0.005–0.006
Vacuum insulation panels (projected)	30–45	0.003–0.005

Source: ORNL, 1997b; Naser et al., 1994.

used in refrigerator walls is urethane foam, or its thermal equivalent, with an R-value of 7. The choice of insulating material and thickness strongly influences the annual energy consumption of a refrigerator, as illustrated in the following example.

Example 6.12

Effect of insulation on refrigerator energy consumption.   The refrigerator in Example 6.5 has a total outside surface area of 5.6 m². The door and walls are insulated with 4.0 cm of urethane foam. The convective conductances along the outside and inside surfaces are 9 and 6 W/m²-K, respectively. Calculate the overall heat transfer coefficient, and use this value to estimate the total heat load through the walls for a design room temperature of 32°C and an interior temperature of 4°C. Also estimate the fraction of time the compressor must run to meet this design room heat load.

Solution:

From Table 6.5, the thermal conductivity of urethane foam is 0.021 W/m-K. The overall heat transfer coefficient is given by Equation (6.35):

$$U_o = \left(\frac{1}{h_{c,o}} + \frac{x}{k} + \frac{1}{h_{c,i}} \right)^{-1}$$

$$= \left(\frac{1}{9} + \frac{0.04}{0.021} + \frac{1}{6} \right)^{-1}$$

$$= 0.46 \text{ W/m}^2\text{–K}$$

From Equation (6.2), the design room heat load is

$$\dot{Q}_{room} = U_o A_o (T_o - T_c)$$

$$= (0.46)(5.6)(32 - 4)$$

$$= 72.1 \text{ W}$$

$$= 260 \text{ kJ/hr}$$

This value is 43 percent of the total design heat load for the refrigerator, which was calculated in Example 6.5 to be 600 kJ/hr, based on a two-minute pull-down time. Thus, to handle only the design room heat load, the compressor would have to run 43 percent of the time.

Effects on Energy Consumption   In practice, the calculation of room heat load for a refrigerator is more complex than in the previous example. A more detailed heat transfer analysis must consider the presence of a freezer in addition to the refrigerator; the use of composite wall materials or cladding around the insulation material; different insulation thicknesses for the refrigerator and freezer compartments; differences in surface area between the outside and inside surfaces; and the use of gasket materials to provide a seal between the refrigerator doors and cabinets.

The results of one detailed analysis of this type are shown in Figure 6.11. This figure shows the total heat flow rate into a refrigerator–freezer as a function of the wall insulation thickness for several different insulating materials. The basic shape of these curves reflects the structure of Equation (6.34), which shows that the room heat load decreases nonlinearly with increasing wall thickness and with materials of lower thermal conductivity.

For a typical refrigerator wall thickness of 3.8 cm, Figure 6.11 shows that switching from a conventional urethane foam insulation (R7) to advanced vacuum insulation panels (R20) would cut the total heat gain in half. This, in turn, would substantially reduce the annual energy consumption and associated environmental impacts. The magnitude of energy savings would be roughly proportional to the reduction in room heat load. At present, however, superior insulation materials of

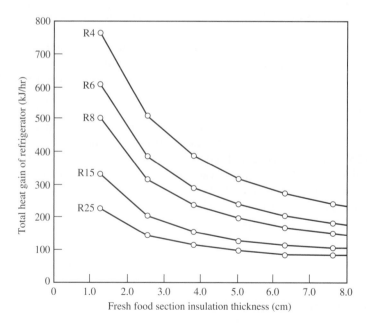

Figure 6.11   Refrigerator heat gain vs. insulation thickness for different insulation R-values. In this case study the freezer insulation thickness was adjusted to give the same heat gain per unit of insulation volume as the fresh food section. Other design assumptions are similar to those in the example problems of this chapter. (*Source:* Based on Naser et al., 1994)

this type are still too expensive, heavy, and difficult to install in mass-produced refrigerators. Nonetheless, such advanced materials indicate what may become feasible with continued research and development.

6.6.5   Energy Impact of CFC Substitutes

As a result of the international ban on CFC production, the refrigerant now being widely used in new refrigerators is HFC-134a. Here we examine the energy and environmental implications of this substitution.

The most direct effect was seen in Table 6.4, which showed that HFC-134a had a coefficient of performance (COP) 7 percent lower than CFC-12 for the same refrigeration cycle. The overall compressor efficiency also is lower because the higher compression ratio reduces both the isentropic efficiency and volumetric efficiency. Repeating the example calculations for total power consumption shows that the refrigeration cycle with HCF-134a requires approximately 20 percent more power—and hence 20 percent more energy—than the CFC-12 design. This means that all of the environmental emissions associated with electric power generation, as illustrated in Example 6.2, would also increase by 20 percent.

The energy efficiency deteriorates even further when one considers that the current urethane foam commonly used for refrigerator insulation has been manufactured using CFC-12 as the blowing agent, which gives the foam its properties. The ban on CFC use means that other blowing agents now must be found. Carbon dioxide gas is one possible substitute, but it produces a poorer insulator that would raise energy requirements by about 10 percent. In the near term, HCFCs are being used as a substitute blowing agent because they produce insulation properties only slightly inferior to CFCs. Although HCFCs are environmentally preferable to CFCs, as indicated in Table 6.1, they still contribute to ozone depletion and global warming, and they are scheduled to be phased out by 2030.

Even if the insulation properties of urethane can be replicated using other materials, the walls of a refrigerator using HFC-134a would have to be approximately 25–30 percent thicker to keep energy consumption at current levels. Indeed, new refrigerators do have thicker walls. But if the energy consumption of refrigerators is to further improve in the face of environmental concerns, new solutions for improved insulating materials and refrigeration cycles must be found. The final section of this chapter briefly highlights some of the trends and developments in these areas.

6.7   TRENDS AND FUTURE TECHNOLOGY

Environmental concerns have forced major design changes in refrigeration technology over the past decade, and that trend is likely to persist in the years ahead. In this section we look at some of the policy actions and technological developments that will shape the future design of household refrigerators.

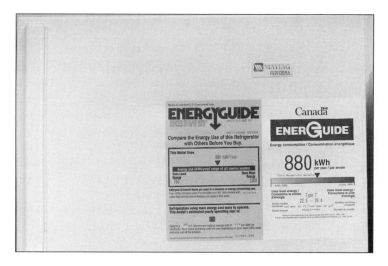

A refrigerator door showing the EnergyGuide labels for the United States and Canada.

6.7.1   Energy Efficiency Standards

Refrigerators built in the early 1970s typically consumed about 2,000 kW-hr/yr, representing the largest electrical use of any household appliance. In 1987 the U.S. Congress established minimum standards of energy efficiency for new refrigerators and other major appliances under the National Appliance Energy Conservation Act. These requirements were subsequently extended by the Energy Conservation Amendments of 1988 and the Energy Policy Act of 1992 to include additional technologies. Household appliances such as dishwashers, clothes washers and dryers, kitchen ranges and ovens, water heaters, and room air conditioners are among the products now included in the Appliance Standards Program administered by the U.S. Department of Energy (DOE). The goal of the program is to continually reduce the energy requirements of newly manufactured products, thus also reducing the environmental impacts of energy use.

For household refrigerators and freezers, the first energy efficiency standards took effect with models built in 1990. These standards were subsequently tightened for appliances built in 1993 or after. Table 6.6 lists the current energy consumption standards, which specify the maximum annual kilowatt-hours of electricity use based on a specified DOE test procedure similar to the calculations illustrated in this chapter. A new energy standard for refrigerators manufactured after July 1, 2001 will require a further 30 percent reduction in maximum energy consumption. The following example illustrates the use of Table 6.6.

Table 6.6   Energy efficiency standards for new refrigerators and freezers.

Product Class[b]	Energy Standards Equation for Maximum Energy Use (kW-hr/yr)[a]		
	Effective Jan. 1, 1990	Effective Jan. 1, 1993	Effective July 1, 2001
1. Refrigerators and refrigerator–freezers with manual defrost	16.3 AV + 316	13.5 AV + 299	8.82 AV + 248.4
2. Refrigerator–freezers—partial automatic defrost	21.8 AV + 429	10.4 AV + 398	8.82 AV + 248.4
3. Refrigerator–freezers—automatic defrost with top-mounted freezer without through-the-door ice service and all refrigerators—automatic defrost	23.5 AV + 471	16.0 AV + 355	9.80 AV + 276.0
4. Refrigerator–freezers—automatic defrost with side-mounted freezer without through-the-door ice service	27.7 AV + 488	11.8 AV + 501	4.91 AV + 507.5
5. Refrigerator–freezers—automatic defrost with bottom-mounted freezer without through-the-door ice service	27.2 AV + 488	16.5 AV + 367	4.60 AV + 459.0
6. Refrigerator–freezers—automatic defrost with top-mounted freezer with through-the-door ice service	26.4 AV + 535	17.6 AV + 391	10.20 AV + 356.0
7. Refrigerator–freezers automatic defrost with side-mounted freezer	30.9 AV + 547	16.3 AV + 527	10.10 AV + 406.0
8. Upright freezers with manual defrost	10.9 AV + 422	10.3 AV + 264	7.55 AV + 258.3
9. Upright freezers with automatic defrost	16.0 AV + 623	14.9 AV + 391	12.43 AV + 326.1
10. Chest freezers and all other freezers except compact freezers	14.8 AV + 223	11.0 AV + 160	9.88 AV + 143.7

[a] AV = Total adjusted volume, expressed in cubic feet (ft³), where AV = fresh food volume in ft³ + (adjustment factor × freezer volume in ft³). The adjustment factors for these product classes are: refrigerator only (no freezer compartment) = 1.0; basic refrigerators = 1.44; refrigerator–freezers = 1.63; freezers = 1.73.

[b] An additional eight categories of compact refrigerators and freezers have separate standards effective July 1, 2001.

Source: CFR, 1998.

Example 6.13

Efficiency standards for a refrigerator–freezer.   Calculate the maximum annual energy consumption permitted for new 0.5 m³ (18 ft³) refrigerator–freezers manufactured in 1990 and 1993, based on a unit with automatic defrost, top-mounted freezer, and no through-the-door ice service. Assume the volume of the fresh food compartment is 0.36 m³ and the freezer is 0.14 m³, as in Example 6.5.

Solution:

From Table 6.6, the applicable standards for maximum annual energy consumption are

$$1990 \text{ model} = 23.5 \text{ AV} + 471$$
$$1993 \text{ model} = 16.0 \text{ AV} + 355$$

The formula for adjusted volume (AV) in Table 6.6 requires units of ft³:

$$\text{Fresh food volume} = (0.36 \text{ m}^3)(35.31 \text{ ft}^3/\text{m}^3) = 12.71 \text{ ft}^3$$
$$\text{Freezer volume} = (0.14 \text{ m}^3)(35.31 \text{ ft}^3/\text{m}^3) = 4.94 \text{ ft}^3$$

Thus

$$AV = \text{(Fresh food volume)} + \text{(Adjustment factor} \times \text{Freezer volume)}$$

$$= 12.71 \text{ ft}^3 + (1.63 \times 4.94 \text{ ft}^3)$$

$$= 20.76 \text{ ft}^3$$

The maximum energy consumption standards are then

1990 Model: $E_{max} = (23.5)(20.76) + 471$

$$= 959 \text{ kW-hr/yr}$$

1993 Model: $E_{max} = (16.0)(20.76) + 355$

$$= 687 \text{ kW-hr/yr}$$

For this example the 1993 model requires 28 percent less energy than the 1990 model. Relative to refrigerator designs of 1970s, energy consumption has been cut by nearly two-thirds. Federal law also now requires all new refrigerators to carry a bright yellow *Energy Guide* label showing the actual annual energy consumption based on the DOE test procedure. The best top-mounted refrigerator–freezers currently available on the market use about 25 percent less energy than the 1993 standard. Appliances that are more efficient than the DOE standards often carry an *Energy Star*[R] label to alert consumers to the energy benefits. Figure 6.12 shows that the appliance efficiency standards are indeed succeeding in reducing the average energy consumption of individual units.

6.7.2   The Fridge of the Future

Researchers and appliance manufacturers are at work designing more efficient refrigerators. One project has been carried out by the DOE's Oak Ridge National Laboratory (ORNL) in conjunction with the Appliance Research Consortium (ARC), a subsidiary of the Association of Home Appliance Manufacturers. The result was a "fridge of the future" that uses about half as much energy as today's refrigerator–freezers—the one kilowatt-hour per day machine.

The new design added a series of energy efficiency improvements to a standard 20 ft³ baseline unit built in 1996. The new technology features included addition of polyurethane foam to the door panels to double the insulation thickness; addition of vacuum insulation panels around the freezer section to reduce heat transfer; replacement of AC motors with more efficient DC motors; and replacement of automatic defrost control with an adaptive defrost that operates only when needed (ORNL, 1997a).

Table 6.7 compares the daily energy consumption of two advanced refrigerator designs to the current baseline design. The fully modified design (called Unit C) uses 45 percent less energy than the baseline unit and 54 percent less than the DOE-mandated maximum for this size unit. Because the addition of vacuum insulation panels adds significantly to the initial cost of the unit, a modified unit without the

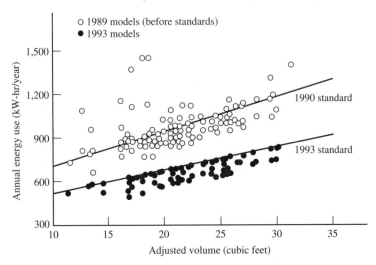

Figure 6.12   Energy efficiency improvements for refrigerator–freezers, showing the reduction in annual energy use after adoption of energy efficiency standards. (*Source:* LBL, 1995)

Table 6.7   Energy consumption of advanced refrigerator designs.

Refrigerator–Freezer Design	Energy Consumption (kW-hr/day)	Percent Run Time[a]	Manufacturer Cost Increase ($)[b]	Simple Payback Time (years)[c]
A Baseline unit	1.68	44.2	—	—
B Unit A with 2-in.-thick doors, high-efficiency compressor, low-wattage condenser fan, and adaptive defrost scheme	1.16	47.6	53.38	6.6
C Unit B with vacuum insulation around freezer section and larger evaporator	0.93	36.5	134.33	11.4

[a] Percentage of time compressor runs each hour.
[b] Cost to consumers is approximately twice this value.
[c] Based on cost to consumers.
Source: ORNL, 1997b.

panels (called Unit B) was also tested. This unit would still be 44 percent more efficient than the 1993 standard and 31 percent more efficient than the actual baseline unit.

Energy efficiency gains translate directly into environmental improvements. For instance, if all 120 million refrigerators in Example 6.2 reduced their annual energy consumption to the level of Unit B in Table 6.7, emissions of SO_2, NO_x, and CO_2 would be less than half the values calculated in Example 6.2. That would be the equivalent of shutting down 22 large (500 MW) power plants across the United States. Although the advanced Unit B refrigerator is higher in initial cost than the baseline unit, the energy saved each year more than pays for this cost over the life of the unit. Chapter 13 takes a closer look at the economics of environmental improvements, including an example based on the data in Table 6.7.

6.8 CONCLUSION

The general approach used in this chapter to analyze the environmental design of refrigerators can be applied to all other types of products that engineers design and build. The engineering analysis begins from a clear understanding of the design objectives and the environmental implications of current designs. Alternative designs that reduce environmental impacts follow from a detailed analysis of the technology in question.

Chapter 7 takes a more comprehensive look at the environmental assessment process, including a life cycle analysis of alternative technology designs. This approach helps ensure that solutions to one environmental problem do not create or aggravate others. This broader perspective complements the type of analysis presented in this chapter by considering not only the use of a particular technology but also its manufacture and disposal.

6.9 REFERENCES

AFEAS, 1998. *Alternative Fluorocarbons Environmental Acceptability Study.* Washington, DC. www.afeas.org/prodsales_download.html.

ASHRAE, 1993. *1993 ASHRAE Handbook, Fundamentals SI Edition.* American Society of Heating, Refrigerating and Air Conditioning Engineers, Atlanta, GA.

CFR, 1998. Code of Federal Regulations. *Federal Register,* 10 CFR Ch. II, January 1, 1998.

IPCC, 1996. *Climate Change 1995, The Science of Climate Change,* J.T.Houghton, et al., Editors, Intergovernmental Panel on Climate Change, Cambridge University Press.

LBL, 1995. *CBS Newsletter, Summer 1995.* Center for Building Science, Lawrence Berkeley Laboratory, Berkeley, CA.

Naser, S.F., G.A. Keolian, L.T. Thompson, and J. Handt (ed.), 1994. *Open-Ended Problem: The Design of a CFC-Free, Energy-Efficient Refrigerator.* National Pollution Prevention Center for Higher Education, University of Michigan, Ann Arbor, MI.

ORNL, 1997a. "The Fridge of the Future." Buildings Technology Center, Oak Ridge National Laboratory, Oak Ridge, TN. www.ornl.gov/ORNL/BTC/adv-rf-tech.html.

ORNL, 1997b. "Powder-Evacuated Panel Insulation." Energy Efficiency and Renewable Energy Program, Oak Ridge National Laboratory, Oak Ridge, TN.

WMO, 1999. *Scientific Assessment of Ozone Depletion: 1998.* Report No. 44, World Meteorological Organization, Geneva.

6.10 PROBLEMS

6.1 Relative to CFCs, is it possible for non-CFC refrigerants to worsen the problem of global warming at the same time they lessen the problem of stratospheric ozone depletion? Discuss the basis for your answer.

6.2 Why is the coefficient of performance (COP) an important parameter for assessing the environmental impacts of a refrigerator design?

6.3 Table 6.3 listed six design parameters affecting the quantity of refrigerant needed. Do any of these parameters also affect the annual energy consumption of a refrigerator? For each parameter discuss *qualitatively* how an *increase* in its value would affect the amount of electricity used and the associated environmental emissions from power generation.

6.4 Revisit the list of design parameters in Table 6.3.

  (a) What values of these variables were used to calculate the annual energy requirement in Example 6.11?

  (b) If the design volume, V, had been 10 percent smaller, by what percentage would the annual energy use (and associated environmental emissions) have changed? Assume all other design parameters are unchanged.

6.5 Repeat Problem 6.4 for a 10 percent decrease in the design value of the refrigerator compartment temperature, T_c. Again keep all other parameters at their original values.

6.6 Repeat Problem 6.4 for a 10 percent decrease in the design value of the outside room temperature, T_o. Again keep all other parameters at their original values.

6.7 Repeat Problem 6.4 for a 10 percent decrease in the design value of the pull-down time, Δt_{pd}. Again keep all other parameters at their original values.

6.8 Repeat Problem 6.4 for a 10 percent decrease in the design value of the evaporator temperature, T_1. Again keep all other parameters at their original values.

6.9 Repeat Problem 6.4 for a 10 percent decrease in the design value of the condenser temperature, T_4. Again keep all other parameters at their original values.

6.10 Summarize the results of Problems 6.4 to 6.9 (or whichever subset of those problems you have done). Which design parameter has the greatest influence on energy use and associated environmental impacts? Which parameter has the least influence?

6.11 In Example 6.6 the temperatures of the evaporator and condenser were selected to be $-20°C$ and $40°C$, respectively. If you wanted to design the most energy-efficient refrigerator possible to meet the rest of the conditions specified in that example, what values of evaporator and condenser temperatures would you choose? Explain the basis for your answer. Then discuss other factors that affect this decision.

6.12 Room air conditioners operate on the same principle as a refrigerator. The room to be cooled is analogous to the interior of the refrigerator compartment, and the outdoor air is the surroundings.

  (a) Calculate the design heat load for a room whose dimensions are 6 m × 4 m × 3 m if it is desired to cool the room from 32°C to 20°C in 15 minutes. Assume the room air has a density of 1.2 kg/m³ and a specific heat of 1.0 kJ/kg-K.

(b) Compare your result to the design heat load for the refrigerator in Example 6.5.

(c) If the two devices use the same type of refrigerant, can you tell which device will have the greater impact on the environment? Discuss your thinking and conclusions about this.

6.13 Calculate the refrigeration effect and COP based on HFC-134a as the refrigerant instead of CFC-12. Assume the same refrigerator design conditions as in Examples 6.5 and 6.6. Discuss the environmental implications of your results compared to the CFC design. (Note: Tables of refrigerant properties can be found in most thermodynamics textbooks and library reference such as ASHRAE, 1993.)

6.14 Analyze the overall impact of several engineering design improvements for the refrigerator compressor in Example 6.10. Assume the isentropic efficiency can be increased to 75 percent; the compressor clearance volume reduced to 4 percent; the loss factor improved to 0.95; and the motor efficiency increased to 88 percent.

(a) What would be the percentage reduction in annual electricity consumption?

(b) Assuming the compressor runs 40 percent of the time, what would be the annual consumer cost savings based on the average U.S. residential electricity price of 8.3 cents/kW-hr? Also estimate the national cost savings if half the refrigerators in the country (60 million units) had these improved compressors.

6.15 One way to improve the efficiency of a refrigeration cycle is to superheat the refrigerant in the evaporator and subcool it in the condenser. To quantify this improvement, revisit the illustrative problems in the text (Examples 6.6 to 6.11). This time assume the saturated vapor is superheated by 5°C (that is, to an evaporator exit temperature of -10°C at 0.15 MPa pressure). Assume the enthalpy gain from superheating equals the enthalpy loss from subcooling in the condenser (a heat exchanger known as an *intercooler* accomplishes this energy transfer). Draw a sketch of this cycle on a pressure–enthalpy diagram. Then calculate the new values of

(a) Refrigeration effect.

(b) Refrigerant mass flow rate.

(c) Total refrigerant mass.

(d) Ideal compressor work.

(e) Coefficient of performance.

(f) Actual compressor power requirement.

What is the overall improvement in energy efficiency?

6.16 Equation (6.35) gave the overall thermal conductance for a one-dimensional wall, which has the same surface area on both sides. A more realistic geometry for a refrigerator compartment would be a three-dimensional box like the one sketched on the following page, where the outside surface area, A_o, is different from the inside surface area, A_i. In this case, a more general form of Equation (6.35) is

$$U_o = \left(\frac{1}{h_{c,o}} + \frac{xA_o}{kA_i} + \frac{A_o}{h_{c,i}A_i} \right)^{-1}$$

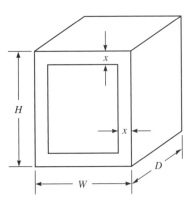

(a) Will the value of U_o calculated from this equation be greater or smaller than the one-dimensional value found from Equation (6.35)? State the reasoning behind your conclusion.

(b) Based on the dimensions shown in the sketch (outside dimensions $H \times W \times D$ and a uniform wall thickness x), write a general equation for the total external surface area, A_o.

(c) Write a similar equation for the total internal surface area, A_i.

(d) Calculate the overall thermal conductance for a refrigerator whose dimensions are $H = 100$ cm, $W = 70$ cm, $D = 60$ cm, and $x = 5$ cm. Assume $k = 0.021$ W/m-K, $h_{c,o} = 9$ W/m²-K, and $h_{c,i} = 6$ W/m²-K.

(e) With the dimensions and assumptions in part (c), what is the percentage difference in the overall heat transfer rate based on the three-dimensional conductance equation compared to the one-dimensional case that assumes $A_i = A_o$? How does this affect the calculation of refrigerator energy use and resulting environmental impacts?

6.17 Suppose the refrigerator in Examples 6.10 and 6.11 were designed with the best current vacuum powder insulation instead of urethane foam.

(a) What would be the annual energy savings (kW-hrs)?

(b) What would be the percentage reduction in average air pollutant emissions associated with the refrigerator's operation?

6.18 Calculate and compare the maximum energy use standards in Table 6.6 for Product Classes 2 to 7 based on an 18 ft³ refrigerator–freezer built in the year 2000. Assume the fresh food section occupies 13 ft³ and the freezer 5 ft³. How much difference is there in the allowable energy use among the six different classes of refrigerator design? Explain (as best you can) the factors that may account for these differences.

6.19 Choose *any two* of the refrigerator–freezer product classes in Table 6.6 and plot the maximum allowable energy use for a new unit as a function of time from 1990 to 2005. Assume an adjusted volume (AV) of 21 ft³. What is the overall percentage reduction in energy use? Is there any difference in the percentage improvement between the two product classes?

6.20 Visit a local appliance dealer with a showroom of refrigerator–freezers carrying the yellow *Energy Guide* label. Obtain the following information for at least four different makes and models of refrigerator–freezers:

(a) The volumes of the fresh food and freezer compartments.

(b) The advertised annual energy use on the *Energy Guide* labels.

(c) The types of refrigerant used.

Calculate the corresponding energy use standards from Table 6.6, and compare them to the values on the labels. Summarize your findings in a brief report. Include any suggestions you might have for improving the consumer information displayed on the *Energy Guide* label.

chapter

7

Environmental Life Cycle Assessments*

*This chapter was written by Cliff I. Davidson.

7.1 INTRODUCTION

In previous chapters we have explored both the benefits and the adverse environmental effects of technology. As engineers we recognize that technology can also help reduce these adverse effects. However, the products and processes we deal with are complex, and the environmental consequences can be far-reaching. How can we ensure that we are doing the best possible job in balancing the benefits of technology with reduced environmental damage?

Although we are just beginning to understand the complexities of this balance, we have a tool that can help us with this challenging task. This is known as *life cycle assessment,* or simply *LCA.* We have seen in the past few chapters that a product such as an automobile, a battery, a power plant, or a refrigerator can affect the environment in many ways. Making the product requires extraction of raw materials, which takes energy and tools. The raw materials need to be refined and put together to form the finished product, which must be packaged and shipped to consumers. Consumers need energy to use the product. Finally, the product must be disposed of at the end of its useful life. Each of these stages has environmental effects. In principle, an LCA gives us a way to account for these effects when we first design a product so that we may minimize environmental consequences associated with its entire life cycle. Although we will focus on life cycle assessment of consumer products in this chapter, an LCA can also account for environmental effects when we design an industrial process, like generation of electrical power or treatment of drinking water.

Chapter 1 introduced the concept of LCA as providing a big picture of how engineering decisions affect the environment. It also introduced the concepts of *green design* and *industrial ecology,* which seek to minimize or eliminate the overall environmental consequences of engineering design decisions. Only recently has the importance of considering this big picture been realized. Many decades ago, people tolerated pollution and enjoyed unrestricted use of natural resources. When the problems of pollution were first recognized, dilution was thought to be the answer: Factory chimneys were raised, and longer pipes were installed for sewage outfalls. Dilution proved inadequate as populations and industries grew; as a result, "end-of-pipe" treatment methods were developed to capture pollutants or change them chemically just before they were emitted into the environment. Electrostatic precipitators for controlling air emissions and wastewater treatment plants to reduce water pollution are examples.

The new environmental ethic goes beyond these concepts. If we can look at the big picture, we can design a product or process so that fewer pollutants are created in the first place rather than controlling them after they form. Furthermore, we can make engineering decisions that decrease the use of scarce natural resources. This will enable us to make progress toward *sustainable development*—development that meets our current needs without compromising the ability of future generations to meet theirs (WCED, 1987). LCA can help us in our efforts to achieve sustainable development.

Because LCA is newly evolving, formal analysis procedures are also evolving. Furthermore, many of the detailed procedures depend on the specific processes involved in a particular industry. However, we can describe some of the general principles on which life cycle assessment is based and provide examples of application of the method. In this chapter we begin by discussing the concepts behind life cycle assessment, including a description of the major steps that comprise an LCA. We then address each of these major steps in a separate section. Finally, we draw conclusions from these steps and mention where further work is needed. Throughout the chapter we provide real-world examples of how the method can help us make sound engineering decisions as we attempt to obtain the benefits of technology while minimizing adverse environmental effects.

7.2   PRINCIPLES OF LIFE CYCLE ASSESSMENT

Environmental attitudes have changed over time: What used to be considered reasonable behavior is no longer considered acceptable. Bans on cigarette smoking in public areas, laws on recycling, and standards for automobile gasoline consumption (miles per gallon) are but a few examples of these new attitudes. Another example is the millions of dollars spent to clean up sites where hazardous wastes were legally dumped many years ago, before we understood the consequences of such behavior. But how long will current attitudes remain unchanged? Can we extrapolate into the future from past experience so that we can avoid spending large sums of money to undo past mistakes, as we are doing now? The most important overall principle behind life cycle assessment is to provide a framework so that decisions made today will, to the extent possible, be viewed many years from now as the "right" decisions from the standpoint of environmental impacts.

What kinds of decisions are we talking about? For the most part, these are choices made by companies or choices that affect companies that produce things for us. Examples of decisions include what materials to use, where to order these materials from, what types of energy to use in production, how to package the products, how to ship the products to customers, and how to dispose of wastes. Such decisions are based on wide-ranging criteria. Minimizing cost and maximizing product quality used to be the primary factors in these decisions. Other factors such as maximizing safety, durability, and consumer appeal have also been important in years past. Now we have added a criterion: reducing environmental effects. An LCA provides the framework for incorporating environmental effects into the decision-making process.

7.2.1   Making Decisions about Product Design

Who makes the decisions? In general, managers at various levels within a company are responsible for the final decisions. But managers need input from people with varied experience. Engineers provide many of the answers to the questions just listed, often as different design options with advantages and disadvantages for each. Specialists in the purchasing department provide information on raw materials used in manufacturing, whereas professionals in waste management consider the various options for handling wastes at each stage in the process. Accountants consider costs involved at each step. Marketing and advertising professionals provide input on what might appeal to users of the final products. Specialists in regulation report on what must be done to ensure compliance with myriad laws. The list can go on; usually many individuals must provide information before a final decision is made. When we add a new criterion to the decision-making process—minimizing environmental effects—input from several additional people may be needed.

In its simplest form, the concept of a *life cycle* is adapted from nature. Decaying plants and animals serve as nutrients for other plants and animals—all wastes are used for something. By contrast, in many human activities raw materials serve as inputs to make a product, the product is used, and then it is thrown away without thought as to how the waste products may be useful. Figure 7.1 shows that a product can be considered to have a life cycle as well. Designing products to fit into a cycle solves two problems: It minimizes adverse impacts of wastes, and it reduces the use of natural resources to make new products. We will see that designing products to avoid wastes altogether is difficult. However, we can make progress by using an LCA to help us redesign products to reduce both natural resource use and generation of wastes.

7.2.2 Steps in a Life Cycle Assessment

Figure 7.2 shows the three major steps in an LCA framework. The first step, inventory analysis, is by far the best understood of the major steps. In fact, it is the only step where established procedures have been adopted on a reasonably wide scale. The inventory is based on the scope and possible uses of the LCA, as identified in the next section. Essentially, the inventory lists and quantifies the inputs and outputs of the processes at each stage in the life cycle. A typical input is the number of kilograms of steel used to make a product; an output might be the mass flow rate of sulfur dioxide gas released during manufacturing of the product.

The second step is impact analysis. Each input and output identified in the inventory has some potential environmental impact, and the goal of this second step is to determine these impacts. Examples include the amount of land needed to obtain the iron ore in manufacturing a certain quantity of steel, or the number of people in a city who become ill from the sulfur dioxide gas released during manufacturing. Determining such numbers can be extremely difficult. As a result, several approaches have been developed to enable qualitative comparisons rather than precise quantification. Processes that affect large areas of the globe, such as CFC releases that deplete stratospheric ozone (Chapter 11) or greenhouse gas emissions (Chapter 12), are considered a higher priority than emissions with more localized effects. Releases of highly toxic pollutants are of greater concern than releases of less toxic materials. Impact analysis is typically the most difficult step in an LCA, mainly because the adverse effects of resource use and environmental releases are not well understood. These issues will be discussed in greater detail in Section 7.4.

The third step is improvement analysis. Here we identify what can be done to reduce environmental impacts by changes in product or process design. This is the key step for engineering design decisions. Examples include reducing the amount of land needed to mine iron ore by adding recycled scrap steel to the raw materials, or

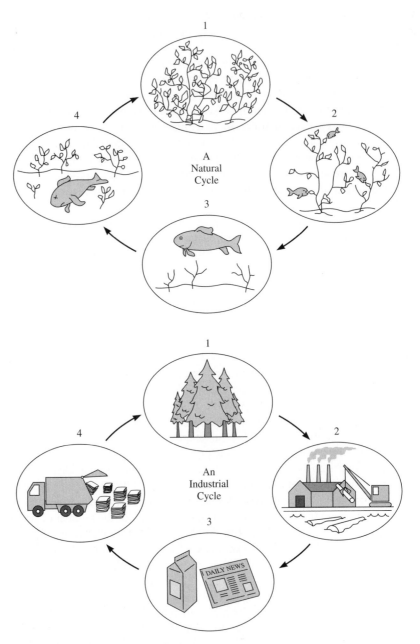

Figure 7.1   In nature, all waste products are used. Plants such as water starwort grow in freshwater lakes (1) and serve as food for fish such as grass carp (2, 3). When the carp die (4), their bodies decay and provide nutrients for another generation of starwort (1). We can design industrial processes to have a cycle. Trees are grown (1) and harvested to make paper (2). The paper products (3) are collected after use to be composted (4). The compost then serves as nutrient material for the next generation of forest (1).

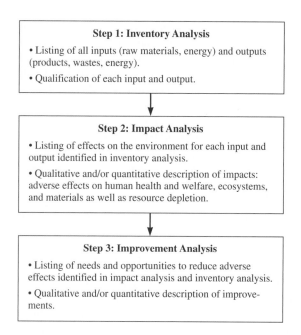

Step 1: Inventory Analysis

• Listing of all inputs (raw materials, energy) and outputs (products, wastes, energy).

• Qualification of each input and output.

Step 2: Impact Analysis

• Listing of effects on the environment for each input and output identified in inventory analysis.

• Qualitative and/or quantitative description of impacts: adverse effects on human health and welfare, ecosystems, and materials as well as resource depletion.

Step 3: Improvement Analysis

• Listing of needs and opportunities to reduce adverse effects identified in impact analysis and inventory analysis.

• Qualitative and/or quantitative description of improvements.

Figure 7.2   Major steps in a life cycle assessment.

changing the product manufacturing process to release less sulfur dioxide. It is often possible to conduct a limited improvement analysis based on either of the first two steps. Inventory analysis can identify which processes have the greatest environmental burdens; impact analysis can identify where in a life cycle the most adverse effects can occur. In either case, the results can guide research on where to look for improvements that can significantly reduce environmental damage. Improvement analysis is discussed in more detail in Section 7.5.

7.2.3   Scope of a Life Cycle Assessment

Figure 7.2 shows that an LCA should account for all inputs, outputs, and environmental effects of carrying out an industrial process or making a product. This can be difficult, partly because of problems in defining the boundaries of the analysis. For example, acquiring raw materials for manufacturing will have environmental effects in terms of resources needed and environmental discharges. If we are studying effects of making paper, tree cutting provides the raw material, and this requires equipment such as power saws. But making power saws requires steel, which itself demands tools for mining iron ore. How far back do we go? This question must be answered before beginning the analyses in Figure 7.2.

Several issues must be considered when establishing the scope of the LCA. These include the motivation for the study, the intended use of the results, the resources available to conduct the study, and the point of view of the analysis.

There may be several motivations for conducting a life cycle assessment. A company may need to satisfy new legal requirements or may decide to go beyond regulations to set an industry standard for others to follow. Alternatively, a company may wish to collect baseline environmental data to which proposed improvements can be compared over a period of years. New technology just becoming available may require an inventory analysis to determine its feasibility. Furthermore, reducing waste often results in cost savings, which may provide added motivation to conduct the analysis.

Linked to the motivation is the intended use of the results. A company may want to identify which processes in its production line pose the greatest environmental burden. This can be the first step in proposing production changes to reduce the burden. For companies that have already installed new technology, the results of an LCA may determine if the technology is meeting expectations. LCA findings can also be used to market products that are "environmentally friendly."

The financial and personnel resources available will greatly affect the scope of the analysis. Gathering data can be time-consuming and expensive; it is necessary to ensure that the benefit will be worth the effort. Several sources of data may already be available, such as open literature (journals, books, and conference proceedings), websites for organizations and government agencies, or government archives. However, care must be taken when using data collected from processes that are somewhat different from the specific processes being studied: The conditions under which the data were obtained must be taken into account. Furthermore, the quality of the data is vitally important.

Finally, linked to all these issues is the point of view of the analysis. An engineer who is working for a manufacturing company may have a perspective different from that of a consumer buying the product. The consumer, in turn, may have a perspective different from that of a government regulator with responsibility to protect public health and welfare. Throughout this chapter, we discuss the analysis from the point of view of an engineer whose job is to consider design aspects of the product in order to balance the interests of the company with broader societal interests.

Having briefly discussed the three steps in an LCA and defined the scope of analysis, we now consider each of the three steps in detail. We begin with inventory analysis, examining some real-world examples of how the analysis can be applied.

7.3   INVENTORY ANALYSIS

7.3.1   Major Components of an Inventory Analysis

Usually the first step in an LCA is to construct a flow diagram that includes the major categories of interest and defines the scope of the study. Figure 7.3 shows an example of a general flow diagram with four stages typically used in an inventory analysis. Each stage encompasses a number of possible processes. Raw materials acquisition involves such activities as mining of ores, harvesting of trees, and extraction of oil from wells. Depending on how the results are to be used, this first stage could also include constructing the mines, planting the trees, and digging the oil wells.

The second stage, manufacturing, embodies several processes. Examples include processing iron ore to make iron, processing wood to make pulp, and processing

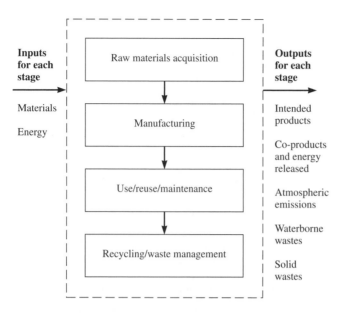

Figure 7.3   General flow diagram for inventory analysis, the first step of life cycle assessment.

crude oil to make petroleum products. The manufacturing category also includes all the processes used to create the final product from these refined materials. Furthermore, packaging the product as well as transportation to the consumer are usually lumped into this second stage.

The third stage involves use of the product by the consumer. All maintenance and repair of the product throughout its useful life are included here.

Finally, disposal of the product at the end of its useful life is the fourth and final stage. This usually means landfilling or incineration, but might also involve dismantling the product and recycling its parts.

Throughout all four stages, the inventory analysis quantifies the amount of each material and amount of energy used as inputs at each stage. Furthermore, the analysis quantifies the outputs, including the intended products, unintended by-products (which may be useful for something, as yet undetermined), energy released, and wastes to the atmosphere and to water as well as solid wastes.

7.3.2   Case Study of a Computer Housing

Consider the various components of a desktop computer. The hard drive contains a motor, magnetic disk, and associated circuitry, while the floppy and CD-ROM drives contain additional motors and circuitry. Boards inside the computer have integrated circuits and various other electrical components. The power supply inside the computer may contain transformers with significant amounts of copper wire. The monitor includes a cathode ray tube, light sources, and beam-focusing equipment. In

Shown here is the plastic housing for IBM's IntelliStation E Pro central processing unit, made of 100 percent recycled content PC/ABS resin. This new housing represents a breakthrough application of recycled resin in the computer industry, motivated by the benefits found from life cycle assessments for materials management.

some cases, a liquid crystal detector may be used instead of a cathode ray tube. The keyboard includes a metal frame, springs, mechanical linkages, and circuitry besides the keys. The mouse includes metal and plastic parts. There are also various cables and plugs connecting the computer system components. Clearly, a lot of materials had to be put together to provide the final product.

Conducting an inventory analysis on a computer is a formidable task. Rather than attempt this for an entire computer, let us consider just one part of the machine: the single piece of plastic that serves as the housing for a typical desktop computer. Note that the choice of plastic for the housing was a design decision made previously—this was chosen from among other options. For simplicity, we will confine the scope of analysis by assuming that the choice of plastic over other possible materials is outside the boundaries of the current analysis. We will, however, consider this issue in the problems at the end of the chapter.

According to the manufacturer, this part weighs 1.72 kg and is made of polyvinylchloride (PVC) plastic coated with copper paint (Brinkley et al., 1996). The housing is fastened onto the computer and remains in place to protect the internal circuitry throughout the computer's useful life.

Where do the materials for the computer housing come from, how are they processed to make the housing, and what disposal options are available? We now answer these questions by expanding the boxes of Figure 7.3 to arrive at detailed flow diagrams for the computer housing.

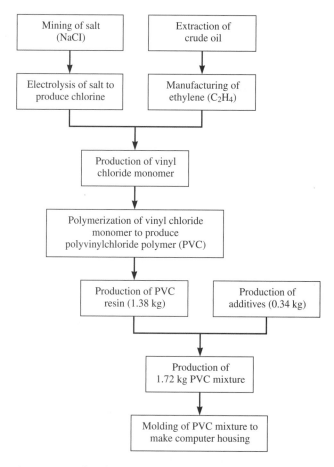

Figure 7.4   Flow diagram for producing the PVC computer housing.

PVC is a *polymer* or a chain of atoms, in this case carbon, hydrogen, and chlorine. It is made by combining ethylene and chlorine to form molecules of vinyl chloride, known as a *monomer*. The monomer molecules are then aggregated (*polymerized*) as shown in Figure 7.4. The PVC must be mixed with additives to stabilize the compound before it is poured into the mold to produce the final part. Note that Figure 7.4 includes only the principal materials involved in making the housing. Minor materials, energy, packaging, and transportation are not included. Thus a significant fraction of the inputs and environmental discharges are not accounted for in this simplified diagram.

After removal from the mold, the PVC housing is coated with copper paint to provide electrical shielding. An additional flow diagram is needed to describe the flow of materials for the paint, shown in Figure 7.5. The figure indicates two categories of raw materials: copper ore and those involved in the production of paint. Both categories of raw materials are processed to make the paint, which is applied

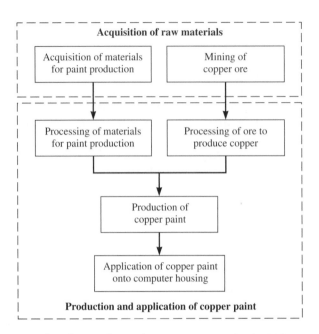

Figure 7.5   Flow diagram for producing copper paint for the PVC computer housing.

onto the PVC computer housing. As with Figure 7.4, this flow diagram includes only major materials, omitting numerous minor materials as well as energy.

Once the housing is completed, it must be packaged and shipped to the computer assembly plant. The finished computer with housing attached is then packaged and transported to stores or to the consumer. Packaging in both cases involves manufacturing of paper and cardboard, illustrated in Figure 7.6. The figure shows that wood and two chemicals, chlorine and sodium hydroxide, are used as principal inputs to make pulp, and the pulp is processed to make paper and cardboard.

After its useful life is over, the computer is either recycled or disposed of, and the housing can have several possible fates. We will assume for the present that the housing is disposed of along with the computer in a landfill, where it remains indefinitely. Other options will be considered when we discuss improvement analysis.

Transportation is needed at several stages in the life cycle of the computer housing. Figure 7.4 implies that raw materials to make the PVC must be brought to manufacturing facilities, whereas Figure 7.5 suggests that raw materials for making copper paint must be shipped to the paint production facilities, from which the paint must be shipped to the location of the computer housings. Figure 7.6 suggests the need to transport raw materials used in paper production to the pulp and paper mills. The finished paper for packaging must then be shipped to the computer assembly plant. Disposal of the paper requires additional shipping. The computers must be shipped to consumers, and the computers or computer parts must be shipped to recycling plants, incineration facilities, or landfills. Several modes of transportation can be used for these separate steps. However, we assume here that most of the shipments are carried by truck.

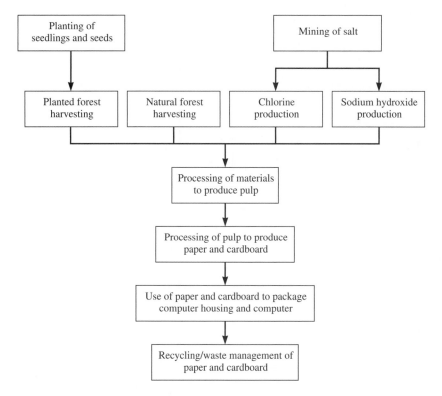

Figure 7.6   Flow diagram for paper and cardboard used for packaging the computer housing.

Assessing the environmental effects of shipping by truck requires accounting for manufacturing of the truck, production and use of fuel, and disposal of the vehicle after its useful life is over. Figure 7.7 shows some of the major steps in this assessment.

Combining Figures 7.4 through 7.7 results in a detailed flow diagram for an inventory analysis of the computer housing, shown in Figure 7.8. Besides the processes shown explicitly in the figure, the effects of transportation from Figure 7.7 are implicit in virtually every component of Figure 7.8. Comparing Figure 7.3 and Figure 7.8 shows that each of the four major stages in the inventory analysis has been taken into account in this case study. We now quantify the analysis by applying data to the inventory flow diagrams.

7.3.3   Quantitative Analysis of the Computer Housing

We have seen that data for an inventory analysis includes inputs (materials and energy) and outputs (intended products, unintended co-products, releases of energy, and releases of waste to the environment). We now consider these inputs and outputs

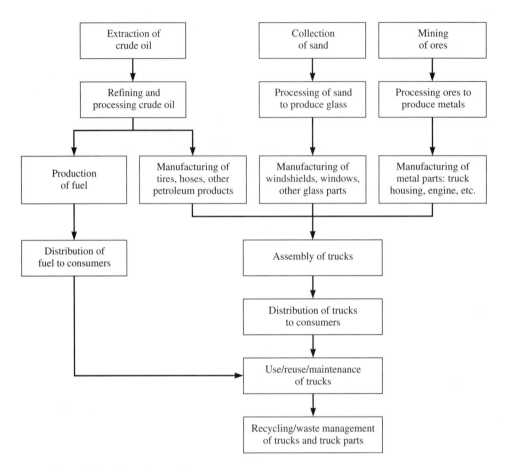

Figure 7.7   Flow diagram for transportation.

applied to Figure 7.4, the flow diagram for manufacturing the PVC computer housing. We also consider the effects of transportation from Figure 7.7.

In principle, one can envision a matrix based on the general flow diagram showing the stages in an inventory analysis. Each column represents one stage or one of
several detailed processes within a stage, such as processing crude oil to make PVC,
or molding the PVC into the product shape. Each row represents a single input or
output, such as the amount of energy used as input, or the quantity of hydrocarbons
released to the air. A simplified conceptual matrix is shown in Table 7.1.

Consistent with the concept of a matrix, data for environmental burdens
involved in manufacturing the computer housing are summarized in Table 7.2. A
number of inputs and outputs are listed. Industrywide generic data are available for
acquisition of raw materials and all production processes leading up to the manufacture of PVC resin (APME, 1994). These data have been used to compute the
inputs and outputs for all the steps in Figure 7.4 up to the box labeled "Production of
PVC Resin." The values have been computed for 1.38 kg, the amount of resin in the
computer housing, and are shown in column 1 of the table. (Data in Table 7.2 and in

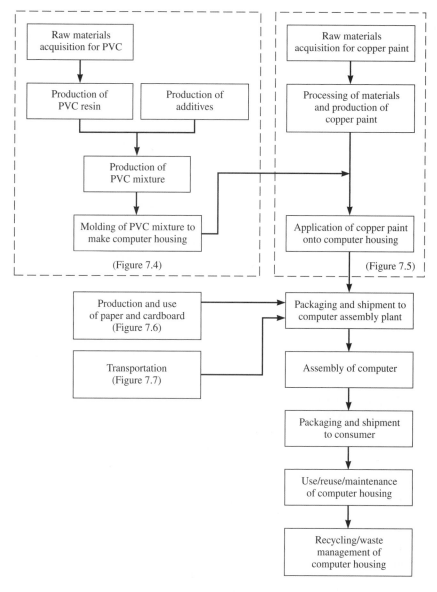

Figure 7.8 Flow diagram for inventory analysis of the computer housing, combining Figures 7.4 through 7.7.

Table 7.1   Conceptual matrix for an inventory analysis.

Process Element	Acquire Raw Materials	Process Raw Materials	Make the Product	Ship to Consumer	Use of Product by Consumer	Management of Wastes
Inputs						
Raw materials						
Energy						
Outputs						
Emissions to air						
Effluents to water						
Solid waste to land						

Table 7.2   Environmental burdens associated with an inventory analysis for manufacturing a PVC computer housing. Inputs of raw materials and outputs of wastes are shown.

		Units	1. All Processes Up to and Including Producing PVC Resin (1.38 kg)	2. Molding the PVC Mixture (Resin + Additives = 1.72 kg)	3. Total: Producing PVC Resin plus Molding the PVC Mixture	4. Complete Production of Housing, Including Resin, Additives, and Molding (1.72 kg)	5. Example 7.1 Calculation: Amount Attributed to Production of Additives (0.34 kg)
Inputs	**Raw Materials**						
	Crude oil (in ground)	kg	0.65	0.016	0.67	0.94	0.27
	Coal (in ground)	kg	0.27	0.26	0.53	0.81	0.28
	Lignite (in ground)	kg	0	0.031	0.031	0.049	0.018
	Natural gas (in ground)	kg	0.8	0.027	0.83	1.2	0.37
	Limestone	kg	0.021	0	0.021	0.18	0.16
	Sand	kg	0.0014	0	0.0014	0.0014	0
	NaCl	kg	0.96	0	0.96	0.93	0
	Bauxite	kg	0.0003	0	0.0003	0.00034	0.00004
	Iron ore	kg	0.0005	0	0.0005	0.00054	0.00004
	Ilmenite ore	kg	0	0	0	0.095	0.095
	Total water	liters	27	0.0088	27	55	28
	Energy	MJ	89	12	100	137	37
Outputs	**Air Emissions**						
	Particulate matter	g	5.4	0.51	5.9	7.3	1.4
	CO_2	g	2,400	720	3,100	4,700	1,600
	CO	g	3.4	0.13	3.5	5.4	1.9
	SO_x	g	18	5.9	24	31	7
	NO_x	g	21	3.1	24	30	6
	N_2O	g	0	0.058	0.058	0.4	0.34
	NH_3	g	0	0.00017	0.00017	0.00024	0.0022
	Cl_2	g	0.0014	0	0.0014	0.0014	0
	HCl	g	0.33	0.13	0.46	0.57	0.11

Table 7.2 (*continued*)

	Units	1. All Processes Up to and Including Producing PVC Resin (1.38 kg)	2. Molding the PVC Mixture (Resin + Additives = 1.72 kg)	3. Total: Producing PVC Resin plus Molding the PVC Mixture	4. Complete Production of Housing, Including Resin, Additives, and Molding (1.72 kg)	5. Example 7.1 Calculation: Amount Attributed to Production of Additives (0.34 kg)
CH_4	g	0	0.23	0.23	2.6	2.4
Total hydrocarbons	g	26	7	33	49	16
Chlorinated organics	g	0.7	0	0.7	0.7	0
Aldehydes	g	0	0.0032	0.0032	0.012	0.0088
H_2S	g	0	0	0	0.0012	0.0012
Other organic	g	0	0.0062	0.0062	0.015	0.0088
Metals	g	0.0041	0	0.0041	0.0043	0.0002
Water Effluents						
BOD5	g	0.11	0	0.11	0.2	0.09
COD	g	1.5	0.00015	1.5	3.5	2
Chlorides	g	58	0	58	58	0
Chlorinated organics	g	0.0041	0	0.0041	0.0041	0
Dissolved organics	g	1.9	0	1.9	1.9	0
Dissolved solids	g	0.58	0.11	0.69	2.3	1.6
Suspended solids	g	2.7	0.0017	2.7	4.3	1.6
Oil	g	0.069	0.0022	0.071	0.12	0.049
Hydrocarbons	g	0	0	0	0.0012	0.0012
Iron (Fe2+ and Fe3+)	g	0	0	0	6.4	6.4
Fluorides	g	0	0.00081	0.00081	0.0017	0.00089
Metals	g	0.27	0	0.27	0.38	0.11
Nitrates	g	0	0.00019	0.00019	0.00054	0.00035
Nitrogen—NH_3	g	0	0.00038	0.00038	0.0043	0.0039
Nitrogen—Organic	g	0.0041	0	0.0041	0.0056	0.0015
Ammonium hydroxide	g	0	0	0	0.00019	0.00019
Sodium ions	g	6.6	0.00013	6.6	8.4	1.8
Sulfates	g	2.1	0.00017	2.1	7.4	5.3
Methanol	g	0	0	0	0.0077	0.0077
Phenol	g	0	0	0	0.00013	0.00013
Fibers	g	0	0	0	0.0091	0.0091
Calcium	g	0	0	0	0.6	0.6
Magnesium	g	0	0	0	2.6	2.6
Aluminum	g	0	0	0	1.4	1.4
Manganese	g	0	0	0	0.38	0.38
Solid Waste						
Hazardous	kg	0.0048	0	0.0048	0.0051	0.0003
Landfilled	kg	0	0.00053	0.00053	0.00053	0
Mineral	kg	0.08	0	0.08	0.083	0.003
Nonhazardous chemicals	kg	0.015	0	0.015	0.015	0
Slags and flyash	kg	0.017	0	0.017	0.017	0
Unspecified	kg	0.0028	0.17	0.17	0.36	0.19
Recovered	kg	0	0.027	0.027	0.044	0.017

Source: Original data taken from Brinkley et al., 1996.

subsequent tables in this chapter are generally shown to two significant figures. In most circumstances, greater accuracy is not justified, although there are some instances where three or four significant figures are used for bookkeeping purposes.)

Figure 7.4 shows that PVC resin is combined with chemical additives to improve the durability of the PVC, and this results in a high-quality, commercial-grade PVC mixture that contains 1.38 kg of resin and 0.34 kg of additives. The PVC mixture is then molded into the shape of the computer housing. Although separate data are not available for the production of additives or production of the PVC mixture in Figure 7.4, we do have information on the last box, namely the process of molding 1.72 kg of the PVC mixture. These data are provided by an injection molding company and are given in column 2 of Table 7.2.

Finally, data are available from a computer manufacturing company for the sum total of inputs and outputs for all of Figure 7.4 for a specific list of industrial plants, as shown in column 4. The relation between the flow diagram and Table 7.2 is shown in Figure 7.9.

Example 7.1

Calculating environmental burdens of a product.   Let us assume that the values in Table 7.2 represent consistent measures and can be combined, even though the data are from different industrial plants. Furthermore, let us focus on the production of PVC resin and production of additives, ignoring environmental burdens of the box labeled "Production of 1.72 kg PVC mixture." For the conditions in Table 7.2, determine the inputs and outputs for making 0.34 kg of PVC additives.

Solution:

If the box indicated in the problem statement is ignored, Figure 7.9 shows that making a computer housing can be considered in simplified form as three steps, each with environmental burdens. These steps are the production of 1.38 kg of PVC resin, production of 0.34 kg of additives, and molding the PVC mixture (resin plus additives) to make the housing. We know the environmental burdens for the first step (column 1 in Table 7.2) and the third step (column 2 in the table). We also know the total environmental burden for all three steps (column 4). We can therefore determine the environmental burden for the second step—making the additives—as follows:

Value for additives = Value in column 4 − (Value in column 1 + Value in column 2)

The sum of values from columns 1 and 2 is shown in column 3. Thus the value for additives is simply the difference between the values in columns 3 and 4. The result is shown in column 5 of Table 7.2, which lists the inputs and outputs of producing 0.34 kg of additives, as desired.

The environmental burdens calculated in Example 7.1 can be used in various engineering design decisions. For example, although the additives improve the quality of the PVC, their production may cause greater environmental consequences than an equivalent amount of PVC resin production. This is illustrated in the following problem.

Example 7.2

Comparing environmental burdens of two different materials.   Consider making a 1.72 kg computer housing from PVC resin alone. Calculate the environmental burdens associated with inputs needed to make the new housing. For which raw materials are the environmental bur-

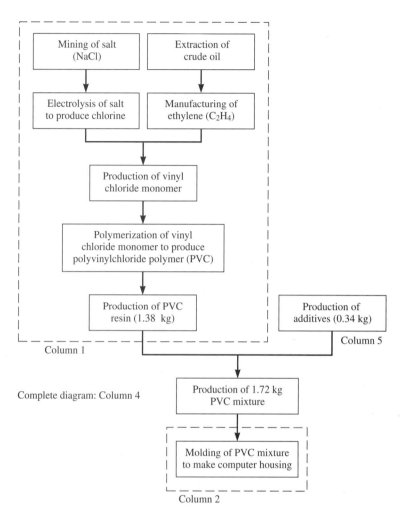

Figure 7.9 Flow diagram for producing the PVC computer housing shown in relation to the columns in Table 7.2.

dens greater compared with the original housing? By visual inspection of the outputs in Table 7.2, can you identify whether the PVC resin housing is responsible for more environmental discharges compared with the original housing?

Solution:

We note that making the new housing has two steps, each with an environmental burden: producing 1.72 kg of resin and molding the resin. The first environmental burden is found by multiplying each value in column 1 of Table 7.2 by 1.72/1.38, or 1.25. The second environmental burden is found in column 2. Thus an alternate way to calculate the total burden is as follows:

Value for new housing = 1.25 × (Value in column 1) + Value in column 2

Table 7.3   Inputs needed to make a computer housing for Example 7.2.

Substance	Units	1. All Processes Up to and Including Producing PVC Resin (1.38 kg), Identical to Table 7.2, Column 1	2. Molding PVC Resin (1.72 kg), Assumed Same as Molding PVC Mixture in Table 7.2, Column 2	3. Example 7.2 Calculation: Complete Production of Housing Made Entirely of PVC Resin (1.72 kg)	4. Complete Production of Housing Made of 1.38 kg Resin and 0.34 kg Additives (Copied from Table 7.2, Column 4)
Raw Materials					
Crude oil (in ground)	kg	0.65	0.016	0.83	0.94
Coal (in ground)	kg	0.27	0.26	0.6	0.81
Lignite (in ground)	kg	0	0.031	0.031	0.049
Natural gas (in ground)	kg	0.8	0.027	1	1.2
Limestone	kg	0.021	0	0.026	0.18
Sand	kg	0.0014	0	0.0017	0.0014
NaCl	kg	0.96	0	1.2	0.93
Bauxite	kg	0.0003	0	0.00038	0.00034
Iron ore	kg	0.0005	0	0.00063	0.00054
Ilmenite ore	kg	0	0	0	0.095
Water	liters	27	0.0088	34	55
Energy	MJ	89	12	123	137

The results are given in Table 7.3. The values show that the new computer housing has smaller environmental burdens except for sand, NaCl, bauxite, and iron ore. Thus for most of the raw materials, the additives increase the environmental burdens. The high-quality PVC mixture uses more resources than the PVC resin.

To consider the second question in the problem statement, we note that visual inspection of column 1 in Table 7.2 suggests the values are much smaller than those in column 4. Even when multiplied by 1.25 and added to column 2, the values shown in column 3 for almost all air emissions and water effluents are smaller than values in column 4. The situation is less obvious for the solid waste categories, where most values are roughly equal between the two housings. Overall, the results indicate that extra environmental burdens are associated with both inputs and outputs for production of the additives.

Let us now consider the role of transportation in an inventory analysis of the PVC computer housing. Figure 7.7 shows the complexity of accounting for transportation. To simplify matters, we focus on emissions of one species—airborne particulate matter—in the following example.

Example 7.3

Pollutant emissions from transportation.   Trucks used for large shipments have trailers with dimensions 12 m long by 3.6 m wide by 3.0 m high. Each computer housing, with packaging, measures 0.5 m by 0.4 m by 0.4 m. Assume that each housing must be shipped a total of 1,000 km during its manufacturing, use, and disposal. If a truck emits 0.6 grams of particulate matter per kilometer of travel (Lowenthal et al., 1994), calculate the total amount of particulate matter released per housing during shipment. Compare with the total amount released from manufacturing according to Table 7.2.

Solution:

The volume of the trailer is 12 m $\times$ 3.6 m $\times$ 3.0 m = 130 m^3, while the volume of one computer housing with packaging is 0.5 m $\times$ 0.4 m $\times$ 0.4 m = 0.08 m^3. Therefore, each trailer truck can carry 130 m^3/0.08 m^3 = 1,600 housings. The total amount of particulate matter emitted per housing for 1,000 km of travel can thus be calculated as

$$\text{Particulate matter} = (0.6 \text{ grams/km}) \times (1,000 \text{ km/truck}) \times (1 \text{ truck/1,600 housings})$$

$$= 0.4 \text{ grams per housing}$$

This compares with Table 7.2, column 4, showing that 7.3 grams of particulate matter are released during production of the housing. Thus the amount of particulate matter released during shipment of the housing is about 5 percent of the amount released from its manufacture based on the previous assumptions. The calculation assumes that the emissions during manufacturing and waste management of the truck are small compared with the emissions during its use; this assumption is supported by current data (MacLean and Lave, 1998). Note that the values in Table 7.2 do not include the copper paint.

Another perspective on environmental burdens involves energy use by transportation. This aspect is applied to the computer housing problem in the following example.

Example 7.4

Energy consumption for transportation. A trailer truck can travel about 2.11 km per liter of diesel fuel (5 miles/gallon). If the energy content of the diesel fuel is 40.8 megajoules (MJ) per liter, determine the total amount of energy used for transportation of a computer housing during its manufacture, use, and disposal based on 1,000 km of travel. Compare with the energy used in the production of the housing from Table 7.2.

Solution:

$$\text{Total energy} = (40.8 \text{ MJ/liter}) \times (1 \text{ liter/2.11 km}) \times (1,000 \text{ km/truck})$$

$$\times (1 \text{ truck/1,600 housings}) = 12.1 \text{ MJ/housing}$$

This compares with 137 MJ needed to produce one housing according to Table 7.2, column 4. Thus transportation uses roughly 9 percent of the energy consumption of manufacturing the housing. This calculation again assumes the energy used during manufacturing and waste management of the truck is small compared with the energy used to operate it.

These calculations, conducted as part of an inventory analysis in an LCA, indicate that making PVC computer housings requires significant use of natural resources and produces environmental discharges. But what are the impacts of the resource use and discharge of wastes? That question is now explored through impact analysis, the second step of an LCA.

7.4   IMPACT ANALYSIS

Analyzing the impacts of resource use and environmental discharges can be complex. Nevertheless, there are ways to organize a study of impacts so that straightforward information can be obtained to guide engineering decisions. In this section we first consider several categories of impacts. Then we examine ways of prioritizing the severity of the various impacts. Finally, we provide some examples of how to quantify impacts for use in a life cycle assessment.

7.4.1   Categories of Impacts

For convenience, we will consider four categories of environmental impacts in this section. These include depletion of natural resources, effects on human health, effects on ecosystems, and impairment of human welfare.

Depletion of resources occurs when materials and energy are needed as inputs to a process. We are more concerned about nonrenewable resources, such as minerals and fossil fuels, than renewable resources such as agricultural crops and wind energy. Some resources, such as forests, are renewable only after long time periods. Note that the impacts of resource depletion may be difficult to quantify: How can one assess the value of a forest that is destroyed? The commercial value of the lumber may be far less than the perceived value of the forest to those who used it as a recreation area. Other resources, such as agricultural land damaged by strip mining, may have a well-defined value, and thus the impacts can be more easily quantified. Depending on the resource, it may be preferable to describe the impacts qualitatively rather than quantitatively.

In the context of resource depletion, we can envision two types of industrial processes: those that have the possibility of being sustainable, and those that are not sustainable even with proper management. An example of the first type is manufacturing in which a mineral resource is changed from one form to another. If the energy used in this process is renewable and if the product can be recycled or reused, then progress has been made toward sustainability. An example of the second type of process is the use of a resource that is irreversibly changed and cannot be reused— for example, combustion of a fossil fuel.

The next category, direct effects on human health, is usually considered the most important. Outputs from an industrial process to air, water, or land may result in health effects if not properly controlled. Several lists of industrial outputs with the potential to cause health effects have been assembled; one example is the Toxics Release Inventory (TRI) of the U.S. Environmental Protection Agency (EPA). As discussed in Chapter 2, the TRI includes many different chemicals and chemical groups for which industries are asked to report their emissions. Development of the TRI has been a major step forward in attempts to quantify outputs for use in assessing environmental impacts. However, the TRI has numerous weaknesses, such as the fact that the total weights of all toxic species emitted from a plant are summed by the EPA in reporting the data. The usefulness of a single number representing the sum of all toxic species, some of which are highly toxic and others much less toxic, is questionable.

An enormous amount of research has been conducted on health impacts of toxic chemicals. For example, hundreds of studies have examined the effects of inhaling toxic gases, hundreds more the effects of inhaling airborne particles, and still hundreds more the effects of ingesting contaminated water. But because there are so many chemical species and so many different effects, we do not yet understand the links between exposure to toxic species and the resulting health problems. Current research (like Horvath et al., 1995) is attempting to rank the severity of health effects from different chemicals reported in the TRI. At best, we can conduct only a very rough impact analysis based on human health. Chapter 14 discusses the health risks of environmental pollutants at greater length.

The third category, ecosystem effects, includes damage to plants and animals living in all types of environments. These include aquatic systems such as rivers, lakes, and oceans, as well as terrestrial environments such as meadows, forests, and deserts. Certain ecosystems are somewhat resilient to pollutants, whereas others are highly sensitive. An example of a sensitive ecosystem is a lake that has poor buffering capacity, which means that acidic pollutants will not be neutralized by natural chemicals in the water. Thus the acids can harm fish and aquatic plants. Our understanding of ecosystem effects is even poorer than that of human health effects.

Finally, effects on human welfare include issues other than health and ecosystem impacts. Examples include changes in climate, destruction of recreation areas, generation of unpleasant odors, corrosion of structures, and impairment of visibility due to haze. Note that many of these examples can also cause effects on plants and animals, such as extinction of species caused by global warming and destruction of recreation areas that are also natural habitats. Human welfare, which is concerned mainly with quality of life issues, is given a lower priority than human health. As with the other categories, our understanding of the effects of industrial outputs on welfare is incomplete.

Given the complexity of environmental effects and varied human perceptions of the impacts in the categories just described, it is obviously difficult to set priorities for these impacts. Yet such priorities are extremely important as input to policy decisions. We now consider some of the methods to rank these various impacts.

7.4.2   Ranking Environmental Impacts

The U.S. Environmental Protection Agency (EPA) established criteria in 1990 for ranking adverse environmental effects. The agency then applied these criteria to several types of environmental problems to come up with a relative ranking of the problems in terms of severity. The EPA's criteria and their application to environmental problems have been summarized by Graedel and Allenby (1995).

1. The physical extent of the impacted area is important. This could be expressed as the population of the region affected, the size of the ecosystem experiencing adverse effects, the number of trees destroyed, or some other measure of the physical extent of damage.

2. The danger posed by the environmental change is of considerable concern. Highly toxic chemicals are ranked above those that are less toxic, and environmental changes that result in severe health effects such as cancer are most feared.

3. The extent of exposure is an additional criterion. For example, some substances in water are highly toxic if they reach living cells of aquatic animals, but they may not constitute a hazard simply because they are attached to suspended particles that are too large to reach the cells. On the other hand, small flyash particles released from coal combustion may contain only trace amounts of toxic metals such as lead and cadmium, but may be hazardous to health because the toxic material on the outside of the particles can be readily absorbed into the bloodstream when inhaled into the lungs. Thus the extent of exposure is high for these particles.

4. The final criterion is the penalty for making the wrong decision. If a pollutant is allowed to enter the environment and is later discovered to be more toxic than previously thought, will the damage persist? A chemical with a long residence time in the environment is likely to be of greater concern than one that will react and disappear quickly and thus have adverse effects over a shorter period.

Note that these criteria may overlap to some extent: A chemical with a long residence time has a greater opportunity to be carried further and thus affect a larger physical area than one with a short residence time. Based on these criteria, the EPA ranked 13 major environmental problems, as shown in Table 7.4.

The ranking of these 13 environmental problems was based on opinions of experts using the criteria just given. Although the four criteria attempt to provide a scientific basis for deciding the rankings, the ultimate decision on the severity of these problems depends on human values. Recognizing the importance of bringing values into the decision process, some attempts have been made to survey scientists,

Table 7.4   EPA ranking of 13 major environmental problems.

Relatively High Priority	**Medium Priority**	**Relatively Low Priority**
• Habitat alteration and destruction	• Herbicides and pesticides	• Oil spills
• Species extinction	• Pollution in rivers, lakes,	• Groundwater pollution
• Stratospheric ozone loss	and other surface waters,	• Radionuclides
• Global climate change	especially toxic chemicals,	• Acid runoff to surface waters
	nutrients, biochemical oxygen	• Thermal pollution
	demand, and turbidity	
	• Acid deposition	
	• Air toxics and smog	

Source: Based on USEPA, 1990.

as well as policymakers without scientific backgrounds, as to the relative importance of different environmental problems. An example of results of such a survey is shown in Figure 7.10. The figure compares the problems thought to be severe at present with problems expected to be severe 25 years from now. Accidental chemical releases, pollution of the oceans, and metals and toxic chemicals in freshwater are of special concern because they are believed to be serious problems both currently and in the future. Note that several natural hazards are included for comparison, such as cyclones, droughts, floods, and earthquakes.

7.4.3   Quantification of Impacts

The rankings just discussed show that some adverse environmental effects are preferred over others, and this information can be used to make design decisions even in the absence of quantifiable impacts. However, in some situations quantification is possible. One example is global warming, where emissions of greenhouse gases from an industrial process are known. Each greenhouse gas has a *global warming*

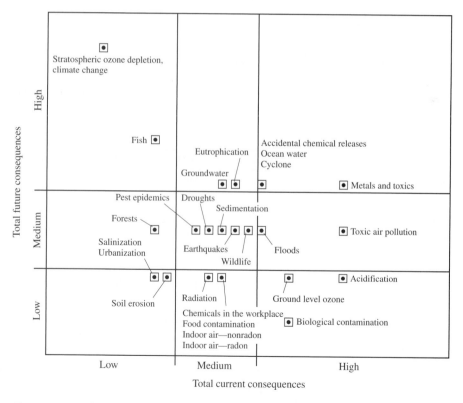

Figure 7.10   Plot comparing the severity of environmental problems expected 25 years from now with the severity of environmental problems today, according to surveys. (*Source*: Norberg-Bohm et al., 1995)

potential (GWP) that is expressed in equivalent units of carbon dioxide (CO_2). Based on a 100-year time horizon, nitrous oxide (N_2O) has a global warming potential of 310, meaning that each gram of N_2O causes the same amount of potential global warming as 310 grams of CO_2 added to the atmosphere. Methane (CH_4) has a global warming potential of 21. Chapter 12 discusses global warming issues and the GWP in greater detail.

Example 7.5

Global warming as an example of impact analysis.   Consider Figure 7.9, which shows the process of making PVC plastic for a computer housing. Using data for emissions of CO_2, N_2O, and CH_4 from Table 7.2, determine the overall global warming potential of each of the following:

(a) Making 1.38 kg of resin.

(b) Making 0.34 kg of additives.

(c) Molding the PVC mixture to form the housing.

(d) Manufacturing the complete computer housing made of 1.38 kg resin and 0.34 kg additives.

Express your answers in terms of total equivalent emissions of CO_2 in grams. Compare the potential global warming impact of these various steps. What do you conclude about which steps contribute most to this impact?

Solution:

The emissions of all three greenhouse gases for production of 1.38 kg of resin are given in column 1 of Table 7.2, and the emissions for production of 0.34 kg of additives are given in column 5 of the same table. Emissions for the molding process are given in column 2. The total global warming potential for each process, and the overall total, are shown in Table 7.5.

Table 7.5   Global warming potential (GWP) for Example 7.5.

Product/Greenhouse Gas	CO_2	N_2O	CH_4	Total*
Global warming potential (100-year horizon)	1	310	21	
Emissions from making 1.38 kg PVC resin, grams				
(from Table 7.2, column 1)	2,400	0	0	
Equivalent emissions of CO_2, grams	2,400	0	0	2,400
Emissions from making 0.34 kg additives, grams				
(from Table 7.2, column 5)	1,600	0.34	2.4	
Equivalent emissions of CO_2, grams	1,600	100	50	1,750
Emissions from molding PVC mixture, grams				
(from Table 7.2, column 2)	720	0.058	0.23	
Equivalent emissions of CO_2, grams	720	18	4.8	740
Emissions from complete production of housing, grams				
(from Table 7.2, column 4)	4,700	0.4	2.6	
Equivalent emissions of CO_2, grams	4,700	120	55	4,880

* Totals rounded to nearest 10 grams.

The values show that producing the resin has the greatest global warming potential, followed by the additives and then the molding process.

Life cycle assessments are frequently used to compare different options as well as to evaluate individual products. The next example compares the global warming potential of two alternatives for the computer housing used in earlier examples.

Example 7.6

Comparing the global warming impacts of two different materials.   Calculate the overall global warming potential of making the new computer housing described in Example 7.2, where the additives are replaced by resin in the PVC mixture. Compare the overall global warming potential of the new housing with that of the original housing considered in Example 7.5.

Solution:

We saw in Example 7.2 that environmental burdens for the new housing can be calculated from Table 7.2 as follows:

$$\text{Value for new housing} = 1.25 \times (\text{Value in column 1}) + \text{Value in column 2}$$

The results of these calculations for the three greenhouse gases are shown in Table 7.6. The result shows that manufacturing of the new housing has an overall global warming potential of 3,740, which is 23 percent smaller than 4,880 for the original housing.

Table 7.6   Global warming potential for Example 7.6.

Product/Greenhouse Gas	CO_2	N_2O	CH_4	Total*
Global warming potential (100-year horizon)	1	310	21	
1.25 × (value in column 1, Table 7.2)	3,000	0	0	
Equivalent emissions of CO_2, grams	3,000	0	0	3,000
Value in column 2, Table 7.2	720	0.058	0.23	
Equivalent emissions of CO_2, grams	720	18	4.8	740
Housing made from PVC resin, total	3,720	0.058	0.23	
Equivalent emissions of CO_2, grams	3,720	18	4.8	3,740
Housing made from PVC mixture, total (Table 7.5)				
Equivalent emissions of CO_2, grams	4,700	120	55	4,880

* Totals rounded to nearest 10 grams.

7.5   IMPROVEMENT ANALYSIS

The final step of life cycle assessment uses the inventory and impact results to identify where improvements may be made. These improvements often involve trade-offs; in some cases saving raw materials may be a higher priority than saving energy. The opposite may be true in other cases. The ultimate decisions depend on availability of certain materials and availability of different types of energy, which are often location-specific.

Although improvement analysis may be applied to either an industrial process or a product, we focus here on application to a product, namely the computer housing. We first consider a new method of providing electrical shielding, and we determine if this method offers an improvement over conventional shielding by copper paint. Then we consider options for better waste management of the computer housing.

7.5.1   Improving Electrical Shielding of the Computer Housing

The same categories of environmental impacts introduced in Table 7.2 for inventory analysis can be used to assess whether a change in computer housing design represents a net improvement. The proposed change is to insert a steel plate inside the PVC housing for electrical shielding rather than applying copper paint. Figure 7.11 shows a flow diagram for manufacturing the steel plate. Two options are included: Option A involves making the steel from iron ore and coal, whereas Option B involves acquiring scrap steel that is melted and processed to make new steel parts.

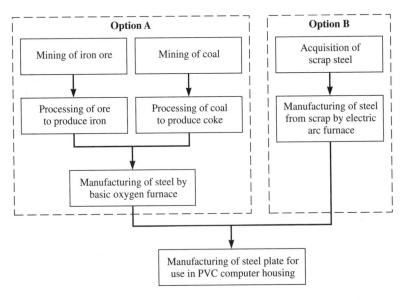

Figure 7.11   Flow diagram showing two options for producing steel plate for the PVC computer housing.

In this analysis we will consider only Option A. Based on this option, the environmental burdens for manufacturing a steel plate and also for manufacturing and applying copper paint are shown in Table 7.7.

Table 7.7 Environmental burdens associated with an inventory analysis for manufacturing copper paint and also a steel plate, two options for shielding the PVC computer housing. In the last column, "s" indicates that the steel plate is preferred, whereas "c" indicates that copper paint is preferred.

	Substance	Units	Steel Plate	Copper Paint	Difference (Copper − Steel)	
Inputs	*Raw Materials*					
	Crude oil (in ground)	kg	0.051	0.09	0.039	s
	Coal (in ground)	kg	0.1	0.12	0.02	s
	Lignite (in ground)	kg	0.012	0.014	0.002	s
	Natural gas (in ground)	kg	0.011	0.1	0.089	s
	NaCl	kg	0	0.0003	0.0003	s
	Iron ore	kg	0.99	0	−0.99	c
	Copper ore	kg	0	2.9	2.9	s
	Water	liters	68	0.59	−67	c
	Energy	MJ	22	13	−9	c
Outputs	*Air Emissions*					
	Particulate matter	g	23	0.4	−23	c
	CO_2	g	1,889	505	−1,400	c
	CO	g	1.5	0.21	−1.3	c
	SO_x	g	8	6.9	−1.1	c
	NO_x	g	4.3	2.3	−2	c
	N_2O	g	0.037	0.09	0.053	s
	NH_3	g	0.00089	0.0005	−0.00039	c
	HCl	g	0.051	0.06	0.009	s
	CH_4 (methane)	g	0.1	0.5	0.4	s
	Total hydrocarbons	g	15	110	96	s
	Chlorinated organics	g	0	0.00039	0.00039	s
	Aldehydes	g	0.0029	0.0023	−0.0006	c
	H_2S	g	0	0.00029	0.00029	s
	Other organic	g	0.0048	0.004	−0.0008	c
	Lead	g	0	0.025	0.025	s
	Metals	g	0	0.00012	0.00012	s
	Water Effluents					
	BOD5	g	0.00027	0.004	0.0037	s
	COD	g	0.0008	0.07	0.069	s
	Chlorides	g	0	0.005	0.005	s

(continued)

Table 7.7 *(continued)*

Substance	Units	Steel Plate	Copper Paint	Difference (Copper − Steel)	
Dissolved organics	g	0	0.0011	0.0011	s
Dissolved solids	g	0.56	0.3	−0.26	c
Suspended solids	g	0.0009	0.027	0.026	s
Oil	g	0.0075	0.013	0.0055	s
Hydrocarbons	g	0	0.009	0.009	s
Fluorides	g	0.00031	0.0004	0.00009	s
Metals	g	0	0.023	0.023	s
Nitrates	g	0	0.00016	0.00016	s
Nitrogen—NH^3	g	0.00014	0.0006	0.00046	s
Nitrogen—organic	g	0	0.0008	0.0008	s
Methanol	g	0	0.008	0.008	s
Other organic	g	0	0.0009	0.0009	s
Solid Waste					
Hazardous	kg	0	0.009	0.009	s
Landfilled	kg	0	0	0	—
Mineral	kg	0	0.7	0.7	s
Nonhazardous	kg	0	0	0	s
Slags and flyash	kg	0	2.2	2.2	s
Unspecified	kg	0.38	0.08	-0.3	c
Recovered	kg	0.01	0.024	0.014	s

Source: Brinkley et al., 1996.

Example 7.7

Improvement analysis of manufacturing the computer housing.   Using Table 7.7, determine whether electrical shielding by a steel plate or by copper paint is preferable when the most important environmental priority is

(a) Reducing total use of oil, coal, and natural gas.

(b) Reducing energy consumption.

(c) Reducing emissions of particulate matter to the atmosphere.

(d) Reducing water effluents.

(e) Reducing total solid wastes.

Solution:

Comparing values for the steel plate and the copper paint in Table 7.7 provides answers to each of the conditions posed. To facilitate the comparison, the difference in values between the two materials is computed on the right side of the table, along with the preferred material. The results show the following:

(a) The steel plate consumes less oil, coal, and natural gas.

(b) The copper paint uses less energy.

(c) The copper paint emits fewer particulates to the air.

(d) For nearly every category of water effluents, the steel plate has lower values.

(e) The steel plate has smaller total solid wastes; these are dominated by high slag and ash wastes for the copper paint.

7.5.2 Improving Waste Management of the Computer Housing

Table 7.8 indicates inputs and outputs for three disposal options: placing the housing in a landfill, burning the housing in an incinerator, and recycling the housing. Note that the first option has the potential of contaminating groundwater or surface waters, whereas the second option releases airborne contaminants. The second option may also require disposal of the ash from incineration, with the potential for contamination of groundwater or surface waters. The original calculations were performed for a 2.26 kg PVC housing of a computer monitor (Brinkley et al., 1995 and 1996). These original values have been corrected using a multiplication factor of 1.72 kg/2.26 kg to construct Table 7.8 because we are interested in the inputs and outputs for a 1.72 kg PVC housing.

Several assumptions are necessary here. For example, it is assumed that the recycled PVC is as good as virgin PVC for making computer housings; it has been demonstrated that this is nearly the case. It is also assumed that heat produced by incinerating the PVC can be used for other purposes such as generating electricity, although in practice most incinerators are not connected to such generators. Energy recovery has important implications. If we had to add energy to incinerate the PVC, this would be a net positive quantity in the value of energy needed as input. However, because the PVC releases a substantial amount of energy when it burns, we can subtract this amount of energy from the energy needed for other purposes in the PVC incineration process, such as transportation of the PVC to the incinerator. Thus some of the values in the incineration column in Table 7.8 are negative, indicating that the amount subtracted exceeds the energy necessary for incineration.

Similarly, recycling the PVC means that fewer raw materials are needed for making new PVC. The reduction in raw materials can be subtracted from the amounts of raw materials used to recycle the PVC, such as materials used to make the trucks that collect the PVC for recycling. As a result, most of the input values in Table 7.8 are negative. Because recycling PVC also reduces emissions that would have been produced from making new PVC, we can subtract emissions values from the emissions necessary in the recycling process. An example of the latter is emissions from trucks carrying old PVC housings to the recycling plant. The reductions in emissions are generally greater than the amounts of emissions necessary in the recycling process, so virtually all of the outputs in Table 7.8 in the recycling column are negative.

The PVC entering a landfill is assumed to be inert; no energy is released from this PVC. Thus there are no negative values in the landfill column. Furthermore, because the PVC is inert, virtually all of the emissions in this column are from transportation of the housings to the landfill. The only exception is the solid waste value representing the housing itself, namely 1.72 kg of PVC.

Table 7.8   Inputs and outputs for three disposal options for the PVC computer housing.

	Substance	Units	Landfill	Incineration	Recycling
Inputs	**Raw Materials**				
	Crude oil	kg	0.027	0.019	−0.84
	Coal	kg	0.00015	−0.51	−0.51
	Natural gas	kg	0.000076	0.003	−1.1
	Limestone	kg	0	1.1	−0.18
	NaCl	kg	0	0	−0.91
	Water	liters	0.0053	−0.0061	−40
	Total primary energy	MJ	32	15	−91
Outputs	**Air Emissions**				
	Particulates	g	0.11	26	−6.4
	CO_2	g	91	1,800	−3,600
	CO	g	0.31	0.84	−4.3
	SO_x	g	0.12	−9.9	−24
	NO_x	g	0.91	−3.2	−24
	NH_3	g	0.00053	0.011	0
	Cl_2	g	0	0	−0.0014
	HCl	g	0	230	−0.41
	Hydrocarbons	g	0.24	−11	−37
	Other organic	g	0	−0.015	−0.7
	Water Effluents	g			
	BOD5	g	0.00015	0.00015	−0.2
	COD	g	0.00046	0.00053	−3.5
	Chlorides	g	0	0	−57
	Dissolved solids	g	0.32	0.37	−1.2
	Suspended solids	g	0.00015	−0.003	−4.2
	Oil	g	0.0038	0.0053	−0.11
	Sulfates	g	0	0	−7.6
	Nitrates	g	0	−0.0003	0
	Sodium ions	g	0	0	−8.4
	Metals	g	0	0	−0.37
			0	0	0
	Solid Waste				
	Hazardous chemicals	kg	0	0	−0.0076
	Landfilled PVC	kg	1.72	0	0.015
	Slag and ash	kg	0	1.3	−0.015
	Other	kg	0	−0.33	00.27

Source: Based on data from Brinkley et al., 1995 and 1996.

Although these values are the best available, they are only approximations. The values for landfilling and incineration are based on mathematical models using reasonable assumptions, whereas the values for recycling are based on measurements (Brinkley et al., 1995 and 1996).

Table 7.2 provided inputs and outputs for manufacturing a PVC computer housing. Table 7.7 listed additional inputs and outputs for providing effective shielding of the housing, either by using a metallic copper paint or inserting a steel plate. Finally, Table 7.8 listed inputs and outputs for different options of disposal of the housing. For each specific input and output, we can sum the values in these tables to provide a total environmental burden for that input or output. For example, Table 7.2 shows that the complete production of a PVC computer housing requires 0.94 kg of crude oil, whereas Table 7.7 indicates that 0.09 kg of crude oil are needed for making and applying copper paint to the housing. Table 7.8 shows that landfilling the housing at the end of its useful life requires 0.027 kg of crude oil, mainly for transportation to the disposal site. The sum of these three values—1.06 kg—is the total amount of crude oil needed for making and disposing of the housing. Note that some of the categories in Table 7.2 are not listed in Table 7.7 or Table 7.8 due to lack of an environmental burden in that category.

We are now ready to consider the total environmental impacts of several options for manufacturing and disposal of PVC computer housings.

Example 7.8

Improvement analysis of the computer housing.   One difficulty with the use of copper paint for shielding is that the paint must be removed before the housing can be recycled. The cost of doing this is unacceptably high at present, so that landfilling and incineration are the only realistic options. A PVC housing with steel plate shielding, in contrast, can be readily recycled. Given this situation, determine the environmental burdens associated with the following three options for making and disposing of a computer housing:

Option 1: Shielding by copper paint, disposal by landfilling.

Option 2: Shielding by copper paint, disposal by incineration.

Option 3: Shielding by steel plate, disposal by recycling.

Use Tables 7.2, 7.7, and 7.8 to calculate these burdens, and identify which option is preferred.

Solution:

The results of the calculation are shown in Table 7.9, which shows that Option 3 is the best in virtually all categories of inputs and outputs. For inputs, the only exceptions are greater amounts of iron ore and water needed to make the steel plate. For outputs, the recycling values are less than or comparable to the other disposal methods with the exception of nitrates and a couple of categories of solid waste, where Option 2 is best. Overall, the results indicate that use of the steel plate shielding with recycling is by far the environmentally superior method of dealing with discarded housings at the end of their useful lives.

Table 7.9   Environmental burdens for Example 7.8.

	Substance	Units	Option 1 PVC, Copper, Landfill	Option 2 PVC, Copper, Incinerate	Option 3 PVC, Steel, Recycle
Inputs	*Raw Materials*				
	Crude oil (in ground)	kg	1.06	1.05	0.15
	Coal (in ground)	kg	0.93	0.42	0.4
	Lignite (in ground)	kg	0.063	0.063	0.061
	Natural gas (in ground)	kg	1.3	1.3	0.11
	Limestone	kg	0.18	1.3	0
	Sand	kg	0.0014	0.0014	0.0014
	NaCl	kg	0.93	0.93	0.02
	Bauxite	kg	0.00034	0.00034	0.00034
	Iron ore	kg	0.00054	0.00054	0.99
	Ilmenite ore	kg	0.095	0.095	0.095
	Copper ore	kg	2.9	2.9	0
	Water	liters	56	56	83
	Energy	MJ	182	165	68
Outputs	*Air Emissions*				
	Particulate matter	g	7.8	34	24
	CO_2	g	5,300	7,000	3,000
	CO	g	5.9	6.4	2.6
	SO_x	g	38	28	15
	NO_x	g	33	29	10
	N_2O	g	0.49	0.49	0.44
	NH_3	g	0.0034	0.014	0.0033
	Cl_2	g	0.0014	0.0014	0
	HCl	g	0.63	230	0.21
	CH_4	g	3.1	3.1	2.7
	Total hydrocarbons	g	160	160	64
	Chlorinated organics	g	0.70	0.70	0.70
	Aldehydes	g	0.014	0.014	0.015
	H_2S	g	0.0015	0.0015	0.0012
	Other organic	g	0.019	0.004	−0.68
	Lead	g	0.025	0.025	0
	Metals	g	0.0044	0.0044	0.0043
	Water Effluents				
	BOD5	g	0.20	0.20	0.00027
	COD	g	3.6	3.6	0.0008
	Chlorides	g	58	58	1.0
	Chlorinated organics	g	0.0041	0.0041	0.0041

(continued)

Table 7.9 (*continued*)

Substance	Units	Option 1 PVC, Copper, Landfill	Option 2 PVC, Copper, Incinerate	Option 3 PVC, Steel, Recycle
Dissolved organics	g	1.9	1.9	1.9
Dissolved solids	g	2.9	3.0	1.7
Suspended solids	g	4.3	4.3	0.10
Oil	g	0.14	0.14	0.017
Hydrocarbons	g	0.010	0.010	0.0012
Iron (Fe2+ and Fe3+)	g	6.4	6.4	6.4
Fluorides	g	0.0021	0.0021	0.00201
Metals	g	0.40	0.40	0.01
Nitrates	g	0.0007	0.0004	0.00054
Nitrogen—NH_3	g	0.0049	0.0049	0.00044
Nitrogen—organic	g	0.0064	0.0064	0.0056
Ammonium hydroxide	g	0.00019	0.00019	0.00019
Sodium ions	g	8.4	8.4	0
Sulfates	g	7.4	7.4	−0.2
Methanol	g	0.016	0.016	0.0077
Phenol	g	0.00013	0.00013	0.00013
Fibers	g	0.0091	0.0091	0.0091
Other organic	g	0.0009	0.0009	0
Calcium	g	0.6	0.6	0.6
Magnesium	g	2.6	2.6	2.6
Aluminum	g	1.4	1.4	1.4
Manganese	g	0.38	0.38	0.38
Solid Waste				
Hazardous	kg	0.014	0.014	−0.0025
Landfilled	kg	1.8	0.00053	0.015
Mineral	kg	0.78	0.78	0.083
Nonhazardous chemicals	kg	0.015	0.015	0.015
Slags and flyash	kg	2.2	3.5	0.002
Unspecified	kg	0.44	0.11	0.47
Recovered	kg	0.068	0.068	0.054

7.6 CONCLUSION

In this chapter we have considered a new relationship between industrial processes and environmental preservation. Rather than attempting to fix the damage after it occurs, the new environmental ethic calls for engineers to include environmental considerations right from the start. In other words, engineers will incorporate green design in the early planning of products and processes. Although in its infancy, life cycle assessment is the principal tool likely to be used in this effort.

We have seen that LCA includes three fundamental steps: inventory analysis, impact analysis, and improvement analysis. Of these three, only the first step is developed to an appreciable extent. We lack information on the effects of resource depletion and environmental discharges, making it difficult to conduct rigorous impact analysis. Furthermore, we lack detailed data on possible substitutions for most industrial processes that could help us perform a more rigorous improvement analysis. These areas are fertile ground for environmental research.

Overall, it is likely that environmental life cycle assessments and their component parts are likely to grow in use as we enter a new era of environmental protection and a concern for sustainability. There will be opportunities for engineers to make substantial contributions in this area for some time.

7.7   REFERENCES

Allen, D.T., and N. Bakshani, 1992. "Environmental Impact of Paper and Plastic Grocery Sacks." *Chemical Eng. Edu.,* Spring, pp. 82–86.

APME, 1994. "Eco-Profiles of the European Polymer Industry, Report 6: Polyvinyl Chloride." Technical Paper from the Association of Plastics Manufacturers in Europe, April.

Brinkley, A., J. R. Kirby, I. L. Wadehra, J. Besnainou, R. Coulon, and S. Goybet, 1995. "Life Cycle Inventory of PVC: Disposal Options for a PVC Monitor Housing." *Proceedings of the IEEE International Symposium on Electronics and the Environment,* May 1–3, Orlando, Florida, pp. 145–151.

Brinkley, A., J. R. Kirby, I. L. Wadehra, J. Besnainou, R. Coulon, and S. Goybet, 1996. "Life Cycle Inventory of PVC: Manufacturing and Fabrication Processes." *Proceedings of the IEEE International Symposium on Electronics and the Environment,* May 6–8, Dallas, Texas, pp. 94–101.

Ciambrone, D. F., 1997. *Environmental Life Cycle Analysis.* Lewis Publishers, Boca Raton, Florida, pp. 9–13.

Graedel, T. E., 1998. *Streamlined Life Cycle Assessment.* Prentice Hall, Upper Saddle River, New Jersey, pp. 78–79.

Graedel, T. E., and B. R. Allenby, 1995. *Industrial Ecology.* Prentice Hall, Englewood Cliffs, New Jersey, pp. 151–52.

Horvath, A., C. T. Hendrickson, L. B. Lave, F. C. McMichael, and T. S. Wu, 1995. "Toxic Emissions Indices for Green Design and Inventory." *Environ. Sci. and Tech.,* vol. 29, pp. 86A–90A.

Lowenthal, D. H., B. Zielinska, J. C. Chow, J. G. Watson, M. Gautam, D. H. Ferguson, G. R. Neuroth, and K. D. Stevens, 1994. "Characterization of Heavy-Duty Diesel Vehicle Emissions." *Atoms. Environ.,* vol. 28, pp. 731–43.

MacLean, H. L., and L. B. Lave, 1998. "A Life-Cycle Model of an Automobile." *Environ. Sci. and Technol.* July 1, pp. 322A–330A.

Norberg-Bohm, V., W. C. Clark, B. Bakshi, J. Berkencamp, S. A. Bishko, M. D. Koehler, J. A. Marrs, C. P. Nielsen, and A. Sagar, 1995. "International Comparisons of Environmental Hazards: Development and Evaluation of a Method for Linking Environmental Data with the Strategic Debate on Management Priorities." *Risk Assessment for Global*

Environmental Change. R. Kasperson and J. Kasperson, eds., United Nations Press, New York.

USEPA, 1990. *Reducing Risk: Setting Priorities and Strategies for Environmental Protection.* FAB-EC-90-211, Science Advisory Board, United States Environmental Protection Agency, Washington, DC.

WCED, 1987. *Our Common Future.* World Commission on Environment and Development, Oxford University Press, Oxford, U.K.

7.8   PROBLEMS

7.1   We saw in Figure 7.1 that environmental impacts could be reduced by modeling industrial processes after natural cycles. For example, products could be designed so that their parts could be reused or recycled rather than discarded at the end of their useful lives. Although it is difficult to design closed-cycle industrial processes as in Figure 7.1, we can minimize environmental effects by recycling as shown in Figure 7.12, using aluminum as an example. The figure shows that bauxite ore containing aluminum is first excavated. The ore is then processed to obtain relatively pure aluminum. The aluminum is formed into sheets, bars, ingots, and other shapes before being shipped to production plants. The final aluminum is used to make several products; aircraft, boats, bicycles, and containers for food and beverages are illustrated as some of the major uses. The figure also shows that aluminum products may be recycled and used as inputs to make aluminum stock once again. Note that recycling reduces the amount of bauxite ore mined and also the amount of ore that needs to be processed.

In this problem, choose *two* other common materials for which a similar diagram can be drawn. For each material, answer the following questions:

(a)   What are the most prevalent products using this material?

(b)   What specific steps in the overall industrial process are reduced or entirely eliminated by recycling this material?

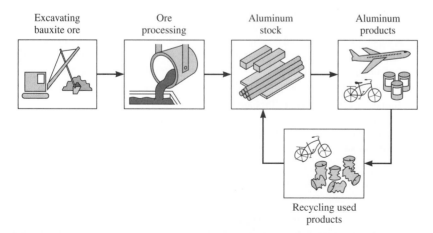

Figure 7.12   Life cycle assessment for aluminum, including recycling.

(c) What natural resources are preserved?

You may need to consult other sources of information to answer these questions.

7.2   Most of the flow diagrams in this chapter begin with the raw materials needed as inputs to the final product. Choosing this step as the starting point is somewhat arbitrary, as discussed in Section 7.2.3. One can go back further in the process by considering the tools needed to obtain the raw materials or the inputs needed to make these tools, for example. List the raw materials identified in this chapter associated with making, packaging, and transporting the PVC computer housings. What tools are used to obtain each raw material? What inputs are needed to make these tools? You may need to consult other sources of information to answer these questions.

7.3   In Section 7.3.2, we mentioned that the choice of PVC material for the computer housing was a design decision made prior to the analyses in this chapter. Now consider the option of making a computer housing entirely from steel. Assume that the environmental burdens for manufacturing a complete steel housing are 10 times the burdens associated with making a steel plate for the PVC housing, shown in Table 7.7. Compare the inputs needed for three types of computer housings: PVC with additives, PVC resin, and steel. The first two housings require a steel plate, which has environmental burdens that must be added in. Which inputs (raw materials) have maximum values for the housing made of PVC with additives? Which inputs are maximum for the PVC resin? The steel housing?

7.4   Extend Table 7.3 by calculating the following outputs for a housing made entirely of PVC resin. Also calculate these outputs for a housing made entirely of steel. For which outputs does the PVC resin housing have the maximum values? Repeat for the PVC with additives and the steel housing.

Outputs to air: particulate matter, CO_2, CO, SO_x, NO_x, N_2O, and CH_4.

Outputs to water: dissolved solids, suspended solids, and oil.

7.5   Assume that your source of iron ore for any of the three housings in Problem 7.3 is from an open pit mine, where the land must be restored to its original condition when the mine is exhausted. This is an extremely expensive process. How might this fact affect your choice of computer housing material?

7.6   Consider shipping the computer housings by trailer truck as described in Example 7.3. Each computer housing without packaging measures 45 cm × 40 cm × 20 cm. If the computer housings could be shipped without packaging, determine the mass of particulate matter emitted per housing for the same conditions as in Example 7.3. What fraction of particulate matter would be emitted from transportation as compared with the emissions from manufacturing the housing?

7.7   Based on your answers to Problem 7.6, one might argue that it would be worthwhile minimizing packaging to reduce environmental effects associated with transportation. Assume that on average 10 percent of the unpackaged housings in a shipment are damaged and must be discarded.

(a) Based on particulate emissions from both manufacturing and transportation of the housings, would it be more beneficial to ship packaged or unpackaged housings? Ignore emissions from manufacturing the packaging material.

(b) Another transportation company using the same size trailer trucks claims they are more careful in handling the shipment. What is the maximum breakage rate that would make it worthwhile to ship unpackaged housings based on particulate emissions?

7.8 In Examples 7.5 and 7.6 we calculated the global warming potential (GWP) for two types of computer housings: one with a PVC mixture that includes resin plus additives, and one with PVC resin only. The calculations were based on emissions of airborne CO_2, N_2O, and CH_4.

(a) Using the same three airborne gases, calculate the GWP for these two types of housings that also include the steel plate insert.

(b) Calculate the GWP for a computer housing made entirely of steel.

(c) Consider your answers to Problems 7.3, 7.4, and 7.8(a) and (b). Which of the three types of computer housings has the most detrimental environmental effects? Discuss your answer.

7.9 Consider energy inputs and air pollution outputs from manufacturing paper and plastic (polyethylene) grocery sacks. We can compare these inputs and outputs by noting that the carrying capacity of 60,790 plastic sacks is equal to that of 30,395 paper sacks (Graedel, 1998). Data are available from three independent studies:

Material	Allen and Bakshani, 1992	Graedel, 1998	Ciambrone, 1997
Plastic			
60,790 sacks			
Energy (MJ)	42,000	42,000	36,000
Air pollution emissions (kg)	33	35	27
Paper			
30,395 sacks			
Energy (MJ)	52,000	52,000	41,000
Air pollution emissions (kg)	89	90	22

(a) What do you conclude about the environmental effects of making plastic versus paper sacks?

(b) Discuss factors that might be responsible for differences in the data among these three studies.

7.10 We have seen in this chapter that life cycle assessment can be a valuable tool for minimizing the environmental damage caused by manufacturing, using, and disposing of products. Yet LCA is not in widespread use. Discuss several reasons for this, and explain whether you feel LCA will play a more prominent role in environmental protection in the future.

MODELING ENVIRONMENTAL PROCESSES

chapter

8

Controlling Urban Smog*

*The principal author of this chapter was Spyros N. Pandis.

8.1 INTRODUCTION TO URBAN AIR POLLUTION

Many cities in both industrialized and developing countries are immersed in a cloud of polluted air for extended periods of the year. This pollution is commonly called *smog*—a term coined in 19th-century England combining the words *smoke* and *fog.* Today the term *smog* refers to the mixture of numerous gaseous and particulate pollutants found in the atmosphere as a result of human activities.

When many people live in the relatively small area of a city, their activities generate enough pollutants to create smog. Such urban air pollution problems appeared even in ancient Rome, whose buildings were blackened by carbon particles emitted during the burning of wood. In the 13th century, King Edward I of England issued a proclamation against the use of coal. The "foul air" of London and other English cities was described by John Evelyn in *Fumifugium: The Inconvenience of the Air and Smoke of London Dissipated,* published in 1661. These problems became a lot more common after the industrial revolution, both in the major cities of Europe and in the United States. The situation was captured well by the poet Percy Shelley, who wrote in the early 19th century, "Hell is a city a lot like London, a populous and smoky city."

Indeed, a number of deadly air pollution episodes occurred in London, most notably the episode of December 1952, in which approximately 4,000 people died during five days of air stagnation and fog over London. At about the same time, in October 1948, the first major smog episode in the United States occurred in Donora, Pennsylvania, an industrial river-valley town near Pittsburgh. Almost half of Donora's population fell sick, with 20 deaths attributed to the episode.

8.1.1 London Smog

The London air pollution episodes were characterized by the combination of extended fog periods and high concentrations of smoke from coal burning for residential heating and other energy needs. As a result, people were breathing air containing fog droplets with high sulfuric acid concentrations (from the sulfur impurities in coal), plus a lot of smoke particles. All of these events took place during the winter with little or no sunlight. This type of urban pollution has been named *London smog.* It is characterized by high concentrations of particulate matter emitted from fuel combustion and industrial processes. Typically, the concentration of sulfur dioxide from fuel combustion also is high. Stagnant air allows the concentration of these pollutants to reach dangerous levels that reduce visibility and imperil human health. This is also the type of smog that affected Donora, Pennsylvania, in 1948, as well as New York City and other urban areas a decade or two later. Today this type of smog is most prevalent in populous cities of eastern Europe and Asia, including Moscow, Sofia, Bombay, and Beijing.

8.1.2 Los Angeles Smog

A second type of smog characterizes urban air pollution in most industrialized countries today. In these big cities the mostly gaseous pollutants emitted by cars, industry, and other anthropogenic activities mix together in the atmosphere and start reacting chemically, sparked by energy from the sun. The result is the creation of *secondary pollutants* that are often more dangerous than their precursors. Ozone is an example of such a secondary pollutant. This type of pollution is known as *photochemical smog,* or *Los Angeles smog,* after the city where it was first discovered.

In this case the word *smog* is really a misnomer because photochemical smog is not a mixture of smoke and fog. Rather, the Los Angeles type of smog includes a whole suite of pollutants, especially hydrocarbons and oxides of nitrogen, plus sulfur dioxide, carbon monoxide, and particles emitted by vehicles and industrial sources. In addition, there are secondary compounds such as ozone, secondary particulate matter, and various toxic species. The smog becomes apparent because of the secondary particulate matter formed in the atmosphere, which reduces visibility as seen in Figure 8.1.

Pollutants are referred to as *primary* if they are emitted directly into the atmosphere from identifiable sources and *secondary* if they are produced in the atmosphere from reactions of the primary pollutants. Some pollutants are partially primary

Figure 8.1   Poor visibility due to urban air pollution. The loss of visibility is due to small particles in the atmosphere that are either emitted directly or formed from chemical reactions of other air pollutants.

and partially secondary. For example, atmospheric particulate matter includes particles emitted directly to the atmosphere (such as soot particles from combustion) plus secondary particulate matter formed from chemical reactions of emitted gases. Examples include SO_2 and unburned gasoline vapors, both of which react in the atmosphere to form products like sulfuric acid and organic substances that condense onto other available particles. The Los Angeles type of smog also harms human health and welfare. As discussed in Chapter 2, its effects include a variety of human respiratory illnesses, as well as damage to materials and vegetation.

8.2   ACHIEVING AIR QUALITY GOALS

The primary objective of this chapter is to understand how pollutants produced by human activities affect the quality of the air we breathe. Based on this understanding, we can then identify effective actions or control measures that can achieve desired air quality levels. To begin, let us examine some of the measures used to quantify pollutant concentrations.

8.2.1   Units of Measurement

A variety of units are used to quantify the concentration or partial pressure of air pollutants in the atmosphere. Because pollutant concentrations are typically very small relative to the major constituents of the atmosphere (oxygen and nitrogen), one of the most useful units of measure for gases is *parts per billion per volume (ppbv,* or usually just *ppb*). If the concentration of a gaseous species is 1 ppb, there is only one molecule of that substance for every billion molecules of air. Its partial pressure is then 10^{-9} atm.

Example 8.1

Units of gaseous concentration in the atmosphere.   Air in the atmosphere is roughly 21 percent oxygen by volume. What is the oxygen concentration in ppb?

Solution:

A concentration of 21 percent by volume means that 21 out of 100 molecules in the atmosphere are oxygen molecules. Therefore, out of 1 billion molecules, 0.21 billion, or $0.21 \times 10^9 = 210$ million, are oxygen. So the oxygen concentration in the atmosphere is 210 million ppb.

As we shall see, an important contribution to modern urban smog is made by gaseous (volatile) *organic compounds,* which are molecules containing carbon plus other elements. Organic compound concentrations in the atmosphere are expressed in units of *equivalent parts per billion of carbon (ppbC)*. To calculate the concentration of an organic compound in ppbC, one needs to multiply the molecular concentration in ppb by the number of carbon atoms in the organic molecule. For example, butane (C_4H_8) has four carbon atoms, so 1 ppb of butane is equivalent to 4 ppbC of butane.

Although gases are usually measured in terms of volumetric concentration, *particles* in the air—which can be solids or liquids—are measured on the basis of their mass concentration. Again, because the amounts are relatively small, the typical unit used for particle concentration is *micrograms of particle mass per cubic meter of air,* commonly written as $\mu g/m^3$. In some cases the concentration of a gaseous substance also may be expressed in terms of mass concentration; but for the most part (including the examples used in this chapter) mass concentration units are reserved for particulate matter, whereas volume concentrations are used for gases.

8.2.2   Air Quality Standards

Health scientists who study urban smog have explored the relationships between air pollution and human health. This information has been used by policymakers to arrive at *ambient air quality standards* that protect human health and welfare by defining the permissible concentrations of atmospheric pollutants. Chapter 2 discussed the U.S. national air quality standards for various pollutants. For example, the maximum permissible concentration of ozone (O_3)—a key measure of Los Angeles–type smog—is a maximum one-hour concentration of only 120 ppb and a maximum 8-hour concentration of just 80 ppb.

In many urban areas of the United States, as well as in large cities elsewhere around the world (like Athens, Mexico City, and Tokyo), measured ambient ozone levels routinely exceed these health-related standards. Thus the key question is what emission control measures are needed to improve air quality so as to achieve the desired ambient air quality standards. The answer to this question is not at all straightforward because the creation of ozone is a complicated problem.

8.2.3   Sources of Emissions

A variety of urban activities cause emission of pollutants to the atmosphere, either directly (such as when we drive our cars, barbecue, mow lawns, or paint houses) or indirectly (as when we heat or cool our homes or buy consumer products). Typical daily emissions to the atmosphere from a large urban city as a result of these activities are shown in Table 8.1. This table highlights emissions of volatile organic compounds (VOCs, mostly hydrocarbons), nitrogen oxides (NO_x), and carbon monoxide—all of which (as we shall soon see) play a critical role in the formation of atmospheric ozone.

Some emission sources, like surface coatings and landfills, emit large quantities of one pollutant (VOCs) and little else, whereas other sources, like automobiles, emit a variety of compounds. Hundreds of different source types are lumped into some of the groups in Table 8.1 (like miscellaneous stationary sources, other solvent use, and off-road vehicles). Furthermore, the volatile organic compounds (VOCs) category shown in Table 8.1 actually represents a complicated mixture of many different organic compounds, some of which are listed in Table 8.2 for a typical urban area. Some of these compounds are more chemically reactive than others. For example, VOCs in the tailpipe exhaust of an automobile are much more reactive than the VOC emissions from a landfill, which contain mainly methane. These chemical differences also play an important role in the formation of ozone and urban smog.

Table 8.1 Daily emissions of pollutants in Los Angeles.

Source	VOCs (Tons/Day)	NO$_x$ (Tons/Day)	CO (Tons/Day)
Stationary Combustion			
Electric utility combustion	11.2	106.1	11.2
Industrial combustion	28.1	99.1	26.3
Commercial fuel combustion	0.1	3.6	0.03
Manufacturing (other) combustion	3.1	39.5	181
Residential and agricultural combustion	16.9	53.3	127
Waste burning and incineration	0.1	0.4	0.3
Other Stationary Sources			
Landfills	778	0.0	0.0
Surface coating (solvent use)	195	1.9	0.4
Other solvent use	162	0.2	0.03
Petroleum refining	2.8	35.3	1.1
Petroleum processes, storage, and transfer	100	11.3	13.6
Industrial processes	26.1	2.6	18.1
Miscellaneous stationary sources	440	3.8	119
Transportation Sources			
On-road vehicles	581	662	5,002
Off-road vehicles	26.8	67.1	172.6
Railroads	3.8	15.3	5.8
Ships	22.2	16.0	88.6
Aircraft	18.1	16.3	84.5
Total	2,415	1,134	5,851

Source: Based on Russell et al., 1983.

8.2.4 The Role of Engineers

Engineers are heavily involved in trying to make some sense out of this complicated system of a polluted atmosphere. Indeed, there may be hundreds of potential choices for reducing air pollutant emissions in order to improve air quality. For instance, one can reduce automobile emissions by changing engine design, changing fuel, expanding public transportation, or adding exhaust catalysts. One can also place additional controls on industrial emissions to remove particulate matter, SO$_2$, or NO$_x$, or introduce new solvents for paints and industrial processes. All of these choices have a cost to society. How should one proceed? Which set of solutions should one pick?

How does one evaluate proposals for improving air quality, such as the suggestion to replace current gasoline-based automobiles with new cars burning a mixture of 80 percent methanol and 20 percent gasoline? Engineers have designed engines that can use this fuel, but is it a good idea to replace all cars in Los Angeles with

Table 8.2   Major atmospheric VOCs in Atlanta, Georgia.

Species	Concentration (ppbC)	Species	Concentration (ppbC)
Isopentane	19.8	Meta,para-ethyl-toluene	3.6
Butane	16.9	3-methyl-pentane	3.4
Toluene	14.7	Ethylene	3.0
Para-cymene	11.0	Methyl-cyclopentane	2.9
Pentane	9.4	Ethylbenzene	2.8
Benzene	8.8	Ortho-xylene	2.8
Meta,para-xylene	7.6	3-methyl-hexane	2.6
2-methylpentane	5.9	2,3-dimethyl-pentane	2.5
Cyclohexane	5.2	1,4-diethyl-benzene	2.4
Ethane	5.0	Iso-butene	2.2
Undecane	4.9	1,2,4-trimethylpentane	2.2
Propane	4.8	Iso-butyl-benzene	2.2
Isobutane	4.8	2-methyl-2-butene	1.8
Isoprene	4.6	1,3,5-trimethylbenzene	1.8
Acetylene	4.3	Cyclopentane	1.6
Hexane	3.8	Propene	1.5
		Total Measured	**197**

Source: NRC, 1992.

these new ones? Will this effectively achieve the desired air quality goals? This would be an expensive experiment: It would require replacing 20 million or so cars with new ones using a different technology. However, the choice is still there. Should we go in this direction? If so, is it possible that reducing emissions will not really make things better, and could even make the air pollution problem worse? Engineers, with their grounding in science and mathematics and their ability to work with complex systems, are in a unique position to contribute answers to such questions.

The rest of this chapter will examine the basics of the modern urban atmospheric system and will demonstrate how engineers can provide valuable guidance on the selection of effective air pollution control strategies. This chapter also will provide insights as to the causes of urban smog, which have important implications for engineers who design commercial products and processes that emit air pollutants.

We will begin by looking at how pollutants accumulate in the atmosphere, starting with pollutants that do not react chemically to produce other species. Then we will consider the more complex problem of atmospheric chemical reactions that produce secondary pollutants like ozone. This understanding is essential in order to devise emission control programs that can improve urban air quality.

8.3 ACCUMULATION OF POLLUTANTS IN AN URBAN AREA

Valuable insights can be gained by describing mathematically the accumulation of pollutants over an urban area. Sophisticated computer models are available, but let us attempt to first construct a simple representation of the system from basic principles.

As a first approximation, the atmosphere over an urban area can be modeled as a well-mixed box with horizontal dimensions Δx and Δy (the dimensions of the city) and height H, as in Figure 8.2. This type of model is known as a *box model*. Let us assume that the wind speed is u and the wind direction is parallel to the x dimension of the city.

The choice of the horizontal dimensions of the box is relatively straightforward, but what about the height H? The troposphere is the lowest layer of the atmosphere and contains most of the atmospheric gases. Normally, the temperature in the troposphere decreases with altitude. Although the troposphere extends up to about 10 km, it is not usually well mixed above urban areas. So typically the *mixing height, H,* is about 1,000 m over an urban area.

Under some circumstances the mixing height can be much lower. In particular, air pollution episodes are usually accompanied by a meteorological phenomenon called a *temperature inversion.* Under inversion conditions a layer is formed above the ground where the temperature actually *increases* with altitude, as sketched in Figure 8.3. This layer of warm air acts as a lid that prevents the mixing of air below with air above. Therefore, the presence of a temperature inversion restricts the dilution of pollutants and sets the stage for the buildup of high concentrations of pollutants during an air pollution episode. The altitude at the bottom of the inversion layer is called the *inversion height.* For typical air pollution episodes it is around 500 m.

Our first task is to calculate the concentration of pollutants in the urban atmosphere. These concentration levels can then be compared to air quality standards to determine whether emission control measures are needed, and if so, how much. We will consider three cases with increasing levels of complexity. First we will look at chemically nonreacting (inert) pollutants at steady state. Next we will look at the buildup of these pollutants over time. Finally, we will examine chemically reacting

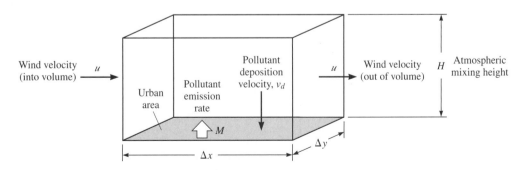

Figure 8.2 A box model for the atmosphere of an urban area.

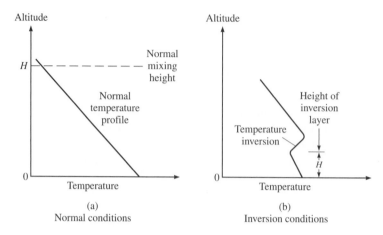

Figure 8.3   Atmospheric inversions and mixing height. Under normal conditions the temperature of the lower atmosphere falls with increasing altitude, and there is good vertical mixing within a height of roughly 1,000 m. Inversion conditions alter the temperature profile and substantially reduce the mixing height.

pollutants and their buildup over time. In each case the concentration over the region is assumed to be spatially uniform because air within the box is well mixed.

8.3.1   Nonreacting Pollutants under Steady State Conditions

Let us start with the simplest case of a pollutant that does not participate in any atmospheric chemical reactions. Such pollutants include various types of particulate emissions, such as lead from auto exhaust and other types of solid or liquid particles from industrial processes and fuel combustion. Even some gaseous pollutants like SO_2 often may be approximated as chemically nonreacting over limited spatial and temporal scales. Our goal is to estimate the urban air concentration of such pollutants for different emission rates and meteorological conditions.

The first step is to identify the sources of the pollutant. In general, there are two types of sources: direct emissions to the atmosphere from sources within the city, and the influx of pollutants from other areas upwind of the city. The second step is to identify the *sinks* of the pollutant. Again, there are two general types: deposition of pollutants onto the ground within the city, and atmospheric flow out of the city carried by the wind. To simplify the system a little further, let us assume that the sources and sinks are in balance—that is, the system is in steady state. The mass balance for the pollutant can then be written as

(Mass emission rate into the urban air volume)

    + (Mass flow rate into the volume with the wind)

= (Mass removal rate into the ground)

    + (Mass flow rate out of the volume with the wind)   (8.1)

Note that the units of all these terms are mass per unit time.

To calculate the average concentration, we need to translate this equation into a mathematical expression. Let c be the uniform concentration of the pollutant over the city in units of mass per volume of air (for example, $\mu g/m^3$), let M be the mass emission rate (units of mass per time), and let R (mass per time) be the mass removal rate. Furthermore, let c° be the background concentration of pollutant in the air just outside the urban area (units of mass per volume). Then the mass flow rate of pollutant into the urban area will be proportional to the wind speed, the cross-sectional area of the box perpendicular to the wind direction, and the background pollutant concentration. In mathematical terms,

$$[\text{Pollutant mass flow rate into the volume with the wind flow}] = u(H\,\Delta y)c^\circ \quad (8.2)$$

Let us check the units to make sure that our calculation is correct:

$$uH\Delta yc^\circ[=]\frac{(\text{length})}{(\text{time})}(\text{length})(\text{length})\frac{(\text{mass})}{(\text{length})^3}[=]\frac{(\text{mass})}{(\text{time})}$$

Everything appears to be in order. Following the same approach, we can express the mass flow *leaving* the box as

$$[\text{Pollutant mass flow rate out of the volume with the wind flow}] = uH\Delta yc \quad (8.3)$$

Substituting these expressions into the overall mass balance gives

$$M + uH\Delta yc^\circ = R + uH\Delta yc \quad (8.4)$$

The overall mass emission rate, M, can be written as a function of the emission rate per surface area in the city, m (units of mass per surface area per time):

$$M = m\Delta x\,\Delta y \quad (8.5)$$

The same simplification can be used for the removal rate, R, defining the removal rate per surface area, r (units of mass per surface area per time) as

$$R = r\Delta x\,\Delta y \quad (8.6)$$

This removal rate per surface area is usually described as the product of a *dry deposition velocity*, v_d, and the concentration of the species:

$$r = v_d c \quad (8.7)$$

The dry deposition velocity is a parameter used to describe how rapidly the species is removed from the atmosphere by deposition onto the ground.

Substituting these expressions into our mass balance and rearranging the terms gives

$$\frac{m}{H} = \frac{v_d}{H}c + \frac{u}{\Delta x}(c - c^\circ) \quad (8.8)$$

This equation expresses mathematically the balance between the emission sources of pollutant and its removal by deposition and flow out of the system.

Example 8.2

Residence time of pollutants in an urban area.   What is the physical meaning of the term $(\Delta x/u)^{-1}$ on the right side of the mass balance equation?

Solution:

The first hint comes from examining the units of the term. $\Delta x/u$ has units of time, so this is a time quantity. It is the ratio of the length of the box to the prevailing wind speed, u. So $\Delta x/u$ measures the *residence time* of air over the urban area, and $(\Delta x/u)^{-1}$ is the reciprocal of this time.

Using the insight from Example 8.2, we can define *residence time, τ,* as an indicator of how long a moving air mass stays over an urban area:

$$\tau = \frac{\Delta x}{u} \tag{8.9}$$

Using the residence time, the mass balance equation can be rewritten as

$$\frac{m}{H} = \frac{v_d}{H}c + \frac{(c - c^\circ)}{\tau} \tag{8.10}$$

Solving for the pollutant concentration gives

$$c = \frac{m\tau + Hc^\circ}{v_d\tau + H} \tag{8.11}$$

Example 8.3

Lead concentrations in an urban area.   The city of Leadville (dimensions 120 × 120 km) is characterized by an average daily emission rate of 5,000 kg/d of lead into the atmosphere from auto exhaust and industrial processes. The background concentration of atmospheric lead is 0.1 $\mu g/m^3$. During a typical hot summer day in Leadville the mixing height is 500 m and the average wind speed is 2 m/s. A typical lead deposition velocity for similar meteorological conditions is 0.05 cm/s.

(a) What is the steady state concentration of lead in the urban air?

(b) What (if anything) must Leadville citizens do to achieve an atmospheric lead concentration of 0.4 $\mu g/m^3$?

(c) What if the air quality target is 0.15 $\mu g/m^3$?

Solution:

(a) The first step is to apply our simple model to determine the steady state lead concentration over the city. We will use hours for time, meters for length, and micrograms for mass. The residence time of air pollutants over the city is then

$$\tau = \Delta x/u = (120{,}000 \text{ m}) \, / \, (2 \text{ m/s}) = 60{,}000 \text{ s} = 16.7 \text{ hr}$$

The emission rate of lead per unit surface area is

$$m = M/(\Delta x \, \Delta y) = (5{,}000{,}000 \text{ g/d})/(120{,}000 \text{ m})^2 = 350 \times 10^{-6} \text{g/m}^2\text{-d}$$
$$= 15 \, \mu g/m^2\text{-hr}$$

The lead deposition velocity in meters per hour is

$$v_d = 0.05 \text{ cm/s} = 5 \times 10^{-4} \text{ m/s} = (5 \times 10^{-4} \text{ m/s}) \times (3{,}600 \text{ sec/hr}) = 1.8 \text{ m/hr}$$

Substituting all the values into our equation for the steady state pollutant concentration gives

$$c = \frac{(15 \times 16.7) + (500 \times 0.1)}{(1.8 \times 16.7) + 500} = 0.57 \ \mu g/m^3$$

(b) This value exceeds the desired air quality level of 0.4 $\mu g/m^3$. Looking at the variables of the equation determining the lead concentration, we can see that the only variable the citizens of Leadville can readily control is the lead emission flux, m. Solving Equation (8.11) for m gives

$$m = v_d c + \frac{(c - c^\circ)}{\tau} H \qquad (8.12)$$

Substituting the target concentration, c, into this equation, we find that to reduce the lead concentration to 0.4 $\mu g/m^3$, an emission flux of 9.7 $\mu g/m^2$-hr is necessary. Comparing this to the current emission flux of 15 $\mu g/m^2$-hr, we find that in order to reduce the atmospheric lead concentration by 0.17 $\mu g/m^3$ (approximately 30 percent), the emissions must be reduced by 5.3 $\mu g/m^2$-hr (approximately 35 percent).

(c) If the target atmospheric concentration is 0.15 $\mu g/m^3$ (a 73 percent reduction), then by using Equation (8.12) we find that $m = 1.8 \ \mu g/m^2$-hr. In this case the lead *emissions* must be reduced by almost 90 percent. This huge decrease in emissions is necessary because of the background lead concentration in the area. Note from the mass balance that if Leadville somehow eliminated all lead emissions ($m = 0$), the atmospheric concentration of lead in the city air would be

$$c = \frac{H c^\circ}{v_d \tau + H} = \frac{500 \times 0.1}{1.8 \times 16.7 + 500} = 0.09 \ \mu g/m^3$$

This is slightly lower than the background concentration because some of the background lead brought in by the wind settles onto the ground in Leadville.

Example 8.3 illustrates that if an air quality target is close to the background concentration, then large emission reductions are necessary for the affected urban area. One could argue that in such cases reducing the emissions of other nearby upwind sources would more effectively reduce the atmospheric concentration in the affected urban area. The following example illustrates this point.

Example 8.4

Local versus regional contributions to urban pollution.   Using the data in Example 8.3, what is the percentage contribution to the ambient lead concentration in Leadville from sources upwind of the city? How will this percentage change as lead emissions in Leadville are reduced in the future?

Solution:

Using Equation (8.11), the concentration of lead can be rewritten as the sum of two terms:

$$c = \frac{\tau}{v_d \tau + H} m + \frac{H}{v_d \tau + H} c^\circ$$

The first term is proportional to the lead emission flux in Leadville and is the local contribution. The second term depends on the background concentration, c^o, and represents the contribution of upwind sources to the pollution problem in Leadville. The second term was calculated in Example 8.3 and was found to equal 0.09 $\mu g/m^3$. As was noted, this is a little less than the background concentration because some of these incoming lead particles are deposited on the ground as they are transported over Leadville by the wind.

The factor multiplying m in the above equation depends only on the residence time, deposition velocity, and mixing height. Based on values of $\tau = 16.7$ hr, $v_d = 1.8$ m/hr, and $H = 500$ m (from Example 8.3), this factor is equal to 0.03 hr/m. So the concentration of lead will be given by

$$c \ (\mu g/m^3) = 0.03 \ m + 0.09$$

The fraction, f, of this concentration contributed by upwind sources will therefore equal

$$f = \frac{0.09}{c} = \frac{0.09}{0.03m + 0.09}$$

These results, shown graphically in Figure 8.4, indicate that if the emissions in Leadville are reduced to below 3 $\mu g/m^2$-hr, most of the lead in the city will be the result of transport from other areas.

A Simplified Case   If a nonreacting pollutant has very low background concentrations ($c^o \approx 0$) and very low deposition velocity ($v_d \approx 0$), Equation (8.11) describing its steady state concentration simplifies to

$$c = \frac{m\Delta x}{Hu} \tag{8.13}$$

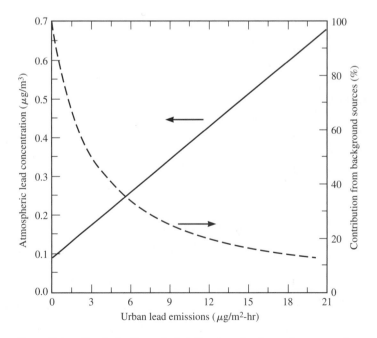

Figure 8.4   Results for Example 8.4: Fraction of pollution from upwind sources.

This simple equation presents some important insights for urban air pollution. The atmospheric concentration of primary pollutants increases as the emission rate per surface area (m) increases and as the size of the urban area (Δx) increases. This means that in large cities pollutant concentrations will tend to be higher than in smaller areas. These two parameters in the numerator (emission rates and city dimensions) are the ones over which we have control.

Equation (8.13) shows that high concentrations also are expected for periods with low mixing height (H) and low wind speed (u). These two parameters are determined by nature; we can do little about them other than try to take them into account during the development of cities. Certain areas like southern California are characterized by frequent periods of pleasant weather, but also low mixing heights and low wind speeds. For such an area, because the terms in the denominator are small, the only way to avoid a buildup of pollutants is to decrease the emission rate, m. The other option is to keep the city size small, but for large urban areas like Los Angeles it's already too late to make this choice.

Control Strategies   Let us look a little more closely at the factors affecting air pollutant emissions. In general, the air pollutant emission rate of a region depends on the type and level of human activity, as reflected in the size of the population and the magnitude of economic activity. This dependency can be roughly approximated by the following expression:

Air pollutant emission rate $=$

(Population) $\times$ (Economic activity per person) $\times$

(Pollutant emissions per unit of economic activity) $\qquad$ (8.14)

The insight from this equation is that emission reductions require one or more of the following: reducing the population of the area, reducing the economic activity per person, or reducing the pollutant emissions per unit of economic activity. The most desirable approach is to reduce the pollutant emissions per unit of economic activity. This is the purpose of environmental control technology and green design activities. However, improvements in this area often are offset by increases in population (southern California is a good example) and economic prosperity. The design of emission control strategies to achieve air quality objectives must take account of all such factors affecting future emission rates. Chapter 15 discusses some of the methods used to forecast environmental emissions.

For simplicity, we will continue in this chapter to focus only on the relationships between emissions and atmospheric concentrations using a box model that assumes uniform concentrations everywhere within a region. The use of more advanced models that account for geographical variations will be left to more specialized courses and texts (like Seinfeld and Pandis, 1998). Nonetheless, the simple box model provides substantial insight into the design of an effective emission control strategy, as illustrated in the next example.

Example 8.5

Emission control strategies for an urban area. A chemically nonreacting pollutant is emitted at a rate of $m = 200$ $\mu g/m^2$-hr within a large urban area. Its background concentration is $c^o = 0.5$ $\mu g/m^3$, and its deposition velocity is 0.001 m/s. The city has dimensions of roughly 100 × 100 km and is characterized by an average wind speed of 3 m/s and an average mixing height of 1,000 m.

(a) What is the steady state concentration of the pollutant over the city?

(b) During the worst period of the year the average wind speed is only 1 m/s and the mixing height is 300 m. What is the steady state concentration of the pollutant over the city during this period?

(c) You want to make sure that the concentration of the pollutant during the worst period of the year does not exceed 3 $\mu g/m^3$. How much should you reduce the maximum concentration? What reduction of emissions is necessary to achieve this goal?

(d) You would like to keep the pollutant concentrations on an average day below 1.5 $\mu g/m^3$ and for the worst period below 3 $\mu g/m^3$. Let us assume that the major sources of the pollutant are automobiles and heavy industry. What might you propose to do?

Solution:

(a) At steady state, the average pollutant concentration is given by Equation (8.11):

$$c = \frac{m\tau + Hc^o}{v_d \tau + H} = \frac{m(\Delta x/u) + Hc^o}{v_d(\Delta x/u) + H}$$

The expression on the right side simply incorporates the definition of τ from Equation (8.9). This form of the equation displays the wind velocity, u, explicitly. So another way of rewriting this equation is

$$c = \frac{m\Delta x + uHc^o}{v_d\Delta x + uH}$$

Substituting the values given for this problem and expressing all rates on a *per hour* basis gives

$$c = \frac{(200 \times 100,000) + (10,800 \times 1,000 \times 0.5)}{(3.6 \times 100,000) + (10,800 \times 1,000)} = 2.3 \ \mu g/m^3$$

So emissions in the urban area will cause the average pollutant concentration to increase by a factor of 4.6 over the background value.

(b) Using the formula developed in part (a) and substituting $u = 3,600$ m/hr and $H = 300$ m, we find that

$$c = \frac{(200 \times 100,000) + (3,600 \times 300 \times 0.5)}{(3.6 \times 100,000) + (3,600 \times 300)} = 13.9 \ \mu g/m^3$$

Under these extreme meteorological conditions (stagnant air, low mixing height), the steady state concentration of the pollutant is predicted to be almost 30 times above the background level.

(c) A reduction in the maximum concentration of 13.9 $\mu g/m^3$ to 3 $\mu g/m^3$ is a reduction of 78.4 percent in concentration. Solving for the emissions as a function of concentration, we find

$$m = v_d c + \frac{uH}{\Delta x}(c - c^o)$$

Substituting the values of the problem and the desired concentration of $c = 3\ \mu g/m^3$, we find

$$m = (3.6 \times 3) + \left(\frac{3{,}600 \times 300}{100{,}000}\right)(3 - 0.5) = 37.8\ \mu g/m^2\text{-hr}$$

The actual emission rate is given to be $200\ \mu g/m^2$-hr. A reduction in emissions of 81.1 percent is therefore required to achieve the desired concentration target.

(d) Let us first calculate the emission reduction required for the average day with a concentration target of $1.5\ \mu g/m^3$. We find that for the average meteorological conditions the emissions must be reduced to

$$m = (3.6 \times 1.5) + \frac{10{,}800 \times 1{,}000}{100{,}000}(1.5 - 0.5) = 113\ \mu g/m^2\text{-hr}$$

So reducing the concentration from 2.3 to $1.5\ \mu g/m^3$ (a 35 percent reduction) would require reducing emissions from 200 to $113\ \mu g/m^2$-hr (a 43 percent reduction).

From this analysis, requiring a reduction in emissions by 81 percent (needed for the worst period) would satisfy both targets. This is one possible solution. However, this requirement may be costly and may be necessary for only a few days per year. A more economical solution may be to reduce emissions by 43 percent or more on average, and then make other efforts to achieve another 38 percent reduction on the worst periods of the year. These additional reductions could come by lessening industrial activity and automobile emissions during air pollution episodes (for instance, asking people to avoid driving if not necessary). If this additional reduction cannot be easily achieved, a larger reduction for the whole year would be necessary to achieve both air quality concentration targets.

8.3.2 Nonreacting Pollutants under Dynamic Conditions

In the previous discussion we assumed that the sources and sinks of the pollutant were in equilibrium (steady state). The result of this assumption was that we could calculate the concentration of the pollutant with a simple algebraic equation. In the more realistic case where the sources and sinks are not equal, a more complex mass balance is necessary. In this mass balance we describe the change over time of the total mass of the pollutant in the atmosphere above the area of interest. Again we assume the pollutant does not react chemically with other substances. The resulting mass balance is:

Rate of change of mass = (Emission rate into the volume)

− (Removal rate to the ground)

+ (Mass flux into the volume with the wind)

− (Mass flux out of the volume with the wind)　(8.15)

Let c be the concentration of the species of interest (in mass per volume of air). Then the total mass of the species, M_t, in the box volume, V, will be equal to $M_t = Vc =$

$\Delta x\,\Delta y\,Hc$. The rate of change of mass is then just the time derivative of this expression:

$$\text{Rate of change of mass} = \frac{dM_t}{dt}$$

$$= \frac{d}{dt}(\Delta x \Delta y Hc) = \Delta x \Delta y H \frac{dc}{dt} \tag{8.16}$$

Here we have implicitly assumed that the mixing height H and the horizontal dimensions of the area remain constant. The remaining terms in this equation have already been discussed in the previous section. So we can now substitute these terms into the mass balance expression of Equation (8.15) to find

$$\Delta x \Delta y H \frac{dc}{dt} = M - R + uH\Delta y c^\circ - uH\Delta y c \tag{8.17}$$

After we divide both sides by $\Delta x \Delta y H$, this equation simplifies to

$$\frac{dc}{dt} = \frac{1}{H}\frac{M}{\Delta x \Delta y} - \frac{1}{H}\frac{R}{\Delta x \Delta y} + \frac{u}{\Delta x}(c^\circ - c) \tag{8.18}$$

We can simplify further by recalling that $m = M/\Delta x\,\Delta y$ is the emission rate per unit area and $r = R/\Delta x\,\Delta y = v_d\,c$ is the removal rate per unit area. So finally

$$\frac{dc}{dt} = \frac{m}{H} - \frac{v_d}{H}c - \frac{u}{\Delta x}(c - c^\circ) \tag{8.19}$$

This equation describes the evolution of the concentration of an urban pollutant with time. The steady state case that we looked at previously corresponds to setting $dc/dt = 0$ in this equation.

The terms on the right side of Equation (8.19) correspond to the changes in the concentrations of pollutant as a result of emissions (m/H), dry removal ($v_d\,c/H$), and flows into ($u\,c^\circ/\Delta x$) and out of ($u\,c/\Delta x$) an urban area. The effects of meteorology are included in the parameters H (mixing height or inversion height), u (wind speed), and implicitly in the deposition velocity v_d (studies show that the higher the wind speed, the higher the deposition velocity). For an urban area $c > c^\circ$, so the flow term, $-u\,(c-c^\circ)/\Delta x$ (also known as the *advection* term), is negative, which tends to decrease the rate of change in concentration of the species. Note that the higher the wind speed, the larger this negative term and the slower the concentration changes.

Example 8.6

Buildup of pollutants over an urban area.   Fogs are usually accompanied by mixing heights as low as 100 m (or even lower) and low wind speeds. If these conditions persist for several days, as happened during the London fog of 1952, the pollutant concentrations can reach dangerous levels. Let us examine this buildup of pollutant concentrations over time.

During the winter, an urban area (dimensions 100 × 100 km) is characterized by average particulate matter (PM) concentrations of 50 $\mu g/m^3$ on normal days. The average mixing height is 1,000 m and the wind speed 5 m/s. A typical deposition velocity for PM is 0.1 cm/s, and the background PM concentration is 10 $\mu g/m^3$.

(a) Calculate the average PM emission rate for this urban area.

(b) A persistent fog episode occurs during this winter period. The mixing height is reduced to 100 m and the wind speed to 0.5 m/s. What will be the PM concentrations after one, two, and three days? Assume that the concentration at the beginning of the fog episode equals the average concentration for the area on a normal day.

(c) What will be the steady state PM concentration, and when will it be reached?

Solution:

(a) As we saw in the previous examples, the steady state emission rate will be given by

$$m = v_d c + \frac{uH}{\Delta x}(c - c^o)$$

In this case

$v_d = 0.1 \text{ cm/s} = 10^{-3} \text{ m/s} = 3.6 \text{ m/hr}$ $H = 1,000 \text{ m}$

$c = 50 \ \mu g/m^3$ $c^o = 10 \ \mu g/m^3$

$u = 5 \text{ m/s} = 18,000 \text{ m/hr}$ $\Delta x = 100,000 \text{ m}$

Therefore, the mass emission rate per unit area is

$$m = (3.6 \times 50) + \frac{18,000 \times 1,000}{100,000}(50 - 10)$$

$$= 7,400 \ \mu g/m^2\text{-hr}$$

To find the total mass emission rate, simply multiply by the area of the urban region:

$$M = m \, \Delta x \Delta y = (7,400 \ \mu g/m^2\text{-hr}) \times (100,000 \text{ m}) \times (100,000 \text{ m}) (24 \text{ hr/d})$$

$$= 1.8 \times 10^{15} \ \mu g/d = 1,800 \text{ tonnes/d}$$

(Recall that 1 tonne = 1 metric ton = 1,000 kg = $10^{12} \ \mu g$)

(b) During the fog episode, the evolution of the concentration of PM will be described by Equation (8.19) derived in the previous section:

$$\frac{dc}{dt} = \frac{m}{H} - \frac{v_d}{H}c - \frac{u}{\Delta x}(c - c^o)$$

The given values of u and H for this case are

$$u = 0.5 \text{ m/s} = 1800 \text{ m/hr, and } H = 100 \text{ m}$$

Substituting these values into the above equation, along with the values of m, v_d, and c^o, the rate of change of the pollutant concentration becomes

$$\frac{dc}{dt} = \frac{7,400}{100} - \frac{3.6}{100}c - \frac{1,800}{100,000}(c - 10)$$

$$= 74.18 - 0.054 \, c$$

or

$$\frac{dc}{dt} + 0.054c = 74.18$$

The solution to this first-order differential equation can be found using standard mathematical methods such as the integrating factor method. (If you are not familiar with solutions to differential equations, don't worry; it will come later in your studies. More details also appear in Chapter 11.) If c_i is the initial PM concentration when the fog starts, then the result is

$$c(t) = 1,374 + (c_i - 1,374)e^{-0.054t}$$

Assuming the initial concentration c_i is equal to the average concentration for the region, then $c_i = 50 \ \mu g/m^3$, and the solution for concentration as a function of time is

$$c(t) = 1,374 - 1,324e^{-0.054t}$$

The pollutant concentration within the fog layer at different times is shown below and plotted in Figure 8.5. Note also that the residence time of pollutants in the area is

$$\tau = \Delta x/u = 100,000/1,800 = 55.5 \ hr = 2.3 \ days$$

Time (Hours)	Concentration ($\mu g/m^3$)
0	50
6	416
12	681
24	1,012
48	1,275
72	1,347
96	1,367

The concentration increases rapidly under these conditions, reaching over 1,000 $\mu g/m^3$ after one day (24 hours). These are extremely high concentrations of PM. For comparison, the current daily national air quality standard for PM is 260 $\mu g/m^3$. PM concentrations above 1,000 $\mu g/m^3$ for several days would be similar to levels that caused

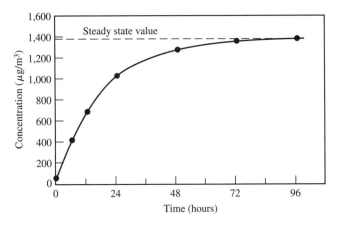

Figure 8.5   Concentration results for Example 8.6.

thousands of deaths in London in the 1950s. However, such meteorological conditions (a three-day fog episode in an urban area) are extremely rare.

(c) As time increases, the exponential term on the right side of the solution becomes smaller. Eventually the concentration approaches its steady state value of 1,374 $\mu g/m^3$, as calculated directly using Equation (8.9). However, because the exponential term never exactly reaches zero (except at $t = \infty$), the concentration of PM never reaches the steady state value. After two days it is 93 percent of this value, and after three days it is 98 percent of this value. So for all practical purposes the concentration has reached its steady state after two to three days. Note that this time is approximately equal to the residence time of pollutants over the area. This is not a coincidence; the residence time is a good approximation of the time it takes to reach steady state for well-mixed nonreactive systems.

A note of caution is in order regarding the use of steady state or equilibrium values. It takes time—often days—to reach these pollutant concentrations, as Example 8.6 illustrates. So using steady state values for short periods can result in significant errors. For example, if the fog episode lasted only six hours but we used the steady state value to calculate the maximum PM concentration, we would calculate 1,374 $\mu g/m^3$. Our dynamic solution indicates that the correct value is only 416 $\mu g/m^3$. It takes much longer than six hours in this case to reach the steady state.

8.3.3 Reactive Pollutants under Dynamic Conditions

Until now we've assumed there are no chemical reactions involving the pollutants of interest. However, some pollutants participate in chemical reactions in the atmosphere—as reactants, as products of reactions, or as both. In this case additional terms must be added to the mass balance equation to describe the overall change in pollutant concentrations. The new terms must reflect the *production* of a pollutant via chemical reactions, as well as its *destruction* via other chemical reactions. The overall mass balance then must be written as

Rate of change of mass $=$ (Emission rate into the volume)

$-$ (Removal rate to the ground)

$+$ (Chemical production rate)

$-$ (Chemical destruction rate)

$+$ (Mass flux into the volume with the wind)

$-$ (Mass flux out of the volume with the wind)   (8.20)

Let P be the chemical production rate of a pollutant and D its chemical destruction rate, both in units of mass per time *per unit volume*. Then the overall production rate of a pollutant over an urban area will equal $P\Delta x \Delta y H$, and the overall destruction rate will equal $D\Delta x\Delta yH$. Substituting these new terms into the mass balance equation, along with the previous expressions from Equation (8.9), gives

$$\Delta x\Delta y\,H\frac{dc}{dt} = M \ + \ P\Delta x\Delta y\,H \ - \ D\Delta x\Delta y\,H \ - \ R \ + \ uH\Delta yc^\circ \ - \ uH\Delta yc$$

After we divide both sides by $\Delta x \, \Delta y \, H$, this equation simplifies to

$$\frac{dc}{dt} = \frac{1}{H}\frac{M}{\Delta x\Delta y} \ + \ P - D - \frac{1}{H}\frac{R}{\Delta x\Delta y} \ + \ \frac{u}{\Delta x}\left(c^\circ - c\right)$$

or, following the same steps as before,

$$\frac{dc}{dt} = \frac{m}{H} + P - D - \frac{v_d}{H}c - \frac{u}{\Delta x}\left(c - c^\circ\right) \tag{8.21}$$

This equation is more general than the earlier equations for nonreacting systems. It is used in complex atmospheric air quality models, where a separate equation is written for each pollutant or chemical species in the atmosphere. The terms P and D are calculated based on the specific reactions involving each species and their corresponding rates. Such simulations of atmospheric chemistry require the solution of systems of nonlinear differential equations, something that is well beyond the scope of this chapter.

One of the most important examples of urban air pollution involving a dynamic, chemically reacting system is the case of ozone. Chapters 2, 3, and 5 noted that ozone in urban areas is formed by chemical reactions in the atmosphere involving hydrocarbon gases and nitrogen oxides. Chapter 2 (Figure 2.2) also showed that more Americans are exposed to unhealthy levels of ozone than any other air pollutant. Now we take a more careful look at this pollutant in the remainder of this chapter. The goal is to understand how it is formed and what we can do to reduce ozone levels in areas where pollution exceeds air quality standards.

8.4 OZONE IN THE ATMOSPHERE

Ozone (O_3) is a reactive gas composed of three oxygen atoms, in contrast to normal molecular oxygen (O_2). Most of the earth's atmospheric ozone (85–90 percent) is found in the *stratosphere*—the portion of the atmosphere between about 10 and 45 kilometers (km) altitude. In the stratosphere, ozone plays a crucial role in absorbing ultraviolet radiation emitted by the sun. The remaining 10–15 percent of atmospheric ozone can be found in the *troposphere*—the lower part of the atmosphere extending from the earth's surface to an altitude of approximately 10 km.

Ozone is a trace atmospheric gas with natural concentrations in the clean troposphere of approximately 30–40 parts per billion (ppb). If the entire atmospheric ozone volume were collapsed to a pressure of one atmosphere, it would form a layer only 3 mm thick. Because ozone exists in such small concentrations in the atmosphere, human activities are more apt to cause changes in its concentration.

8.4.1 Urban Ozone Levels

High ozone concentrations near the surface of the earth are usually associated with urban areas and periods of high temperature and sunlight. Although it first gained notoriety as a pollutant in Los Angeles smog, high levels of ozone are now observed in most urban areas, including most of the eastern United States during the summer months. The daily maximum ozone concentration occurs between noon and 5 P.M. in most central or downtown urban areas. The concentration then decreases during the night and often becomes zero at ground level. For locations downwind of these urban centers, peaks in ozone concentrations are observed even during the night.

The highest levels of ozone occur during air pollution episodes. Figure 8.6 shows the ozone concentration during a three-day episode at Montague, Massachusetts, which is characterized by two daily peaks. The first peak occurs in the early afternoon and is due to ozone produced locally. A second peak, often larger, is evident in the evening due to ozone transported from other locations. As we will see later, ozone production requires a lot of sunlight, so ozone peaks in the late afternoon or evening indicate that ozone produced somewhere else earlier in the day is arriving in the area of concern, transported by the wind.

Measurements like the ones in Figure 8.6 suggest that ozone in the eastern United States is both a local and regional problem. In other words, ozone is both produced from local emissions and transported long distances from cities upwind of an affected area. As a result, a blanket of ozone often covers the eastern United States, with concentrations exceeding 80 ppb on geographical scales of over 1,000 km. During the "ozone season" from about May to October, these high concentrations of ozone are present for several hours per day, persist for several days during smog episodes, and occur in both urban and rural areas.

Despite efforts at the local, regional, and federal levels to control ozone during the last 25 years, ambient ozone concentrations in urban, suburban, and rural areas of the United States continue to be a major environmental and health concern. Ninety-eight counties across the United States did not meet the ozone national air quality standard (maximum one-hour average concentration of 120 ppb) in 1990. These *nonattainment areas* are shown in Figure 8.7 together with the degree of

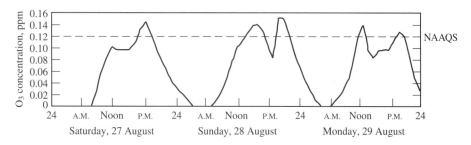

Figure 8.6   Measured ozone concentrations during a three-day episode at Montague, Massachusetts. Data show locally generated midday peaks and transported late peaks. (*Source:* Martinez and Singh, 1979)

Figure 8.7   Ozone air quality nonattainment areas in the United States in 1991. (*Source:* NRC, 1992)

violation of the standard. Areas are characterized by the U.S. Environmental Protection Agency (EPA) as "extreme" if the ozone levels exceed the standard by more than 133 percent, "severe" from 50 to 133 percent above the standard, "serious" from 33 to 50 percent, "moderate" from 15 to 33 percent, and "marginal" from 0 to 15 percent. It is clear from Figure 8.7 that high ozone concentrations are a widespread problem.

8.4.2   Ozone Health Effects

High ozone concentrations exceeding roughly 100 ppb have been shown to be associated with adverse effects on the human respiratory system. The effects are more significant for people who exercise. Acute effects, such as changes in lung capacity and function, have been observed in some people after exposures of less than an hour. In some cases the effects persist for many hours or days. High ozone concentrations also can injure trees and agricultural crops as well as damage certain materials. For further discussion of these effects, see Chapter 2.

8.4.3   Bad Ozone versus Good Ozone

The ozone discussed in this chapter is ozone in the troposphere formed from human activities at and near the surface of the earth. This is sometimes referred to as *bad ozone* because it causes the adverse health effects just described. In contrast, natural ozone high in the stratosphere—sometimes called *good ozone*—has beneficial effects because it absorbs some of the sun's dangerous ultraviolet radiation. Chapter 11 discusses how stratospheric ozone is being depleted by a class of humanmade chemicals known as *halocarbons,* leading to the *ozone hole* you may have heard about. Chemically, there is no difference between ozone in the stratosphere and ozone in the troposphere. But because the two atmospheric layers do not easily mix (again, see Chapter 11), they represent different environmental problems.

8.5   FORMATION OF OZONE IN URBAN AREAS

Ozone is formed in the atmosphere from the reaction of an oxygen atom (O) with an oxygen molecule:

$$O_2 + O \rightarrow O_3 \quad \text{(Reaction 1)} \tag{8.22}$$

There are a lot of oxygen molecules available in the atmosphere (about 21 percent of air molecules are oxygen), so the oxygen atom is the key reactant. Single oxygen atoms are extremely energetic substances that can be formed by certain chemical reactions. One way this oxygen atom can be formed is by the breakup of an NO_2 molecule by sunlight:

$$NO_2 + (\text{Sunlight}) \rightarrow NO + O \quad \text{(Reaction 2)} \tag{8.23}$$

Small amounts of NO_2 are naturally present in the atmosphere, but larger amounts are introduced in urban areas from the combustion of fuels in automobiles, power plants, and other sources (see Chapters 2, 3, and 5 for more details about sources of NO_2). Unlike many other molecules, the atomic structure of NO_2 allows it to absorb the energy in sunlight. So the ozone formation process starts when the absorbed sunlight gives the NO_2 molecules extra energy, causing them to split into two pieces: an NO molecule and an energetic O atom. The O atom carries most of the energy and immediately searches for a reaction partner. The first thing it notices is a lot of N_2 molecules (about 79 percent of air is N_2). However, these nitrogen molecules are inert and are not interested in reacting with anything. The next best thing are the oxygen molecules, which are much more reactive than nitrogen. The oxygen atom rapidly finds an oxygen molecule, and they combine to form ozone.

The reactions shown by Equations (8.22) and (8.23) represent the most significant path for ozone formation in the troposphere. The two basic ingredients are NO_2 and sunlight. However, if this were the only process going on, ozone would accumulate in the atmosphere continuously, with concentrations growing higher each day. We know that doesn't happen, so we must look at the sinks as well as sources of this environmental pollutant. What happens to the ozone that is formed in the troposphere?

Because ozone itself is chemically reactive, it also looks for other molecules to react with. There are several possibilities, but one of the most attractive is NO. Recall that both NO and NO_2 are emitted from various combustion processes. NO combines with ozone-producing NO_2 and oxygen:

$$NO + O_3 \rightarrow NO_2 + O_2 \quad \text{(Reaction 3)} \tag{8.24}$$

Notice that this reaction is in a sense the opposite of the two previous reactions; it uses the products of those reactions to re-form NO_2 and O_2. The net effect thus appears to be zero buildup of O_3. But this is misleading. What is important is how *fast* these reactions occur. The first two reactions produce ozone, whereas the third reaction destroys it. The faster Reactions 1 and 2 are, the more ozone will be produced. But the faster Reaction 3 is, the lower the ozone levels. What determines these rates of reaction, and what are their relative speeds?

8.5.1   The Photochemical Cycle

We know from the study of chemistry that the rates of these reactions are proportional to the product of the reactant concentrations. For example, the rate of Reaction 3, which destroys ozone, is equal to

$$R_3 = k_3[NO][O_3] \tag{8.25}$$

where k_3 is the *reaction rate constant* and is a function of temperature only. The square brackets denote the concentration of the species in units of ppb. The units of the reaction rate R_3 are ppb/min. Therefore, the units of the rate constant are ppb-min. Laboratory studies of this reaction show that at 298 K, $k_3 = 0.03$/ppb-min. So if there are 100 ppb of ozone and 50 ppb of NO, the rate of destruction of ozone will be $R_3 = (0.03)(100)(50) = 150$ ppb/min. This reaction is fairly rapid; it would take less than a minute to destroy 100 ppb of ozone if enough NO is available to do the job. High NO availability thus will tend to quickly destroy the available ozone.

Let us a look next at the pathway for ozone production. Everything starts with Reaction 2, called the *photolysis* of NO_2. This reaction is fundamentally different from Reaction 3 because sunlight is one of the reactants. One could write the reaction rate as

$$R_2 = k \,[\text{Sunlight intensity}][NO_2] \tag{8.26}$$

but usually the sunlight intensity is incorporated into the rate constant, k, so the reaction rate is

$$R_2 = k_2 \,[NO_2] \tag{8.27}$$

where k_2 has units of min^{-1} and depends on the sunlight intensity. In turn, the sunlight intensity depends on location, time of year, time of day, cloud cover, the existence of haze, and a host of other factors. Experimentally, we find that k_2 varies from zero at night to a maximum value of around 0.3/min at noon on clear summer days for a midlatitude location like Los Angeles. For an NO_2 concentration of 100 ppb, for this maximum sunlight, the rate of photolysis is then $R_2 = (0.3)(100) = 30$ ppb/min.

Therefore, this is also a fairly rapid reaction. This reaction rate indicates that the available NO_2 will break down to NO and O in just a few minutes.

Finally, the rate of ozone production from Reaction 1 will equal

$$R_1 = k_1 [O_2][O] \tag{8.28}$$

In this case it is not necessary to evaluate k_1 numerically if we are interested in only the final equilibrium concentration of ozone. One can show that at equilibrium the ozone concentration resulting from this system of three reactions is

$$[O_3] = \frac{k_2 [NO_2]}{k_3 [NO]} \tag{8.29}$$

This expression is called the *photostationary state relationship*. Note that the equilibrium concentration of ozone is proportional to the $[NO_2]/[NO]$ ratio. The higher the ratio, the higher the equilibrium ozone concentration. Also observe that the ozone concentration is proportional to the photolysis constant k_2, which depends on the sunlight intensity. A derivation of this relationship is left as a problem at the end of the chapter.

The photostationary state relationship is useful, but to find the equilibrium ozone concentration we still need to know the equilibrium values of the NO and NO_2 concentrations. In an urban area, combustion sources like cars and power plants emit mainly NO and just a little NO_2 (see Chapters 3 and 5 for details); initially the equilibrium ozone concentration will be low because NO_2 is low. However, NO is gradually oxidized to NO_2 in the atmosphere, while some NO_2 dissociates to produce more NO. So over time the ratio of NO to NO_2 changes. But for this simple reaction system, each time a molecule of NO reacts, a molecule of NO_2 is produced, and vice versa; the *sum of their concentrations* (defined as NO_x) remains constant. If the initial concentrations of ozone, NO, and NO_2 are $[O_3]_0$, $[NO]_0$, and $[NO_2]_0$ respectively, then the resulting equilibrium concentration of ozone will be given by

$$[O_3] = -\frac{1}{2}\left([NO]_0 - [O_3]_0 + \frac{k_2}{k_3}\right) +$$

$$+ \frac{1}{2}\left[\left([NO]_0 - [O_3]_0 + \frac{k_2}{k_3}\right)^2 + \frac{4k_2}{k_3}\left([NO_2]_0 + [O_3]_0\right)\right]^{1/2} \tag{8.30}$$

Example 8.7

Atmospheric ozone concentration from NO_x photooxidation. Use Equation (8.30) to investigate the equilibrium ozone concentration as a function of the initial concentration of ozone and oxides of nitrogen.

(a) Assume that $[O_3]_0 = 0$ and that $[NO_2]_0/[NO_x]_0 = 0.2$. Also assume that $k_2/k_3 = 10$ ppb. Calculate the ozone concentration as a function of initial $[NO_x]_0$ for values between 0 and 1,000 ppb. What do you observe?

(b) Repeat the calculation in part (a) for $[NO_2]_0/[NO_x]_0$ ratios of 0.6, 0.8, and 1.0. What differences do you observe?

(c)   Ambient concentrations of NO_x, even in polluted areas, typically are less than 500 ppb, whereas maximum ozone concentrations can exceed 300 ppb. Can the mechanism in Equation (8.30) explain these observed ozone concentrations?

Solution:

(a)   By definition, $[NO_x]_o = [NO]_o + [NO_2]_o$. Because $[NO_2]_o = 0.2\,[NO_x]_o$ we find that

$$[NO]_o = [NO_x]_o - [NO_2]_o$$

$$= [NO_x]_o - 0.2\,[NO_x]_o$$

$$= 0.8\,[NO_x]_o$$

Substituting these values into Equation (8.30) gives

$$[O_3] = (-0.5)(0.8[NO_x]_o + 10$$

$$+ \; 0.5\sqrt{(0.8[NO_x]_o + 10)^2 + (4)(10)(0.2[NO_x]_o)}$$

This function is plotted in Figure 8.8 for $[NO_x]_o$ levels of 0 to 1,000 ppb. We observe that the predicted ozone concentration increases from zero to 2.5 ppb. These are extremely low levels of ozone.

(b)   Repeating the same procedure for other ratios of initial concentration yields the Figure 8.8 results for the predicted ozone. The maximum O_3 values are found for the case where all the available NO_x exists initially as NO_2 (ratio of 1.0). For this extreme case, the equilibrium ozone concentration can reach values as high as 95 ppb for the highest NO_x levels of 1,000 ppb.

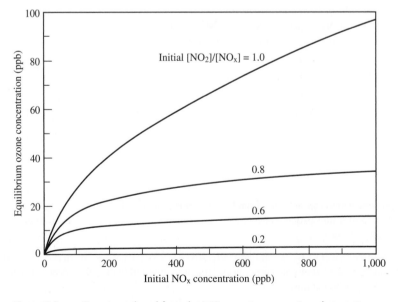

Figure 8.8   Ozone produced from the NO_x reactions assuming photostationary state equilibrium.

As the availability of NO_2 decreases (lower ratios of NO_2 to NO_x), the ozone concentrations decrease rapidly and remain below 40 ppb for all conditions shown in Figure 8.8. For $[NO_2]_0 = 0$, no ozone is formed because there is no means to produce the atomic oxygen needed to form ozone.

(c) As noted in the text, most NO_x is initially emitted in the form of NO, not NO_2. Under these conditions the equilibrium ozone concentrations predicted by Equation (8.30) are below 40 ppb. Therefore, this cycle is unable to predict the much higher ozone concentrations observed in polluted areas.

Based on the results of Example 8.7, we can conclude that chemical reactions other than the three discussed so far are important in polluted areas where high ozone concentrations occur. We will discuss these reactions next.

8.5.2 The Role of Hydrocarbons

Our previous discussion indicates that NO_x emissions alone cannot produce significant amounts of ozone, especially because NO_x is emitted mainly in the form of NO. The photostationary state relationship indicates that ozone formation requires a high initial NO_2/NO ratio. So some other means of converting the available NO to NO_2 is needed to produce high ozone levels. This is where two other ingredients of photochemical smog, *hydroxyl radicals (OH)* and *hydrocarbons,* come into play.

We saw that photochemical reactions in the atmosphere can lead to the dissociation of NO_2 and the formation of an excited oxygen atom that is very reactive. Not only do these O atoms react with plentiful oxygen molecules, they also tend to react with other atmospheric species to form *free radicals*—molecular fragments containing an odd number of electrons. Because they are missing an electron, free radicals react readily trying to gain the missing electron and achieve a more stable state. We saw that O can react with O_2 to form ozone. The same O can react with water vapor, which is also abundant in the atmosphere, to form two hydroxyl radicals:

$$O + H_2O \rightarrow OH \cdot + OH \cdot \qquad (8.31)$$

The chemical symbol denoting a free radical is a centered dot following the chemical formula, indicating that an unpaired electron is present.

The hydroxyl radical is one of the most important species in atmospheric chemistry. This free radical does not react with molecular oxygen, but it does react with most other species in the atmosphere, including hydrocarbons.

There are hundreds of different hydrocarbon compounds in the atmosphere as a result of natural processes and human activities (like driving cars that emit automotive exhaust). The hydroxyl radical attacks practically all of these compounds, initiating their chemical oxidation. This hydrocarbon chemistry was a mystery until the middle of the 20th century, when high ozone levels emerged as a major air pollution problem in southern California. But after 40 years of research, the main processes that produce urban smog are now fairly well understood.

The reactions that follow the initial attack of a hydroxyl radical on a large hydrocarbon molecule are quite complex but follow similar patterns as for smaller molecules. So to examine these general patterns, let us consider the attack of an OH· free radical on the simplest hydrocarbon, methane (CH_4). Our insights from the methane oxidation will be useful for understanding the role of hydrocarbons in ozone production.

The hydroxyl radical attack on methane results in the removal of a hydrogen atom and the formation of water and an organic radical:

$$CH_4 + OH\cdot \rightarrow CH_3\cdot + H_2O \tag{8.32}$$

This reaction is the first step in the atmospheric oxidation of methane. By destroying methane (and other hydrocarbons), the hydroxyl radical (now converted to stable water vapor) prevents hydrocarbons from accumulating in the atmosphere. It thus acts as a cleaning agent, which is the good news. The bad news is that this reaction also initiates a new chain of radical reactions. In this case the victim of the attack, the methane molecule, is now missing one electron and has become a free radical itself. The new *methyl radical* ($CH_3\cdot$) is also reactive and will look for another molecule to react with. Oxygen is once again the best candidate. In this reaction yet another radical (*methyl peroxy*) is formed:

$$CH_3\cdot + O_2 + M \rightarrow CH_3O_2\cdot + M \tag{8.33}$$

M in this reaction refers to another air molecule (nitrogen or oxygen) needed to absorb some of the energy released in the reaction. The methyl peroxy radical, $CH_3O_2\cdot$, can react with a variety of species, but its reaction with NO is extremely important for the creation of photochemical smog:

$$CH_3O_2\cdot + NO \rightarrow CH_3O\cdot + NO_2 \tag{8.34}$$

This reaction creates one more free radical (called the *methoxy radical*), but it also converts one molecule of NO to NO_2. The methoxy radical then reacts with oxygen to form *formaldehyde* and the *hydroperoxyl radical*:

$$CH_3O\cdot + O_2 \rightarrow HCHO + HO_2\cdot \tag{8.35}$$

The overall result of these last four reactions can be written as

$$CH_4 + OH\cdot + NO \rightarrow HCHO + HO_2\cdot + NO_2 \tag{8.36}$$

This result shows that the photochemical oxidation of hydrocarbons leads to the conversion of NO to NO_2. By converting one species that destroys ozone (NO) to one that produces it (NO_2), hydrocarbons provide the fuel for the NO_x photochemical cycle "engine" that produces ozone.

The reaction chain continues, with the formaldehyde (HCHO) reacting with OH· to form CO and eventually CO_2. At that point, all of the hydrocarbon "fuel" has been burned away. In the meanwhile, however, the hydrocarbons have played a critical role in the formation of ozone. Hydrocarbons by themselves cannot produce ozone, but combined with NO_x and sunlight they form a potent mixture that results in higher ozone formation than predicted by the photostationary state relationship discussed earlier.

The reaction in Equation (8.36) can be generalized for any hydrocarbon, RH (where R is a chemist's shorthand notation for the "residual" fragment of a pure hydrocarbon molecule that has been stripped of a hydrogen atom; for instance, CH_3, C_2H_5, and C_3H_7 are examples of residuals):

$$RH + OH \cdot + NO \rightarrow RCHO + HO_2 \cdot + NO_2 \qquad (8.37)$$

where RCHO is the corresponding *aldehyde.* Modern air quality mathematical models simulate the chemistry of hundreds of different hydrocarbons using hundreds of reactions like the one shown above. The rates of these different reactions are used to calculate the chemical production and destruction terms (P and D, respectively) in the mass balance equation for each substance [such as Equation (8.21) derived in Section 8.3.3]. The overall result of these hundreds of (often competing) reactions determines the rate and magnitude of predicted ozone concentrations for a given region. Additional details are presented in more advanced textbooks (like Seinfeld and Pandis, 1998).

8.5.3   Photochemical Smog and Meteorology

The meteorological conditions of a region (sunlight, temperature, wind speed, and other parameters) also directly affect the formation of ozone. In general, episodes of high ozone concentrations are associated with slow-moving, high barometric pressure weather systems. These high-pressure systems are usually indicated on weather maps and TV weather reports with an H (the symbol L is used for low-pressure areas). High-pressure weather systems promote high ozone concentrations because

1. These systems are characterized by air falling from higher altitudes of the atmosphere. This phenomenon creates a pronounced inversion of the normal atmospheric temperature profile, in which temperature normally decreases with height in the troposphere. As discussed earlier (see Figure 8.3), in an inversion layer the air temperature *increases* with height, so cooler air below cannot mix with warmer air above. This inversion layer acts as a lid that contains pollutants in a shallow layer of the atmosphere, as low as a few hundred meters.

2. Wind speeds associated with high-pressure systems typically are low. Therefore, pollutants stay longer over urban areas and accumulate in the atmosphere.

3. Clear skies, sunshine, and warm conditions usually accompany high-pressure systems, accelerating the photochemical formation of ozone. The sinking air motion in areas of high pressure compresses and warms the air, which tends to evaporate any water droplets present, leaving the skies comparatively free of clouds.

Warm temperatures, clear skies, and light winds thus offer the most favorable conditions for high ozone concentrations. In the eastern United States and in Europe, the worst ozone episodes occur in the early summer with the longest periods of daylight. In other locations, like Los Angeles, the weather is dominated by a persistent high-pressure system throughout most of the year. These conditions are ideal for the photochemical production of ozone.

These episodes end when a new weather front brings cooler, cleaner air over the area. Near these low-pressure regions, the air tends to rise rather than fall. This upward motion leads to expansion and cooling of the air and the formation of clouds and storms if the relative humidity at ground level is high. The stable inversion layer also breaks up, allowing greater mixing and dilution of polluted air with cleaner air aloft.

8.6   CONTROLLING OZONE FORMATION

As we saw, ozone formation in an urban area depends on the availability of both nitrogen oxides and hydrocarbons. The term *volatile organic compounds (VOCs)* also is frequently used to include all vapor-phase organic species in the atmosphere except CO and CO_2. However most VOCs are hydrocarbon compounds.

To control the formation of ozone in the troposphere, we must first understand how ozone levels depend on the concentrations of VOCs and NO_x in the atmosphere. Based on this understanding, we can then appropriately reduce the emissions of these pollutants from various types of sources, such as those listed earlier in Table 8.1.

We begin by looking separately at the effects of VOC and NO_x concentrations on ozone formation. Then we consider what happens when both pollutants vary simultaneously.

8.6.1   Effect of VOC Concentration on Ozone Formation

VOCs (mostly hydrocarbons from human activities) serve as the fuel for the formation of ozone. Increasing their concentration leads to faster and more complete conversion of NO to NO_2 and further ozone formation.

There are a number of ways to study the impact of VOC availability on ozone levels. One of the first approaches was to simulate the atmosphere by a set of dynamic mass balance equations similar to Equation (8.21) in Section 8.3.3. To solve these equations, we first need to describe the chemical transformations taking place (terms P and D in the equations) for all the relevant VOCs. Because these equations assume the atmosphere above an urban area behaves as a well-mixed box (that is, the same concentration at all locations), the corresponding models are again known as box models.

These mathematical models attempt to simulate the ozone concentration during the day. To do this, they use observations for the diurnal variation of temperature, wind speed, mixing height, and sunlight. Temperature affects all chemical reactions (they speed up when the temperature increases), and sunlight affects the photochemical reactions (such as the dissociation of NO_2 discussed in Section 8.5.1). The results of these model predictions are consistent with actual observations: Ozone is formed as the day progresses and reaches a maximum level in the afternoon.

A mathematical box model is used to calculate ozone concentrations for different initial conditions, keeping the atmospheric NO_x concentration constant and varying the VOC concentrations. This reveals the dependence of O_3 concentration on VOC levels for a given level of NO_x. Such a diagram is shown in Figure 8.9 for the case of $[NO_x] = 100$ ppb.

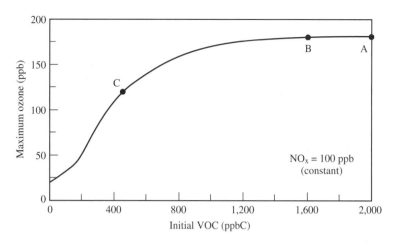

Figure 8.9 Predicted maximum ozone concentration as a function of VOC level for constant NO_x.

For zero VOCs, the available NO_x can produce only a small amount of ozone, around 20 ppb in this case, similar to the results found in Example 8.7. Higher VOC concentrations cause more ozone to be formed. For instance, with 800 ppbC of VOCs, the maximum ozone concentration becomes 150 ppb. However, adding more VOCs to the system after this point has a smaller effect. The ozone concentration reaches a plateau above 1,200 ppbC of VOCs; adding more hydrocarbons then has no effect on the system.

Let us examine what this curve means for an urban area with concentrations of 100 ppb of NO_x and 2,000 ppbC of VOCs (point A in Figure 8.9). According to Figure 8.9, this region will have a maximum daily ozone concentration of 180 ppb, which is 50 percent higher than the national air quality standard of 120 ppb.

In general, the concentrations of VOCs in the urban atmosphere will be approximately proportional to the emission rates of anthropogenic sources such as vehicles, power plants, and other industrial processes. Suppose we decide to reduce these VOC emissions by 20 percent, decreasing the VOC concentration in the atmosphere to 1,600 ppbC (point B). From Figure 8.9, we see that there will be no reduction of ozone concentrations for this strategy. The environment in this case is already saturated with VOCs, so reducing them only a little has a negligible effect; there are still plenty of VOCs left to convert the available NO to NO_2, forming ozone. In this case, to reduce ozone to 120 ppb (the air quality standard), we must reduce the VOC concentration to approximately 450 ppbC (point C), corresponding to a 77 percent reduction in VOC emissions.

8.6.2   Effect of NO_x Concentration on Ozone Formation

Oxides of nitrogen play a dual role: They are both the engine for ozone formation and are also capable of converting ozone back to oxygen (through the NO + O_3 reaction) when NO concentrations are high enough. The mathematical box model is again used to show how ozone levels are affected by different NO_x concentrations, fixing the VOC concentration at a constant value, say 600 ppbC. By varying the NO_x concentration one can generate Figure 8.10.

Once more, as the NO_x concentration increases the ozone concentration also increases. However, in this case a maximum exists at an NO_x concentration of around 80 ppb. If NO_x increases further, the reaction of NO with ozone destroys more ozone than is created, and the maximum ozone concentration decreases.

This complex behavior has some unexpected results for strategies to control urban ozone levels. Let us assume that current conditions in a region correspond to point A in Figure 8.10, with 120 ppb of NO_x and 600 ppbC of VOCs. The maximum ozone concentration then will be 120 ppb, just equal to the national standard. Assume that atmospheric NO_x concentrations are again approximately proportional to the emissions from human activities. If we reduce NO_x emissions by an additional 33 percent (to be on the safe side or to allow for future growth in emissions), the atmospheric NO_x concentration would then drop to 80 ppb (point B). However, even though we have reduced the concentration of a key ingredient for ozone formation, Figure 8.10 shows that the ozone concentration will *increase* from 120 ppb to 150 ppb! The emission reduction strategy not only failed to reduce ozone levels, it actually increased the ozone concentration significantly.

This complex behavior was not well understood many years ago when ozone control strategies were first developed. The result in some cases was counterintuitive—reducing emissions actually made things worse!

8.6.3   Ozone Isopleth Diagrams

To capture this complexity, the dependence of ozone production on the availability of VOC and NO_x for a given area is frequently represented by an *ozone isopleth dia-*

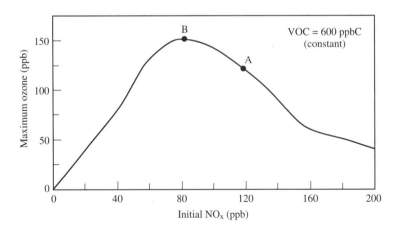

Figure 8.10   Predicted maximum ozone concentration as a function of NO_x level for constant VOC.

gram. An *isopleth* is a line of *constant concentration.* In this case, the plot is a contour map showing the maximum ozone concentrations achieved as a function of different VOC and NO$_x$ levels. The diagram is generated by plotting the predicted ozone maximum obtained from many mathematical model simulations similar to the ones just described. In this approach (known as the *EKMA technique,* for empirical kinetic modeling approach), the initial concentrations of VOC and NO$_x$ are varied while all the other variables are held constant.

Such a plot is shown in Figure 8.11. This approach again assumes that (1) the urban area can be represented as a well-mixed box and (2) the initial concentrations of VOCs and NO$_x$ are proportional to their source emission rates. On this diagram, an area with 600 ppbC of VOCs and 100 ppb of NO$_x$ would correspond to point A on the diagram. The predicted ozone concentration in this case is 143 ppb. Each isopleth (line of constant O$_3$ concentration) shows the combination of VOC and NO$_x$ concentrations needed to produce that ozone level. Isopleth diagrams differ from one urban region to another because of differences in source emissions, meteorology, and other factors.

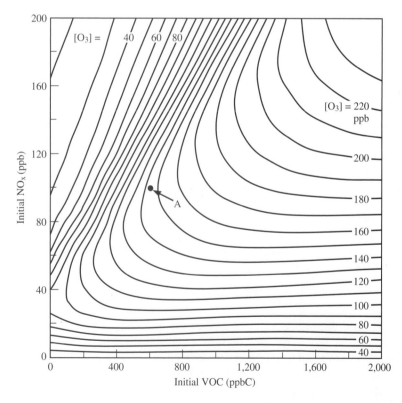

Figure 8.11   Example of an ozone isopleth diagram for an urban area. (*Source:* Jeffries and Crouse, 1990)

Example 8.8

Effects of NO$_x$ versus VOC controls on regional ozone levels.   Assume the ozone concentrations in an urban area are described by the isopleth plot of Figure 8.11, and the current state of the atmosphere is at point A— that is, $[VOC]_o = 600$ ppbC, $[NO_x]_o = 100$ ppb, and $[O_3]_o = 143$ ppb. Estimate the effect on the atmospheric ozone concentration of the following changes in emissions:

(a)   A 20 percent reduction of VOC emissions (no change in NO$_x$).

(b)   A 20 percent reduction of NO$_x$ emissions (no change in VOCs).

(c)   A 20 percent increase of VOC emissions (no change in NO$_x$).

(d)   A 20 percent increase of NO$_x$ emissions (no change in VOCs).

Assume in all cases that atmospheric concentrations of VOCs and NO$_x$ are proportional to emissions.

Solution:

(a)   Assuming that the initial concentrations used to generate Figure 8.11 are proportional to the urban emissions, a 20 percent reduction of VOC emissions corresponds to a new VOC concentration of

$$[VOC] = (1 - 0.2)\,[VOC]_o = (0.8)(600) = 480 \text{ ppbC}$$

So after the reduction in emissions, the area will have 480 ppbC of VOCs and 100 ppb of NO$_x$. These values will produce 125 ppb of ozone, according to Figure 8.11. Therefore, a 20 percent reduction in VOC emissions will reduce atmospheric ozone by 13 percent.

(b)   If the NO$_x$ emissions are reduced by 20 percent, the new NO$_x$ concentration will be $(0.8)(100) = 80$ ppb. The VOC concentration will remain at 600 ppbC. Using Figure 8.11, we see that the new ozone concentration will increase by 1 ppb to 144 ppb. In this case, reducing NO$_x$ emissions by 20 percent leads to a 0.7 percent *increase* in atmospheric ozone.

(c)   Following the same procedure, the new VOC concentration will be $(1.2)(600) = 720$ ppbC. The point with $[VOC] = 720$ ppbC and $[NO_x] = 100$ ppb corresponds to an ozone concentration of approximately 155 ppb. So a 20 percent increase in VOC emissions increases the ozone level by 8 percent.

(d)   Increasing NO$_x$ emissions by 20 percent will result in $[NO_x] = (1.2)(100) = 120$ ppb. For $[VOC] = [VOC]_o = 600$ ppbC, the new ozone concentration is 130 ppb. So in this case *increasing* the NO$_x$ emissions by 20 percent *decreases* the regional ozone concentration by 9 percent.

Of the four strategies evaluated, the 20 percent reduction in VOCs would achieve the greatest reduction in ozone.

Part (d) of the previous example raises the intriguing possibility of controlling O_3 levels by *increasing* the emissions of a key ingredient in its formation, NO$_x$. If ozone were our only concern, this counterintuitive strategy would indeed be effective in reducing atmospheric ozone levels. However, the region would now be characterized by higher NO$_x$ concentrations. This means higher concentrations of atmospheric NO$_2$, which is also a major air pollutant with adverse health effects (see the discussion in Chapter 2).

This example underscores the complexity of atmospheric chemistry and the need for careful analysis of the problem before decisions are made about the best way to achieve air quality standards. By the same token, strategies that work well for one area (like Los Angeles) may not work at all in other areas (like Atlanta). Communicating this complexity to laypersons is one of the challenges facing engineers who work in this area. The next example illustrates a case in point.

Example 8.9

Should we increase NO$_x$ emissions? Your grandmother talks to you:

I was watching TV yesterday and I heard a strange-looking guy—I think that he was a spokesperson for something—saying that cleaning up the emissions of cars may *increase* this ozone thing in the air and make the smog problem worse. It doesn't make sense to me. I mean, if what he said was true, then we could make our car emissions dirtier and the air would get cleaner! You are my smartest grandchild. You're going to this university where they charge your parents a fortune for tuition. What do you think?

Explain to your grandmother if the statement she heard could be true, and discuss her ideas about reducing air pollution. She is not an engineer, so avoid all the terminology used by experts.

Solution:

Many different answers are possible. Here is one:

What I've learned, Grandma, is that nature sometimes works in complicated ways. We put a lot of different pollutants into the air from cars and factories and even the dry cleaning store in the neighborhood. Everything would be OK if we just stopped emitting *all* the pollution that we emit today. But depending on the situation, reducing only one of the pollutants could sometimes make this ozone thing worse, like the guy said. But then if we *increased* the pollution from cars like you said, there would be other problems.

Do you remember the stories you used to tell me, Grandma, about the farm where you grew up? You had problems with snakes, and you also had problems with rats infesting the barn. But when you started getting rid of the snakes, your rat problem became a lot worse. The rats started multiplying without the snakes around to eat them. You even thought about bringing more snakes onto the farm to get rid of the rats, but then you thought that this was not such a great idea. The snakes caused a lot of other problems you weren't too crazy about.

So you see, Grandma, it's the same story with pollution. By reducing only one of the bad things out there, like certain types of car emissions (or snakes), its possible to make the ozone problem a little worse (like the rats). But at the same time, if we increased the car emissions, it would create other problems that could be even worse. Ozone isn't the only pollutant that makes your eyes water and your lungs hurt. Other stuff that cars put into the air can do the same things, and they cause other problems as well. So if we made car emissions dirtier, the ozone might actually go down a little (like the rats), but then we'd have other problems to deal with (more snakes).

The real trick, Grandma, is to figure out the right combination of things we need to do to solve the smog problem without causing other problems. That's what I'm learning about at my university. It's neat stuff. (Even with all the chemistry!)

Ozone isopleth diagrams like the one in Figure 8.11 are a critical element in devising effective emission control strategies for regions where current or projected ozone levels exceed air quality standards. The following section explains how to make best use of these diagrams.

8.6.4   Control Strategy Regimes

Ozone isopleth plots usually can be separated in two regions by a *ridge line.* The ozone ridge line identifies the maximum ozone concentration that can be achieved at a given VOC level, allowing the NO_x level to vary. Figure 8.12 shows the ridge line for Figure 8.11. The ozone ridge line separates the regions of low VOC-to-NO_x ratio (close to the upper left corner of Figure 8.12) and high VOC-to-NO_x ratio (close to the lower right corner of the diagram). The region of the diagram below the ridge line is described as being "NO_x-limited." In this regime the ozone concentration is relatively insensitive to changes in VOC levels but responds readily to reductions in NO_x. In other words, in this regime there are plenty of hydrocarbons available (plenty of fuel) but not enough NO_x for the production of ozone. Hence it is NO_x-

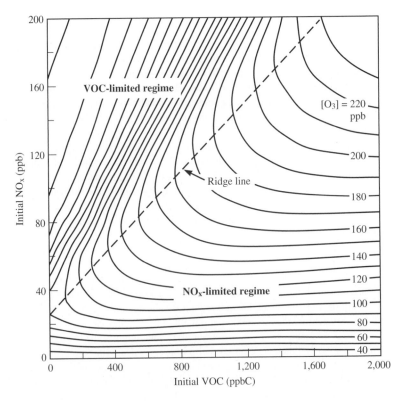

Figure 8.12   Control strategy regimes of an ozone isopleth diagram.

limited. The region of the plot above the ridge line is described as being "VOC-limited." In this regime the ozone concentration is sensitive to reductions in VOC concentration; but decreases in NO_x concentration cause an *increase* in ozone levels.

Determining where the current concentrations for a particular urban area fall on this plot is extremely important for the design of an effective control strategy to reduce ozone levels. The next two examples illustrate this point by examining the situation for different initial values of VOC and NO_x concentrations.

Example 8.10

Ozone control strategies for an urban area. An urban area is characterized by a VOC concentration of 1,800 ppbC and an NO_x concentration of 60 ppb. Assume that Figure 8.11 describes the dependence of ozone concentrations on the NO_x and VOC availability for this area. Also assume (as before) that atmospheric concentrations of NO_x and VOC are proportional to their respective source emission rates.

Discuss the advantages and disadvantages of the following strategies to achieve ozone levels at or below the current national standard of 120 ppb:

(a) An NO_x emission control program.

(b) A VOC emission control program.

(c) A control program for both NO_x and VOC emissions.

Solution:

(a) The current air quality for this urban area lies near the lower right corner of the isopleth diagram of Figure 8.11, which is the NO_x-limited region. According to the diagram, the current ozone concentration is a little above 140 ppb. In this case, a control program to reduce NO_x emissions will be very effective in reducing ozone levels. Assuming atmospheric concentrations are proportional to emissions, a 25 percent reduction in NO_x emissions will reduce the NO_x concentration from 60 to 45 ppb. This will reduce the ozone levels to less than 120 ppb, assuming VOC levels remain constant at 1,800 ppbC.

(b) A control program to reduce VOC levels will have practically no effect on ozone concentrations for this region; it could even have a small negative impact until VOC levels are reduced below 800 ppbC (a 56 percent reduction in concentration and emissions). Only after this large reduction would further decreases in VOC emissions begin to lower the ozone levels. To get down to 120 ppb would require an over 80 percent reduction in VOCs (350 ppbC or less). In fact, for this area even if the anthropogenic VOC emissions were reduced to zero, there would still be 55 ppb of ozone according to Figure 8.11. The reason is that the natural hydrocarbons emitted by trees, together with the available NO_x, are sufficient to produce this relatively high level of ozone.

(c) A control strategy reducing both NO_x and VOCs in this urban area is not better than the strategy of just reducing the NO_x. The ozone isopleths are nearly horizontal in the part of the diagram around the current VOC and NO_x concentrations. So there is no additional benefit from reducing VOCs by modest amounts. For example, reducing both NO_x and VOCs by 50 percent has the same effect on ozone as reducing only the NO_x by 50 percent. Because this urban area is in the NO_x-limited regime, controlling NO_x is the most promising strategy for reducing ozone levels to meet the federal air quality standards.

The preceding example illustrated the use of an ozone isopleth diagram to analyze emission reduction requirements to meet health-based air quality standards for ozone. In a NO_x-limited area (below the ozone ridge line), the most effective approach is to reduce NO_x emissions only. Such control strategies are now being adopted in many areas of the eastern United States. The next example illustrates a different situation that may be found in some urban areas.

Example 8.11

Ozone control strategies for a different urban area.   Repeat the analysis of Example 8.10 for an urban area where the current atmospheric concentration of NO_x is 160 ppb and the VOC concentration is 800 ppbC. Assume that Figure 8.11 again represents the ozone isopleth diagram for this region.

Solution:

(a) This area lies in the upper part of the isopleth diagram above the ridge line, which is the VOC-limited region. Based on Figure 8.11, the current ozone concentration is 125 ppb. In this case, reducing NO_x a modest amount (while keeping VOCs constant) will actually *increase* the ozone concentration to as high as 165 ppb! Only after NO_x has been reduced to 45 ppb (a 72 percent reduction) will further reductions start paying off. To get the ozone levels down to the 120 ppb standard (just 4 percent below the current 125 ppb), the NO_x concentration must fall to 40 ppb. This means that a NO_x-only control strategy would require a 75 percent reduction in emissions (assuming again that emissions are proportional to atmospheric concentration).

(b) In contrast, a strategy focused only on reducing VOCs will have a huge effect on ozone for this area. For example, going from 800 ppbC to 600 ppbC (a 25 percent reduction) will reduce ozone to less than 80 ppb (a 36 percent reduction). To meet the 120 ppb ozone standard, a VOC reduction of only 4 percent is needed.

(c) A strategy of reducing both VOCs and NO_x will have varying effects on ozone, depending on the relative reductions of the two pollutants. Figure 8.11 shows that ozone isopleths near the point of current NO_x and VOC levels are approximately straight lines with a slope of 1 ppb NO_x per 5 ppbC VOC. So if 1 ppb of NO_x is reduced for every 5 ppbC of VOCs, the area will simply move along a 125 ppb isopleth, and the ozone concentration will remain unchanged. This strategy would at least reduce the high NO_x concentrations in the area without increasing the ozone.

If the VOC and NO_x levels are reduced by less than the 5-to-1 ratio, ozone will *increase*. On the other hand, if the VOC-to-NO_x reductions exceed the 5-to-1 ratio, the ozone concentrations will also be reduced. This strategy is not as effective in reducing ozone levels as VOC reductions alone, but it has the advantage of reducing NO_x concentrations at the same time. Because of the high current NO_x levels, a reduction in NO_x may indeed be desired or needed to achieve air quality standards for NO_2 (see Chapter 2 for details).

8.6.5   Ozone Formation Potential of Hydrocarbons

In our discussion so far, all hydrocarbon compounds have been lumped together and included under VOCs. However, not only are there hundreds of unique hydrocarbon species in the atmosphere, but different species are emitted from different sources. Before ozone chemistry was well understood, lumping all hydrocarbons together was a convenient way to represent their role. Today, however, researchers are seeking better ways to identify the ozone formation potential of different species in order to devise more effective environmental regulations and control technologies.

The Carbon Mass Approach   One way to distinguish between and regulate different hydrocarbon or VOC emissions is to focus only on the carbon mass emitted and neglect the subsequent chemistry. This is the approach we have been using to quantify hydrocarbon concentrations in ppbC. This technique assumes that larger molecules are, in general, more reactive than small molecules. The advantage of this method is that it is simple and straightforward. One needs merely to compare the total carbon emissions for various hydrocarbons. For example, one molecule of m-xylene (C_8H_{10}) has the same mass of carbon as four molecules of ethanol (C_2H_5OH). So for a carbon-based strategy, reducing the m-xylene concentration by 1 ppb is considered equivalent to reducing the ethanol concentration by 4 ppb. However, there are some problems with this method.

We now know that individual hydrocarbon or VOC compounds react at differing rates in the atmosphere because the reaction rate constants with the OH radical are different. Therefore, some VOCs are not very reactive and participate marginally in ozone production, whereas others are major players in ozone production. Of course, there are also hydrocarbons that have intermediate reactivities. Thus the shortcoming of the carbon mass–based approach is that if the resulting hydrocarbon reductions (based only on carbon mass) happen to be unreactive species, the net change in ozone will be marginal.

The Reactive Organic Gas Approach   To incorporate differences in chemical reactivity, two categories of VOCs initially were defined by government agencies for regulatory purposes: unreactive and reactive species. The latter are usually referred to as *reactive organic gases* (*ROGs*). These are the compounds most responsible for ozone formation. In contrast, compounds like methane and carbon monoxide are considered to be unreactive for ozone formation in urban areas. (The term *unreactive* is probably misleading; these compounds do react and contribute, though slowly, to ozone formation in the troposphere.)

Deciding which VOCs are unreactive was not easy. In the mid-1980s the U.S. Environmental Protection Agency (EPA) decided to consider the reactivity of ethane (C_2H_6) with the OH radical as the dividing line between reactive and unreactive species. So ethane and all VOCs that react with the OH radical more slowly than ethane were considered unreactive, and all others were considered reactive organic gases.

To compare different ROGs, the carbon mass approach was then applied to the ROGs alone. This combined approach assumes that reactive compounds with the greatest carbon mass contribute most to ozone formation. Although this grouping into two categories was an improvement over the earlier carbon-based approach, more refined approaches were subsequently proposed, as described next.

A Chemical Reactivity Approach   A better way of comparing the ozone-forming potential of different hydrocarbons was to develop a reactivity scale that could be used as a weighting factor for individual species. The first such scale proposed and used by air pollution control agencies was based on the rate constant k_{OH} for reactions with the OH radical. Values of this reaction rate constant for a number of VOCs are given in Table 8.3. You can see that the value of k_{OH} spans nearly six orders of magnitude across the set of VOCs that are listed, with methane being the least reactive and isoprene the most reactive. The typical OH radical concentration in an urban area during the daytime in summer is about $5–10 \times 10^6$ molecules/cm^3.

Suppose you wanted to compare the ozone-forming potential of m-xylene and ethanol. According to the k_{OH} reactivity scale, 1 ppb of m-xylene is equivalent to

Table 8.3   OH rate constants (k_{OH}) and maximum incremental reactivity (MIR) values for selected VOCs at 298 K.

Compound	$10^{12} \times k_{OH}$ (cm^3 molecule per sec)	MIR (g O$_3$ formed per g VOC emitted)
Methane	0.006	0.0016
Carbon monoxide	0.21	0.065
Acetone	0.22	0.49
Ethane	0.25	0.32
Methanol	0.94	0.65
Propane	1.1	0.57
2-Butanone	1.1	1.4
Benzene	1.2	0.81
n-Butane	2.4	1.18
Methyl tert-butyl ether	2.9	0.73
Ethanol	3.3	1.7
2,2,4-Trimethylpentane	3.6	1.34
Toluene	6.0	5.1
Ethene	8.5	8.3
n-Octane	8.7	0.69
Ethyl tert-butyl ether	8.8	2.2
Formaldehyde	9.4	6.6
Acetaldehyde	16	6.3
m-Xylene	24	14.2
Propene	26	11.0
1,2,4-Trimethylbenzene	32	5.3
o-Cresol	42	2.5
a-Pinene	54	3.9
trans-2-Butene	64	13.2
Isoprene	101	9.3

Source: NRC, 1999.

24/3.3 = 7.3 ppb of ethanol. As this example shows, the advantage of the k_{OH} scale is its simplicity: The reaction constants are known, and the calculation of equivalent emissions is straightforward. In addition, these weighting factors do not depend on the geographic area of concern.

The disadvantage of this scale is that it considers only the reaction rate of the first step of the oxidation process of VOCs in the atmosphere. The remaining chemical reactions that lead to the conversion of NO to NO_2 and the formation of ozone are neglected. These processes also are very important. In fact, we now know that the number of ozone molecules that are formed for every VOC molecule that reacts with OH is highly variable and depends on the chemical structure of the specific VOC.

An Improved Reactivity Scale   In an effort to account for the overall chemistry of ozone formation, a more useful definition of VOC reactivity is the *incremental reactivity* (*IR*). This is defined as the amount of ozone formed per unit mass of a specific VOC added to or subtracted from the overall VOC mixture in a given air mass. Mathematically, this can be written as

$$\text{Incremental reactivity (IR)} = \frac{\Delta[O_3]}{\Delta[VOC]} \ (\text{for } \Delta[VOC] \to 0) \qquad (8.38)$$

The IR value of a given compound depends on the chemical properties of the atmosphere in the urban area of concern. So it depends on the VOC/NO_x ratio in the area and also on the amounts and chemical composition of other VOCs that are present.

A number of different IR scales have been proposed to deal with the dependency of incremental reactivity on local environmental conditions. One of the most popular is the *maximum incremental reactivity* (*MIR*) scale. The MIR of an organic compound is the IR value under conditions where the IR value has its maximum. This usually occurs for low VOC/NO_x ratios (upper left corner of the isopleth diagram of Figure 8.12), where the chemistry is VOC-limited. The California Air Resources Board proposed in 1990 to use the MIR scale for regulatory applications, especially to assess the environmental impacts of alternative automotive fuels. Since then, the MIR scale has been used extensively for ozone-related policy making in the United States.

Table 8.3 shows the MIR values for a selected set of organic compounds. These values are determined from state-of-the-art air quality models. The absolute value of the MIR for a particular compound can vary by about 20–60 percent because of different environmental conditions. However, the relative values (ratios of MIR) usually vary by less than 30 percent. Table 8.3 shows that some compounds, like carbon monoxide, methane, and ethane, have very low incremental reactivities, consistent with the earlier methods used for ranking ozone-forming potential. Other compounds, like formaldehyde, ethene, isoprene, propene, and m-xylene, have increasingly higher MIRs. Values above 5 are considered to mark the most potent VOCs for ozone formation. Note also that the compounds with the highest MIR values do not necessarily have the highest reactivity on the k_{OH} scale. This difference reflects the importance of other chemical reactions that produce ozone.

Example 8.12

Comparing reactivity scales for ozone formation potential. An air pollution source in an urban area emits the following mass of pollutants per day: 1.3 kg CO, 2.0 kg ethene, and 1.5 kg acetaldehyde. A new technology is proposed to reduce the contribution of this source to ozone formation. The expected new emissions per day would be 10 kg CO, 0.3 kg acetone, and 11 kg methanol.

Compare the emissions from the two technologies using each of the four methods for evaluating ozone-forming potential:

(a) The carbon mass approach.

(b) The ROG carbon mass approach.

(c) The k_{OH} reactivity scale.

(d) The MIR reactivity scale.

Based on your results, is it a good idea to use the new technology if you are trying to reduce the ozone levels in the area?

Solution:

(a) In the carbon mass approach, the total carbon emitted by each technology first must be calculated. The carbon contents and mass emissions of the five pollutants involved are shown in this table:

Compound	Formula	Carbon per Mol (g)	Molecular Weight (g)	% Carbon	Emissions (kg/d) Current	Emissions (kg/d) New
Carbon monoxide	CO	$1 \times 12 = 12$	28	42.9	1.3	10
Acetone	C_3H_6O	$3 \times 12 = 36$	58	62.1		0.3
Methanol	CH_4O	$1 \times 12 = 12$	32	37.5		11
Ethene	C_2H_4	$2 \times 12 = 24$	28	85.7	2.0	
Acetaldehyde	C_2H_4O	$2 \times 12 = 24$	44	54.5	1.5	

The current carbon emissions are then

$$(1.3 \times 0.429) + (2 \times 0.857) + (1.5 \times 0.545) = 3.2 \text{ kg C per day}$$

The carbon emissions of the new technology would be

$$(10 \times 0.429) + (0.3 \times 0.621) + (11 \times 0.375) = 8.6 \text{ kg C per day}$$

The new technology would therefore increase the carbon mass emitted by a factor of 2.7. By this measure, ozone is predicted to get *worse* using the new technology.

(b) Recall that ROGs are defined as compounds with k_{OH} values above that of ethane. Carbon monoxide and acetone are not ROGs because their reaction rate constants with OH are less than that of ethane (see Table 8.3). So in this approach, only the emissions of ethene and acetaldehyde are included in the calculation of current total ROG carbon emissions. Thus the current ROG carbon emissions are

$$(2.0 \times 0.857) + (1.5 \times 0.545) = 2.5 \text{ kg C per day}$$

For the new technology, only methanol is a ROG, based on the k_{OH} values in Table 8.3. Thus the ROG carbon emissions of the new technology would be

$$11 \times 0.375 = 4.1 \text{ kg C per day}$$

According to this method, the new technology would increase the ROG carbon emissions by approximately 40 percent. Thus ozone levels again are predicted to be *higher* using the new technology.

(c) Using the k_{OH} reaction rate constants (times 10^{12}) given in Table 8.3 as weighting factors, the current emissions correspond to

$$(1.1 \times 0.21) + (2.0 \times 8.5) + (1.5 \times 16) = 41.3 \text{ reactivity units}$$

The corresponding weighted emissions of the new technology are

$$(10 \times 0.21) + (0.3 \times 0.22) + (11 \times 0.94) = 12.5 \text{ reactivity units}$$

According to this scale, the new technology would result in emissions that are over three times *less* reactive than the current technology. Thus the new technology is predicted to *decrease* the ozone levels.

(d) Finally, using the maximum incremental reactivity (MIR) values from Table 8.3, the overall MIR of the current technology is

$$(1.2 \times 0.065) + (2.0 \times 8.3) + (1.5 \times 6.3) = 26.1 \text{ kg O}_3$$

For the new technology the overall MIR would be

$$(10 \times 0.065) + (0.3 \times 0.49) + (11 \times 0.65) = 7.1 \text{ kg O}_3$$

So the new technology has the potential to *decrease* the ozone formed by nearly a factor of 4.

The first two approaches based on carbon weighting and ROG emissions indicated that the new technology would significantly *increase* ozone levels in the region. However, according to the two reactivity-based scales (especially the MIR), the new technology would *decrease* the production of ozone by 19 kg per day. This scale affords the best current measure of ozone-forming potential, so it would indeed be a good idea to use the proposed new technology to reduce ozone levels, all else being equal. In general, other options and their costs would be assessed first to find the most cost-effective solution.

8.7   CONCLUSION

Urban air pollution is one of the oldest problems of human civilization. For centuries the problem was mainly associated with smoke and other emissions from coal combustion. In the 20th century, the advent of the automobile and a host of industrial processes gave rise to a new and more complex form of urban air pollution in which sunlight triggered a set of complex chemical reactions that created smog and harmful ozone.

Research over the past four decades has slowly but persistently uncovered the mysteries of modern urban air pollution. The relationships between emissions from human activities and the resulting changes in environmental quality are now understood reasonably well. This chapter has attempted to convey those findings in a form that can be used by modern engineers to understand the environmental implications of alternative technological designs. Because modern air quality analysis relies heavily on sophisticated computer models, there are limitations in the types of material that can be presented in a chapter like this. By emphasizing the underlying fundamentals, as well as some realistic applications of the results, this chapter has placed readers in a good position to pursue this subject in more detail.

8.8 REFERENCES

Jeffries, H. E., and R. Crouse, 1990. "Scientific and Technical Issues Related to the Application of Incremental Reactivity." Department of Environmental Sciences and Engineering, University of North Carolina, Chapel Hill, NC.

Martinez, J. R., and H. B. Singh, 1979. "Survey of the Role of NO_x in Nonurban Ozone Formation." EPA/450/4-79-035, U.S. Environmental Protection Agency, Research Triangle Park, NC.

NRC, 1992. *Rethinking the Ozone Problem.* National Research Council, National Academy Press, Washington, D.C.

NRC, 1999. *Ozone-Forming Potential of Reformulated Gasoline.* National Research Council, National Academy Press, Washington, DC.

Russell, A. G., G. J. McRae, and G. R. Cass, 1983. "Mathematical Modeling of the Formation and Transport of an Ammonium Nitrate Aerosol." *Atmospheric Environment* 17, pp. 949–64.

Seinfeld, J. H., and S. N. Pandis, 1998. *Atmospheric Chemistry and Physics: From Air Pollution to Climate Change.* Wiley, New York.

8.9 PROBLEMS

8.1   Based on the emissions inventory for Los Angeles (Table 8.1), what are the three largest sources of VOCs and NO_x in that region? Based on this information, discuss the general feasibility of reducing urban ozone concentrations by focusing emission control efforts on only a small number of major sources.

8.2   The maximum eight-hour average concentration of carbon monoxide (CO) in a downtown urban area is measured to be 15 ppm. The national ambient air quality standard (NAAQS) is 9 ppm. Automobile emissions are believed to be the only major source of CO; it is assumed there is no background contribution from other sources. Assume also that CO can be modeled as chemically nonreactive over the period of interest.

   (a)   Under these conditions, what is the relationship between the mass of CO emissions and the steady state atmospheric concentration of CO for fixed values of the average wind speed, u, mixing height, H, and urban area dimension, Δx?

   (b)   Based on this model, what percentage reduction in vehicle emissions is needed to achieve the NAAQS?

   (c)   New data show that there is actually a measurable background concentration of 1 ppm CO, even when there are no auto emissions. How does this change your analysis for parts (a) and (b)? (Note: The measured maximum CO is still 15 ppm.)

8.3   The city of Pleasantville has asked you to help assess its future compliance with national ambient air quality standards for carbon monoxide (CO). The current maximum eight-hour average concentration in downtown Pleasantville is 8.0 ppm, which is below the national air quality standard of 9 ppm. Your analysis shows that weekday vehicle emissions are the dominant source of CO downtown. You've estimated the following values for current and future CO emissions in the downtown region:

Weekday CO Emissions in Downtown Pleasantville (10^6 g/da)

Source	Current	2010
Passenger cars	2.5	2.4
Other vehicles	0.5	1.1

Your analysis of the downtown air quality data reveals a background CO concentration of 0.5 ppm in the absence of any vehicle emissions. This background is due primarily to distant industrial sources.

(a)   Based on these data and projections, does it appear that downtown Pleasantville will still meet the eight-hour NAAQS of 9 ppm in 2010? Because a preliminary answer is needed quickly, the simplifying assumptions outlined in Problem 8.2 (modeling CO as nonreactive) can be used here as well. However, state clearly the assumptions you make to relate CO mass emissions to atmospheric concentrations.

(b)   New air pollution controls on industrial sources might reduce the future background concentration of CO to 0.25 ppm by 2010. Would this change the results of your analysis? Explain.

8.4   The national air quality standards for fine particulate matter less than 10 microns (PM_{10}) are an annual average concentration of 50 $\mu g/m^3$ and a daily average maximum of 150 $\mu g/m^3$. Which of these two standards is likely to require the greater degree of emissions control in the city of Middleburgh? The city dimensions are 75 km $\times$ 75 km, and the background concentration of PM_{10} in Middleburgh is 30 $\mu g/m^3$. On most days the atmospheric mixing height in the region is 900 m, and the prevailing wind velocity is 2.5 m/sec. For a few days each year, meteorological conditions drop the mixing height to 250 m and the wind speed to 0.8 m/sec. You may assume that the PM_{10} is chemically nonreacting and that its deposition velocity is 0.12 cm/sec.

8.5   Revisit the city of Middleburgh described in the previous problem. The concentration of PM_{10} is 55 $\mu g/m^3$ on a typical day. Suddenly the adverse meteorological conditions outlined in the previous problem set in. How long will it take for the air quality to exceed the daily standard of 150 $\mu g/m^3$? Assume that particulate emission rates remain unchanged during this period.

8.6   (a)   What are the primary emissions from automobiles responsible for the formation of tropospheric ozone?

(b)   Why do peak ozone concentrations occur in the summer months, rather than randomly throughout the year?

(c)   Identify and label the five quantities or variables shown in the following figure.

(d)   Use this figure to show how a reduction in emissions could make the air quality worse instead of better. Label your initial condition as point A and your final condition as point B.

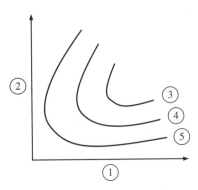

Figure for Problem 8.6

8.7   (a) Explain the role that sunlight plays in the formation of tropospheric ozone.

(b) What environmental effects of tropospheric ozone make it a criteria pollutant subject to regulation?

(c) How do energy-related sources or activities contribute to ozone formation?

(d) What nonenergy sources or activities contribute to ozone formation?

(e) Is it possible for increased levels of tropospheric ozone to make up for the loss of ozone in the stratosphere (that is, the ozone hole)? Explain.

8.8   Use Equations (8.22) to (8.24) (Reactions 1, 2, and 3) to derive the expression for equilibrium ozone concentration given by Equation (8.29). Note that the equilibrium state means that $d[O_3]/dt = 0$. You may also assume that there is no net accumulation of oxygen *atoms* because they are highly reactive. Thus assume that $d[O]/dt = 0$ for this derivation.

8.9   The current state of an urban atmosphere is at point A in Figure 8.11; that is, [VOC] = 600 ppbC, [NO_x] = 100 ppb, and [O_3] = 143 ppb. We want to reduce the ozone level to 100 ppb (safely below the one-hour standard of 120 ppb). Describe *four* emission reduction strategies that will achieve the desired ozone level. Discuss the relative merits (pros and cons) of each strategy.

8.10   The EPA has ordered sizable reductions in NO_x emissions from electric power plants to control ozone levels in the eastern United States. Assume that power plants account for half of the total NO_x emissions in a region where the current air quality concentrations of NO_x and VOC are 100 ppb and 800 ppbC, respectively.

(a) If power plant NO_x emissions are reduced by 80 percent, how much improvement in O_3 levels can be expected based on the isopleth diagram of Figure 8.11? Give your answer both as an absolute value (in ppb) and as a percentage change.

(b) How would your answer to part (a) change if the VOC concentration were 1,200 ppbC?

8.11   The results of a complex urban air quality model have been approximated by the following equation relating peak one-hour ozone concentrations to NO_x and VOC emissions in the region:

$$O_3 \text{ concentration (ppb)} = A + B\sqrt{NO_x}\left\{1 - \exp\left[C\left(\frac{VOC}{NO_x}\right)^D\right]\right\}$$

where

NO_x = fraction of current NO_x emissions

VOC = fraction of current VOC emissions

A, B, C, D = coefficients as shown in the following table

Coefficients for the Ozone Air Quality Model (One-Hour Average, ppb)

Urban Region	A	B	C	D
Bigapplis	3.8	586.8	−0.6293	0.8483
Metropolis	29.6	114.9	−8.2172	1.4938
Gotham	30.8	197.3	−1.4059	0.5496

This table gives the values of the model coefficients A, B, C, and D for three different urban areas. The current ozone concentration in each area corresponds to a fraction of 1.0 for both NO_x and VOC in the equation given.

(a) Select *any two* of the urban regions. For each region calculate the values of peak O_3 produced by NO_x and VOC fractions varying from zero to 1.0 at increments of 0.1 (this will give you a 10 × 10 grid for each region).

(b) Use these results to locate and draw the 120 ppb O_3 isopleth for each region (two separate graphs). Then sketch several additional isopleths at intervals of 10 ppb O_3 above and below the 120 ppb line.

(c) For each region, identify *two* possible strategies you would recommend to achieve the one-hour ozone air quality standard. Will the same strategies work for the two regions? Explain.

8.12 In 1997 the U.S. EPA promulgated a new air quality standard for ozone: a maximum of 80 ppb over an eight-hour averaging time. Since 1970, the standard has been a maximum one-hour average of 120 ppb. Is the new eight-hour standard more or less stringent than the one-hour standard? That is, which standard will require the greater *percentage reduction* in *ambient O_3 concentration* relative to current levels? To answer this question, use the simplified air quality model from Problem 8.11. The model coefficients for one-hour and eight-hour averaging times are shown in the following table for one urban region. Again, the values of NO_x and VOC in this equation vary from 0.0 to 1.0. The current ozone concentration corresponds to a fraction of 1.0 for both VOC and NO_x emissions.

(a) Calculate the current ozone concentration for the two averaging times.

(b) Determine the percentage reduction in ozone level needed to meet each standard.

(c) Evaluate and compare the levels of NO_x and/or VOC emission reductions needed to meet each standard.

Coefficients for Ozone Air Quality Model for the Study Region

O_3 Averaging Time	A	B	C	D
One hour	30.8	197.3	−1.4059	0.5496
Eight hours	30.7	165.6	−1.3242	0.4483

8.13 An alternative motor vehicle fuel has been proposed to help reduce urban ozone concentration. It is claimed that the new fuel will reduce regional emissions of formaldehyde by 500 kg/d, toluene by 100 kg/d, isoprene by 30 kg/d, and propene by 10 kg/d. On the other hand, there will be some increases in other VOCs: an additional 300 kg/d of methanol, 170 kg/d of ethanol, 90 kg/d of ethene, and 30 kg/d of propane. Evaluate the claim that this change will help reduce ozone. Prepare a brief report discussing your results and method of analysis.

chapter
9

PCBs in the Aquatic Environment[*]

*This chapter was written by David A. Dzombak, Richard G. Luthy, and Cliff I. Davidson.

9.1   INTRODUCTION: WHAT ARE PCBs?

PCB stands for *polychlorinated biphenyl*. This category of chemicals includes over 200 compounds with a distinct molecular structure and common properties. PCBs are mixtures of PCB compounds that exist as clear or light yellow oily liquids at room temperature. The qualities of these compounds made them highly useful in industry for almost 50 years beginning in 1929.

One of the most important properties of PCBs is that they are a poor conductor of electricity. They are thus useful as electrical insulators in equipment such as transformers, capacitors, and fluorescent light ballasts. Furthermore, PCBs are not easily ignited and in fact retard flames. This makes them useful as additives to hydraulic fluids and lubricating oils in machinery. Because PCBs are chemically stable and do not decompose easily, they also have been used as additives in pesticides, paints, and sealing compounds. As shown in Figure 9.1, the most significant uses of PCBs have been as insulators in capacitors and transformers.

One example of the usefulness of PCBs can be seen in the aluminum industry. Thousands of different products are made of aluminum; food and beverage containers, wheels for cars and trucks, and structural components for aircraft are among the largest uses. Shaping the aluminum requires rolling or pressing operations that are performed with hydraulic pressure and lubrication. Large volumes of hydraulic oils and lubricating oils are used, and these are often subject to elevated temperatures. In the 1950s a big aluminum manufacturing plant in Texas was destroyed by a fire that began when leaking hydraulic oil was somehow ignited. The result was the loss of a plant valued at hundreds of millions of dollars, loss of jobs and income, and most importantly, the loss of the lives of a number of workers in the fire. As a result of this experience, the aluminum industry searched for chemicals that could be added to hydraulic and lubricating oils that would retard fires. PCBs were identified as the most effective available fire retardant. Soon after the fire in Texas, hydraulic and lubricating oils in aluminum manufacturing plants were replaced with oils contain-

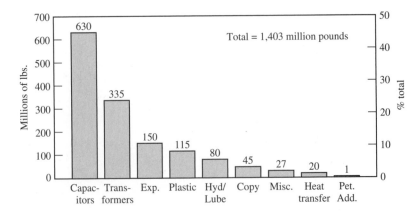

Figure 9.1   Applications of PCBs in the United States based on sales records from 1930 to 1975. (*Source:* Durfee et al., 1976)

ing up to 50 percent PCBs by weight. These oils proved very effective in preventing fires and providing all attendant benefits, including greatly increased worker safety.

Although highly useful in industry, PCBs were subsequently found to have adverse effects on animals and humans. As a result, their production was banned in the United States in 1977. However, their contamination of the environment has persisted. In this chapter we explore PCBs as a case study in contamination of surface waters. We begin by considering the toxicity of PCBs to emphasize the importance of studying these compounds. We also examine typical concentrations of PCBs in the environment and compare these concentrations to national standards. We then consider the chemistry of PCBs. Having established the fundamental properties and chemistry of PCBs, we examine the movement of these compounds through the environment, beginning with release at the source, movement through natural waters, and partitioning among different environmental compartments. Finally, we examine issues involved with managing PCB contamination in aquatic environments and discuss lessons learned from our experiences with PCBs.

9.2   TOXICITY OF PCBs

The properties that made PCB compounds so resilient in industrial processes have also made them long-lasting in the environment. In the 1960s scientists discovered that PCBs tend to accumulate rather than break down over time in soil and water. This aroused concerns that if PCBs were someday found to be toxic, their persistence in the environment would be a serious problem. Then in 1968 a major event focused worldwide attention on PCBs.

Kyushu University Hospital in western Japan started treating people in June 1968 for the skin disease chloracne as well as other symptoms such as fatigue, headaches, swelling of the joints and eyes, and increased skin pigmentation. Within several months, the number of patients with these symptoms increased to epidemic proportions, eventually involving more than 1,600 people. An investigation showed that all the patients had eaten food prepared with a certain brand of rice oil. Inspection of the rice oil plant revealed a leak in a heat exchanger where PCBs and other chemicals were used; the leak had contaminated the rice oil with these chemicals (NRC, 1979). The contamination became known as "Yusho poisoning."

Further research on Yusho poisoning eventually showed that chemicals other than PCBs were the primary cause of the symptoms in this case. Nevertheless, the attention given to PCBs by this event sparked a considerable amount of research on the toxicity of PCBs. This research has shown that PCBs are responsible for chloracne, skin and eye irritation, and increased skin pigmentation. However, these symptoms are found only with very high exposures, such as among workers at PCB manufacturing plants. Furthermore, there are no other acute symptoms such as adverse effects on internal organs.

The human health effect of greatest concern from PCBs is the possibility that these compounds cause cancer. Based on studies with rats and mice, the U.S. Environmental Protection Agency concluded that PCBs are "probable human carcinogens" (Erickson, 1997). Additional studies have been conducted in human populations exposed to PCBs, but these studies are not yet adequate to determine whether PCBs are carcinogenic. Nevertheless, because of the classification as probable human carcinogens, the production of PCBs has been banned in the United States.

9.3   PCBs IN THE ENVIRONMENT

Over the years PCBs have been released to the environment in great quantities. From 1930 to 1975 production of PCBs in the United States has been estimated as 570 million kg. As of 1975 it was estimated that about 68 million kg (12 percent of total production) were mobile in the environment, and about 130 million kg (21 percent) were in landfills or equipment dumps (Durfee et al., 1976; NRC, 1979). Even though U.S. production of PCBs stopped two years later, releases to the environment continued from leaks in structures holding PCB liquids, spills from equipment, and fires that burned PCB-bearing materials. In addition, landfills were poorly designed before the 1980s, so leakage of PCBs from some older landfills has continued.

9.3.1   Fate and Concentration of PCBs

Where in the environment are PCBs found? We generally identify various environmental *compartments* to denote where pollutants accumulate; in the broadest terms, these compartments include the atmosphere, oceans, soil, groundwater, surface waters such as rivers and lakes, and the biosphere, which includes plants and animals. We can consider such compartments globally or within a well-defined geographical area, such as a river system that includes separate compartments like the river water, aquatic plants, aquatic animals, sediments at the bottom, and the atmosphere above the river.

The environmental compartments likely to contain PCBs can be identified by considering the chemical and physical properties of these compounds. For example, based on experiments with a particular mixture of PCB compounds used previously in industry, a maximum of 100×10^{-6} grams of total PCBs was found to dissolve in 1 liter of water (Luthy et al., 1997). This concentration, also expressed as 100 μg/L, is extremely low, suggesting that only tiny quantities of PCBs are found in rivers, lakes, and other bodies of freshwater. But even at these low concentrations, PCBs can be harmful.

Another property of PCBs is that they are denser than water. This is significant in that drops of PCB liquids will sink to the bottom in a body of water, thus contaminating the sediments. The implications for aquatic ecosystems are discussed in Sections 9.6 and 9.7.

Because they contain carbon, PCB molecules attach more readily to other molecules that also contain carbon than to molecules without carbon. This is consistent with the low solubility in water: PCB molecules interact poorly with water molecules. It is also consistent with the fact that many PCB compounds do not evaporate into the air to a significant extent. But because soil particles often contain carbon,

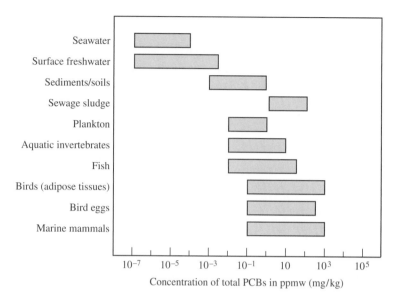

Figure 9.2   Ranges of total PCB concentrations detected in various environmental compartments. (*Source:* Schwarzenbach, et al., *Environmental Organic Chemistry,* ©1993. Reprinted by permission of John Wiley & Sons, Inc.)

PCBs readily attach to soil. In fact, more than 99 percent of the total PCB mass in the environment occurs in soil (Travis and Hester, 1991). The remaining 1 percent of PCB mass is distributed among other environmental compartments.

Concentration ranges measured for total PCBs (the sum of the concentrations of all PCB compounds present) in various environmental compartments are presented in Figure 9.2. Note that these concentrations are in units of mass of PCB (in milligrams) per mass of substance (in kilograms). The units of mg/kg can also be expressed as parts per million by weight, abbreviated ppmw or simply ppm. Because 1 liter of water has a mass of 1 kg, we note that 1 mg/kg is the same as 1 mg/L for. water concentrations. Thus the values in Figure 9.2 for surface freshwater can be compared directly with the maximum soluble concentration in water, which is 100 μg/L (or 10^{-1} mg/L) for a typical PCB mixture, as already noted. The maximum value typically found in surface freshwater is 10^{-3} mg/L according to Figure 9.2, far less than the maximum soluble concentration.

As indicated in Figure 9.2, concentrations of PCBs in animals can be relatively high. In fact, by weight, the concentrations are higher than those in water, air, and soil. This is because certain tissues of animals, like fatty tissues, have chemical characteristics that favor retention of PCBs. When animals, including humans, contact soil, water, and air containing PCBs, some of the PCBs are transferred to these tissues. The same occurs when animals ingest food containing trace amounts of PCBs. All people in the United States have measurable PCBs in certain body tissues (Lucas et al., 1982), primarily as a result of food chain transfer.

PCBs were banned in many other countries after the United States imposed its ban; hence worldwide production of these compounds fell dramatically in the 1980s. Because of lower production and because of slow but steady degradation of PCBs by bacteria, environmental concentrations have declined. Figure 9.3 shows concentration decreases in sediments and fish in Lake Ontario. The figure illustrates that the

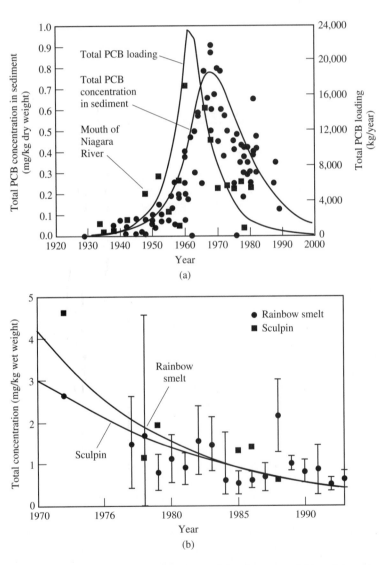

Figure 9.3   Observed total PCB concentrations in (a) sediments, and (b) sculpin and rainbow smelt from Lake Ontario. (*Source:* Reprinted with permission from Gobas et al., *Environ. Sci. Technol., 29*, p. 2043, Copyright 1995, American Chemical Society.)

most rapid decline occurred immediately after the U.S. ban in 1977, with a much slower decline since 1990.

9.3.2 Environmental Standards for PCBs

Because of the presence of PCBs throughout the environment and their suspected toxicity, the U.S. Environmental Protection Agency (EPA) has established standards for the maximum allowable concentrations of polychlorinated biphenyls in water. The standard for drinking water, developed under the authority of the Safe Drinking Water Act, specifies a *maximum contaminant level* or *MCL,* which is the maximum permissible concentration of a contaminant in water that is delivered to users of public water systems. MCLs are based on the likelihood of health risks associated with the contaminants. The technical feasibility and cost of treatment also are considered in establishing primary MCLs. MCLs currently exist for a number of contaminants, including total PCBs (see Chapter 2, Table 2.5, for details). The current MCL for total PCBs in drinking water is 5×10^{-7} grams per liter (g/L), which is equivalent to 0.5 micrograms per liter (μg/L).

The EPA also has developed *water quality criteria* to protect aquatic life in rivers, lakes, and other bodies of water. These bodies of water are known as *receiving waters* because they often receive contaminants from industries, sewage plants, agricultural runoff, and other sources. Just as with MCLs, there are water quality criteria for many contaminants, including PCBs. These water quality criteria, developed under the Clean Water Act, are used by individual states to decide how much of a contaminant should be permitted to enter a particular receiving water. Water quality criteria are based entirely on the adverse effects of contaminants.

Two water quality criteria for PCBs have been developed for the Great Lakes region of the United States and are being used as national recommendations for water quality management. The water quality criterion for protection of human health from a lifetime ingestion of water from a PCB-contaminated water body is 3.9×10^{-6} μg/L. The water quality criterion for protection of wildlife in the food web of a PCB-contaminated water body is 7.4×10^{-5} μg/L. Note that these concentrations are much lower than the MCL for drinking water. These extremely low water quality criteria for PCBs basically imply that any detectable amount of PCBs in water is too high. Given the limitations of available water treatment technology, these criteria also imply that there should be no measurable discharges of any industrial process water containing PCBs.

In addition to the MCL and water quality criteria, there are also limits on the PCB content of fish to protect people eating the fish. PCB concentrations in fish are related to the PCB content of the water where the fish live and to the food web position of the particular species of fish. The U.S. Food and Drug Administration (FDA) has set a tolerance level of 2.0×10^{-3} grams of total PCBs per kilogram (2.0 mg/kg) of the edible portion of all fish. Because many fish have been caught with concentrations above this limit, there has been considerable attention to reducing PCBs in a number of contaminated water bodies, such as the Hudson River in New York, as discussed later in Section 9.7.3.

9.4   CHEMISTRY OF PCBs

We defined PCBs as polychlorinated biphenyls at the beginning of this chapter. Let us now explore the chemistry behind this name.

The word *polychlorinated* simply means that several chlorine atoms are associated with the molecules of this chemical compound. *Biphenyl* requires a bit more explanation. We begin by calling attention to the benzene molecule shown in Figure 9.4, which has great importance in organic chemistry. Benzene (C_6H_6) has six carbon atoms arranged in a ring, with one hydrogen atom attached to each carbon atom. This molecule is a basic building block for many organic compounds that are made for industry-specific uses. When one of the hydrogen atoms is removed from a benzene molecule, it is known as a *phenyl group.* Two phenyl groups linked together are known as a *biphenyl molecule,* pictured in Figure 9.5. A polychlorinated biphenyl is therefore a biphenyl molecule with chlorine atoms substituted for some or all of the hydrogen atoms.

There are 10 hydrogen atoms on the biphenyl molecule, and any or all of these can be replaced by chlorine atoms. With each substitution of a chlorine atom for a hydrogen atom, a new compound is formed. There are 209 different PCB compounds possible, each of which is distinguished by the number and location of chlorine atoms on the biphenyl molecule. These compounds range from molecules containing one chlorine atom and nine hydrogen atoms (monochlorobiphenyl) to a molecule containing 10 chlorines (decachlorobiphenyl). PCB compounds are often considered as groups of molecules containing the same number of chlorine atoms:

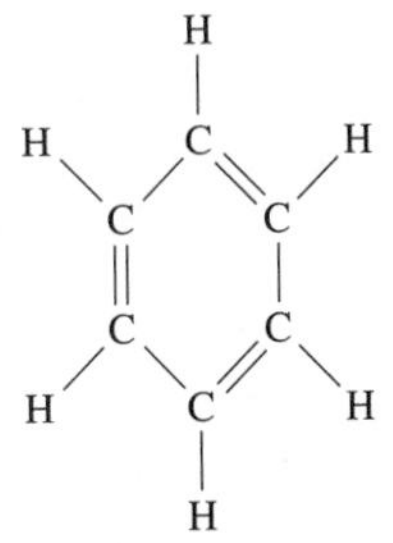

Figure 9.4   The benzene molecule. A single line denotes a single bond, whereas a double line denotes a double bond.

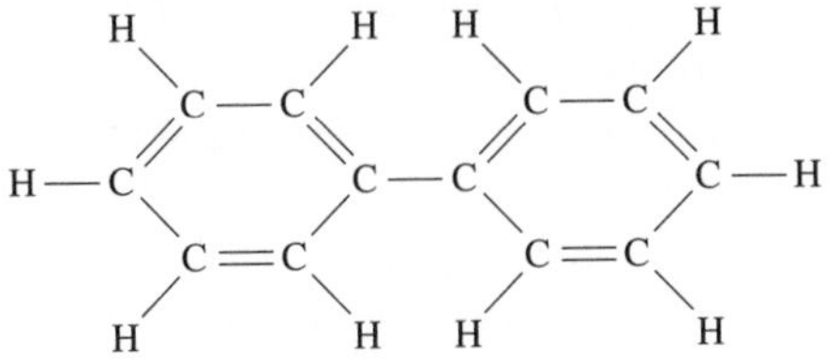

Figure 9.5   The biphenyl molecule.

monochlorobiphenyl, dichlorobiphenyl, trichlorobiphenyl, and so on. Examples of two trichlorobiphenyl molecules are shown in Figure 9.6. There are 24 unique trichlorobiphenyl compounds, known as *isomers*; they have the same atomic weight and the same number of carbon, hydrogen, and chlorine atoms, but different structure owing to the location of chlorine atoms on the biphenyl molecule.

When PCBs were used in industry, they were manufactured to include many different PCB compounds. Typical mixtures included 40 to 60 different PCB compounds.

When PCB liquids contact water, the PCB compounds present in the liquid mixture dissolve in water only to a very limited extent. If a drop of PCB liquid is placed in a beaker of water, for example, there will be no visually noticeable loss of PCB mass. Some small amounts of each compound present will dissolve, however, as can be detected with sensitive measuring equipment. Each compound has a maximum amount that can dissolve in water at a particular temperature. This maximum amount, known as the *aqueous solubility,* is a characteristic property of the compound. The aqueous solubilities measured for PCB compound groups are listed in Table 9.1, which shows that the values are extremely low. Chemists refer to PCBs and similar organic chemicals with very low aqueous solubility as being *hydrophobic* or *water-hating.* Note that these values are for individual compounds, unlike the aqueous solubility of 100 μg/L described previously, which is total dissolved PCBs for a particular commercial mixture of PCBs. The commercial mixture includes some of the higher-solubility compounds as well as low-solubility compounds from Table 9.1.

A common characteristic among PCB compounds is that they do not have an electrical charge associated with them; that is, they are electrically neutral. Other electrically neutral compounds include oils and fats. In contrast, the hydrogen ion H^+ in water is a dissolved chemical species possessing an electrical charge. Uncharged compounds tend to interact with other uncharged compounds, so PCBs dissolve in oils and fats like the fatty tissue of animals. Some soil particles are uncharged; in

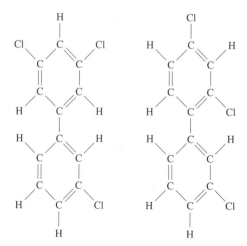

Figure 9.6   Examples of 2 out of the possible 24 trichlorobiphenyl molecules.

Table 9.1   Maximum dissolved concentrations in water for PCB compound groups.

Compound Group	Chlorine Atoms per Biphenyl	Number of Isomers	Aqueous Solubility (mg/L)[a]
Monochlorobiphenyl	1	3	5.90
Dichlorobiphenyl	2	12	2.03
Trichlorobiphenyl	3	24	0.64
Tetrachlorobiphenyl	4	42	0.18
Pentachlorobiphenyl	5	46	0.03
Hexachlorobiphenyl	6	42	0.01
Heptachlorobiphenyl	7	24	0.006
Octachlorobiphenyl	8	12	0.001
Nonachlorobiphenyl	9	3	0.0002
Decachlorobiphenyl	10	1	0.00008

[a] Highest measured solubility value at 25°C for individual compounds in the compound group.
Source: Opperhuizen et al., 1988.

fact, organic soil components have properties similar to oils and fats, and thus PCBs also attach to soil.

The low concentrations of PCBs indicated in Table 9.1 require carefully designed sampling and analytical procedures and sophisticated equipment to measure. This is true for detection of PCBs in all types of environmental samples, as the amounts of

From onboard a ship, this device is used to retrieve samples of sediment from the bottom of Boston Harbor. Test results showed significant concentrations of PCBs.

PCB compounds present in sediment and fish are also very low, typically in the range of 10^{-9} to 10^{-6} gram of PCB per gram of sample. The low concentrations in water, sediment, and fish are measured by removing PCB from the sample using a liquid in which PCBs dissolve readily, such as hexane. A small amount of hexane can extract PCBs from a relatively large volume of water, thus concentrating the small amount of PCB mass present in the sample into a relatively small volume. Samples of hexane can then be analyzed for specific PCB compounds with an analytical technique known as *gas chromatography.* This method provides information on the mass of individual PCB compounds present in the sample.

9.5 RELEASE OF PCBs FROM SOURCES

9.5.1 Pathways of Release

How did PCBs get out into the environment to begin with? As with any widely used liquid, leaks and spills are impossible to avoid. PCB liquid leaking from capacitors, transformers, hydraulic equipment, and other sources found its way into drains and ultimately into wastewater or stormwater flows. These flows were often directed to an industrial or municipal wastewater treatment plant, with final discharge to rivers, lakes, or coastal waters. The released PCB liquids dissolved in part and also became suspended as fine oily droplets in water. Because PCBs are relatively nonbiodegradable and associate strongly with particles of all types, including fine particles that do not settle rapidly, much of the released PCB mass was not captured by treatment plants but rather ended up in the receiving waters.

Another pathway for release of PCBs is through the ground. PCB-bearing liquids were often stored in tanks at industrial plants. Leaks from these tanks, from sumps below the tanks, and from equipment employing PCB liquids discharged PCBs to the ground. Because of the high density of PCB liquids, they typically sank into the ground and reached groundwater. The contaminated groundwater then flowed out to rivers, lakes, and other bodies of water, carrying dissolved and fine particle–associated PCBs with it. Thus, from both wastewater and groundwater discharges, the water environment has been the receptacle for significant amounts of PCB releases.

9.5.2 Example: PCBs in Boston Harbor

PCBs were used in industries in Boston for a long time. A fraction of the PCBs released to the environment in the Boston area made their way to coastal areas. Based on measurements from the late 1970s and early 1980s, the average total PCB concentration in treated wastewater from Boston's municipal treatment facilities at Deer Island and Nut Island was 17 micrograms per liter (μg/L). These discharges, along with those from sewer overflows during wet weather, contaminated Boston Harbor sediments (Figure 9.7) and fish (Figure 9.8).

Concentrations in industrial wastewater discharges have generally been decreasing since PCB production was banned, resulting in lower water concentrations and

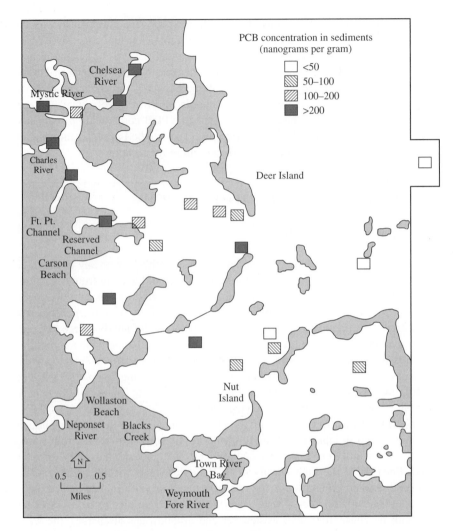

Figure 9.7   PCB distribution in Boston Harbor surface sediments, 1984–1986. (*Source:* Morel and Jones, 1986)

hence a decrease in sediment and fish concentrations. The data in Figure 9.8 illustrate the significant reduction in PCB concentrations that have occurred in winter flounder in Boston Harbor since the mid-1980s.

Although concentrations of PCBs in wastewater discharges have been decreasing, at most locations they have not yet decreased to zero. PCBs continue to be detected in municipal wastewater in Boston, for example. In the discharge permit issued for Boston's upgraded Deer Island Treatment Plant, steps have been taken to reduce PCBs in wastewater entering the plant (MADEP, 1999). The permit requirements include monitoring industries where PCBs are typically found (like waste oil handlers, salvage yards, and metal pressing and forming operations), building dikes

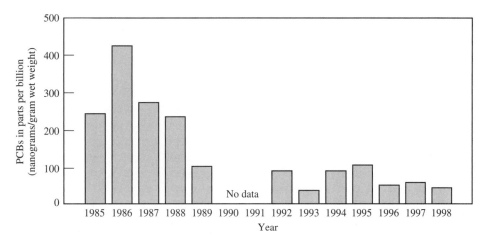

Figure 9.8 PCB concentrations in winter flounder (fillet) caught near Deer Island, Boston Harbor, 1985–1998. (*Source:* MWRA, 1999)

around waste storage areas to prevent further leakage of PCBs, and stopping PCB discharges when they begin.

In addition to inputs of PCBs from wastewater discharges, another important source is release from sediments at the bottom of rivers and other water bodies. PCBs tend to associate with particles, which eventually deposit in the sediments. Over time, the mass of PCBs builds up in sediments. If the discharge of PCBs to receiving waters is reduced, the amounts of PCB in the sediment can become greater than the amounts in the water above it. The contaminated sediment can then serve as a source, releasing PCBs in small quantities to the overlying water. The process of particle deposition to the sediments is illustrated in Figure 9.9. Sediments typically are the largest reservoirs of PCBs in surface water systems because organic matter associated with sediment particles is much more abundant than fish, oils, or other types of organic matter in these systems.

9.6 MOVEMENT OF PCBs IN RECEIVING WATERS

A chemical in a water body typically exists either in dissolved form or associated with particles. Dissolved chemicals tend to be transported with the water, whereas particle-associated chemicals are generally deposited in sediments. As pointed out in Section 9.4, both categories are important for PCBs.

In this section we examine processes that affect the movement of dissolved and particulate contaminants such as PCBs. We focus on rivers because the flow patterns are relatively well defined, although many of the concepts presented here apply to other natural water systems. We consider the chemicals to be nonreactive; that is, they are not subject to any changes by chemical or biochemical reactions. Examination of mixing in complex flows and in chemically reacting systems is outside the

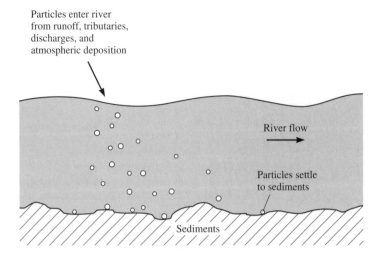

Figure 9.9   Particle deposition to the sediments of a river.

scope of this chapter. Our focus here is on two topic areas: mixing and dilution in rivers, and settling of particles to the river bed.

9.6.1   Mixing and Dilution

Most chemicals are discharged into a river not in concentrated form, but rather diluted in a volume of wastewater. We generally assume the chemical is uniformly mixed in the wastewater. Having entered the river, usually from the side, the wastewater mixes with the river water. For the present, we will focus on locations sufficiently far downstream that the wastewater has mixed uniformly across the entire width and throughout the depth of the river. Let us derive a simple method of calculating the concentration of a chemical at this downstream point.

The volumetric flow of a river, Q_{river}, is typically given in units of cubic feet per second (cfs) or liters per second (L/s). This quantity represents the volume of water moving past a location every second. There is also a volumetric flow of the wastewater, $Q_{wastewater}$, containing the chemical, as shown in Figure 9.10. The figure defines a *control volume,* or a region of constant volume, that includes a length of the river and also the wastewater discharge. We note that the volume of water entering the region must equal the volume of water leaving the region each second:

$$Q_{river-upstream} + Q_{wastewater} = Q_{river-downstream} \qquad (9.1)$$

The concentration of chemical in the river upstream of the discharge point is $C_{upstream}$, usually expressed in milligrams/liter (mg/L). Unless the river is already polluted before the wastewater enters, $C_{upstream}$ is assumed to be zero. The concentration of chemical in the wastewater is $C_{wastewater}$, and the concentration in the river downstream of the discharge is $C_{downstream}$.

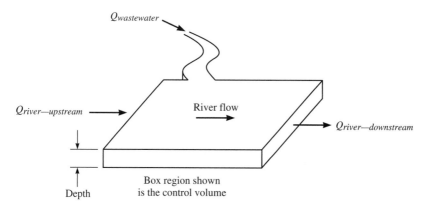

Figure 9.10   Control volume for a length of river illustrating the volumetric flows of water.

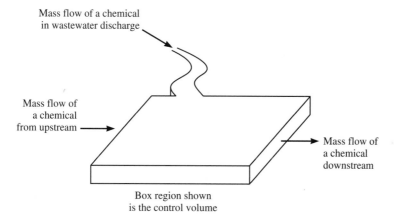

Figure 9.11   Control volume for a length of river illustrating the mass flows of a chemical.

Just as the volume of water entering and leaving the region each second must be equal, so the mass of chemical entering and leaving the region each second must also be equal (assuming the absence of any degradation reaction). This mass balance is illustrated in Figure 9.11. We can write expressions for the mass of chemical entering the region and the mass of chemical leaving the region each second based on this figure:

$$\text{Mass of chemical entering the region per second} =$$

$$Q_{wastewater}C_{wastewater} \ + \ Q_{river-upstream}C_{upstream}$$

$$\text{Mass of chemical leaving the region per second} =$$

$$Q_{river-downstream}C_{downstream}$$

Thus we can write

$$Q_{wastewater}C_{wastewater} \ + \ Q_{river-upstream}C_{upstream} =$$

$$Q_{river-downstream}C_{downstream} \tag{9.2}$$

Note that the units for each term $Q \times C$ are

$$\frac{\text{liters}}{\text{sec}} \times \frac{\text{mg}}{\text{liter}} = \frac{\text{mg}}{\text{sec}}$$

If we assume that $C_{upstream} = 0$ (no initial contamination), Equation (9.3) becomes

$$Q_{wastewater}C_{wastewater} = Q_{river-downstream}C_{downstream}$$

$$C_{downstream} = \frac{Q_{wastewater}}{Q_{river-downstream}} \times C_{wastewater} \qquad (9.3)$$

The situations pictured in Figures 9.10 and 9.11, and defined by the above equations, are termed a *mass balance,* which is an important concept in environmental engineering as well as in many other disciplines. As discussed in Chapter 1, a mass balance simply states that mass is conserved.

Example 9.1

Concentration of PCBs in a river from a wastewater discharge.   Treated wastewater from an industrial facility is discharged into a river at a rate of 8,000 gallons per minute (gpm). The discharge contains a total PCB concentration (sum of all PCB compounds in the water sample, including those on suspended particles) of 0.50 mg/L. The river upstream of the discharge location has no detectable PCBs. The river flow rate is 1,000 cubic feet per second (cfs). What is the total PCB concentration in mg/L at a point downstream of the discharge where the PCBs are completely mixed throughout the width and depth of the river? Assume that particle settling is sufficiently slow that it can be neglected over the time scale of this problem.

Solution:

Equation (9.1) can determine the flow rate of the river downstream of the discharge. However, the wastewater and river flow rates must be expressed in the same units in order to use this equation:

$$Q_{wastewater} = 8{,}000 \,\text{gal/min} \times 0.1337 \,\text{ft}^3/\text{gal} \times 1 \,\text{min}/60 \,\text{sec} = 17.8 \,\text{ft}^3/\text{sec (cfs)}$$

$$Q_{river-downstream} = Q_{river-upstream} + Q_{wastewater} = 1{,}000 \,\text{cfs} + 17.8 \,\text{cfs} = 1{,}017.8 \,\text{cfs}$$

Now we use Equation (9.3) to calculate the PCB concentration at the downstream location:

$$C_{downstream} = \frac{Q_{wastewater}}{Q_{river-downstream}} \times C_{wastewater} = \frac{17.8 \,\text{cfs}}{1{,}017.8 \,\text{cfs}} \times 0.50 \,\text{mg/L} = 0.0087 \,\text{mg/L}$$

The final result has been expressed to two significant figures because one of the terms in the equation, namely 0.50 mg/L, is expressed to only two significant figures. Note that we have maintained a higher number of significant figures for intermediate calculations only.

Now let us assume that we are only a short distance downstream of the discharge point. The wastewater has not had a chance to mix uniformly across the width of the river. However, because rivers are usually fairly shallow compared with their width, we still assume the wastewater has mixed completely over the river depth. For this situation the PCB concentration will be considerably greater because there is less dilution by the river water.

Example 9.2

Concentration of PCBs in a river close to a wastewater discharge.   Assume the river in Example 9.1 has a width of 50 feet. Determine the concentration of PCBs a short distance downstream of the discharge where the wastewater has spread only 10 feet from the shore instead of the full distance across the river.

Solution:

The portion of the river containing the wastewater, downstream of the discharge, has a volumetric flow rate that can be calculated as

$$Q_{river-upstream} = (10 \text{ ft}/50 \text{ ft}) \times 1{,}000 \text{ cfs} = 200 \text{ cfs}$$

$$Q_{river-downstream} = Q_{river-upstream} + Q_{wastewater} = 200 \text{ cfs} + 17.8 \text{ cfs} = 217.8 \text{ cfs}$$

Thus we can use Equation (9.3) to calculate the PCB concentration at the downstream location:

$$C_{downstream} = \frac{Q_{wastewater}}{Q_{river-downstream}} \times C_{wastewater} = \frac{17.8 \text{ cfs}}{217.8 \text{ cfs}} \times 0.50 \text{ mg/L} = 0.041 \text{ mg/L}$$

Note that the concentration is about five times as high as in Example 9.1 because the wastewater is diluted by only one-fifth as much river water in this initial mixing zone.

These examples show that relatively high concentrations of contaminants are found at the point of discharge into the river, but dilution and mixing occur as the contaminants flow downstream. The distance required for complete mixing depends on the river geometry, especially the width and depth, and also the flow properties such as water velocity and volumetric flow rate. Figure 9.12 illustrates the concept of mixing in a river.

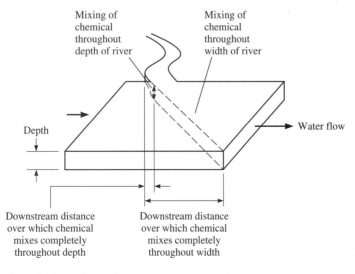

Figure 9.12   Mixing of a chemical in a river after discharge.

9.6.2 Settling of Particles

The time scale for Examples 9.1 and 9.2 was sufficiently short so that settling of particles could be neglected. Now let us consider what happens as we progress further downstream from a discharge point.

Figure 9.13 shows that three vertical forces act on a particle suspended in water: the force of gravity, a buoyant force, and friction. For most particles, the density of the particle is greater than the density of water, and thus the force of gravity is greater than the buoyant force. This means that the particle will accelerate downward. As it begins to move, the force of friction becomes important. The particle continues to accelerate, and the force of friction increases as the particle velocity increases. Eventually the particle reaches a velocity where the forces of friction, gravity, and buoyancy all sum to zero. At this point the particle is moving downward at its *terminal settling velocity,* also called the *settling velocity, V_s.* We can calculate this parameter by equating expressions for the three forces to yield

$$V_s = (g/18\mu)(\rho_s - \rho_w)d^2 \tag{9.4}$$

where g is gravitational acceleration (9.81 m/s^2), μ is the dynamic viscosity of water (1.002×10^{-3} kg/[m-sec] at 20°C), ρ_s is the density of the particle (kg/m^3), ρ_w is the density of water ($1,000$ kg/m^3), and d is the diameter of the spherical particle (m). This expression shows that the settling velocity increases with the square of particle diameter and decreases to zero as particle density approaches the density of water. For typical particles with a density greater than that of water, the particle reaches its settling velocity quickly.

Several sources of particles are found in a river. Rain and snow runoff can deliver particles from nearby land. The sediments at the bottom of the river can be disturbed by water turbulence, stirring up particles. Biological debris from plants and animals can also contribute to the river's loading of particles. Examples of some types of par-

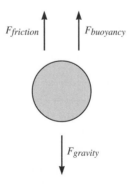

Figure 9.13   Three vertical forces acting on a particle suspended in water.

Table 9.2   Examples of particle types in rivers.

Particle Type	Approximate Density (kg/m³)[a]	Particle Size Range (m)[b]
Plankton and biological debris	1,100	$10^{-9} - 10^{-5}$
Clay and silt	2,000	$10^{-9} - 10^{-5}$
Sand	2,650	$10^{-5} - 10^{-4}$

[a] For reference, the density of water = 1,000 kg/m³.
[b] Corresponds to individual particle sizes.

ticles found in rivers, along with approximate density and range of sizes, are given in Table 9.2. Note that the sizes shown in this table correspond to the diameters of individual particles. Aggregation often occurs in natural water systems, leading to particle aggregates as large as a millimeter or more (10^{-3} m). The density and size of particles is important for their ultimate fate: Small particles of low density can be carried for very long distances because of their slow settling velocities. In contrast, larger, heavier particles settle rapidly and are not transported far downstream.

In cases where there is little or no turbulence in the river flow, the sediments will be undisturbed, and particles will continually deposit on the bottom of the river. Thus particles that settled to the sediments earlier will be buried by more recent deposits.

Example 9.3

Settling velocity of particles.   Calculate the maximum settling velocity for biological debris, clay, and sand particles.

Solution:

For each particle type, the maximum settling velocity will occur for the largest particle of that type, assuming that the different size particles all have the same density. From Table 9.2, the upper end of the size range for biological debris (mostly plant degradation products) is a particle with a diameter of 10^{-5} m (0.01 mm). Considering that the density of biological debris is 1,100 kg/m³ and that of water is 1,000 kg/m³, we can calculate the settling velocity with Equation (9.4):

$$V_s = \frac{g}{18\mu}(\rho_s - \rho_w)d^2 = \frac{9.81 \text{ m/s}^2}{18(1.002 \times 10^{-3} \text{ kg/m-s})}(1,100 - 1,000)\text{kg/m}^3 \, (10^{-5} \text{ m})^2$$

$$= 5.4 \times 10^{-6} \text{m/s}$$

Using a similar approach for clay and sand, the maximum settling velocity can be calculated as 5.4×10^{-5} m/s for clay and 9.0×10^{-3} m/s for sand.

Note that all particles with a density greater than that of water will eventually settle to the bottom of the river, given enough time. If the river depth is D, then the time needed for the particles to settle from the top of the river to the bottom is simply

$$t_b = D/V_s \qquad (9.5)$$

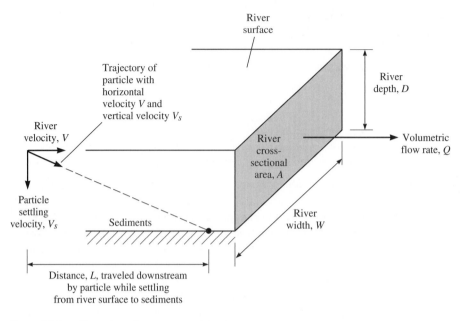

Figure 9.14   Illustration of parameters of the river system.

Over this period of time, the particles will have moved downstream a distance L determined by the river velocity V:

$$L = Vt_b \qquad (9.6)$$

The river velocity can be determined if the cross-sectional area of the river, A, and the volumetric flow rate, Q_{river}, are known:

$$V = Q_{river}/A = Q_{river}/WD \qquad (9.7)$$

Here W is the width of the river. These parameters are illustrated in Figure 9.14.

Example 9.4

Distance downstream for sand particles to settle out.   If the river of Example 9.1 is 10 feet deep, calculate the downstream distance, L, at which all of the contaminants will have settled out. Assume the contaminants are associated with the largest-diameter sand particles from Table 9.2.

Solution:

The maximum-diameter sand particle listed in Table 9.2 is 10^{-4} m, and the settling velocity of a 10^{-4} m diameter sand particle was calculated in the previous example to be 9.0×10^{-3} m/s. The time for this particle to settle from the top of the river to the bottom is given by Equation (9.5):

$$t_b = \frac{D}{V_s} = \frac{10 \text{ ft}}{9.0 \times 10^{-3} \text{ m/s } (1 \text{ ft}/0.305 \text{ m})} = 339 \text{ sec} = 5.65 \text{ min}$$

To determine the downstream distance, L, that the particle will move over this time, we first need to calculate the river velocity according to Equation (9.7):

$$V = \frac{Q_{river}}{A} = \frac{Q_{river}}{DW} = \frac{1,000 \text{ ft}^3/\text{s}}{(10 \text{ ft})(50 \text{ ft})} = 2.0 \text{ ft/s}$$

Finally, Equation (9.6) can be used to determine the desired distance, L:

$$L = Vt_b = (2.0 \text{ ft/s})(339 \text{ s}) = 678 \text{ ft} \approx 680 \text{ ft}$$

Note that the final answer should be given to no more than two significant figures because the settling velocity has been calculated to this many figures.

Thus if we move 680 feet downstream from the point of discharge in Example 9.1, all of the sand particles will have reached the sediments. What about smaller particles? Because the settling velocity is proportional to the square of particle diameter, there can be a significant difference in the distances traveled by small particles and by larger particles. Consider the following example.

Example 9.5

Distance downstream for clay particles to settle out.   Suppose that the contaminant in Example 9.4 is attached to clay particles of diameter 1 μm. How far downstream will the contaminant travel before settling out?

Solution:

From Table 12.2, the density of the clay particle is estimated as 2,000 kg/m^3. A diameter of 1 μm is equivalent to 10^{-6} m. Using Equation (9.4), we have

$$V_s = \frac{g}{18\mu}(\rho_s - \rho_w)d^2 = \frac{9.81 \text{ m/s}^2}{18(1.002 \times 10^{-3} \text{ kg/m-s})}(2,000 - 1,000) \text{ kg/m}^3 (10^{-6} \text{ m})^2$$

$$= 5.4 \times 10^{-7} \text{ m/s}$$

Using this value of V_s, we can apply Equation (9.5) to calculate the amount of time needed for the particle to settle from the top of the river to the bottom:

$$t_b = \frac{D}{V_s} = \frac{10 \text{ ft}}{5.4 \times 10^{-7} \text{ m/s}(1 \text{ ft}/0.305 \text{ m})} = 5.65 \times 10^6 \text{ sec} = 65.4 \text{ days}$$

The distance downstream, L, that the particle travels before depositing is given by Equation (9.6):

$$L = Vt_b = (2 \text{ ft/s})(5.65 \times 10^6 \text{ s}) = 1.13 \times 10^7 \text{ ft} \approx 1.1 \times 10^7 \text{ ft} = 2,100 \text{ miles}$$

This calculation indicates that a 1 μm diameter clay particle settles so slowly that it will not reach the bottom for a very long time. Over this period, it will travel a great distance with the river. In fact, the particle will probably not reach the bottom before the river empties into a lake, ocean, or other body of water. Thus a contaminant associated with this type of particle will behave essentially like a dissolved contaminant, moving with the water rather than settling at a significant rate.

In these examples we have seen that PCBs may travel far from their point of discharge into the receiving waters. The distance traveled is a function of whether the PCBs are dissolved in the water or associated with solid matter such as suspended particles. We now explore quantitatively the physical–chemical process that determines whether PCBs will be dissolved or attached to solid matter, known as partitioning.

9.7   PARTITIONING OF PCBs IN RECEIVING WATER SYSTEMS

Figure 9.2 showed that some environmental compartments have higher concentrations of PCB compounds than others. The term *partitioning* refers to the processes by which the PCBs become distributed among these compartments. The most important compartments for PCBs in rivers are the river water, fish and other aquatic animals, and sediments at the bottom. Here we consider binary systems—that is, systems where the partitioning occurs between two compartments.

The concept of equilibrium is important in understanding partitioning. If a known amount of a PCB is introduced into a river system, such as when a tank ruptures and its contents spill into the river, the PCB will distribute itself among the various environmental compartments. The movement of PCBs is complex during the initial spill, and there will be changes in PCB concentrations in the various compartments. After time, however, equilibrium will be reached and no net exchange of PCBs will occur from one compartment to another; concentrations will not change. Here we focus mainly on equilibrium conditions.

We begin this section by examining the partitioning of PCBs between river water and the sediments under equilibrium conditions. Then we consider partitioning between river water and fish for the same conditions. Finally, we examine as a case study the Hudson River, where conditions are not always at equilibrium.

9.7.1   Partitioning between River Water and Sediments

When introduced into a river system, some of the PCB molecules become associated with suspended and sedimenting particles. This association can be *adsorption,* where the molecules of PCB bond to the surface of a sediment particle, or *absorption,* where the molecules become incorporated inside a sediment particle. It is often difficult to determine whether adsorption or absorption or a combination of the two is taking place, so we often refer to the combined impact of both processes simply as *sorption.*

As long as the concentration of a PCB is less than or equal to its aqueous solubility (as listed in Table 9.1), the amount of PCB sorbed onto sediment particles will be proportional to the amount of PCB dissolved in the river water. The amount of PCB sorbed onto sediment particles can be measured by analyzing the mass of PCB in a sample of sediment obtained from the bottom of a river. We report this as the concentration, $C_{sediment}$, in units of mg PCBs per kg of sediment after the sediment has dried (mg/kg). We also recall from Section 9.3.1 that the concentration of PCBs in water, C_{water}, is expressed as mg PCBs/liter of water (mg/L). The proportionality can thus be expressed as

$$C_{sediment} = Constant \times C_{water} \qquad (9.8)$$

The constant of proportionality in Equation (9.8) contains two parts. The first part is known as the *sediment–water partition coefficient.* This parameter is denoted by K_{oc}, and is a function of only the particular PCB compound. The second part of the proportionality constant is the organic carbon content of the sediment particles, a fraction denoted by f_{oc}. We recall from Section 9.4 that PCB molecules do not have an electrical charge and thus tend to attach themselves to other materials that do not have an electrical charge. Organic carbon in sediment particles is such a material. This organic carbon comes mostly from decaying plant material that was incorporated into the sediment when the particles were originally formed. Note that K_{oc} must have units of liters of water per kilogram of sediment in order for the equation to balance. Equation (9.8) can thus be written as

$$C_{sediment} = K_{oc} f_{oc} C_{water}$$

$$(mg/kg) = (L/kg) \times (Fraction) \times (mg/L) \qquad (9.9)$$

This equation enables us to calculate the concentration of PCBs in the sediment that is in equilibrium with the PCBs in the river water.

Values of K_{oc} are similar for all isomers of a PCB compound group. These values are given in Table 9.3. Values of f_{oc} vary depending on the type of sediment but are usually in the range of 0.01 to 0.1.

Table 9.3   Sediment–water partition coefficients for PCB compound groups.

Compound Group	Chlorine Atoms per Biphenyl	Number of Isomers	log K_{oc}[a]
Monochlorobiphenyl	1	3	4.7
Dichlorobiphenyl	2	12	5.1
Trichlorobiphenyl	3	24	5.5
Tetrachlorobiphenyl	4	42	5.9
Pentachlorobiphenyl	5	46	6.3
Hexachlorobiphenyl	6	42	6.7
Heptachlorobiphenyl	7	24	7.1
Octachlorobiphenyl	8	12	7.5
Nonachlorobiphenyl	9	3	7.9
Decachlorobiphenyl	10	1	n/a

[a] Units of K_{oc} are L/kg; n/a = not available.
Source: Based on Mackay et al., 1992.

Example 9.6

Partitioning of a PCB between river water and sediment.   The concentration of dichloro-biphenyls in river water is found to be 35 μg/L. What is the concentration of dichlorobiphenyls in the sediment at equilibrium with the river water if the sediment particles are 10 percent organic carbon?

Solution:

Dichlorobiphenyls have an aqueous concentration of 35 μg/L, or 0.035 mg/L, according to Table 9.1. If this concentration is present in river water above the sediment, the concentration of dichlorobiphenyls in the sediment can be determined from Equation (9.9) with the appropriate value of K_{oc} from Table 9.3, and $f_{oc} = 0.1$ based on 10 percent organic carbon:

$$C_{sediment} = K_{oc} f_{oc} C_{water} = 10^{5.1}\,\text{L/kg} \times 0.1 \times 0.035\,\text{mg/L} = 440\,\text{mg/kg}$$

The result shows that by weight, the concentration of dichlorobiphenyls in the sediment (440 mg PCB/kg sediment) is much greater than the concentration in the river water expressed in the same units (0.035 mg PCB/L water, equivalent to 0.035 mg PCB/kg water). This suggests that the PCB molecules preferentially attach to organic material in the sediment rather than remaining dissolved in the water. We say that PCBs have a greater *affinity* for organic material in the sediment than they have for water.

In the previous example we were given a concentration in water and asked to determine the concentration in the sediment. Another way to use Equation (9.9) is to begin with a known total mass of PCB in the entire system and determine the amount of PCB in each phase. This is illustrated in Example 9.7.

Example 9.7

Partitioning of a known total mass of PCB in a system.   A tank contains 100 L of water and 0.5 kg of sediment. The sediment has an organic carbon content of 10 percent. If 0.1 mg of dichlorobiphenyls is added to the tank, determine how much of this mass of PCBs will dissolve in the water and how much will become attached to the sediment particles.

Solution:

The total mass of dichlorobiphenyls is 0.1 mg, which is divided between the water and the sediments. We can therefore write

$$M_{total} = M_{water} + M_{sediment}$$

where

$$M_{total} = \text{total mass of dichlorobiphenyls in the system}$$

$$M_{water} = \text{mass of dichlorobiphenyls in the water}$$

$$M_{sediment} = \text{mass of dichlorobiphenyls in the sediment}$$

We know that

$$M_{water} = C_{water} \times 100\,\text{L}$$

and

$$M_{sediment} = C_{sediment} \times 0.5\,kg$$

Thus we can write

$$
\begin{aligned}
M_{total} &= M_{water} + M_{sediment} = (C_{water} \times 100\,L) + (C_{sediment} \times 0.5\ kg) \\
&= (C_{water} \times 100\,L) + (K_{oc}f_{oc}C_{water} \times 0.5\,kg) \\
&= C_{water}[100\,L + (K_{oc}f_{oc} \times 0.5\,kg)] \\
&= C_{water}[100\,L + (10^{5.1}\,L/kg \times 0.1 \times 0.5\,kg)] \\
&= C_{water}(100\,L + 6{,}300\,L) = 0.1\ mg
\end{aligned}
$$

We can solve for M_{water} and $M_{sediment}$

$$C_{water} = \frac{0.1\,mg}{6{,}400\,L} = 1.56 \times 10^{-5}\ mg/L$$

so that

$$M_{water} = 1.56 \times 10^{-5}\,mg/L \times 100\,L = 0.00156\ mg$$

$$C_{sediment} = K_{oc}f_{oc}C_{water} = 105.1\,L/kg \times 100\,L = 0.00156\ mg$$

and

$$M_{sediment} = 0.197\,mg/kg \times 0.5\,kg = 0.0984\ mg$$

We note that of the 0.1 mg dichlorobiphenyls, 1.56 percent is dissolved in the water and 98.4 percent is attached to the sediment.

We can also consider the partitioning of a PCB between water and suspended particles in the water. The suspended particles are treated in the same way as the sediment particles.

Example 9.8

Partitioning of a PCB between water and suspended particles. River water contains suspended particles at a concentration of 20 mg/L. The fraction of organic carbon in the particles is 0.1. A sample of the river water is collected in a bottle and analyzed; the total concentration of dichlorobiphenyls in both dissolved and suspended phases is 0.2 mg/L. What fraction of these PCBs is dissolved in the river water, and what fraction is associated with the suspended particles?

Solution:

Analogous to Example 9.7, we can sum the mass of dichlorobiphenyls dissolved in the water, M_{water}, and the mass attached to suspended particles in the solid phase, M_{solid}, to obtain the total mass of dichlorobiphenyls in a specified volume of river water. We use 1 liter of river water as the volume for convenience:

$$M_{total} = M_{water} + M_{solid}$$

As in Example 9.7, we know that $M_{water} = (C_{water} \times 1\ L)$ and $M_{solid} = (C_{solid} \times 20\ mg)$ because there are 20 mg of suspended particles in the 1 liter volume of river water. We note that

C_{water} has units of mg PCB/L and C_{solid} has units of mg PCB/(kg solids). C_{solid} is analogous to $C_{sediment}$. Thus we can write

$$M_{total} = M_{water} + M_{solid} = (C_{water} \times 1\,L) + (C_{solid} \times 20\,mg\,solids)$$

$$= (C_{water} \times 1\,L) + (K_{oc}f_{oc}C_{water} \times 20\,mg\,solids)$$

$$= C_{water}[1\,L + (K_{oc}f_{oc} \times 20\,mg\,solids)]$$

$$= C_{water}\left[1\,L + \left(10^{5.1}\,\frac{L}{kg\,solids} \times \frac{1\,kg}{10^6\,mg} \times 0.1 \times 20\,mg\,solids\right)\right]$$

$$= C_{water}(1\,L + 0.25\,L) = 0.2\,mg$$

Solving for C_{water} and C_{solid} yields

$$C_{water} = \frac{0.2\,mg\,PCB}{1.25\,L} = 0.089\,\frac{mg\,PCB}{L}$$

$$C_{solid} = K_{oc}f_{oc}C_{water} = 10^{5.1}\,\frac{L}{kg\,solids} \times 0.1 \times 0.16\,\frac{mg\,PCB}{L} = 2{,}014\,\frac{mg\,PCB}{kg\,solids}$$

In 1 liter of water there are 20 mg solids. Thus we have

$$\frac{Mass\ of\ PCB\ in\ solid\ phase}{L} = 2{,}014\,\frac{mg\,PCB}{kg\,solids} \times \frac{20\,mg\,solids}{L} \times \frac{1\,kg}{10^6\,mg}$$

$$= 0.04\,\frac{mg\,PCB}{L}$$

where the PCB is in the dichlorobiphenyl group of compounds. Note that the total mg of PCB in the dissolved and solid phases in 1 liter of river water is $0.16 + 0.04 = 0.2$ mg/L as required.

The previous examples involved only one PCB compound group, dichlorobiphenyls. In Section 9.4 we noted that PCBs were most often used as mixtures of several compounds. We can use the data for each compound separately to determine the distribution of PCB mixtures in different environmental compartments, as shown in the following example.

Example 9.9

Partitioning of PCB mixtures between river water and sediments.   PCBs were sold commercially as mixtures called Aroclors. For example, Aroclor 1242 contained primarily PCBs having two to five chlorine atoms. The aqueous solubility of each prominent PCB compound in Aroclor 1242 has been measured experimentally (Luthy et al., 1997), representing the maximum possible amount of each compound that can dissolve in water. Representative values of the major compounds in the mixture are dichlorobiphenyls, 35 μg/L; trichlorobiphenyls, 45 μg/L; tetrachlorobiphenyls, 15 μg/L; and pentachlorobiphenyls, 5 μg/L. Assume that dissolved PCB compounds in river water have been measured at these maximum possible concentrations. Estimate the total PCB concentration of all four of these compound groups in the sediment. Assume that the sediment has 10 percent organic matter.

Solution:

We can use Equation (9.9) and Table 9.3 for each compound group separately. For dichlorobiphenyls:

$$C_{sediment} = K_{oc}f_{oc}C_{water} = 10^{5.1} \text{ L/kg} \times 0.1 \times 0.035 \text{ mg/L} = 440 \text{ mg/kg}$$

For trichlorobiphenyls:

$$C_{sediment} = K_{oc}f_{oc}C_{water} = 10^{5.5} \text{ L/kg} \times 0.1 \times 0.045 \text{ mg/L} = 1,420 \text{ mg/kg}$$

For tetrachlorobiphenyls:

$$C_{sediment} = K_{oc}f_{oc}C_{water} = 10^{5.9} \text{ L/kg} \times 0.1 \times 0.015 \text{ mg/L} = 1,190 \text{ mg/kg}$$

For pentachlorobiphenyls:

$$C_{sediment} = K_{oc}f_{oc}C_{water} = 10^{6.3} \text{ L/kg} \times 0.1 \times 0.005 \text{ mg/L} = 1,000 \text{ mg/kg}$$

The total PCB concentration in sediment is therefore

$$C_{sediment} = (440 + 1,420 + 1,190 + 1,000) \text{ mg/kg} = 4,050 \text{ mg/kg}$$

9.7.2 Partitioning between River Water and Fish

We noted in the previous section that PCBs attach to organic carbon in sediment particles. In fact, we have seen that far more PCB mass attaches to each kg of sediment than is dissolved in a kg of river water. The same is true when PCBs partition between river water and fish. Because fish tissue contains organic carbon, PCBs tend to accumulate in fish rather than remain dissolved in the water. The uptake of chemicals, such as PCBs into fish tissue, is known as *bioconcentration* because the chemicals from a large amount of water become highly concentrated in a small amount of fish tissue.

Bioconcentration in fish and other aquatic animals is an important process for PCBs in the food chain. For algae and other plants, direct uptake of PCBs from water is the dominant uptake mechanism. For fish and other animals higher on the food chain, uptake by ingestion can be important.

Quantitatively, the amount of bioconcentration occurring in a natural system can be determined using the *bioconcentration factor* or *BCF*, which is analogous to the sediment–water partition coefficient, K_{oc}. The units of *BCF* are the same as those of K_{oc}, as can be seen from the following equations:

$$C_{sediment} = K_{oc}f_{oc}C_{water}$$

$$(\text{mg/kg}) = (\text{L/kg}) \times (\text{Fraction}) \times (\text{mg/L}) \tag{9.10}$$

$$C_{fish} = BCF \times C_{water}$$

$$(\text{mg/kg}) = (\text{L/kg}) \times (\text{mg/L}) \tag{9.11}$$

Bioconcentration factors are tabulated in various reference sources (like Kenaga and Goring, 1980). Typical values of bioconcentration factors for selected PCB compounds are shown in Table 9.4.

Table 9.4   Bioconcentration factors for selected PCB compounds.

Chemical	Bioconcentration Factor (L/kg)[a]
Monochlorobiphenyl	590
Dichlorobiphenyl	215
Trichlorobiphenyl	49,000
Tetrachlorobiphenyl	73,000
Pentachlorobiphenyl	46,000

[a] Values given are for particular PCB compounds in each group.
Source: Kenaga and Goring, 1980.

Example 9.10

Bioconcentration of PCBs in fish.   In Section 9.3.2 we discussed regulatory standards for PCBs and noted that the maximum permissible level of PCBs in fish is 2.0 mg/kg. What concentration of pentachlorobiphenyl is allowable in water so that the PCB fish standard is not exceeded?

Solution:

According to Table 9.4, the bioconcentration factor for pentachlorobiphenyl is 46,000 L/kg. Using this value in Equation (9.10) yields

$$C_{water} = \frac{C_{fish}}{BCF} = \frac{2.0 \text{ mg/kg}}{46,000 \text{ L/kg}} = 4.3 \times 10^{-5} \text{ mg/L}$$

This is an extremely small concentration needed to keep PCB levels low in fish.

Note that the values of *BCF* listed in Table 9.4 for the individual PCB compounds differ by orders of magnitude. Thus mixtures of PCBs containing many compounds will range widely in concentration activity.

Although we have considered two binary systems for partitioning of PCBs, it is clear that a real river system will include several environmental compartments. Consideration of systems with more than two components is outside the scope of this chapter. However, given the tendency of PCBs to associate with organic carbon, it is apparent that PCBs introduced into a river in discharge waters will tend to accumulate in all environmental compartments that contain organic carbon, including various aquatic animals and plants as well as sediments that contain carbon.

We will now examine the effects of PCB contamination of sediments in the Hudson River as a case study, and we'll consider the complex environmental management issues that this problem poses.

9.7.3   PCBs in the Hudson River

We have seen that in typical river systems a large fraction of the PCB mass is associated with sediment particles. The sediments became contaminated when PCBs were in general use. Now that PCBs are no longer manufactured in the United States, we are faced with a new dilemma: Should we allow the huge mass of PCBs in the sediments to remain there, or should we remove the sediments by dredging? There is no clear-cut answer. Some experts argue that dredging actually increases the problem by releasing substantial amounts of PCB into the water during removal.

They further argue that if the sediments were left in place, additional particle deposition (in the runoff from nearby land) would continue to bury the contaminated areas, thus reducing risk over time. Dredging also can harm the sediment ecosystem in the short term. Other experts are concerned that burial of the contaminated sediments by new deposition is not permanent—the PCBs could be released at any time by a heavy river flow, such as during a flood, which erodes the river bottom. They argue that the only sure way to protect future generations from the PCBs is to remove them.

As an example of the problem, we can consider a site in New York State that has been contaminated with PCBs. From the mid-1940s until 1976, PCB wastewater was discharged to the upper Hudson River by capacitor manufacturing plants located in Hudson Falls and Fort Edward, New York (Rhea et al., 1997). PCBs from these plants accumulated in the sediments upstream of a dam at Fort Edward. When the dam was removed in 1973, the sediments were disturbed and PCBs were released. Several older PCB deposit sites were also exposed. The PCBs contaminated a 40-mile stretch of river from Fort Edward to Troy, New York. In 1984 the EPA recommended action to stabilize the older deposit sites, but they recommended no action for the sediment deposits (USEPA, 1984). The stabilization procedures for the older deposits were completed in 1991.

Problems continued, however. Measurements of total PCBs in Hudson River water at Fort Edward showed an increase in 1991 from 0.1 μg/L to 4 μg/L. This was attributed to the collapse of a structure at an old mill that had retained a pool of PCBs adjacent to a Hudson Falls manufacturing site. PCB liquids were found to be discharging directly into the river through fractures within the river bed. The PCB liquids and sludges were removed, and barriers were installed. These efforts decreased the PCB concentrations in the river water to generally less than 0.02 μg/L.

In 1982, 75 percent of the fish in the Hudson River showed PCB concentrations of less than 5 mg/kg of edible portion of fish, which was the Food and Drug Administration's limit at that time. In 1983 the limit was changed to 2 mg/kg of edible portion. Since then, studies in the upper Hudson River have shown PCBs in fish ranging from 1 to 30 mg/kg (USEPA, 1991), and the EPA has been considering further action.

This experience shows that contamination is still an issue in the Hudson River. Several other U.S. rivers have similar problems. Thus, despite the ban on PCBs, releases of these compounds from old sites and structures may increase the concentrations in rivers and contaminate fish and ecosystems. People eating the fish may then be at increased health risk.

Although we have focused on river systems in this chapter, PCBs can contaminate land as well. Just as bioconcentration causes high PCB levels in fish, the same process can result in high concentrations in animals on land. Earthworms ingesting contaminated soil, for example, can have PCB concentrations in their tissue that are much greater than the soil concentrations. Birds eating the contaminated earthworms can develop still higher concentrations. If a carnivore then eats contaminated birds, the carnivore's PCB levels increase further. This process of increasing concentrations of synthetic organic chemicals through the food chain first received public attention in 1962 with the publication of *Silent Spring* by Rachel Carson. Her book emphasized the need for careful scientific studies of how pesticides and other hazardous chemicals move through the environment, and what can be done to minimize the risk of chemicals to ecosystems and people. These studies are ongoing and will continue to provide guidance on how best to balance the benefits of technology while minimizing adverse environmental consequences.

9.8   CONCLUSION

Our experience with PCBs provides important lessons about the need to consider potential environmental and health impacts in selecting chemicals for industrial use. As is evident from this chapter, the discharge of compounds that do not degrade can lead to their widespread distribution in the environment and to possible long-lasting adverse effects. Although engineering properties of PCB compound mixtures were studied in detail during their development, properties relevant to environmental fate and human health were not. PCBs were designed for many uses, for example as fire retardants, insulating fluids in electrical equipment, oils in hydraulic and lubrication equipment, paints and coatings, and other applications. Process and product designers adopted PCBs for use because of their robust properties. These designers, as well as the developers of PCBs, all failed to investigate adequately the toxicity and likely environmental fate of PCB compounds.

Once the environmental and health problems associated with PCBs became clear and production of PCBs was banned in the United States, research efforts were initiated to identify alternative non-PCB liquids for industrial use. For example, high-temperature hydrocarbons and silicone oil were identified for use in transformers and other electrical equipment. These replacements have certain performance characteristics similar to those of PCBs (like insulating capacity), but they have weaker fire safety characteristics (USDOT, 1984). Because PCBs are highly fire resistant and do not produce combustible gases when heated, transformers and electrical equipment using PCBs could be installed indoors and next to buildings with relatively few restrictions. Use of the replacement liquids no longer allows these benefits and has required redesign and retrofitting of electrical equipment with various kinds of fire prevention devices.

As in all design endeavors, there are multiple objectives and trade-offs in the design and selection of chemicals for particular uses. Environmental and health factors should be included in addition to performance characteristics, costs, and safety issues. The U.S. experience with PCBs demonstrates that the cost of not adequately considering long-term environmental and health effects in designing or selecting chemicals can be very high indeed.

9.9 REFERENCES

Carson, R. L., 1962. *Silent Spring,* Houghton Mifflin, New York.

Durfee, R. L., G. Contos, F. C. Whitmore, J. D. Barden, E. E. Hackman, and R. A. Westin, 1976. "PCBs in the United States—Industrial Use and Environmental Distributions." Report EPA 560/6-76-005, U.S. Environmental Protection Agency, Office of Toxic Substances, Washington, DC.

Erickson, M. D., 1997. *Analytical Chemistry of PCBs,* 2nd ed. CRC Lewis Publishers, Boca Raton, FL.

Gobas, F. A. P. C., M. N. Z'Grabben, and X. Zhang, 1995. "Time Response of the Lake Ontario Ecosystem to Virtual Elimination of PCBs." *Environmental Science and Technology,* vol. 29, pp. 2,038–46.

Kenaga, E. E., and C. A. I. Goring, 1980. In *Aquatic Toxicology.* ASTM STP 707, J. G. Eaton, P. R. Parrish, and A. C. Hendricks, eds., American Society for Testing and Materials, Philadelphia, PA.

Lucas, R. M., V. G. Iannacchione, and K. D. Melroy, 1982. "Polychlorinated Biphenyls in Human Adipose Tissue and Mother's Milk." Report EPA-560/5-83-011, U.S. Environmental Protection Agency, Office of Toxic Substances, Washington, DC.

Luthy, R. G., D. A. Dzombak, M. J. R. Shannon, R. Unterman, and J. R. Smith, 1997. "Dissolution of PCB Congeners from an Aroclor and Aroclor/Hydraulic Oil Mixture." *Water Research,* vol. 31, pp. 561–73.

Mackay, D., W. Y. Shiu, and K. C. Ma, 1992. *Illustrated Handbook of Physical–Chemical Properties and Environmental Fate of Organic Chemicals,* vol. 1, Lewis Publishers, Chelsea, MI.

MADEP, 1999. Massachusetts Department of Environmental Protection, "Authorization to Discharge under the National Pollutant Discharge Elimination System, MWRA Publicly Owned Treatment Works, Boston, MA." Permit No. MA0103284, Commonwealth of Massachusetts, Department of Environmental Protection, May 20, 1999.

Morel, F. M. M., and G. J. Jones, 1986. "Pattern of Contaminant Distribution in Boston and Salem Harbors." Report to National Oceanic and Atmospheric Administration, Washington, DC.

MWRA, 1999. Massachusetts Water Resources Authority, "Update on Boston Harbor." http://www.mwra.state.ma.us/harbor/html/bhrecov.htm (accessed July 13, 1999).

NRC, 1979. *Polychlorinated Biphenyls.* National Research Council, National Academy of Sciences, Washington, DC.

Opperhuizen, A., F. A. P. C. Gobas, J. M. D. Van der Steen, and O. Hutzinger, 1988. "Aqueous Solubility of Polychlorinated Biphenyls Related to Molecular Structure." *Environmental Science and Technology,* vol. 22, pp. 638–46.

Rhea, J., J. Connolly, and J. Haggard, 1997. "Hudson River PCBs: A 1990s Perspective." *Clearwaters,* vol. 27, no. 2, Summer, pp. 24–28.

Schwarzenbach, R. P., P. M. Gschwend, and D. M. Imboden, 1993. *Environmental Organic Chemistry.* Wiley-Interscience, New York, NY.

Travis, C. C., and S. T. Hester, 1991. "Global Chemical Pollution." *Environmental Science and Technology,* vol. 25, pp. 815–18.

USDOT, 1984. "Polychlorinated Biphenyls (PCBs) in Transit System Electrical Equipment." Report DOT-TSC-UMTA-84-15, U.S. Department of Transportation, Transportation Systems Center, Cambridge, MA, 1984.

USEPA, 1984. "Feasibility Study, Hudson River PCB Site, New York," vol. 1. Prepared by NUS Corporation, U.S. Environmental Protection Agency, Region II Office, New York, NY.

USEPA, 1991. "Phase 1 Report: Review Copy Interim Characterization and Evaluation, Hudson River PCB Reassessment RI/FS." U.S. Environmental Protection Agency, Region II, New York, NY, August.

9.10 PROBLEMS

9.1 A river 100 feet wide is contaminated for a distance of 20 miles downstream from a former electrical transformer manufacturing facility. The water in the river is five feet deep on average and contains 10 μg/L total PCBs (this includes water and suspended particles). The sediments in the river are contaminated to a depth of six inches, with an average sediment concentration of 2 mg/kg. What is the total mass of PCBs in the water (including suspended particles)? The total mass of PCBs in the sediments (in kilograms)?

9.2 If you consumed two liters of water containing 0.5×10^{-6} g/L total PCBs (the 1999 MCL for PCBs) every day for 70 years, what would be the total mass of PCBs (in kilograms) potentially accumulated in your body? Assume there is no breakdown or excretion of PCBs from your body.

9.3 Of the 36 inches of annual precipitation (rain and snowmelt) that is deposited on five acres of contaminated industrial property, about 50 percent runs off to a stream.

    (a) Considering that the precipitation is deposited uniformly over the five-acre area, what is the average runoff flow from the contaminated property in gallons per day?

    (b) If the total PCB concentration in the runoff flow is 20 μg/L, what will be the total PCB mass loading (kilograms per day) to the stream?

9.4 A river has a flow rate of 500 cubic feet per second (cfs) and a total PCB concentration of 30 μg/L. If a wastewater discharge of 25 cfs has a total PCB content of 300 μg/L, what will be the total PCB concentration downstream of the discharge after it has mixed completely across the width and throughout the depth of the river?

9.5 Silt particles in a river are 10^{-8} m in diameter. If PCBs are attached to biological debris in the river settling at the same velocity as the silt, determine the diameter of the biological debris.

9.6 Two discharge pipes, each with a volumetric water flow of 1 cubic meter/sec, enter the same stream a few miles apart. The first pipe emits 1,000 mg/sec PCBs, whereas the second pipe emits 200 mg/sec. If the concentration of PCBs downstream of the second discharge pipe is 0.1 mg/liter, what is the concentration in the stream between the two pipes? Assume that there is complete mixing.

9.7 Table 9.2 shows that silt grains typically range in size up to 10^{-5} m in diameter. Assume that an agglomerate is formed of 100 silt grains, each of this maximum size. This cluster of silt grains has a roughly spherical shape, and you may assume that there is negligible space between the grains. Determine the settling velocity of such an agglomerate.

9.8 The cluster of sand grains in Problem 9.7 enters the stream of Problem 9.6 at the stream's surface, downstream of the second discharge pipe. If the width of the stream is 5 m and the water velocity is 2 m/sec, determine how far downstream the agglomerate will travel before reaching the bottom.

9.9 If the aqueous phase in a river has a dichlorobiphenyl concentration of 100 μg/L and a trichlorobiphenyl concentration of 50 μg/L, estimate the total PCB concentration (mg/kg) associated with the suspended solids in the river. The river has 50 mg/L of suspended solids, and the solid particles have 2 percent organic carbon. Assume that the river water is in equilibrium with the suspended solids and that dichlorobiphenyl and trichlorobiphenyl compounds are the only PCB compounds present.

9.10 If the aqueous phase in a river has a dichlorobiphenyl concentration of 100 μg/L and a trichlorobiphenyl concentration of 50 μg/L, estimate the total PCB concentration (μg/kg) associated with the fish in the river. Assume that compound partitioning between the river water and the fish is in equilibrium, and that dichlorobiphenyl and trichlorobiphenyl compounds are the only PCB compounds present.

Human Exposure to Toxic Metals*

*This chapter was written by Cliff I. Davidson.

10.1 INTRODUCTION

Metals have been vitally important to human civilization for centuries. They have been the material of choice for countless products, from jewelry and tools in ancient times to computers and spaceships today. In this chapter we explore the release of metals to our environment as a by-product of our working with metals. We begin with a brief history of metallurgy to illustrate how the use of metals has become pervasive in our world. We then consider evidence that working with huge amounts of metals has resulted in their release to the environment, and we demonstrate that this is of concern because of adverse effects. Next we examine human exposure to metals in different environmental media. Finally, we consider the concept of dose–response relationships as a means of estimating the effects of human exposure to metals.

10.2 A BRIEF HISTORY OF METALLURGY

The history of working with metals goes back to prehistoric times. Indeed, one can follow the progress of civilization by examining how metals have been used over the years.

Figure 10.1 shows a time line of the history of metals. Stone Age dwellers hammered metals in their natural form to make ornaments some 8,000 years ago, whereas heating of ores to extract metals, known as *smelting,* began in roughly 4000 B.C. Workers determined as early as 3000 B.C. that adding tin to copper in the ratio of 1 to 10 provided the most durable bronze products, including tools, weapons, and jewelry. The use of leaded paints also began during the Bronze Age, as did coins made of silver. It is interesting that for a time, iron was valued at eight times the value of silver in early Babylonia. Similarly, copper was more valuable than gold for a period in ancient Egypt.

Gradually, iron replaced bronze as the most important metal for tools and weapons. Iron had the potential to be more durable, but it required *carburizing* (heating in the presence of carbon, such as with charcoal) and *quenching* the fire to reduce the heat quickly, both of which were difficult for the ancients. Some of these products of iron that included appreciable carbon content are considered to be the first manufacture of steel.

By the beginning of the Iron Age, manufacturing with iron was dominant in several regions of Asia, including the Caucasus Mountains region, the Middle East, and northern India. Bronze was still used extensively in China at this time, although iron-working became important several hundred years later.

The rise of the Greek Empire greatly increased metal production. Besides the six metals already mentioned, the Greeks isolated mercury and developed its use. Following the decline of Greece, the Romans advanced the manufacture of iron and steel for household tools, farming implements, weapons for their armies, and parts for ships, including anchors and chains. They produced ornaments and coins using gold and silver. They also relied extensively on the use of lead for water pipes, water tanks, and food storage containers; their total production of lead between 200 B.C. and 500 A.D. is estimated at 18 million metric tonnes.

After about 500 A.D., the disintegration of the Roman Empire caused a substantial decline in the production and use of metals in the Western world. It wasn't until

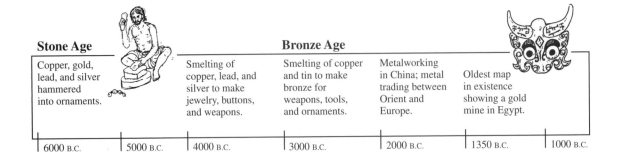

Stone Age

	Bronze Age			
Copper, gold, lead, and silver hammered into ornaments.	Smelting of copper, lead, and silver to make jewelry, buttons, and weapons.	Smelting of copper and tin to make bronze for weapons, tools, and ornaments.	Metalworking in China; metal trading between Orient and Europe.	Oldest map in existence showing a gold mine in Egypt.

| 6000 B.C. | 5000 B.C. | 4000 B.C. | 3000 B.C. | 2000 B.C. | 1350 B.C. | 1000 B.C. |

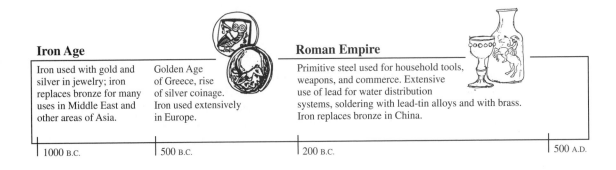

Iron Age

		Roman Empire
Iron used with gold and silver in jewelry; iron replaces bronze for many uses in Middle East and other areas of Asia.	Golden Age of Greece, rise of silver coinage. Iron used extensively in Europe.	Primitive steel used for household tools, weapons, and commerce. Extensive use of lead for water distribution systems, soldering with lead-tin alloys and with brass. Iron replaces bronze in China.

| 1000 B.C. | 500 B.C. | 200 B.C. | 500 A.D. |

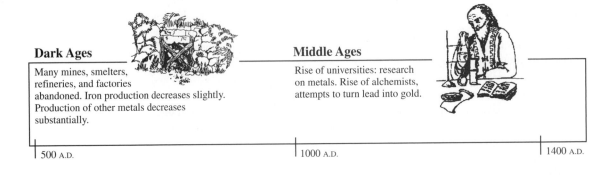

Dark Ages

	Middle Ages
Many mines, smelters, refineries, and factories abandoned. Iron production decreases slightly. Production of other metals decreases substantially.	Rise of universities: research on metals. Rise of alchemists, attempts to turn lead into gold.

| 500 A.D. | 1000 A.D. | 1400 A.D. |

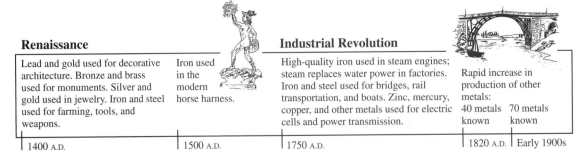

Renaissance

		Industrial Revolution	
Lead and gold used for decorative architecture. Bronze and brass used for monuments. Silver and gold used in jewelry. Iron and steel used for farming, tools, and weapons.	Iron used in the modern horse harness.	High-quality iron used in steam engines; steam replaces water power in factories. Iron and steel used for bridges, rail transportation, and boats. Zinc, mercury, copper, and other metals used for electric cells and power transmission.	Rapid increase in production of other metals: 40 metals known 70 metals known

| 1400 A.D. | 1500 A.D. | 1750 A.D. | 1820 A.D. | Early 1900s |

Figure 10.1   Time line of metalworking (*Sources*: Aitchison, 1960; Nriagu, 1983; Wedeen, 1984; Wertime, 1964)

after 800–900 A.D. that Europe began to emerge from the Dark Ages and production of metals increased once again. By 1100 A.D. new techniques in metalworking developed as part of the rise of universities in Europe. It was also at this time that alchemy became widespread. Although well known for their attempts to change lead into gold, many alchemists in fact conducted worthwhile research on alloying, casting, separating, and plating metals.

Artisans in the Middle Ages and during the Renaissance used metals extensively to produce jewelry, ornaments, and works of art. Many European cities had goldsmiths and silversmiths. Lead became the dominant material for artistic creations on buildings as well as for roof covering and plumbing. Pewter made of lead and tin was widely used for kitchenware, whereas brass and bronze became popular for decorating buildings and making monuments and memorials. Iron and steel were commonly used for farming and household equipment and for implements of war.

In England in the early 1700s, one of the most important discoveries in the history of metalworking was made: An improved process was developed to use coke rather than charcoal to smelt iron. Smelting was no longer limited by the availability of wood to serve as fuel. This led to the beginning of the industrial revolution, in which mechanical power produced by steam replaced conventional water power in factories by the late 1700s. The availability of high-quality iron had a dramatic impact. Boats began to be manufactured of iron, and water distribution systems were constructed with iron pipes. Public transportation by trams on iron rails began in the early 1800s. Steam engines made of iron were used in countless new factories. Although only seven metals had been used extensively from ancient times up to the mid-1700s, the industrial revolution initiated development of new metals. By 1820, 40 metals were known, and by the early 1900s there were 70.

New uses of metals greatly improved lifestyles throughout the 18th and 19th centuries. Long bridges made of iron closed gaps between communities and regions, and railroad construction facilitated travel. Dry cells made with zinc, mercury, and eventually other metals produced electricity that enabled the separation of many metallic elements and served as the forerunner of modern electric power generation. The availability of copper was instrumental for the increased use of electric power. The rise of steelmaking in the late 19th century was vital for the growth of numerous industries, such as automobile manufacturing.

The importance of metals has continued to grow. For example, an alloy of aluminum with copper, magnesium, manganese, and silicon was both lightweight and strong enough to be used in early aircraft. Alloys of nickel and chromium can withstand high temperatures and have been used as heating elements in electric furnaces. It was discovered that germanium and silicon have unique electrical properties and can be used to develop electrical components; a variety of new metals have since been created, and these have been vital to the recent growth of electronics and computers. The construction industry has been transformed by development of many new metal alloys with superior strength. Production of nuclear energy relies on containment vessels made of metals, especially stainless steel, that can withstand the extremely high temperatures of nuclear reactions. Many more examples of the benefits of newly developed metals have been reported (Tylecote, 1992). It is safe to say that a large fraction of the world's population is now much better off as a result of metalworking.

10.3   RELEASE OF METALS TO THE ENVIRONMENT: EVIDENCE OF ADVERSE EFFECTS

Huge amounts of metals are now processed worldwide. Steel ranks first in global production, followed by aluminum, copper, and zinc. Although the massive development of metals has been a key factor in improving our standard of living, such development has not been free of adverse effects. As we mine metals, smelt them, and manufacture products, we release quantities of metal to the environment, and thus we can be exposed to them. Most of these metals are found in low concentrations, so we refer to them as *trace metals*.

Certain metals are essential for our nutrition, and small amounts must be consumed for our well-being. Examples include calcium, copper, iron, magnesium, phosphorus, potassium, sodium, and zinc. Nonessential metals are toxic even at very low concentrations. All metals, whether essential or nonessential, become toxic at sufficiently high concentrations.

The toxic effects of metals vary, although they generally attack key organs in the body and can impair functioning of the nervous system. For example, lead causes learning and behavioral disorders as well as problems in motor coordination. Lead also attacks the kidneys, the reproductive system, and the hematopoietic system, which is responsible for making blood. Nickel and chromium have both been linked to respiratory cancer and are known to cause skin disease. Cadmium impairs functioning of the kidneys; mercury can damage the nervous system and the brain. Exposure to any of these metals can be fatal at sufficiently high levels.

The first written account of metal toxicity dates back to ancient Greece, where in 200 B.C. several effects of lead exposure were described (Nriagu, 1983). Other writings at this time describe miners who covered their bodies and especially their mouths while they worked, suggesting they knew of the toxicity associated with metal dust. The extent of lead usage by the Romans—for their water distribution system and for food storage—has been used to estimate that lead exposure among the Romans was much greater than that today and was a likely cause of excessive illness at the time. Many modern writers have even claimed that excessive exposure to lead contributed to the decline of the empire (Gilfillan, 1965; Nriagu, 1983). There are also reports of toxicity of lead and mercury in China dating to the first century A.D., attributed to early Chinese alchemists who advocated human consumption of these metals to prolong life.

Toxic effects of metals have persisted through the years and are still common today. In the case of lead, several notable sources are responsible for recent reported poisoning cases. Examples include ingestion of whiskey contaminated by lead-soldered joints in pipes of homemade stills; inhalation of lead vapor from backyard automotive battery smelting to recover lead; ingestion of contaminated food stored in leaded containers; and ingestion by children of chips of leaded paint in older housing (NAS, 1980). Mercury poisoning has also been documented in several instances. Hundreds of people in Minamata Bay, Japan, were poisoned by eating fish contaminated with mercury from emissions by a plastics factory. Treatment of wheat and other grains with mercury compounds resulted in several thousand poisoning cases in Iraq and Pakistan.

Other trace metals have been responsible for reported poisoning cases. Water polluted with cadmium from a mine was used to irrigate crops in Toyama Bay, Japan, resulting in numerous fatalities. Arsenic in drinking water poisoned hundreds of people in Taiwan. Antimony in enamel used on food storage vessels has been implicated in poisoning cases. Beryllium, nickel, and manganese dusts have caused disease in occupationally exposed individuals.

How do these exposures to trace metals occur? To understand the problem, we need to examine the pathways by which people are exposed to metals.

10.4 PATHWAYS OF HUMAN EXPOSURE TO TRACE METALS

10.4.1 Distribution of Trace Metals in the Environment

Virtually all metals are found naturally in the environment. For example, the atmosphere over land contains significant amounts of silicon, aluminum, and iron from soil dust, whereas air over the oceans contains high levels of sodium. Surface waters as well as groundwater contain metals from the soil and from biota. Animals and plants also contain metals as a result of their exposure to natural levels of metals in the air, water, and soil around them. All of these media contain nutrient metals as well as tiny amounts of toxic metals such as lead, mercury, and cadmium. Evolutionary processes over millions of years have ensured that our bodies are not harmed by the concentrations of toxic metals typically found in nature.

Human activities associated with metalworking and with other industrial processes have increased the amounts of trace metals in our environment. Cities where leaded gasoline is used can have lead concentrations of several micrograms per cubic meter of air, rather than the few nanograms per cubic meter found naturally. Airborne levels of cadmium, zinc, and other metals near smelters can be elevated hundreds or even thousands of times above natural levels. Water draining from mines may have greatly increased concentrations of metals compared with levels that occurred before the mines were opened, when the earth's crust was undisturbed. Similarly, food stored in improperly prepared metal containers may become contaminated with metal far above the levels found naturally. In all of these cases, metals were removed from their isolated positions within the earth's crust and mobilized to enable them to be used for various implements of human civilization. The process of extracting and mobilizing them has unintentionally released quantities of these metals and made them available for human exposure.

Contamination by trace metals is far from a local phenomenon. Centuries of mining, smelting, and otherwise redistributing metals have resulted in elevated environmental levels worldwide. The levels now found in the most remote areas of the world—even the Arctic and Antarctic—are not necessarily toxic, but they have been documented as much greater than we would expect before industrialization. Thus human activities have altered environmental cycling of metals on a global scale.

The discovery of trace metal pollution throughout the globe has been an incentive to study health effects of metals at relatively low levels once thought to be safe.

For example, recent work by several scientists concerning lead has shown that both the human influence on environmental concentrations and the effects of these concentrations are far greater than previously thought (Davidson, 1999). This, in turn, has prompted development of methods to calculate human exposure to lead and other toxic metals.

It is convenient to focus on four exposure routes, namely air, drinking water, food, and soil or dust. Each of these contains trace metals that provide pathways for human exposure, as shown in Figure 10.2. The last category is included because tiny amounts of soil or house dust are sometimes inadvertently ingested. The quantity that determines whether there are health effects is the amount of a metal that is eventually absorbed by the body. Here we focus on absorption by the bloodstream, which is the main route by which toxic species are spread throughout the body. We can calculate the mass of a trace metal absorbed by the body per unit time from any exposure route as follows:

$$\text{Mass of trace metal absorbed by the body/time} = A_i = C_i U_i f_i \qquad (10.1)$$

where

i = air, water, food, or soil/dust
C_i = concentration of the trace metal in medium i
U_i = uptake rate of air, water, food, or soil/dust
f_i = fraction of trace metal absorbed by the bloodstream

The units of C_i and U_i depend on the exposure route. We now consider each of the four exposure routes to enable us to use Equation (10.1).

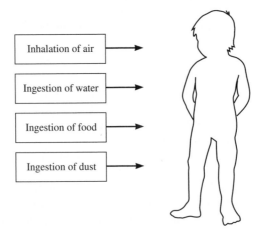

Figure 10.2   The four principal routes of human exposure to trace metals and other environmental pollutants.

10.4.2 Trace Metals in the Air

Trace metals in the air can be in the form of gases or particles. Volatile metals like mercury can be mostly in the vapor phase, although most airborne metals are predominantly associated with solid particles. Both chemical and physical characteristics can affect the uptake and subsequent toxicity of a metal.

One of the most important parameters of an airborne metal is its particle size distribution. We typically characterize airborne particles in terms of their *aerodynamic diameter*. This is a convenient way to account for variations in shape and density: Particles with an aerodynamic diameter of 1 μm will settle through the air and be carried by winds in a manner similar to that of a perfect sphere of diameter 1 μm and a density equal to that of water (1 gram/cm³), regardless of their actual shape and density. Particles with a high density will have an aerodynamic diameter somewhat greater than their physical diameter.

The size distribution depends on how the particles were formed. For example, the high temperatures of coal, oil, or gasoline combustion can vaporize some of the metals contained in the fuel. As the exhaust gases move out of the combustion zone, they cool and the metal vapors condense, producing many tiny particles. The overall size distribution from this source has a bell-shaped curve (a single peak) with most of the particle mass in the range 0.02–0.05 μm aerodynamic diameter. This is illustrated in Figure 10.3 as the *nuclei mode*.

Particles in the nuclei mode are not very stable due to high rates of diffusion for small particle sizes. If airborne concentrations are high, the particles will collide with each other over a period of minutes to form larger particles. They will grow rapidly at first but will eventually reach a stable size. The result is another bell-shaped curve with most of the particle mass in the range 0.2–0.5 μm in diameter, as shown in Figure 10.3. This is known as the *accumulation mode*.

Other metals in the fuel with higher vaporization points will not form gases but rather will be emitted from the combustion zone as particles with a wide range of

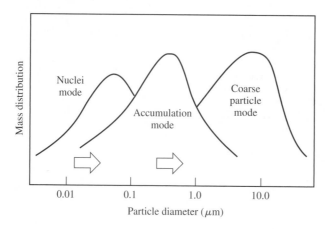

Figure 10.3   The distribution of particle mass in the atmosphere. This is known as the *trimodal* distribution because of the three distinct peaks resulting from a variety of processes.

sizes. Some of these particles will be as small as the condensed particles just described, whereas others may be as large as 10 μm in diameter or even bigger. These particles are in the *coarse particle mode*.

Airborne particles can also result from mechanical processes. For example, soil dust eroded by winds and salt particles generated by bubbles bursting on the ocean surface can be significant. Although a fraction of these particles can have diameters smaller than 1 μm, much of the particle mass is likely to be associated with larger particles, generally 5–10 μm in diameter. Particles bigger than 10 μm are also created, but under usual conditions most of them settle out of the air much more quickly than the smaller particles.

Note that the *x*-axis in Figure 10.3 has a logarithmic scale, so a very wide range of particle sizes is represented. In Section 10.6 we will revisit the bell-shaped curve in somewhat greater detail. Although Figure 10.3 has been drawn to represent all airborne particles, the mass distribution of those particles containing trace metals will be similar.

Example 10.1

Growth of spherical particles.   Spherical particles of 0.02 μm diameter are moving by diffusion and forming larger, more stable particles. How many particles of the former size are needed to create one particle 0.5 μm in diameter in the accumulation mode? How many additional 0.02 μm diameter particles are needed to enable one 0.5 μm diameter particle to grow to 0.7 μm diameter?

Solution:

The volume of a sphere of diameter *d* is given by $(\pi/6)d^3$. If the desired number of 0.02 μm diameter particles is given by *N*, we can write

$$N(\pi/6)(0.02 \; \mu m)^3 = (\pi/6)(0.5 \; \mu m)^3$$

Although the lead content of U.S. gasoline has been greatly reduced, much of the world still uses automotive fuels with high lead concentration, making vehicle emissions a major environmental source of this toxic metal.

Thus $N = 15,600$. Clearly, very large numbers of smaller particles are needed to form one particle in the accumulation mode. For the second question, we have

$$N(\pi/6)(0.02 \ \mu m)^3 = (\pi/6)[(0.7 \ \mu m)^3 - (0.5 \ \mu m)^3]$$

So $N = 52,000$. Note that the number of additional particles is $52,000 - 15,600 = 36,400$. As the particle diameter increases, it takes more $0.02 \ \mu$m diameter particles to enable even slight growth of the accumulation mode particle, due to the relationship between volume and diameter. This is the main reason why particles in the accumulation mode will not continue to grow indefinitely.

In Equation (10.1), the concentration C_{air} is usually expressed as the mass of the trace metal per unit volume of air, such as μg metal/m³ air. The uptake rate, U_{air}, is simply the amount of air inhaled per unit time, expressed as m³ air/day. The value of f_{air} is the fraction of the trace metal entering the body via inhalation that is eventually absorbed into the bloodstream. This fraction is the result of a two-step process: It is determined both by the rate of deposition in the lower lung and also by the subsequent absorption of the metal from the lower lung into the bloodstream. We sometimes assume that essentially all of a metal reaching the lower lung is absorbed into the bloodstream, in which case f_{air} is determined only by deposition in the lower lung. To understand this step, we need to consider the human respiratory system.

Figure 10.4 shows a diagram of the respiratory system, divided into three regions. The nasopharyngeal and tracheobronchial regions, in the upper respiratory system, are not likely to have particles accumulate in them. This is because the walls are lined with a mucous membrane that is constantly in motion as a natural defense mechanism to remove foreign matter that gets into the airways. The material removed in this manner is eventually coughed up or is swallowed and reaches the stomach, where it may be treated in the same way as water, food, or dust. In contrast, particles reaching the pulmonary region are likely to remain there. Figure 10.4 shows that the pulmonary region contains the alveolar sacs where oxygen transfer to the

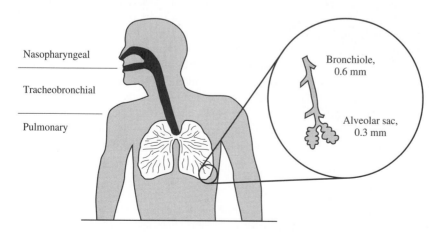

Figure 10.4 The three regions of the human respiratory system.

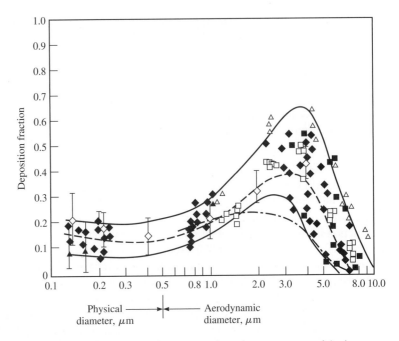

Figure 10.5   Efficiency of particle deposition in the pulmonary region of the human respiratory system as a function of particle diameter. (*Source:* USEPA, 1982)

bloodstream occurs. When particles reach these sacs, a fraction of the toxic material in the particles can also be absorbed into the bloodstream.

The fraction of incoming particles reaching the pulmonary region depends on many factors, among them particle size and breathing rate. Figure 10.5 shows the efficiency of deposition in this region as a function of particle size, presented as physical or aerodynamic diameter. The two solid curves in the figure show lower and upper limits of the experimental data from several human inhalation studies; note that the actual efficiency of deposition is expected to be between these limits. For example, the highest deposition appears at about 3 or 4 μm aerodynamic diameter, where the efficiency varies from roughly 0.3 (or 30 percent) to over 0.6 (or 60 percent). Thus if we know the airborne particle size distribution of a trace metal, we can calculate the value of f_{air} and hence the total amount absorbed by the body.

Figure 10.6 shows the average airborne particle size distribution of lead based on measurements in several cities. The figure shows the fraction of airborne lead concentration in each of four size ranges according to aerodynamic diameter of the particles. We now use this information to calculate the value of f_{air}.

Example 10.2

Deposition of particles in the lung.   Calculate the overall fraction of deposition in the pulmonary region of the human respiratory system for the average size distribution of lead in urban areas.

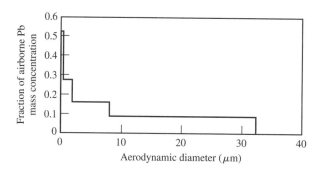

Figure 10.6 Particle size distribution for airborne lead mass, modified from a data summary of 32 distributions for airborne lead in urban areas. (*Source:* Milford and Davidson, 1985)

Solution:

The size distribution of Figure 10.6 shows the fraction of airborne lead mass concentration in the size ranges 0.125–0.5 μm, 0.5–2 μm, 2–8 μm, and 8–32 μm aerodynamic diameter. Note that the size distribution varies a lot in small particle sizes. To allow us to better identify variations in the distribution for small particles, we usually represent the x-axis on a logarithmic scale by converting the given particle aerodynamic diameters d_p to their log equivalents:

d_p (μm)	Log d_p
0.125	−0.903
0.5	−0.301
2	0.301
8	0.903
32	1.51

When we use log d_p as the x-axis rather than d_p, the result is the graph shown in Figure 10.7. Note that the lower end of the distribution now has much better representation. We can make the simplification that all of the lead mass in each size range is associated with the particle diameter at the midpoint of that size range. Because we are using a logarithmic scale, the midpoint of the first size range is found as

$$\log d_p = (log\ 0.125 + log\ 0.5)/2 = -0.602$$

so that

$$d_p = 10^{-0.602} = 0.25\ \mu m$$

We can calculate the midpoint of the other size ranges similarly.

The fraction deposited in each size range, according to the worst-case condition represented by the upper curve in Figure 10.5, is calculated as

Size range 1: 0.125–0.5 μm, midpoint = 0.25 μm, deposition efficiency = 0.2
Size range 2: 0.5–2 μm, midpoint = 1.0 μm, deposition efficiency = 0.3
Size range 3: 2–8 μm, midpoint = 4.0 μm, deposition efficiency = 0.65
Size range 4: 8–32 μm, midpoint = 16 μm, deposition efficiency = 0

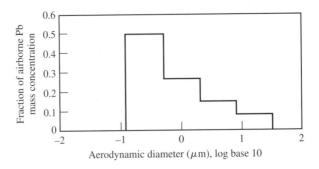

Figure 10.7   Distribution of Figure 10.6 plotted on a logarithmic x-axis.

Thus the overall fraction of deposition for lead can be computed as

$$(0.5 \times 0.2) + (0.27 \times 0.3) + (0.15 \times 0.65) + (0.08 \times 0) = 0.28$$

We stated earlier that the value of f_{air} is assumed to be entirely determined by deposition in the pulmonary region. Thus the calculation just performed leads us to conclude that f_{air} must be 0.28. This approximately agrees with a more rigorous calculation by the U.S. Environmental Protection Agency, which showed that f_{air} was equal to 0.3 (NAS, 1980). Only one significant figure is justified in any case due to the approximate nature of the calculation, so we adopt the value of 0.3 for lead absorbed from inhalation. We are now ready to use Equation (10.1) for the case of inhaled lead. Note that children are more vulnerable than adults to the effects of lead, so we apply Equation (10.1) to the case of an urban child in the next example.

Example 10.3

Deposition of particles in the lung.   Calculate the amount of lead absorbed by the body of a child each day for inhaled particles in an urban area where leaded gasoline is used. The airborne concentration of lead is $C_{air} = 1 \ \mu g$ Pb/m^3. The inhalation rate for a child is taken as $U_{air} = 10$ m^3 air/day.

Solution:

According to Equation (10.1), we can now write

$$\text{Mass of lead absorbed by the body/day from air} = A_{air}$$

$$= C_{air} \times U_{air} \times f_{air} = 1\mu g/m^3 \times 10 \ m^3/day \times 0.3$$

$$= 3 \ \mu g \ lead/day$$

10.4.3   Trace Metals in Water

Both surface waters and groundwater are commonly used for drinking water supplies. In both cases trace metals may be dissolved in the water, or they may be undissolved in the form of suspended particles. Some waters are naturally high in metals. For example, groundwater at many locations contains a high concentration of iron, a nutritive element.

Several categories of human activities can contribute metals to drinking water. Runoff from agricultural lands containing fertilizers or pesticides may add metals to drinking water supplies. Metals from industries may reach the water through effluent streams. Contaminated rain can contribute additional amounts of trace metals. Construction activities can disrupt the earth's crust, releasing quantities of trace metals.

Most people living in urban areas receive their drinking water through a distribution system from a water treatment plant. In some cases pipes made of lead are used in this distribution and can add significant amounts of lead to the drinking water.

The concentration of a trace metal in drinking water, C_{water}, is usually expressed as μg/liter of water, whereas the uptake rate, U_{water}, is expressed as liters of water consumed per day. However, human exposure occurs not only through drinking the water directly, but also through drinking beverages and eating foods made with water containing trace metals. For this reason, the uptake rate for drinking water may indicate only part of the total exposure to trace metals found in water.

The value of f_{water} reflects the absorption of a trace metal through the walls of the stomach and small intestines to the bloodstream. This is different from the factor f_{air} for absorption of an airborne metal from the lower lung to the bloodstream. For example, f_{water} for lead is equal to 0.5, as compared with 0.3 for f_{air}. Medical data are needed in each case to determine the appropriate values of f_{air} and f_{water} for a specific metal. Approximate values are known for these factors for some metals, such as lead, but our knowledge is limited for many other metals.

10.4.4   Trace Metals in Food

There are numerous sources of trace metals in food. Airborne particles containing metals can deposit onto foods grown outdoors, such as grains, fruits, and vegetables. Soldered cans containing food can leach metals into their contents, especially if the foods are acidic. Metal-containing dust can be inadvertently added to foods during processing if the work area is not sufficiently clean. Improperly applied glazes used on ceramic storage containers or plates may contain metals, especially lead, and in some cases this may contaminate food. Recently many of these problems have been minimized in the United States due to changes in canning and food handling practices.

The concentration, C_{food}, of trace metals in foods is usually expressed in μg/gram of food. The uptake rate, U_{food}, is given as grams of food/day; the value of f_{food} represents the absorption of trace metals from food through the walls of the stomach and small intestine, similar to the process for water. Thus we use the same values for f_{food} and f_{water} for a given trace metal.

10.4.5   Dust and Soil

We usually consider *dust* to be settled airborne particles, often indoors. This can include microscopic pieces of human skin and hair, tiny insects such as dust mites, bits of fabric from clothing and furniture, and many other sources created by human activity around the house. In contrast, *soil* is often considered to be outdoors and is generally assumed to be natural with contamination added by human activities. There is some ambiguity in defining these two terms, however: *soil dust* is sometimes defined as airborne soil particles. Both soil and dust can be of concern. High levels of metals may be found in soils near major stationary sources of pollution, such as smelters, or near heavily traveled roads. Soil surrounding houses with flaking leaded paint is likely to be contaminated. Dust inside a house may contain lead and other metals if interior leaded paint has been used and if the paint is flaking. Small amounts of dust may be ingested when we eat food that we have handled with our fingers. Furthermore, bits of soil and dust may be ingested by children who are playing outdoors or indoors.

We normally characterize concentrations of trace metals in soil and dust in terms of μg/gram of soil or dust. The uptake rate, U_{dust} is given as grams of soil or dust ingested per day. Although data are lacking for f_{dust} the value of this factor is sometimes taken as less than the corresponding value for food and water because trace metals associated with soil and dust are likely to be less soluble than metals in food and water.

10.4.6   Quantifying Total Human Exposure

To use Equation (10.1) for all environmental exposure routes, we need representative values of concentration, uptake rate, and absorption of a metal in each medium. Figure 10.8 shows typical values for these parameters for the case of lead. The following example illustrates the use of this information.

Example 10.4

Absorption of lead by the body.   Calculate the amount of lead absorbed by the body each day for water, food, and dust ingested by a child. Values of concentration, uptake rate, and fraction absorbed for the various exposure routes are shown in Figure 10.8.

Solution:

In Example 10.3, we found that the mass of lead absorbed by the body/day from air could be calculated as $A_{air} = C_{air} \times U_{air} \times f_{air} = 1\ \mu\text{g/m}^3 \times 10\ \text{m}^3/\text{day} \times 0.3 = 3\ \mu\text{g lead/day}$. Using Figure 10.8 to calculate the absorption from the other exposure routes, we obtain the following:

$$\text{Mass of lead absorbed by the body/day from water} = A_{water}$$

$$= C_{water} \times U_{water} \times f_{water} = 10\ \mu\text{g/liter} \times 1.4\ \text{liter/day} \times 0.5 = 7\mu\text{g/day}$$

$$\text{Mass of lead absorbed by the body/day from food} = A_{food}$$

$$= C_{food} \times U_{food} \times f_{food} = 0.1\ \mu\text{g/gram} \times 1{,}000\ \text{g/day} \times 0.5 = 50\ \mu\text{g/day}$$

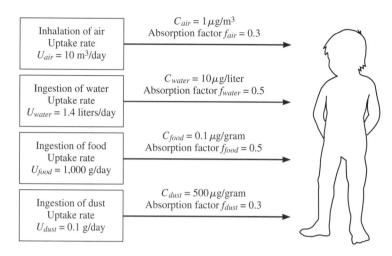

Figure 10.8 Uptake rates for air, water, food, and dust for a child, along with environmental concentrations and absorption factors for lead exposure. Values in the boxes do not depend on which pollutant is being considered, whereas the values along the arrows are specific for lead. (*Source:* Based on NAS, 1980)

Mass of lead absorbed by the body/day from dust $= A_{dust}$

$$= C_{dust} \times U_{dust} \times f_{dust} = 500 \ \mu g/gram \times 0.1 \ g/day \times 0.3 = 15 \ \mu g/day$$

Total mass of lead absorbed by the body/day from all four exposure routes (air, water, food, and dust) $= 3 + 7 + 50 + 15 = 75 \ \mu g/day$.

If the contribution of lead in air is such a small part of the total, why are we so concerned with airborne lead, such as that emitted from automobiles using leaded gasoline? The answer is that lead emitted to the air also affects lead in other environmental compartments. For example, outdoor reservoirs used for drinking water may receive deposited airborne lead. Some of the lead in food is also the result of deposited airborne particles, as previously mentioned. For the case of soil and dust, virtually all of the lead near roadways may be the result of deposited airborne particles.

10.5 TOTAL DOSE OF ABSORBED METALS

We have seen that trace metals present in our environment can be absorbed into our bodies. We have calculated rates at which the metals are absorbed in units of μg of a metal per day. But the effect of a metal depends not only on the rate at which the metal is being absorbed, but also on the total amount of the metal that has been absorbed up to the present. The total amount absorbed is known as the *dose*, denoted by D_{total}, and it is defined simply as

$$D_{total} = A_{total} \times t \qquad (10.2)$$

where A_{total} is the total absorption rate from all exposure routes (μg/day) and t is the time of exposure. In Example 10.4, the total daily dose would be 75 μg.

This simple calculation is sufficient as long as the absorption rate is constant over time. However, there are many reasons why we would not necessarily expect the absorption rate to be constant. For example, the concentration in the environment might change with time, or the uptake rate might vary. The more general expression for dose D_{total} is

$$D_{total} = \int_0^T A_{total}(t)\,dt \qquad (10.3)$$

where the absorption rate varies over time between 0 and T. Consider the following modification to the calculations in the previous section.

Example 10.5

Doses of lead due to inhalation.   Compute the total dose of lead due to inhalation by a child over a period of two years for two situations:

(a) The airborne concentration of lead has a constant value of $C_{air} = 1$ μg/m^3 over the two-year period.

(b) The airborne concentration of lead is $C_o = 1$ μg/m^3 at $t = 0$, but then decreases linearly to 0.8 μg/m^3 over the two-year period according to $C_{air}(t) = C_o(1 - 0.1t)$, where t is measured in years.

Solution:

(a) Constant airborne lead concentration:

$$D_{air} = A_{air} \times t = C_{air} \times U_{air} \times f_{air} \times t$$

$$= 1\mu g/m^3 \times 10\ m^3/day \times 0.3 \times 2\ \text{years} \times 365\ \text{days/year} = 2{,}190\ \mu g\ \text{lead}$$

(b) Time-varying airborne lead concentration:

$$D_{air} = \int_0^T A_{air}(t)\,dt = \int_0^2 C_{air}(t) \times U_{air} \times f_{air}\,dt = C_o \times U_{air} \times f_{air} \int_0^2 (1 - 0.1t)\,dt$$

$$= 1\ \mu g/m^3 \times 10\ m^3/day \times 365\ \text{days/year} \times 0.3 \times \left(t - 0.1\frac{t^2}{2} \right)\Bigg|_0^2$$

$$= 1{,}970\ \mu g\ \text{lead}$$

The same procedure used in Example 10.5 can be employed to find the total dose from multiple exposure routes. The following example illustrates such a calculation.

Example 10.6

Exposure to lead by several pathways. The child in Example 10.5 is also exposed to lead in water, food, and dust over the two-year period. The airborne lead concentration is characterized by $C_{air}(t) = C_o(1-0.1t)$ as before. The lead concentrations in water and food are unchanged over this period and are identical to those in Example 10.4. The concentration of lead in dust is assumed to be proportional to the concentration in air because lead in the dust is a result of settling airborne particles. Determine the exposure of the child to lead from all four exposure routes over the two-year period.

Solution:

$$D_{air} = 1{,}970 \ \mu g \text{ lead as before}$$

$$D_{water} = A_{water} \times t = C_{water} \times U_{water} \times f_{water} \times t$$
$$= 10 \ \mu g/\text{liter} \times 1.4 \ \text{liter/day} \times 0.5 \times 2 \ \text{years} \times 365 \ \text{days/year}$$
$$= 5{,}110 \ \mu g \text{ lead}$$

$$D_{food} = A_{food} \times t = C_{food} \times U_{food} \times f_{food} \times t$$
$$= 0.1 \ \mu g/\text{gram} \times 1{,}000 \ g/\text{day} \times 0.5 \times 2 \ \text{years} \times 365 \ \text{days/year}$$
$$= 36{,}500 \ \mu g \text{ lead}$$

$$D_{dust} = \int_0^2 A_{dust}(t)dt = \int_0^2 C_{dust}(t) \times U_{dust} \times f_{dust} dt$$
$$= \int_0^2 C_o(1-0.1t) \times 0.1 \ \text{gram/day} \times 365 \ \text{days/year} \times 0.3 \ dt$$
$$= 9{,}850 \ \mu g \text{ lead}$$

where $C_o = 500 \ \mu g/\text{gram}$ dust as in Example 10.4. Thus

$$D_{total} = D_{air} + D_{water} + D_{food} + D_{dust}$$
$$= 1{,}970 + 5{,}110 + 36{,}500 + 9{,}850 = 53{,}400 \ \mu g \text{ lead}$$
$$= 0.0534 \ \text{grams lead}$$

10.6   DOSES IN A POPULATION

Each and every person is exposed to a different amount of trace metals: We all breathe air at different locations and different respiration rates, we drink water from various sources and in varying amounts, we eat different foods, and we are exposed to different quantities of dust. Therefore, we would expect a wide variation in the dose of trace metals in our population. Indeed, this turns out to be the case. Some people have surprisingly low doses, whereas others have doses sufficiently high to warrant medical attention. How can we determine the distribution of doses in our population?

The method outlined in Section 10.5—calculating the dose from each exposure pathway—may be satisfactory for estimating the dose of an individual where it is feasible to obtain data on concentrations and uptake rates. However, it is far too tedious to apply this method to an entire population. If we wish to investigate how doses are distributed throughout a population, it is more convenient to measure the concentration of a metal in human tissue, such as the bloodstream. This method works because the human body itself provides an integrator of trace metal dose over time. In the case of lead, for example, it is known that the concentration in a person's blood roughly indicates the dose over the previous few weeks. We measure *blood lead concentration*—often called *PbB*—in units of μg lead per deciliter (dl) of blood, where 1 dl is equal to 0.1 liters, or 100 ml. To illustrate the method, we now consider an example of data for PbB distributed in a population.

Table 10.1 shows the results of a hypothetical study in which 500 people are sampled for PbB. The table lists the number of people who have PbB values in the range 0–5 μg/dl, the number who have values in the range 5–10 μg/dl, and so forth up to the range 40–45 μg/dl. Note that the range 0–5 μg/dl actually means between a value just above 0 and a value just less than 5 μg/dl, assuming the measurements are sufficiently precise. Similarly, the range 5–10 μg/dl means a value just greater than 5 but just less than 10 μg/dl. This reasoning applies to all ranges. Note that PbB can never attain a value of *exactly* 5 μg/dl, because in theory one could continue to improve the precision of the measurement, thus adding significant figures indefinitely.

How can we use the data? First we determine the *fraction* of the total population sampled rather than the actual number of people in each PbB interval. Then we divide the resulting fraction by the range in PbB values, which is 5 μg/dl for each interval. The result for the first range (0–5 μg/dl) is equal to

$$\frac{5 \text{ people}}{(500 \text{ people}) \times 5 \ \mu\text{g/dl}} = .002(\mu\text{g/dl})^{-1}$$

Table 10.1   Distribution of concentrations of lead in the blood for a sample of 500 people. Although these data are hypothetical for illustration only, the characteristics of the distribution have been chosen to approximate the blood lead distribution prevalent in urban children when leaded gasoline use was widespread.

PbB (μg/dl)	Number of People with PbB in the Indicated Range	Value of Probability Density Function (pdf)
0–5	5	0.002
5–10	15	0.006
10–15	25	0.010
15–20	105	0.042
20–25	200	0.080
25–30	105	0.042
30–35	25	0.010
35–40	15	0.006
40–45	5	0.002

Values for the other ranges are computed in the same way and are shown in Table 10.1. These values define the *probability density function,* or *pdf,* for this set of data. A graph of the pdf is shown in Figure 10.9. We use the pdf to perform an important calculation: We can integrate under the curve between any two values in order to obtain the fraction of the population having PbB between those two values. This can help us determine how many people are safe from lead poisoning, how many have PbB values high enough to require monitoring over time, and how many require immediate treatment due to high lead exposure. As an illustration, individuals with PbB greater than 30 μg/dl are considered to be at risk of health effects from lead exposure; those with PbB in the range 20 to 30 μg/dl may need to monitor their PbB levels to ensure they do not exceed 30. The fraction of the population with PbB in the range 20 to 30 μg/dl may be found by integrating as follows:

$$\int_{20}^{30} f(x)dx = \text{Area under the curve in the range 20–30 } \mu\text{g/dl}$$

$$= (0.8 \times 5 \ \mu\text{g/dl}) + (.042 \times 5\mu\text{g/dl}) = 0.61$$

Here we are simply measuring the area of two rectangles in Figure 10.9. The area under all of the rectangles in this figure is equal to 1.0, consistent with the fact that 100 percent of the population must have PbB between zero and the maximum value. Such a calculation is easy when we are working with small numbers of rectangles. However, let us now generalize the curve of Figure 10.9 to the case where we have many tiny ranges of PbB, so that the curve becomes smooth, as shown in Figure 10.10.

What function is represented by the curve of Figure 10.10? The function is called the *normal distribution,* and it has great significance in the field of statistics. The analytical expression for the normal distribution is

$$f(x) = \frac{1}{\sigma\sqrt{2\pi}}\exp\left[-\frac{(x-M)^2}{2\sigma^2}\right] \tag{10.4}$$

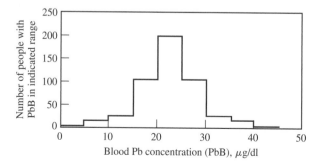

Figure 10.9 Graph of the probability density function whose values are shown in Table 10.1.

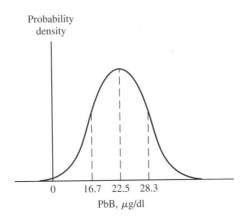

Figure 10.10   Probability density function for the population with data in Table 10.1 and Figure 10.9, plotted as a smooth curve. The result is a normal distribution with $M = 22.5$ μg/dl and $\sigma = 5.8$ μg/dl.

where x in this case is the blood lead concentration, PbB. In Equation (10.4) M is the *mean value* of x and σ is the *standard deviation* of x. The mean value in a normal distribution is the value of x at the peak of the curve. The standard deviation is a measure of the curve's spread. To be precise, a point on the curve located one standard deviation from the mean is at the *point of inflection,* which is where the slope is a maximum, as shown in Figure 10.10. There are two such points, one above and one below the mean.

Using the data of Table 10.1 and performing a detailed statistical analysis, we see that the value of M is 22.5 μg/dl and the value of σ is about 5.8 μg/dl. In principle, we could substitute these values into Equation (10.4) and perform the integration between any two values of PbB to obtain the fraction of the population with PbB in that range. However, the function $f(x)$ cannot be integrated directly; thus we would have to resort to numerical integration. Fortunately, there is a straightforward way to solve this type of problem without the need for numerical integration on a computer.

The method is based on the fact that the shape of the normal distribution for $M = 0$ and $\sigma = 1$ is always the same. So all we need to do is take any variable and make a simple change so that it has these values of M and σ. Consider the distribution of Figure 10.10 for PbB. The curve has $M = 22.5$ and $\sigma = 5.8$. We define a new variable z as follows:

$$z = \frac{x - M}{\sigma} = \frac{x - 22.5}{5.8} \tag{10.5}$$

where x represents the value of PbB. If we substitute various values of x into Equation (10.5), we can generate a new curve similar to Figure 10.10 but with values of z instead of PbB for the x-axis. The new curve looks like the old one but is displaced 22.5 units to the left, so that the mean value is centered on $z = 0$. Furthermore, the curve has its maximum slope at $z = -1$ and $z = +1$. The new curve, shown in Fig-

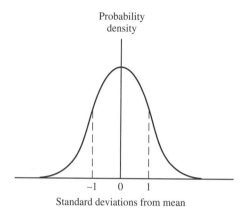

Probability
density

−1   0   1

Standard deviations from mean

Figure 10.11   Normal distribution with $M = 0$ and $\sigma = 1$.

ure 10.11, has $M = 0$ and $\sigma = 1$. We can use a standard table, shown in Table 10.2, for this curve, which essentially performs the numerical integration for us.

Table 10.2 lists values for the function $\Phi\,(z)$:

$$\Phi(z) = \int_{-\infty}^{z} \frac{1}{\sqrt{2\pi}}\exp\left(-\frac{u^2}{2}\right)du \qquad (10.6)$$

which represents the integral of Equation (10.4) with $M = 0$ and $\sigma = 1$. Thus $\Phi\,(z)$ is the fraction of the population having PbB less than z. Note that u is simply a dummy variable of integration to distinguish it from z, which takes on numerical values.

The use of Table 10.2 is illustrated in the next two examples.

Example 10.7

Fraction of population with blood lead concentrations below a specified value.   Use Table 10.2 to determine the fraction of people in the distribution discussed earlier who have blood lead concentrations less than 22.5 μg/dl.

Solution:

According to Equation (10.4),

$$\text{Fraction} = \int_{0}^{22.5} \frac{1}{\sigma\sqrt{2\pi}}\exp\left[-\frac{(x - M)^2}{2\sigma^2}\right]dx$$

where $M = 22.5$ μg/dl and $\sigma = 5.8$ μg/dl.

Figure 10.10 shows that the normal distribution actually extends from $-\infty$ to $+\infty$. So we approximate the above integral as

$$\text{Fraction} = P(x < 22.5) = \int_{-\infty}^{22.5} \frac{1}{\sigma\sqrt{2\pi}}\exp\left[-\frac{(x - M)^2}{2\sigma^2}\right]dx$$

Table 10.2   Values of Φ (z) for a normal distribution.

z	0	1	2	3	4	5	6	7	8	9
0.0	0.5000	0.5040	0.5080	0.5120	0.5160	0.5199	0.5239	0.5279	0.5319	0.5359
0.1	0.5398	0.5438	0.5478	0.5517	0.5557	0.5596	0.5636	0.5675	0.5714	0.5753
0.2	0.5793	0.5832	0.5871	0.5910	0.5948	0.5987	0.6026	0.6064	0.6103	0.6141
0.3	0.6179	0.6217	0.6255	0.6293	0.6331	0.6368	0.6406	0.6443	0.6480	0.6517
0.4	0.6554	0.6591	0.6628	0.6664	0.6700	0.6736	0.6772	0.6808	0.6844	0.6879
0.5	0.6915	0.6950	0.6985	0.7019	0.7054	0.7088	0.7123	0.7157	0.7190	0.7224
0.6	0.7257	0.7291	0.7324	0.7357	0.7389	0.7422	0.7454	0.7486	0.7517	0.7549
0.7	0.7580	0.7611	0.7642	0.7673	0.7703	0.7734	0.7764	0.7794	0.7823	0.7852
0.8	0.7881	0.7910	0.7939	0.7967	0.7995	0.8023	0.8051	0.8078	0.8106	0.8133
0.9	0.8159	0.8186	0.8212	0.8238	0.8264	0.8289	0.8315	0.8340	0.8365	0.8389
1.0	0.8413	0.8438	0.8461	0.8485	0.8508	0.8531	0.8554	0.8577	0.8599	0.8621
1.1	0.8643	0.8665	0.8686	0.8708	0.8729	0.8749	0.8770	0.8790	0.8810	0.8830
1.2	0.8849	0.8869	0.8888	0.8907	0.8925	0.8944	0.8962	0.8980	0.8997	0.9015
1.3	0.9032	0.9049	0.9066	0.9082	0.9099	0.9115	0.9131	0.9147	0.9162	0.9177
1.4	0.9192	0.9207	0.9222	0.9236	0.9251	0.9265	0.9278	0.9292	0.9306	0.9319
1.5	0.9332	0.9345	0.9357	0.9370	0.9382	0.9394	0.9406	0.9418	0.9430	0.9441
1.6	0.9452	0.9463	0.9474	0.9484	0.9495	0.9505	0.9515	0.9525	0.9535	0.9545
1.7	0.9554	0.9564	0.9573	0.9582	0.9591	0.9599	0.9608	0.9616	0.9625	0.9633
1.8	0.9641	0.9648	0.9656	0.9664	0.9671	0.9678	0.9686	0.9693	0.9700	0.9706
1.9	0.9713	0.9719	0.9726	0.9732	0.9738	0.9744	0.9750	0.9756	0.9762	0.9767
2.0	0.9772	0.9778	0.9783	0.9788	0.9793	0.9798	0.9803	0.9808	0.9812	0.9817
2.1	0.9821	0.9826	0.9830	0.9834	0.9838	0.9842	0.9846	0.9850	0.9854	0.9857
2.2	0.9861	0.9864	0.9868	0.9871	0.9874	0.9878	0.9881	0.9884	0.9887	0.9890
2.3	0.9893	0.9896	0.9898	0.9901	0.9904	0.9906	0.9909	0.9911	0.9913	0.9916
2.4	0.9918	0.9920	0.9922	0.9925	0.9927	0.9929	0.9931	0.9932	0.9934	0.9936
2.5	0.9938	0.9940	0.9941	0.9943	0.9945	0.9946	0.9948	0.9949	0.9951	0.9952
2.6	0.9953	0.9955	0.9956	0.9957	0.9959	0.9960	0.9961	0.9962	0.9963	0.9964
2.7	0.9965	0.9966	0.9967	0.9968	0.9969	0.9970	0.9971	0.9972	0.9973	0.9974
2.8	0.9974	0.9975	0.9976	0.9977	0.9977	0.9978	0.9979	0.9979	0.9980	0.9981
2.9	0.9981	0.9982	0.9982	0.9983	0.9984	0.9984	0.9985	0.9985	0.9986	0.9986
3.0	0.9987	0.9990	0.9993	0.9995	0.9997	0.9998	0.9998	0.9999	0.9999	1.0000

The designation $P(x < 22.5)$ stands for the probability of x being less than 22.5, and this extends down to $-\infty$ instead of to zero. But there is no appreciable loss of accuracy because the area under the curve for PbB less than zero in Figure 10.10 is very small.

Equation (10.5) tells us that we can transform the variable x into the variable z, such that $x = 22.5$ corresponds to $z = 0$. This is also apparent if we compare Figure 10.10 and Figure 10.11. Similarly, the value of x located one standard deviation above the mean in Figure 10.10

8

is 22.5 + 5.8 = 28.3, and we note that $x = 28.3$ in Figure 10.11 corresponds to $z = 1.0$ in Figure 10.11. Thus $P(x < 22.5) = P(z < 0)$, and the equation for the desired fraction is

$$\text{Fraction} = \int_{-\infty}^{22.5} \frac{1}{\sigma\sqrt{2\pi}} \exp\left[-\frac{(x-M)^2}{2\sigma^2}\right] dx = \int_{-\infty}^{0} \frac{1}{\sqrt{2\pi}} \exp\left[-\frac{u^2}{2}\right] du$$

But this is just Equation (10.6) with $z = 0$. So we read the desired fraction from Table 10.2, noting that $P(z < 0) = \Phi(z = 0)$. The value of $\Phi(z = 0)$ is 0.5000. This tells us that half of the population will have PbB less than 22.5 μg/dl. The result is expected because the area under the normal distribution curve to the left of $x = M$ (that is, $x < M$) corresponds to half the total area, or 0.5.

Another useful calculation involves finding the number of people with blood lead levels between two specified values. This is illustrated in the next example.

Example 10.8

Fraction of population with blood lead concentrations between two values. Use Table 10.2 to determine the fraction of people in the distribution discussed earlier who have blood lead concentrations between 20 and 30 μg/dl.

Solution:

Equation (10.5) tells us that $x = 20$ μg/dl corresponds to

$$z = \frac{20 - 22.5}{5.8} = -0.43$$

whereas $x = 30$ μg/dl corresponds to

$$z = \frac{30 - 22.5}{5.8} = 1.29$$

Thus we need to integrate the standard normal distribution between -0.43 and $+1.29$, shown in Figure 10.12. To use Table 10.2, we note the following equality:

$$\text{Fraction} = \int_{-0.43}^{1.29} \frac{1}{\sqrt{2\pi}} \exp\left[-\frac{u^2}{2}\right] du$$

$$= \int_{-\infty}^{1.29} \frac{1}{\sqrt{2\pi}} \exp\left[-\frac{u^2}{2}\right] du - \int_{-\infty}^{-0.43} \frac{1}{\sqrt{2\pi}} \exp\left[-\frac{u^2}{2}\right] du$$

$$= P(z < 1.29) - P(z < -0.43) = \Phi(1.29) - \Phi(-0.43)$$

The value of $\Phi(1.29)$ is given in Table 10.2 as 0.9015. The values of Φ for negative arguments like $\Phi(-0.43)$ are not given. However, we note that the area under the curve between $z = -\infty$ and $z = -0.43$, which is denoted by $\Phi(-0.43)$, is the same as the area under the curve between $z = 0.43$ and $z = \infty$ due to symmetry. We further note that $\Phi(\infty) = P(z < \infty)$ represents the area under the entire curve, which is 1.0. Thus

$$\Phi(-0.43) = \Phi(\infty) - \Phi(0.43) = 1.0 - 0.6664 = 0.3336$$

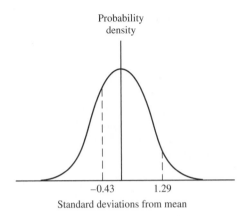

Figure 10.12   Normal distribution with $M = 0$ and $\sigma = 1$ used in Example 10.8.

The desired fraction is then

$$0.9015 - 0.3336 = 0.5679 \cong 0.57$$

The difference between this value and the value of 0.61 calculated previously for Figure 10.9 is due to the difference between a smooth curve such as Figure 10.10 and a curve of discrete data such as Figure 10.9.

It is clear that the normal distribution has its greatest utility when we integrate under the curve. What type of plot do we obtain if we graph the integral of the normal curve—for example, a plot of $\Phi(z)$ versus z? This is another important graph in the field of statistics, known as the *cumulative density function* or *cdf*. The shape of the cdf is interesting: At the far left side of the bell-shaped curve, the area under the curve is accumulating slowly because the curve is near the x-axis. By the time we reach the midpoint, however, we are adding lots of area, so that the value of the cdf increases rapidly. Continuing to the right, we are accumulating less area, and thus the cdf is no longer increasing as fast. Eventually the cdf levels off at a value approaching 1.0. We can plot this type of graph by integrating under the curve of Figure 10.10, as shown in Figure 10.13. We could also integrate under the curve of Figure 10.11 to obtain a similar graph: Values along the x-axis would become values of z ranging from $-\infty$ to $+\infty$, with $z = 0$ corresponding to the steepest point at the center of the curve. In the next section we consider dose–response curves using the concept of the cdf.

10.7   RESPONSE TO A DOSE

We have seen that the dose of lead can be represented by the concentration of lead in a person's blood, and that these concentrations are widely distributed in the pop-

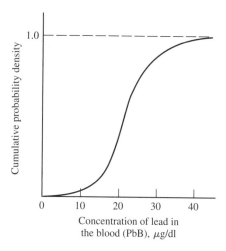

Figure 10.13 Cumulative density function for PbB corresponding to Figure 10.10.

ulation. The same is true of other trace metals. How can we evaluate the harmful effects of a person's dose of a trace metal?

This must be determined empirically; that is, we must observe the effects for various doses and draw conclusions about these effects from the observations. One way to evaluate effects is to experiment with animals. Another way is through epidemiology research, where we identify effects in human populations that may be exposed to high levels of a pollutant. For example, epidemiologists may study people living near a major pollution source. Both animal studies and epidemiological research must have control populations that are not exposed to the pollutants of concern, so that the effects of the pollutants can be isolated from other factors.

Results of these types of studies have demonstrated an important finding: Even if the dose of metal—or other toxic pollutant—is identical in many members of a population (either animals or people), the effects may be quite variable. For example, assume animals in a laboratory experiment are all given an identical low dose of a toxic chemical. Most animals will not be affected by the chemical, although a small fraction may show a response such as illness, death, or some other effect. When the dose is increased, a larger fraction of animals will show a response. As the dose becomes high, most of the animals will have a response, but nevertheless a few may still be unaffected by the chemical. Finally, when the dose is increased sufficiently, all of the animals will show a response. This phenomenon may be illustrated with a classic *dose–response curve,* which has the same shape as Figure 10.13. Values of the dose are represented on the *x*-axis, whereas values of the response are shown on the *y*-axis. Indeed, both Figure 10.13 and the corresponding dose–response curve are integrals of a normal distribution. The normal distribution corresponding to Figure 10.13 was illustrated in Figure 10.10, namely the fraction of population having PbB in each range of values. The probability distribution corresponding to the dose–response curve is simply the fraction of population having a response to each range of dose. In the case of PbB, one response of interest is learning disability: Lead is known to affect learning in children.

Dose–response relationships are important in setting regulations for toxic pollutants such as trace metals. They are also important for estimating various types of health risks, as discussed later in Chapter 14. In many cases, the shape of a dose–response curve differs from the normal distributions assumed here. Based on available dose–response relationships we often try to establish a maximum safe level below which there are no effects, known as the *threshold level*. One way to illustrate this is to examine the case of a National Ambient Air Quality Standard (NAAQS).

As noted in Chapter 2, the NAAQS for lead was established by the EPA in 1978. This standard is based on measurements of learning ability in children along with measurements of their PbB values. Such studies suggest that learning ability is impaired even at low PbB levels, although the impairment increases rapidly as PbB rises above 30 μg/dl. For this reason, the EPA selected 30 μg/dl as the maximum permissible level of PbB for the bulk of the population of children (99.5 percent)—essentially a threshold level.

To illustrate the calculation performed by the EPA in setting this NAAQS, we assume that the hypothetical distribution of Figure 10.10 applies to PbB in children in the years before 1978. According to the previous paragraph, only 0.5 percent of the population of children are permitted to have PbB above 30 μg/dl. The results of Example 10.8 show that the fraction of children with PbB in this range before 1978 was actually much greater:

$$P(x < 30) = P(z < 1.29) = \Phi(1.29) = 0.9015$$

Thus

$$P(x > 30) = 1 - \Phi(1.29) = 0.0985$$

This means that about 10 percent of the childhood population had PbB levels above 30 μg/dl. EPA reasoned that PbB would be decreased if ambient airborne lead were decreased. In order to achieve $P(x > 30) = 0.005$, the mean value M would have to decrease from its existing value of 22.5 to a lower value:

$P(x < 30) = 0.9015$ in pre-1978 data, but it is desired that $P(x < 30) = 0.995$

According to Table 10.2, $\Phi(z) = 0.995$ implies $z = 2.575$. Therefore

$$z = \frac{x - M}{\sigma}$$

so that

$$M = x - \sigma z = 30 \,\mu g/dl - (5.8 \,\mu g/dl \times 2.575) = 15 \,\mu g/dl$$

Thus the distribution of blood lead concentrations in children must be moved from $M = 22.5$ μg/dl to $M = 15$ μg/dl. The result is shown in Figure 10.14. The curve has been moved to the left just the correct distance so that $P(x > 30 \,\mu g/dl) = 0.005$ as desired.

Once the desired mean value of PbB was established, the EPA noted through separate calculations that the total contribution to PbB from ingestion of water,

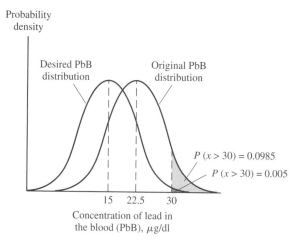

Probability
density

Desired PbB
distribution

Original PbB
distribution

$P(x > 30) = 0.0985$

$P(x > 30) = 0.005$

15   22.5   30

Concentration of lead in
the blood (PbB), μg/dl

Figure 10.14   Shifting the distribution of blood lead concentrations downward as a basis for
the National Ambient Air Quality Standard for lead.

food, and dust was 12 μg/dl. Therefore, out of the total desired PbB value of 15
μg/dl, only 3 μg/dl should be allowed from inhalation of air. Finally, the EPA noted
from other studies that when the airborne lead concentration is increased by 1
μg/m^3, the value of PbB increases by 2 μg/dl. Thus the final value of the NAAQS
is 1.5 μg/m^3.

We saw earlier that a child living in an urban area where leaded gasoline is used
would have a total absorption of 75 μg lead/day. This is essentially responsible for
the mean value of PbB equal to 22.5 μg/dl in Figure 10.10. The EPA determined
that the airborne lead concentration must fall substantially so that the mean PbB
would fall to 15 μg/dl. A substantial fraction of the final value, 3 μg/dl out of 15
μg/dl, was attributed to inhalation of airborne lead. Yet airborne lead, even in a city
where leaded gasoline is used, is responsible for only 3 μg lead/day out of 75 μg
lead/day. The difference in relative importance of airborne lead occurs partly
because PbB is affected by ingestion of deposited airborne lead in addition to
inhalation. There are also different assumptions inherent in the estimate of 75 μg
lead/day absorption as compared with the estimate of the fraction of 15 μg/dl PbB
due to airborne lead. Indeed, surveys since the 1970s have shown that PbB has been
reduced much further than anticipated as a result of switching to unleaded gasoline.

It is clear that a critical step in establishing the lead standard was the calculation
of the desired mean value of 15 μg/dl for PbB. This demonstrates the importance of
the calculations described earlier. In actuality, the distribution of PbB values is log-
normal rather than normal, meaning that the curve has a symmetric bell shape only
when the PbB values on the x-axis are plotted on a logarithmic scale. We used a hypothetical normal distribution here for simplicity to illustrate the calculation. Nevertheless, the approximation of a normal distribution for PbB works well to illustrate the
point, and indeed the method used by the EPA is very similar to that just outlined.

10.8   CONCLUSION

In this chapter we have seen that metalworking has been a vital part of advances in technology to improve our standard of living. We have also seen that trace quantities of metal released to the environment can have adverse effects, especially on human health. To reduce the risk associated with inhalation and ingestion of toxic metals, regulations have been established. The National Ambient Air Quality Standards (NAAQS), established under the 1970 Clean Air Act, require all regions of the United States to demonstrate that airborne lead concentrations will not exceed the standards. This involves consideration of allowable pollutant emissions, background concentrations (resulting from sources upwind), and meteorology. Industries also must report their pollutant releases to air, water, and land through the Toxic Release Inventory (see Chapter 2 for details). New regulations adopted in 1990 further require industries that emit hazardous air pollutants to demonstrate that they are using maximum achievable control technology (MACT) to reduce emissions. This new approach avoids the controversies associated with the need to identify a single threshold value of air quality; it also encourages industries to use the best control technologies available. Furthermore, many industries are taking initiatives to reduce pollutant releases through product and process modification rather than using add-on control technologies. These new methods are more likely to achieve the largest gains in environmental quality.

10.9   REFERENCES

Aitchison, L., 1960. *A History of Metals,* vols. 1 and 2. Interscience Publishers, New York.

Davidson, C. I., 1999. *Clean Hands: Clair Patterson's Crusade against Environmental Lead Contamination.* Nova Science Publishers, Commack, New York.

Gilfillan, S. C., 1965. *Journal of Occupational Medicine,* vol. 7, pp. 53–60.

Milford, J. B., and C. I. Davidson, 1985. "The Sizes of Particulate Trace Elements in the Atmosphere—A Review." *Journal of the Air Pollution Control Association,* vol. 35, pp. 1,249–60.

NAS, 1980. *Lead in the Human Environment.* National Academy of Sciences, National Academy Press, Washington, DC.

Nriagu, J. O., 1983. *Lead and Lead Poisoning in Antiquity.* John Wiley and Sons, New York.

Schroeder, H. A., 1973. *The Trace Elements and Man.* Devin-Adair Company, Old Greenwich, Connecticut.

Shaw, G. E., 1991. *Physical Properties and Physical Chemistry of Arctic Aerosol.* Elsevier, London, p. 127.

Tylecote, R. F., 1992. *A History of Metallurgy.* 2nd ed. The Institute of Materials, Brookfield, Vermont.

USEPA, 1982. *Air Quality Criteria for Particulate Matter and Sulfur Oxides.* EPA 600/8-82-029a, U.S. Environmental Protection Agency, Washington, DC.

Wedeen, R. P., 1984. *Poison in the Pot: The Legacy of Lead.* Southern Illinois University Press, Carbondale, Illinois.

Wertime, T. A., 1964. "Man's First Encounters with Metallurgy." *Science,* vol. 146, pp. 1,257–67.

10.10 PROBLEMS

10.1 Many toxic metals are discussed in this chapter: mercury, cadmium, nickel, chromium, and lead, to name a few. However, we have focused our quantitative analysis on lead because it is by far the most widely studied toxic metal and one of the most widely studied environmental pollutants.

  (a) Discuss several reasons why lead has received so much attention in scientific studies.

  (b) Consider the steps leading to establishment of the National Ambient Air Quality Standard for lead. List some of the problems with each step. Which of these problems are the most serious?

10.1 In this chapter we learned that a sizable number of small particles are needed to enable a big particle to grow only slightly bigger. Here we explore the same idea by considering coagulation of small particles. Assume that spherical particles of diameter 0.05 μm are coagulating in the atmosphere to form 0.1 μm diameter particles.

  (a) By what fraction does the number of particles decrease if all of the 0.05 μm particles form 0.1 μm particles?

  (b) If the atmosphere contains equal numbers of 0.1 μm and 0.3 μm diameter particles, determine how much more volume is associated with the 0.3 μm particles, expressed as a percentage.

  (c) Assume spherical particles of both sizes are composed entirely of the same toxic material. What does the calculation in part (b) imply about the effects of human uptake of equal numbers of 0.1 and 0.3 μm diameter particles?

10.3 Inhalation of airborne lead can be an important pathway contributing to the total lead burden in the body. A size distribution measurement of airborne lead particles in a downtown area shows that there are 1.7 μg/m^3 in the size range 0.3–1.2 μm, 0.2 μg/m^3 in the size range 1.2–4.8 μm, and 0.1 μg/m^3 in the size range 4.8–19.2 μm.

  (a) If these lead particles are inhaled, what fraction of the airborne lead is deposited in the pulmonary region of the lung?

  (b) What is the total amount of lead absorbed each day by a child who breathes this lead at an inhalation rate of 10 m^3/day?

10.4 An adult inhales 20 m^3 air/day, ingests two liters of water/day, and consumes 2,000 g of food/day. Ingestion of dust is considered negligible. How much greater is the absorption of lead per day for an adult compared with that for a child? Use the absorption factors and concentrations of lead shown in Figure 10.8. Why are children more vulnerable to the effects of lead than adults? To answer this question, consider the specific health effects of lead presented in Section 10.3.

10.5 When there are several pathways of human exposure to lead, we must focus our attention on those pathways with the greatest contribution to total exposure. Assume we can ignore a certain pathway only if its contribution is less than 2 percent of the total lead absorbed by a child. Consider the decrease in airborne lead according to Example 10.5: $C_{air}(t) = C_o (1 - 0.1t)$. Assume this equation applies from $t = 0$ to $t = 10$ years, at which time $C_{air}(t)$ has reached zero concentration. Also assume the concentration in dust is proportional to the airborne concentration as in Example 10.6.

  (a) For what value of t can we begin to ignore the inhalation pathway?

(b) For what value of t can we begin to ignore ingestion of dust?

(c) What will be the total dose of lead over the full 10-year period if inhalation and dust ingestion are ignored once their contribution falls to less than 2 percent of the total amount absorbed? Compare this value with the true total dose that accounts for inhalation and ingestion over the 10-year period. How much difference is there between these values (in percentage)?

10.6 Of the 0.1 μg/g of lead in food, assume 80 percent of this amount is a constant, whereas 20 percent results from deposited airborne lead. The amount of deposited airborne lead in food varies in proportion to the airborne lead concentration, which varies over time as given in Problem 10.5. Write an expression for the amount of lead absorbed by a child as a function of time, and compute the total dose from food over the 10-year period. How does this compare with the total dose of lead from food, assuming constant exposure over this period (Example 10.6)?

10.7 Section 10.6 points out that members of a population usually have exposures that vary widely, for example due to variations in lead concentration in food. Consider a population of children where the average value of C_{food} is known for each child. Assume the values of C_{food} for this population follow a normal distribution with $M = 0.1$ μg/g and $\sigma = 0.015$ μg/g. We are most concerned with those children having the highest values of C_{food}, possibly leading to high values of lead absorption by the body.

(a) What fraction of the children have C_{food} above 0.13 μg/g?

(b) What fraction of the children have C_{food} in the range 0.07 to 0.13 μg/g?

10.8 The distribution of values of C_{food} in Problem 10.7 gives rise to variations in the amounts of lead absorbed by the children, A_{food}. One can consider each value of C_{food} in the distribution being used to compute one value of A_{food} using Equation (10.1): $A_{food} = C_{food} \times U_{food} \times f_{food}$. Assume each child in the distribution of Problem 10.7 consumes exactly 1,000 g of food/day and has $f_{food} = 0.5$. The distribution of values for A_{food} must also follow a normal distribution because each value of C_{food} is multiplied by a constant ($U_{food} \times f_{food}$). The mean value of A_{food} in this distribution, denoted by $\bar{A}_{food}$, is given by

$$A_{food} = C_{food} \times U_{food} \times f_{food} = 0.1\ \mu\text{g/g} \times 1{,}000\ \text{g/day} \times 0.5 = 50\ \mu\text{g/day}$$

where $\bar{C}_{food}$ is the mean value of C_{food}. The standard deviation is given by

$$\sigma(\text{for } A_{food}) = \sigma(\text{for } C_{food}) \times U_{food} \times f_{food}$$

$$= 0.015\ \mu\text{g/g} \times 1{,}000\ \text{g/day} \times 0.5 = 7.5\ \text{g/day}$$

Thus we have a normal distribution for A_{food} defined by $M = 50$ μg/day and $\sigma = 7.5$ μg/g.

(a) What fraction of the children have A_{food} above 65 μg/g?

(b) What fraction of the children have A_{food} in the range 35–65 μg/g?

(c) Public health authorities maintain a close watch on the fraction of the population with the highest amounts of absorbed lead. What value of A_{food} is exceeded by the highest 1 percent of the population?

10.9 A medical study involves human subjects who are representative of certain segments of the population in terms of blood lead concentration, PbB. Assuming the normal distribution discussed in the chapter is accurate for the population of children ($M = 22.5$ μg/dl, $\sigma = 5.8$ μ/dl), determine the range of PbB values (minimum and maximum) that should be used for each of the following segments of the population:

(a) The top 1 percent of PbB values in the population.

(b) The lowest 3 percent of PbB values in the population.

(c) The middle quarter (25 percent) of the population, centered about the mean.

10.10 Consider estimating the fraction of the population of children with PbB precisely at 22.5 μg/dl, the mean value in the distribution of Problem 10.9. What happens when you attempt to use an integral to evaluate this fraction? Explain the problem in terms of accuracy in measuring PbB.

CFCs and the Ozone Hole*

*The principal author of this chapter was Cliff I. Davidson.

11.1 INTRODUCTION: THE PROBLEM OF OZONE DEPLETION

Several trace gases in the atmosphere play a role in blocking ultraviolet radiation from the sun, which can damage our immune systems, cause cataracts in our eyes, induce various forms of cancer, and damage agricultural crops and other plants. The most important of these trace gases is ozone (O_3), found in the stratosphere far above the earth's surface. Chemically, this stratospheric ozone is the same molecule as the tropospheric ozone at the surface of the earth, which contributes to urban air pollution, as discussed in Chapter 8. However, we can consider ozone in the stratosphere as "good" ozone that protects us from ultraviolet radiation, in contrast to "bad" ozone in the troposphere, which can affect human health.

Because of its importance, several research programs have measured the amounts of stratospheric ozone in different locations at different times of the year. In one of these programs, the British Antarctic Survey discovered a problem in the early 1980s: It appeared that stratospheric ozone concentrations over Antarctica were greatly reduced for two months of the year (Farman et al., 1985). These data showed that a rapid decline had been occurring there since the mid-1970s, as summarized in Figure 11.1. Recent information from the World Meteorological Organization (WMO) shows that certain parts of the stratosphere around the world are now experiencing depletion of ozone. For example, concentrations at 30–60 degrees latitude of both Northern and Southern Hemispheres (a zone referred to as the *midlatitudes*) have been decreasing at a rate of 3–6 percent every 10 years since 1979. Stratospheric ozone is not disappearing everywhere, however. For instance, there has been no significant decrease in concentration over the tropics (between 20° north and 20° south latitude), according to the WMO.

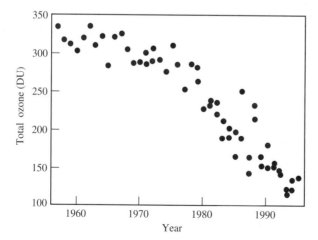

Figure 11.1 Antarctic ozone concentration (in Dobson Units) in October of each year from 1955 to 1995. The data points include both ground-based and satellite observations. Total ozone levels declined by about 50 percent over this period. (*Source:* NASA, 2000)

The depletion of stratospheric ozone is now known to be caused by a class of humanmade chemicals known as *chlorofluorocarbons* or *CFCs*. Certain chemicals similar to CFCs, such as other halocarbons, also contribute to this problem. In this chapter we investigate the reasons for this loss of ozone. We begin by discussing the formation and destruction of stratospheric ozone under natural conditions. Then we describe chlorofluorocarbons and other halocarbons and how they are used. Next we describe how CFCs perturb the natural ozone cycle, considering the chemical reactions occurring in the midlatitudes and also the special case of ozone loss over Antarctica. We then quantify the CFC problem using a simple mass balance model. The model allows us to calculate CFC concentrations in the atmosphere for different emission rates; it also enables us to determine the amount of ozone lost at various elevations. In the final section we discuss what is currently being done to address the problem of stratospheric ozone depletion.

11.2   THE NATURAL OZONE LAYER

Why does ozone occur naturally in the upper atmosphere? How much is there, and how is it destroyed? To answer such questions, we must first look at the structure of the atmosphere and the role of solar energy in the formation of ozone.

11.2.1   The Structure of the Atmosphere

The atmosphere is composed of several layers with differing characteristics. In the context of ozone depletion, we are most interested in the two lowest layers. These are the *troposphere,* which extends from the earth's surface to an altitude of roughly 10 kilometers, and the *stratosphere,* which occupies the interval 10–45 kilometers above the surface, as shown in Figure 11.2. The boundary between the troposphere and the stratosphere is known as the *tropopause.*

The troposphere is characterized by large-scale turbulence and mixing. Essentially all weather phenomena occur in this layer. An important characteristic is that the temperature decreases with altitude, starting from an average of 15°C at the earth's surface and reaching −55°C at the tropopause. This occurs because the weight of the air causes differences in pressure over a range of heights, and the laws of thermodynamics (such as the ideal gas law) show that pressure changes are accompanied by temperature changes. The thickness of the troposphere is not the same everywhere; it is as high as 16 kilometers in the tropics and as low as 9 kilometers in the polar regions.

The temperature structure of the stratosphere is quite different than that of the troposphere. As we shall see shortly, this is mainly because of the presence of ozone, which absorbs solar energy and releases it as heat. Ninety percent of the ozone is found in the stratosphere, and for that reason the stratosphere is also called the *ozone layer.* The temperature of the air remains roughly constant from the tropopause to a height of about 20 kilometers. It then increases with height, reaching 0°C at the top of the stratosphere where the incoming solar intensity is strongest. The fact that the

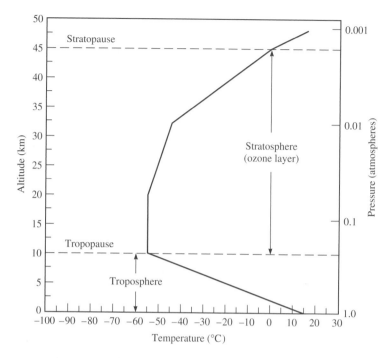

Figure 11.2   The two lowest layers of the atmosphere, showing the temperature profile of the troposphere and the stratosphere. Most atmospheric mass is concentrated in the troposphere, but ozone is found mainly in the stratosphere.

stratosphere is warmer than the troposphere limits mixing between the two regions: Warm air is less dense than cool air, so the warm air of the stratosphere remains above the cooler air of the troposphere. Note that particles and gases can be brought into the stratosphere either by direct injection (such as aircraft emissions or large volcanic eruptions) or by a circulation pattern near the tropics that produces limited mixing of tropospheric and stratospheric air masses.

11.2.2   Ultraviolet Radiation from the Sun

Incoming solar radiation is classified according to wavelength. Figure 11.3 shows the different forms of radiation, from the shortest wavelengths, which contain the most energy, to the longer wavelengths, which contain much less energy. Ultraviolet radiation, which is invisible, is so named because it occurs next to violet in the visible light spectrum. Wavelengths are commonly measured in units of micrometers ($1 \ \mu m = 10^{-6}$ m), but in the ultraviolet range the units of nanometers (1 nm $= 10^{-9}$ m) are often preferred. Thus a wavelength of $0.2 \ \mu m$ is equivalent to 200 nm.

Three ranges of ultraviolet radiation of interest in this chapter are shown in Figure 11.4. UV-C (200 to 280 nm) has the shortest wavelength of the three categories and is the most damaging. However, this radiation is entirely absorbed by oxygen

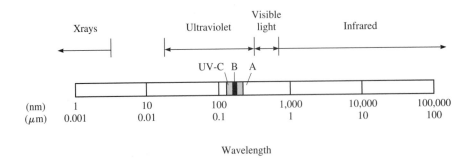

Figure 11.3   The spectrum of electromagnetic radiation.

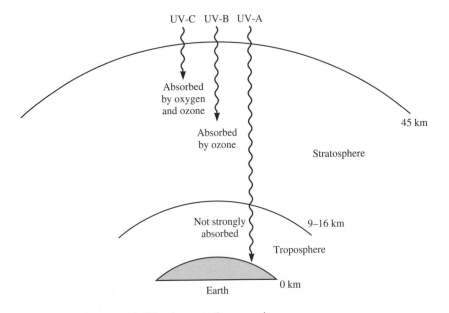

Figure 11.4   Absorption of UV radiation in the atmosphere.

and ozone molecules in the stratosphere and therefore never reaches the earth's surface. UV-B (280 to 320 nm) is a potentially very harmful form of radiation that is mostly absorbed by ozone molecules in the stratosphere. UV-A (320 to 400 nm) is the least damaging form of UV radiation. The structure of oxygen, ozone, and other atmospheric molecules is such that radiation in the UV-A range is not as strongly absorbed as radiation of shorter wavelengths. Thus UV-A normally reaches the earth in the greatest quantity. When stratospheric ozone is destroyed, however, greater quantities of UV-B can reach the earth's surface, and this is the principal concern associated with ozone loss. Chapter 12 on global warming describes in greater detail the absorption characteristics of ozone and oxygen that influence transmission of UV radiation through the atmosphere.

11.2.3 Formation and Destruction of Ozone

The process of creating ozone in the stratosphere consists of several steps:

$$O_2 + hv \rightarrow O + O \qquad \text{Creation of oxygen atoms } (\lambda < 240 \text{ nm}) \qquad \text{(Reaction 1)}$$

$$O + O_2 \rightarrow O_3 \qquad \qquad \text{Formation of ozone} \qquad \text{(Reaction 2)}$$

$$O_3 + hv \rightarrow O_2 + O \quad \text{Absorption of UV } (240 \text{ nm} < \lambda < 320 \text{ nm}) \text{ (Reaction 3)}$$

$$O + O_3 \rightarrow 2O_2 \qquad \qquad \text{Destruction of ozone} \qquad \text{(Reaction 4)}$$

In the first step, oxygen molecules absorb UV radiation with wavelengths less than 240 nm (UV-C). This causes each O_2 molecule to split into two oxygen atoms as described by Reaction 1, a step known as *photolysis*. The term hv in this equation refers to input solar energy—in this case UV-C radiation. The variable h is Planck's constant, whereas v is the frequency of the radiation. The single oxygen atoms are highly reactive. In the second step, shown as Reaction 2, they combine quickly with molecular oxygen to form ozone. The resulting ozone molecules then absorb UV radiation between 240 nm and 320 nm (UV-B and UV-C). This causes them to split apart, once again resulting in an oxygen molecule and a single oxygen atom (Reaction 3). This third step protects us from harmful effects: Dangerous UV radiation in the range 240–320 nm is absorbed by the ozone and is thus prevented from reaching the earth's surface. The products of Reaction 3 can then combine to form ozone again, as described by Reaction 2.

Besides Reaction 3, ozone can be destroyed by reaction with an oxygen atom as described by Reaction 4. This is known as *recombination*. The amount of ozone in the stratosphere is thus determined by a balance between the creation of ozone through photolysis and the destruction of ozone, according to the four reactions. The overall process, discovered by Chapman (1930), is known as the *Chapman mechanism*. A more detailed description of this mechanism is given by Elliott and Rowland (1987) and Rowland (1991).

The net result of these reactions is a variation of ozone concentration with height. Concentrations are low in the troposphere, increase to a peak value in the mid-stratosphere, and then decrease above this height. The variation in ozone with altitude is responsible for the temperature profile shown in Figure 11.2: UV radiation is continually being converted to heat as a result of the destruction and reformation of ozone.

Example 11.1

Chemistry of ozone formation and destruction. Confirm that there is no buildup of ozone, molecular oxygen, or atomic oxygen in the sequence of chemical reactions just described.

Solution:

This can be shown by doubling Reaction 2 to account for both of the oxygen atoms in Reaction 1 and then adding Reactions 1–4 together. The net result is

$$O_2 + 2(O + O_2) + O_3 + (O + O_3) \rightarrow (O + O) + 2O_3 + (O_2 + O) + 2O_2$$

Combining terms yields

$$3O_2 + 3O + 2O_3 \rightarrow 3O_2 + 3O + 2O_3$$

which indicates a balance.

The concentration of ozone in the stratosphere varies naturally with latitude, with season, and from day to day. The variation with latitude is primarily due to differences in solar radiation. Ozone is produced in largest quantities over the equator because this is where the sun's radiation is the most intense. However, stratospheric winds move the ozone away from the equator toward the poles, so that the ozone layer is generally thickest over the poles and thinnest around the equator. The concentrations near the equator are approximately the same year-round. Away from the equator, seasonal variations can be as much as 10–20 percent of the annual average, with the lowest concentrations in winter (Rowland, 1989).

11.2.4   Measurement of Ozone Concentrations

There are several ways to report the concentration of ozone in the air. Four common ways are concentration expressed as a fraction, as parts per million (ppm), as micrograms per cubic meter, or as Dobson Units (DU). For the first of these, concentration is expressed simply as

$$C_{O_3} \text{ (expressed as a fraction)} = \frac{\text{molecules of ozone}}{\text{molecules of air}} = \frac{\text{moles of ozone}}{\text{moles of air}}$$

$$= \frac{n_{O_3}}{n} \qquad (11.1)$$

where the number of molecules of air is simply the sum of molecules of all the constituents of air, including nitrogen, oxygen, and the trace constituents. Fractional concentrations of ozone are typically very small, so we express concentration in parts per million:

$$C_{O_3} \text{ (expressed as ppm)} = \frac{\text{molecules of ozone}}{10^6 \text{ molecules of air}}$$

$$= \frac{\text{moles of ozone}}{10^6 \text{ moles of air}}$$

Note that

$$\frac{10^{-6} \text{ moles ozone}}{1 \text{ mole air}} = \frac{1 \text{ mole ozone}}{10^6 \text{ moles air}} = 1 \text{ ppm ozone} \qquad (11.2)$$

Thus we multiply the fractional concentration by 10^6 to obtain concentration expressed as ppm.

We can also express the concentration in grams of ozone/m³ air, which is related to the fractional concentration through the *ideal gas law,* which relates the properties of a pure substance at low pressures (such as in the atmosphere):

$$n = \frac{pV_{air}}{R_o T} \tag{11.3}$$

where p = pressure of the air (atmospheres), V_{air} = volume of the air (m³), R_o = universal gas constant (m³ atmosphere/mole K), and T = absolute temperature (K). Therefore we can write

$$n_{O_3} = C_{O_3} \text{ (expressed as a fraction) } n$$

$$= C_{O_3} \text{ (expressed as fraction)} \frac{pV_{air}}{R_o T} \tag{11.4}$$

Note that the mass of ozone is equal to $n_{O_3} M_{O_3}$, where M_{O_3} is the molecular weight of ozone. Therefore

$$\text{Mass of ozone } = n_{O_3} M_{O_3}$$

$$= C_{O_3} \text{ (expressed as a fraction)} \frac{pV_{air}}{R_o T} M_{O_3} \tag{11.5}$$

We can now use this equation to express the ozone concentration in terms of mass per unit volume:

$$C_{O_3} \text{ (concentration expressed as g/m}^3) = \frac{\text{Mass of ozone}}{V_{air}}$$

$$= C_{O_3} \text{ (expressed as a fraction)} \frac{pM_{O_3}}{R_o T} \tag{11.6}$$

Example 11.2

Expressing the concentration of ozone in various units. The fractional concentration of ozone in clean tropospheric air is about 1×10^{-8} at 1 atmosphere pressure and $T = 0°C$. Calculate this concentration in ppm and also in g/m³.

Solution:

$$C_{O_3} \text{ (expressed as a fraction)} = 1 \times 10^{-8} \frac{\text{moles of ozone}}{\text{mole of air}}$$

$$C_{O_3} \text{ (expressed in ppm)} = 10^6 \times C_{O_3} \text{ (expressed as a fraction)}$$

$$= 1 \times 10^{-2} \text{ ppm} = 0.01 \text{ ppm}$$

$$C_{O_3} \text{(expressed in g/m}^3\text{)} = C_{O_3} \text{(expressed as a fraction)} \times \frac{pM_{O_3}}{R_o T}$$

$$= 10^{-8} \times \frac{(1 \text{ atm}) \times (48 \text{ g/mole})}{\left(8.21 \times 10^{-5} \dfrac{\text{m}^3\text{atm}}{\text{mole-K}} \right) \times 273 \text{ K}}$$

$$= 2.14 \times 10^{-5} \text{ g/m}^3$$

Note that 2.14×10^{-5} g/m^3 is equivalent to 21.4 μg/m^3. We often convert g/m^3 to μg/m^3 for simplicity in expressing the concentrations.

Concentrations that are expressed as fractions, ppm, and g/m^3 are most useful close to the ground where air density, pressure, and temperature are roughly constant. At high altitudes, however, properties of the air may vary greatly, and it is more convenient to express the ozone concentration using Dobson Units. This method is named after the scientist who designed an instrument to measure ozone. The instrument reports the total number of ozone molecules occupying a vertical column of air stretching from the earth's surface to the top of the stratosphere. One DU is defined as 0.01 mm thickness of pure ozone at 0°C and one atmosphere pressure. Ozone concentrations can vary naturally from more than 500 DU in polar regions in the spring to less than 250 DU at the equator. The typical column of ozone is approximately 300 DU. We can measure stratospheric ozone concentrations using ground-based instruments, rockets, satellites, and high-altitude balloons.

Unlike the measures of concentration in Example 11.2, the number of Dobson Units tells us the total amount of ozone in a vertical column of air without providing any information about how the ozone is distributed with height. Because most of the ozone in a vertical column is present in the stratosphere, this measurement provides a straightforward way of assessing stratospheric ozone. The following three examples illustrate some of the physical concepts behind the Dobson Units method of reporting concentrations.

Example 11.3

Relation between Dobson Units and number of molecules per cm².   Calculate the number of molecules of ozone per cm^2 of the earth's surface for 1 DU, based on the definition of DU just given and the ideal gas law.

Solution:

$$\text{Number of moles of ozone } = n_{O_3} = \frac{p_{O_3} V_{O_3}}{R_o T_{O_3}}$$

$$\text{Number of molecules of ozone } = n_{O_3} N_a = \frac{p_{O_3} V_{O_3}}{R_o T_{O_3}} N_a$$

where P_{O_3}, V_{O_3}, and T_{O_3} refer to the pressure, volume, and temperature of pure ozone, and N_a is Avogadro's number. So the number of molecules of ozone per unit area of the earth's surface is given by

$$\frac{n_{O_3} N_a}{A} = \frac{p_{O_3} N_a}{R_o T_{O_3}} \times \frac{V_{O_3}}{A}$$

$$= \frac{p_{O_3} N_a}{R_o T_{O_3}} \times h_{O_3}$$

where A is the area of the earth's surface and h_{O_3} is the height of a layer of pure ozone. Substituting values, we obtain

$$\frac{p_{O_3} N_a}{R_o T_{O_3}} \times h_{O_3} = \frac{1 \text{ atm} \times 6.023 \times 10^{23} \text{ molecules/mole}}{8.21 \times 10^{-5} \frac{m^3 \text{ atm}}{\text{mole K}} \times 273 \text{ K}} \times 0.01 \text{ mm} \times \frac{1 \times 10^{-3} \text{ m}}{\text{mm}}$$

$$= 2.69 \times 10^{20} \frac{\text{molecules of ozone}}{m^2 \text{ of the earth's surface}}$$

$$= 2.69 \times 10^{16} \frac{\text{molecules of ozone}}{cm^2 \text{ of the earth's surface}}$$

The next example illustrates the relationship between Dobson Units and the more familiar units of parts per million used to express the concentration of air pollutants in the troposphere.

Example 11.4

Relation between Dobson Units and parts per million. Concentrations of ozone at ground level on a smoggy day can reach 0.2 ppm. If this concentration of ozone is mixed up to a height of 1,000 m, express the ozone concentration in DU. Assume air is an ideal gas with uniform values of $T = 0°C$ and $p = 1$ atmosphere over this height. Compare your answer with the typical value of 300 DU for the entire atmospheric column, including the stratosphere.

Solution:

We know that 1 DU corresponds to 2.69×10^{16} molecules of ozone per cm^2 of the earth's surface. Thus we need to calculate the number of molecules of ozone per cm^2 in the column of air up to 1,000 m height, and divide by 2.69×10^{16} to get the number of DU.

The number of air molecules in a volume 1 cm × 1 cm × 1,000 m (which has a total volume of 0.1 m^3) is calculated as

$$nN_a = \frac{pV_{air} N_a}{R_o T}$$

$$= \frac{1 \text{ atm} \times 0.1 \text{ m}^3 \times 6.023 \times 10^{23} \text{ molecules/mole}}{8.21 \times 10^{-5} \frac{m^3 \text{ atm}}{\text{mole K}} \times 273 \text{ K}}$$

$$= 2.69 \times 10^{24} \text{ molecules of air}$$

This is exactly 10^8 times greater than the number of ozone molecules for 1 DU calculated in Example 11.3. Such a result is expected: 1 DU corresponds to a thickness of 0.01 mm of ozone, whereas this calculation assumes a thickness of 1,000 m of air, which is 10^8 times greater. Both ozone and air are considered ideal gases, so the thicknesses of the layers of ozone and air can be compared.

The number of ozone molecules in the 1,000 m column of air is calculated as

$$n_{O_3} N_a = C_{O_3} n\, N_a$$

$$= (0.2 \times 10^{-6})\,(2.69 \times 10^{24})$$

$$= 5.38 \times 10^{17}\ \text{molecules of ozone}$$

This is the number of molecules of ozone present in a column of air that occupies an area of 1 cm^2 of the earth's surface and is 1,000 m high. The concentration in Dobson Units is therefore

$$5.38 \times 10^{17}\ \frac{\text{molecules of ozone}}{\text{cm}^2\ \text{of the earth's surface}} \times \frac{1\text{DU}}{2.69 \times 10^{16}\ \dfrac{\text{molecules of ozone}}{\text{cm}^2\ \text{of the earth's surface}}}$$

$$=\ 20\ \text{DU}$$

This is much smaller than the typical value of 300 DU for ozone in a column that extends upward to the top of the stratosphere. Thus the amount of ozone close to the ground, even on a smoggy day, is much smaller than the amount of ozone in the entire atmospheric column, most of which is in the stratosphere.

The large decrease in pressure from the earth's surface to the top of the stratosphere requires a different approach to determining average concentration near the surface using the ideal gas law. The following example illustrates this point.

Example 11.5

Avoiding complications of varying atmospheric pressure.   Determine the concentration of ozone in ppm averaged over the column from the ground to the top of the stratosphere for a typical value of 300 DU. Assume that there are a total of 1.1×10^{44} molecules of air in the troposphere and stratosphere together, and that the average radius of the earth is 6,370 km.

Solution:

We cannot simply use the ideal gas law as we did for Example 11.4. That problem assumed a constant pressure and temperature over a height of 1,000 m; we must now consider the full atmospheric column, where the temperature and pressure vary greatly. If we can determine the total number of ozone molecules per cm^2 of the earth's surface (extending up to the top of the stratosphere) and similarly determine the total number of air molecules per cm^2 up to this height, then the ratio of these two quantities is the desired average concentration. Using the information given in the problem,

$$\left(\frac{2.69 \times 10^{16}\ \dfrac{\text{molecules of ozone}}{\text{cm}^2\ \text{of the earth's surface}}}{1\ \text{DU}} \right) (300\ \text{DU})$$

$$= 8.07 \times 10^{18} \frac{\text{molecules of ozone}}{\text{cm}^2 \text{ of the earth's surface}}$$

Because A is the area of the earth's surface given by $4\pi r^2$, we can write

$$\frac{1.1 \times 10^{44} \text{ molecules of air}}{4\pi \ (6,370 \text{ km})^2 \times 10^{10} \frac{\text{cm}^2}{\text{km}^2}} = 2.15 \times 10^{25} \frac{\text{molecules of air}}{\text{cm}^2 \text{ of the earth's surface}}$$

$$\frac{8.07 \times 10^{18} \dfrac{\text{molecules of ozone}}{\text{cm}^2 \text{ of the earth's surface}}}{2.15 \times 10^{25} \dfrac{\text{molecules of air}}{\text{cm}^2 \text{ of the earth's surface}}} = 3.75 \times 10^{-7} \frac{\text{molecules of ozone}}{\text{molecule of air}}$$

$$= 0.375 \text{ ppm}$$

Note that we cannot strictly compare the value of 0.375 ppm of ozone, which represents a gross average over the entire column, with the concentration of 0.2 ppm in only the lowest 1,000 m of the atmosphere. The full column concentration is influenced by the scarcity of air molecules in the stratosphere: Most of the 1.1×10^{44} molecules of air in the atmosphere are found in the troposphere. It is, however, feasible to use the previous calculation to compare the total number of ozone molecules per cm^2 in the entire atmospheric column with the total number of ozone molecules per cm^2 in the lowest 1,000 m on a smoggy day.

For the full atmospheric column with 300 DU, there are

$$8.07 \times 10^{18} \frac{\text{molecules of ozone}}{\text{cm}^2 \text{ of the earth's surface}}$$

For the lowest 1,000 m where the ozone concentration is 0.2 ppm, there are

$$5.38 \times 10^{17} \frac{\text{molecules of ozone}}{\text{cm}^2 \text{ of the earth's surface}}$$

Thus the number of molecules of ozone per cm^2 in the entire atmospheric column is 15 times as large as the number of ozone molecules in the lowest 1,000 m on a smoggy day. This is the same as comparing the value of 300 DU in the entire atmospheric column with 20 DU in the lowest 1,000 m, calculated in Example 11.4.

Having established the mechanisms of formation and destruction of the stratospheric ozone layer under natural conditions, we now provide a background for considering the influence of anthropogenic activities on this layer. The next section introduces the chemicals responsible for damage.

11.3 CHLOROFLUOROCARBONS (CFCS) AND HALOCARBONS

Chlorine (Cl) belongs to a class of chemicals called *halogens* that also includes fluorine (F), bromine (Br), and iodine (I). When one or more of these elements combine with carbon, the resulting compounds are known as *halocarbons*. CFCs are the most well-known type of halocarbons. Here we take a closer look at these compounds.

11.3.1   What Are CFCs?

Early work in the 1970s by American scientists Sherwood Rowland and Mario Molina had shown that chlorine atoms can destroy stratospheric ozone. These investigators, along with Paul Crutzen of Germany, further showed that a class of chemicals known as *chlorofluorocarbons* were responsible for the chlorine atoms found in the stratosphere. As discussed in Chapter 6, chlorofluorocarbons were developed in the 1930s to replace other refrigerants such as sulfur dioxide, ammonia, and methyl chloride. The new compounds were thought to be an ideal solution to the growing demand for refrigerant chemicals: CFCs are odorless, colorless, nonflammable, noncorrosive, and nontoxic, unlike the chemicals they replaced. They are also extremely stable, so once produced they could be used for years.

Soon it was recognized that CFCs were effective in other applications. In recent years CFCs have been used as industrial solvents for cleaning semiconductors (21 percent of total CFC usage), as blowing agents for foam and insulation manufacturing (26 percent), and as propellants for aerosol spray cans (28 percent), in addition to use as refrigerants (23 percent). Table 11.1 lists the applications of several CFCs as well as other halocarbon compounds, including two important chlorinated compounds, carbon tetrachloride (CCl_4) and methyl chloroform (CH_3CCl_3). The table also shows their 1992 atmospheric concentrations, which are measured in units of

Table 11.1   Characteristics of CFCs and other ozone-depleting compounds.

Chemical Name	Chemical Formula	Main Uses or Sources	Atmospheric Concentration (ppt) in 1992	Atmospheric Lifetime, τ (Years)
CFC-11	CCl_3F	Aerosol propellant; foam blowing	268	50
CFC-12	CCl_2F_2	Aerosol propellant; refrigerant; foam blowing	503	102
CFC-113	$C_2Cl_3F_3$	Solvent	82	85
CFC-114	$C_2Cl_2F_4$	Aerosol propellant; refrigerant	20	300
Carbon tetrachloride	CCl_4	Solvent	132	42
Methyl chloroform	CH_3CCl_3	Solvent	135	4.9
Halon-1211	$CBrF_2Cl$	Fire retardant	7	20
Halon-1301	$CBrF_3$	Fire retardant	3	65
Halon-2402	$C_2Br_2F_4$	Fire retardant	0.7	20
HCFC-22	CHF_2Cl	Aerosol propellant; refrigerant; foam blowing; solvent; fire retardant	100	12.1
HCFC-141b	$C_2H_3FCl_2$	Foam blowing; solvent	2	9.4
HCFC-142b	$C_2H_3F_2Cl$	Foam blowing; solvent	6	18.4

Source: IPCC, 1996.

parts per trillion (ppt) by volume, and the atmospheric lifetime, τ, a parameter we will discuss a little later in this chapter.

The expansion of CFC production increased emissions of these compounds to the atmosphere, mainly through leaks, spills, and disposal. Unfortunately, the very same characteristic that made CFCs so useful—their stability—also allowed them to remain in the atmosphere for long periods. These chemicals thus had time to make their way up to the stratosphere, where they could destroy substantial quantities of ozone.

Because of their role in the destruction of stratospheric ozone, CFCs are now recognized as one of the most serious environmental problems on a global scale. In recognition for their work in identifying the problem, Rowland, Molina, and Crutzen were awarded the Nobel Prize in chemistry in 1995. As we will see in Chapter 12, CFCs are also powerful greenhouse gases that play a role in global warming as well as ozone depletion.

11.3.2 The Naming Convention for CFCs

You may have wondered why CFCs and other halocarbons have such a bewildering array of numbers associated with them. Here we reveal the formula behind the naming convention for CFCs. A CFC molecule is created by substituting chlorine and fluorine atoms for the four hydrogen atoms in methane (CH_4). For example, CCl_2F_2 (CFC-12) has two chlorine and two fluorine molecules substituted, as illustrated in Figure 11.5.

There are several ways to refer to compounds in this class. CFC-12, Freon-12, and R-12 are all common names for CCl_2F_2. Freon is the brand name trademarked by the DuPont company. The letter R was originally short for refrigerant but often indicates CFC compounds used for other purposes. The numbers after the label are related to the number of carbon, hydrogen, fluorine, and chlorine atoms in the compound. All compounds can be represented by a three-digit number, which may include 0 as the first digit, using the following rules:

- The number of carbon atoms equals the hundreds digit plus one.
- The number of hydrogen atoms equals the tens digit minus one.
- The number of fluorine atoms equals the ones digit.
- The number of chlorine atoms is derived by difference using the fact that each carbon atom has four atomic bonds.

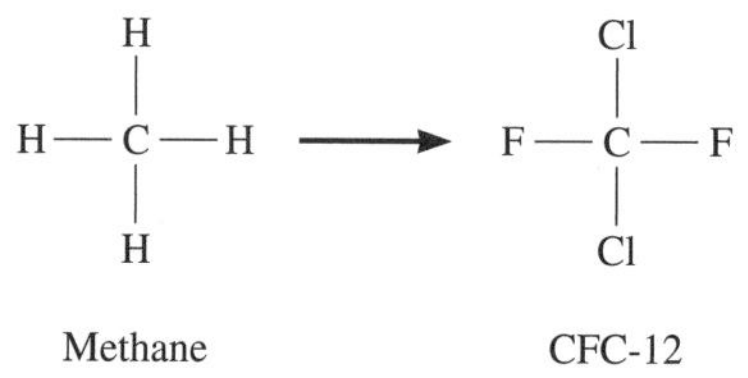

Methane   CFC-12

Figure 11.5 The dichlorodifluoromethane (CFC-12) molecule is created by substitution of two chlorine and two fluorine atoms for the four hydrogen atoms in methane.

This convention does not specify the chemical structure of other refrigerants, such as ammonia, or combinations of halocarbons, like CFC-500. Some compound names include a letter *a* or *b* after the number (such as CFC-134a), which refers to the locations of the atoms in the molecule structure.

Example 11.6

Structure of a chlorofluorocarbon.   How many atoms of carbon, hydrogen, fluorine, and chlorine does CFC-12 have? How many atoms of each does R-290 have? Write the chemical formulas for these compounds.

Solution:

CFC-12:

The three-digit number corresponding to 12 is actually 012:

$$C: 0 + 1 = 1$$
$$H: 1 - 1 = 0$$
$$F: 2$$

The number of chlorine bonds can be found from the number of carbon bonds minus the number of other bonds (two fluorine in this case). The one carbon atom has four bonds. Thus

$$Cl: 4 - 2 = 2$$

The formula is therefore CF_2Cl_2, or CCl_2F_2 as in the previous text.

R-290:

$$C: 2 + 1 = 3$$
$$H: 9 - 1 = 8$$
$$F: 0$$

There are three carbon atoms bonded together, which leaves eight bonds available for other molecules.

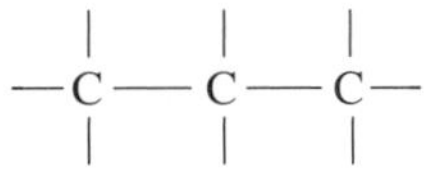

$$-\overset{\displaystyle |}{\underset{\displaystyle |}{C}} - \overset{\displaystyle |}{\underset{\displaystyle |}{C}} - \overset{\displaystyle |}{\underset{\displaystyle |}{C}} -$$

The number of chlorine atoms can be found from the number of available carbon bonds minus the number of bonds taken up by hydrogen and fluorine: $Cl = 8 - 8 = 0$. Thus there are no Cl atoms or F atoms, and the formula is C_3H_8. This is the chemical propane, which is a pure hydrocarbon that can also be used as a refrigerant (see Chapter 6).

11.4   CFC DESTRUCTION OF STRATOSPHERIC OZONE

Earlier we saw that the creation and destruction of ozone molecules in the stratosphere are in a delicate balance under natural conditions. When CFCs are emitted at the earth's surface they slowly disperse, and some of them eventually reach the stratosphere. It takes an average of three to five years for a CFC molecule to travel from the earth's surface to the lower stratosphere. The molecule can then diffuse upward to reach the upper portions of the stratosphere (over 25 km height), where global levels of ozone depletion are seen. A CFC molecule can take from 40 to 100 years to reach this height after being emitted at the earth's surface.

11.4.1 Mechanisms of Ozone Destruction by CFCs in the Midlatitudes

Why is it necessary for CFCs to reach high altitudes to react chemically? The reason is that these molecules can be dissociated only by solar radiation with wavelengths shorter than 230 nm (UV-C). Radiation with such short wavelengths cannot penetrate the stratosphere to reach the troposphere because oxygen and ozone at high altitudes absorb all radiation with wavelengths shorter than 293 nm.

At high elevations in the stratosphere short-wave radiation is abundant, and CFCs can undergo photolysis to release a chlorine atom. This photolytic reaction is similar to those described in Chapter 8 in which the energy of solar photons (hv) breaks apart certain types of molecules. For example, fluorotrichloromethane (CFC-11) experiences the following reaction:

$$CCl_3F + hv \rightarrow Cl + CCl_2F, \text{ at } \lambda < 230 \text{ nm (25–40 km)} \qquad \text{(Reaction 5)}$$

The chlorine released from this reaction can then react with ozone to form chlorine monoxide (ClO):

$$Cl + O_3 \rightarrow ClO + O_2 \qquad \text{(Reaction 6)}$$

Reaction 6 leads to lower ozone levels not only by destroying ozone directly, but also through bypassing the photolysis of ozone that creates atomic oxygen (Reaction 3). By preventing the creation of atomic oxygen, Reaction 6 further hinders the reformation of ozone (Reaction 4) that is necessary to keep the cycle in balance.

Adding to the problem is the instability of the chlorine monoxide formed in Reaction 6. If chlorine monoxide were stable, chlorine would have little effect on the concentration of ozone because ozone is 1,000 times more abundant than current atmospheric concentrations of chlorine monoxide. However, chlorine monoxide is not stable and in fact reacts with atomic oxygen

$$ClO + O \rightarrow Cl + O_2 \qquad \text{(Reaction 7)}$$

Reaction 7 shows that ClO depletes the atomic oxygen by which ozone is formed. More importantly, it also releases the chlorine atom, which allows for the repetition of Reactions 6 and 7. The net result of Reactions 6 and 7 is the same as Reaction 4. Thus chlorine acts as a catalyst to increase the ozone destruction rate. The maximum ozone depletion from these chlorine reactions occurs in the upper stratosphere near 40 km. This is because ozone molecules are scarce above that altitude as a result of the chemistry discussed earlier in section 11.2.3. Below that altitude, the oxygen atoms required for Reaction 7 are scarce.

Example 11.7

Destruction of ozone by chemical reactions involving chlorine.   Show that the chlorine chemical reactions decrease ozone concentrations without destroying the chlorine molecule.

Solution:

Adding Reactions 6 and 7 yields

$$O_3 + O + ClO + Cl \rightarrow 2O_2 + ClO + Cl$$

Thus ozone is destroyed and chlorine is recycled, enabling it to destroy more ozone.

Over about a month, a chlorine atom in the stratosphere will destroy approximately 100,000 ozone molecules before it eventually reacts with the small amount of natural atmospheric methane found in the stratosphere. This reaction forms HCl:

$$Cl + CH_4 \rightarrow HCl + CH_3 \qquad \text{(Reaction 8)}$$

The HCl molecule is eventually transported down to the troposphere, where it is washed out by rain in a matter of months. The net result is that small amounts of chlorine can destroy much greater amounts of ozone.

The CCl_2F in Reaction 5 is soon photolyzed to give two more chlorine atoms and a fluorine atom. The chlorine atoms follow the path just described. The fluorine atom combines with hydrogen to form HF, which is inert in the atmosphere and does not harm ozone. Figure 11.6 summarizes these processes graphically.

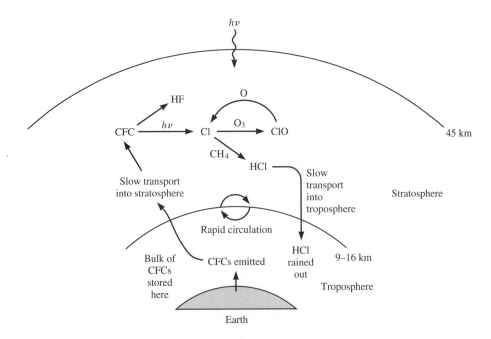

Figure 11.6   The fate of a CFC molecule in the atmosphere.

The fact that chlorine can act as a catalyst to decrease ozone is suggested by observed increases in measured chlorine concentrations that have accompanied the measured decreases in ozone. Historically, chlorine has had a concentration of less than one part per billion by volume (1 ppb) in the stratosphere, mostly from methyl chloride; this compound is emitted from biological ocean processes and from chemical processes in biomass burning. However, chlorine concentrations in the stratosphere have been rising steadily, from 1.5 ppb in 1970 to 2.5 ppb in 1980 and to 3.6 ppb in 1991. Without controls, it was estimated that chlorine concentrations would reach 9 ppb by the year 2000.

11.4.2   Mechanisms of Ozone Destruction by CFCs in the Antarctic

Ozone levels over Antarctica in late August, which is winter in the Southern Hemisphere, are naturally low (about 300 DU) due to the low temperatures. Historically, ozone levels have slowly increased as the sunlight increased during September and October. However, starting in the late 1970s, it was observed that about 95 percent of the ozone was destroyed in the lower stratosphere at altitudes between 15 and 20 km. Above 25 km the decreases were smaller; in 1994 about 60 percent of the ozone above this height was destroyed. Ozone levels returned to more normal values in November.

This loss of stratospheric ozone over the Antarctic region became widely known as the *ozone hole*. Figure 11.7 depicts the size and depth of this "hole" by showing the average loss of ozone in the early 1990s compared to the early 1970s. The geographic

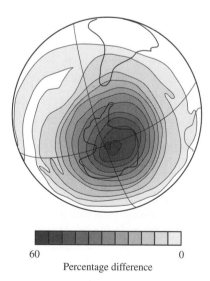

60   0

Percentage difference

Figure 11.7   The ozone hole over Antarctica. This figure shows the average percentage loss in ozone (measured in Dobson Units) from the early 1970s to the early 1990s. (*Source:* NASA, 2000)

extent of the ozone hole extends across the entire continent. At the core of the hole ozone losses exceed 50 percent, as seen earlier in Figure 11.1.

The massive ozone depletion at 15–20 km in Antarctica occurs at a much lower altitude than noted earlier for ozone depletion over the entire globe (40 km). Furthermore, the depletion above 25 km over Antarctica is much larger than for the rest of the globe. This indicates that different processes are responsible. Although the amount of chlorine over the Antarctic is similar to the amounts elsewhere, several natural factors, including geography, wind, temperature, and sunlight, contribute to this problem. The loss of ozone over Antarctica can be explained by three steps: the creation of active chlorine due to the presence of polar stratospheric clouds, photolysis of this active chlorine, and finally the catalytic destruction of ozone.

During the Antarctic winter, strong stratospheric winds create a *polar vortex*— a mostly circular wind pattern that isolates the Antarctic from the rest of the atmosphere. The polar vortex is extremely cold, below $-80°C$. Under these conditions many clouds appear in the stratosphere, commonly referred to as *polar stratospheric clouds* or *PSCs*. Because little sunlight is available for photolysis during the winter, most of the Antarctic chlorine is either in the form of chlorine nitrate ($ClONO_2$, formed from natural concentrations of NO_2 and ClO) or HCl. These species are reservoirs for chlorine and do not directly reduce ozone concentrations. However, the polar stratospheric clouds create ice crystals that cause the formation of Cl_2 and $HOCl$, as shown in Reactions 9 and 10 here. These species build up during the winter when polar stratospheric clouds are prevalent (McCormick et al., 1982).

Polar stratospheric clouds at 15–20 km:

$$ClONO_2 + HCl \rightarrow Cl_2 + HNO_3 \qquad \text{(Reaction 9)}$$

$$ClONO_2 + H_2O \rightarrow HOCl + HNO_3 \qquad \text{(Reaction 10)}$$

The second step is photolysis of the chlorine in Reactions 9 and 10. This begins in late August when the Antarctic starts to receive sunlight. The UV radiation triggers reactions with the Cl_2 and $HOCl$ formed throughout the winter to produce large amounts of chlorine and chlorine monoxide:

$$HOCl + h\nu \rightarrow OH + Cl \qquad \text{(Reaction 11)}$$

$$Cl_2 + h\nu \rightarrow 2Cl \qquad \text{(Reaction 12)}$$

$$Cl + O_3 \rightarrow ClO + O_2 \qquad \text{(Reaction 13)}$$

Finally, the third step is the catalytic cycle that enables chlorine monoxide to destroy ozone. This was shown earlier by Reaction 7. ClO thus plays a key role in the Antarctic just as it does in the midlatitudes. However, the process that enables ClO to destroy ozone at 40 km over most of the globe cannot be responsible for the depletion in the Antarctic. The process at 40 km uses free oxygen atoms, which are abundant only in the upper stratosphere; in contrast, the Antarctic ozone hole forms in the lower stratosphere. The true mechanism of Antarctic ozone destruction is somewhat complicated. However, the key points are that it depends on the UV radiation that becomes abundant in the lower stratosphere in the spring, it depends on the formation of ClO, and it depletes ozone without consuming chlorine. The net result is

$$2O_3 \rightarrow 3O_2 \qquad \text{(Reaction 14)}$$

Thus there is a long buildup of Cl_2 and HOCl during the winter (Reactions 9 and 10), followed by production of Cl and ClO by sunlight in the spring (Reactions 11–13), which leads to massive ozone destruction (Reaction 14). This mechanism is believed responsible for about 70 percent of the Antarctic ozone loss. As the stratosphere heats up during the spring, the polar vortex dissipates and allows ozone-rich air from outside the hole to mix with ozone-depleted air inside the hole. The result is that the Antarctic ozone hole disappears by late November.

11.5   QUANTIFYING OZONE DESTRUCTION BY CFCS: THE MASS BALANCE MODEL

The previous discussion implies the need to reduce and eventually eliminate CFC emissions in order to maintain natural ozone concentrations in the stratosphere. But how long will it take for the ozone levels to recover? What are the implications of reducing emissions slowly versus quickly? To answer such questions, we need mathematical models for predicting the concentrations of CFCs and ozone for different hypothetical situations.

Several types of mathematical models are used in the field of air quality engineering. One of the simplest but most useful types is the mass balance model, which was introduced in Chapter 8 to describe urban air pollution. As before, we consider four processes to predict the amount of pollutant mass in a box of specified volume:

- The flow of pollutant mass into the box.
- The flow of pollutant mass out of the box.
- Emissions of the pollutant within the box.
- Loss of the pollutant within the box (such as loss by chemical reaction or deposition on surfaces).

In general, this type of model can be used both indoors and outdoors; the "box" can be a room of a building, an entire building, or a region of the atmosphere (like the air over a city, as in Chapter 8). It is usually assumed that mixing within the box is complete and instantaneous so that the concentration of the pollutant is the same everywhere in the box. The basic equation is

Rate of change of pollutant mass in the box over time =
  (Rate of pollutant flow into the box)
  − (Rate of pollutant flow out of the box)   (11.7)
  + (Rate of emissions within the box)
  − (Rate of loss of pollutant within the box)

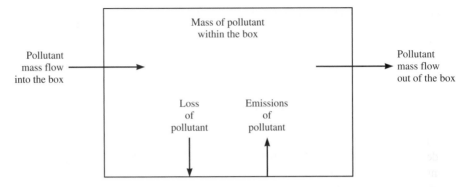

Figure 11.8 Mass balance for a pollutant in a "box" that can represent compartments such as a room, a building, a region of the atmosphere, or the entire earth's atmosphere. Mixing within the box is assumed to be complete and instantaneous.

The various terms in Equation (11.7) are illustrated in Figure 11.8. Each term in the equation has units of mass per time, such as grams/second. Note that we must account for initial conditions if there is any mass of pollutant in the box at the beginning of the period of interest.

In the context of this chapter, we are interested in the amounts of CFC and ozone present in the atmosphere of an entire hemisphere or even both hemispheres. Thus, in this case, the whole atmosphere is modeled as a huge box.

11.5.1 Calculating Amounts of CFC in the Atmosphere

Applying Figure 11.8 to CFCs in the atmosphere, we note that there is no flow of CFCs into the box or out of the box from the surroundings (there are no CFCs in outer space). However, there are emissions of CFCs at the earth's surface and losses of CFCs within the box. For this situation we can define the following terms in Equation (11.7):

Rate of change of CFC mass in the box over time = $\dfrac{dm}{dt}$

CFC flow into the box = 0

CFC flow out of the box = 0

Emissions within the box = P

Removal of CFCs within the box = $R = \beta m$

The last term assumes that the rate of removal is proportional to the CFC mass in the box: The rate of removal will be greatest when there is a large amount of CFC. The proportionality constant is β.

The resulting equation is thus

$$\frac{dm}{dt} = P - R = P - \beta m$$

which can be rewritten as

$$\frac{dm}{dt} + \beta m = P \tag{11.8}$$

where m is a variable representing the CFC mass in the atmosphere (grams), t is a variable representing the time (in years), and P is the CFC emission rate in grams/year. Although the value of P may change over time, the simplest case is when P is constant.

Equation (11.8) is a first-order linear differential equation, and it can be solved to determine how the mass of the CFC varies with time. This type of equation is commonly encountered in engineering. It is essentially a statement that the rate of change of a variable is proportional to the value of the variable.

In order to solve this equation, we must actually find two solutions, namely the *homogeneous* solution and the *particular* solution. The total solution to the differential equation is the sum of these two. The homogeneous solution is found by solving the equation

$$\frac{dm}{dt} + \beta m = 0 \tag{11.9}$$

Equation (11.9) can be solved by rearranging terms and integrating. The result is

$$m_{homogeneous} = K e^{-\beta t} = K e^{-t/\tau} \tag{11.10}$$

where K is a constant of integration and τ is a new proportional constant defined as $1/\beta$. We introduce τ in place of β solely for convenience because it has the units of time (whereas β had units of time^{-1}). The constant τ is called the *atmospheric life-time* or *residence time*. Numerical values of τ have been tabulated for different compounds, as shown earlier in Table 11.1. We will discuss the significance of this parameter later in the chapter. First, let us finish solving Equation (11.8).

The particular solution can be found by a number of different methods. We can merely set dm/dt equal to zero and solve the equation to obtain the particular solution equal to a constant. Thus we have

$$0 + \frac{1}{\tau} m_{particular} = P \tag{11.11}$$

$$m_{particular} = P\tau \tag{11.12}$$

The total solution is thus

$$m(t) = m_{homogeneous} + m_{particular}$$

$$= K e^{-t/\tau} + P\tau \tag{11.13}$$

The next step is to solve for the constant K in Equation (11.13) by applying boundary conditions. It is convenient here to use the initial concentration:

$$m(t = 0) = m_o \tag{11.14}$$

where m_0 is the initial mass of the CFC at the start of the period of interest. Therefore, at $t = 0$ we have

$$m_0 = Ke^0 + P\tau = K + P\tau \tag{11.15}$$

$$K = m_0 - P\tau \tag{11.16}$$

We then substitute this expression for K into Equation (11.13) to obtain the final solution for the mass of the CFC as a function of time:

$$m(t) = (m_0 - P\tau)e^{-t/\tau} + P\tau$$

$$= P\tau(1 - e^{-t/\tau}) + m_0 e^{-t/\tau} \tag{11.17}$$

Equation (11.17) shows that if m_0 is zero, then $m(t)$ starts out as zero and rises exponentially. Even if m_0 is large, its effect becomes negligible as t increases due to the negative exponential, and we are left with $m(t) = P\tau$ in steady state as $t \to \infty$. The value of $m(t)$ as t approaches infinity is called the "steady state solution," termed m_{ss}. We can thus set $m_{ss} = P\tau$ in Equation (11.17):

$$m(t) = m_{ss}(1 - e^{-t/\tau}) + m_0 e^{-t/\tau} \tag{11.18}$$

where

$$m_{ss} = P\tau$$

This is our final expression for the total mass of a CFC in the atmosphere as a function of time. This value also can be converted into a concentration based on the molecular weight of the compound. Using a similar approach to that used earlier for the concentration of ozone, we can write

$$C(\text{expressed as a fraction}) = \frac{\text{molecules of CFC}}{\text{molecules of air}} = \frac{mN_a}{M_i N_m} \tag{11.19}$$

where m is the total mass of the CFC in grams, N_a is Avogadro's number (equal to 6.023×10^{23} molecules/mole), M_i is the molecular weight of the CFC, and N_m is the number of molecules of air in the atmosphere, equal to 1.1×10^{44} as in Example 11.5. We can then write Equation (11.18) in terms of concentration as

$$C(t) = C_{ss}(1 - e^{-t/\tau}) + C_0 e^{-t/\tau} \tag{11.20}$$

where C_{ss} and C_0 are the steady state and initial values of concentration, respectively, given by

$$C_{ss} = \frac{m_{ss} N_a}{M_i N_m} \quad \text{and} \quad C_0 = \frac{m_0 N_a}{M_i N_m} \tag{11.21}$$

Note that besides expressing concentration as a fraction, we can also express it as parts per million (ppm), parts per billion (ppb), or parts per trillion (ppt). As with ozone, we can report the concentration in ppm simply by multiplying the concentration expressed as a fraction by 10^6. To report the concentration in ppb or ppt, we multiply the concentration expressed as a fraction by 10^9 or 10^{12}, respectively.

Example 11.8

Steady state mass and concentration for a constant emission rate. The production of CFC-11 throughout the 1960s averaged about 1.2×10^{11} g/year. We assume that $t = 0$ corresponds to the beginning of 1960 and that $m_o = C_o = 0$. We also assume that all of the compound produced each year is emitted to the atmosphere in the same year. If the emission rate of CFC-11 in the 1960s continued indefinitely, calculate the steady state mass and steady state concentration of this CFC. How long would it take for the concentration to reach 98 percent of steady state?

Solution:

The mass of CFC-11 in the atmosphere at steady state can be found by noting that $m_{ss} = P\tau$ in Equation (11.19), where $\tau = 50$ years according to Table 11.1. Also recall that P is the CFC mass emissions into the box, which is given as 1.2×10^{11} g/yr. Thus,

$$m_{ss} = P\tau = 1.2 \times 10^{11} \text{g/year} \times 50 \text{ years} = 6.0 \times 10^{12} \text{ g}$$

To find the concentration at steady state, we first calculate the molecular weight of CFC-11 using the molecular weight of carbon (12 grams/mole), fluorine (19 grams/mole), and chlorine (35 grams/mole). The chemical formula for CFC-11 is CCl_3F, so the molecular weight is 136 grams/mole. We can then use Equation (11.21)

$$C_{ss} = \frac{m_{ss}N_a}{M_i N_m} = \frac{(6.0 \times 10^{12} \text{ grams})(6.023 \times 10^{23} \text{ molecules of CFC/mole CFC})}{(136 \text{ grams/mole})(1.1 \times 10^{44} \text{ molecules of air})}$$

$$= 0.241 \times 10^{-9} = 0.241 \text{ ppb}$$

We should express the answer to two significant digits because the input data are expressed only to this number of significant digits. Thus the steady state concentration is 0.24 ppb. However, we continue to use three significant digits in subsequent calculations to avoid cumulative rounding error. Equation (11.20) with $C_o = 0$ can be used to write the expression for concentration as a function of time

$$C(t) = C_{ss}(1 - e^{-t/\tau}) + C_o e^{-t/\tau}$$

$$= (0.241)(1 - e^{-t/50})$$

A graph of this equation is shown in Figure 11.9. Note that time must go to infinity in order for the concentration to reach steady state. Thus, as a practical matter, we typically look for a good approximation to steady state. In order to determine the time it takes for the concentration to reach 98 percent of its steady state value, we use Equation (11.20) with $C_o = 0$ and $C(t) = 0.98 \, C_{ss}$, rearranged to solve for time

$$C(t) = C_{ss}(1 - e^{-t/\tau})$$

so that

$$t = -\tau \times \ln\left(\frac{C_{ss} - C(t)}{C_{ss}}\right) = -50 \text{ years} \times \ln\left(\frac{C_{ss} - 0.98C_{ss}}{C_{ss}}\right)$$

$$= -50 \text{ years} \times \ln(0.02)$$

$$= 196 \text{ years}$$

Expressed to two significant figures, the result is 200 years.

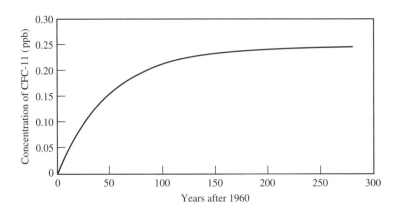

Figure 11.9   Concentration of CFC-11 as a function of year based on the conditions of Example 11.8.

Example 11.8 shows that the airborne concentration of CFC-11 approaches a steady state value after a sufficient number of years. How can the concentration approach a steady state if emissions continue to inject CFC-11 into the atmosphere? Indeed, these emissions cause the airborne concentration of CFC-11 to keep rising. However, the loss rate of CFC-11 also increases as the airborne concentration increases, as we have seen from Equation (11.8). Eventually, the airborne concentration gets so high that the loss rate exactly balances the emission rate; at this point the airborne concentration is at steady state.

Example 11.9

Concentration as a function of time for variable emission rates.   The production of CFC-11 increased significantly beginning in 1970; the average over the next two decades was 3.1×10^{11} g/year. Use this value and the emissions from Example 11.8 to write expressions for CFC-11 concentration as a function of time for two periods: $0 < t < 10$ years (1960–1969) and $t > 10$ years (beginning January 1970). Assume the emissions for $t > 10$ years continue indefinitely. Graph $C(t)$ for $0 < t < 50$ years.

Solution:

As in the previous example, we assume that $t = 0$ corresponds to the year 1960 (with CFC concentration = 0). The general expression for concentration as a function of time in period 1 (1960–1969) is

$$C_1(t) = C_{1ss}(1 - e^{-t/50}) + C_{1o}e^{-t/50}$$

where $C_{1ss} = 0.241$ ppb and $C_{1o} = 0$ as in Example 11.8.

The general expression for concentration as a function of time in period 2 (1970 and later) is

$$C_2(t) = C_{2ss}(1 - e^{-(t-10)/50}) + C_{2o}e^{-(t-10)/50}$$

where the exponential terms incorporate $(t - 10)$ to reflect time from the beginning of period 2. The steady state concentration in period 2 is calculated as

$$C_{2ss} = \frac{m_{ss}N_a}{M_iN_m} = \frac{P\tau N_a}{M_iN_m}$$

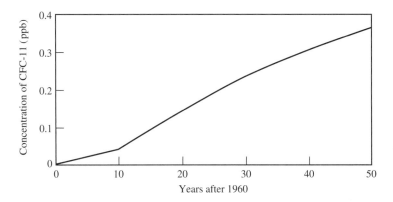

Figure 11.10   Concentration of CFC-11 as a function of year for the conditions of Example 11.9.

$$= \frac{3.1 \times 10^{11}\,\text{g/year} \times 50\,\text{years} \times 6.023 \times 10^{23}}{136\,\text{g/mole} \times 1.1 \times 10^{44}} = 0.624 \times 10^{-9}$$

$$= 0.624\,\text{ppb}$$

The initial CFC concentration at the beginning of period 2, C_{2o}, is equal to the final concentration at the end of period 1 ($t = 10$ years), calculated using the expression from Example 11.8

$$C_{2o} = C_1(t = 10\,\text{years}) = 0.241(1 - e^{-10/50}) = 0.0437\,\text{ppb}$$

Substituting into the general equation for concentration in period 2, we obtain

$$C_2(t) = 0.624(1 - e^{-(t-10)/50}) + 0.0437e^{-(t-10)/50}$$

$$= 0.624 - 0.580e^{-(t-10)/50}$$

This equation applies for $t > 10$ years. The complete solution, including both $C_1(t)$ and $C_2(t)$, is shown in Figure 11.10 out to $t = 50$ years, as requested in the problem statement. Note that the graph gives a concentration of 0.250 ppb in 1992 compared with the actual value of 0.268 ppb (268 ppt) for CFC-11 listed in Table 11.1. The agreement is reasonable, despite the rough approximations in emissions and assumptions in the mass balance.

11.5.2   Calculating Amounts of Ozone Depletion from CFCs

Calculating the amount of ozone depletion can be quite complex. However, a simple relationship can be used to compute the approximate ozone depletion based on changes in the chlorine monoxide airborne concentration (Elliott and Rowland, 1987). The equation does not describe the more complicated problem of the Antarctic hole,

Table 11.2   Values needed to determine ozone depletion from Equation (11.22). Concentrations are 1986 values given in molecules of each species per molecule of air.

Altitude (km)	Concentration (as a Fraction)		
	NO_2	ClO	$4ClO/NO_2$
20	1×10^{-9}	5×10^{-12}	0.02
30	5×10^{-9}	1×10^{-10}	0.08
40	5×10^{-9}	5×10^{-10}	0.4

Source: Based on Elliot and Rowland, 1987.

nor can it predict the spatial variability of the ozone depletion with latitude or longitude. The relationship is

$$\frac{O_{3,f}}{O_{3,i}}(z) = \frac{1 + \dfrac{4ClO_i}{NO_2}(z)}{1 + \dfrac{4ClO_f}{NO_2}(z)} \tag{11.22}$$

The i indicates initial conditions at the beginning of the period of interest, and f indicates the value of the variable during the final year in the calculation. Note that the equation depends on the altitude, z. We assume that the concentration of ClO is directly proportional to the CFC concentration. Ozone chemistry also is affected by the trace amounts of nitrogen dioxide (NO_2) that are found in the stratosphere. (Recall from Chapter 8 that NO_2 is a criteria air pollutant that can also accumulate in the troposphere and react chemically to produce urban smog.) NO_2 concentrations in the stratosphere are assumed to remain constant over time, so that the ratio $4ClO_i/NO_2$ also increases by the same amount as CFC concentrations. The measured concentration values for NO_2, ClO, and $4ClO_i/NO_2$ as a function of altitude can be found from Table 11.2 for the year 1986. This year is sometimes taken as the base year for calculating future ozone depletions.

The following example illustrates the use of the data in Table 11.2.

Example 11.10

Percentage ozone depletion at different altitudes.   Determine the percentage depletion of ozone at altitudes of 20, 30, and 40 kilometers between 1986 and some future time when steady state CFC-11 concentration is achieved. Use the results of Example 11.9 for both the 1986 and the steady state CFC-11 concentrations.

Solution:

Recalling that $t = 0$ corresponds to the year 1960, the concentration of CFC-11 in 1986 can be computed as

$$C(t = 26 \text{ years}) = 0.624 - 0.580e^{-(26-10)/50} = 0.203 \text{ ppb}$$

Thus we can write

$$\frac{C_{final}}{C_{initial}} = \frac{C_{ss}}{C(1986)} = \frac{0.624 \text{ ppb}}{0.203 \text{ ppb}} = 3.07$$

According to the previous discussion, we assume that the concentration of ClO increases in the same ratio as the concentration of CFC-11. Therefore

$$\frac{4ClO_f}{NO_2} = 3.07 \frac{4ClO_i}{NO_2}$$

We can now use Equation (11.22) with the final column of Table 11.2 as follows:

$$\frac{O_{3,f}}{O_{3,i}}(20 \text{ km}) = \frac{1 + (0.02)}{1 + 3.07(0.02)} = 0.96 \Rightarrow 4\% \text{ depletion}$$

$$\frac{O_{3,f}}{O_{3,i}}(30 \text{ km}) = \frac{1 + (0.08)}{1 + 3.07(0.08)} = 0.87 \Rightarrow 13\% \text{ depletion}$$

$$\frac{O_{3,f}}{O_{3,i}}(40 \text{ km}) = \frac{1 + (0.4)}{1 + 3.07(0.4)} = 0.63 \Rightarrow 37\% \text{ depletion}$$

Note that the amount of depletion calculated in Example 11.10 increases as altitude increases, consistent with earlier discussion. The total depletion would actually be greater than calculated here because other CFCs (especially CFC-12) also contribute to ozone depletion. For illustration, we will continue with calculations specific to CFC-11.

It is useful to determine the total ozone depletion at all heights. To do this, we must account for the number concentration of ozone molecules as a function of height. Note that the amount of ozone lost at 40 km is actually smaller than that lost at lower heights, despite the high percentage loss, because there were originally fewer ozone molecules per cm^3 at high altitudes. For the base year 1986, the number concentration of ozone molecules below 15 km can be approximated by 1.0×10^{12} molecules/cm^3, the number concentration between 15 and 25 km by 4.0×10^{12} molecules/cm^3, and the number concentration above 25 km by (Elliot and Rowland, 1987)

$$n = 4.0 \times 10^{12} \times e^{-(z-25)/7} \tag{11.23}$$

This information can be used as input data for estimating total column losses.

Example 11.11

Overall ozone depletion throughout the atmospheric column. Estimate the total percentage of ozone depletion over the atmospheric column in steady state, based on CFC-11. Use the average of losses at 10, 20, 30, and 40 km as the basis for the calculation. Assume the values at 10 km represent the troposphere where there is no depletion, and use Equation (11.23) for the number concentration of ozone molecules above 25 km.

Table 11.3   Concentration of ozone molecules in number/cm³ at different altitudes based on the conditions in Example 11.11.

z (km)	Original Number per cm³	Fraction Remaining	Final Number per cm³	O₃ loss (Number/cm³)
10	1.0×10^{12}	1.00	1.0×10^{12}	0.0
20	4.0×10^{12}	0.96	3.8×10^{12}	0.2×10^{12}
30	2.0×10^{12}	0.87	1.7×10^{12}	0.3×10^{12}
40	4.7×10^{11}	0.63	3.0×10^{11}	0.17×10^{12}
Average	$\mathbf{1.87 \times 10^{12}}$		$\mathbf{1.70 \times 10^{12}}$	

Solution:

The original numbers of ozone molecules at 10 and 20 km are 1.0×10^{12} and 4.0×10^{12}, respectively, according to the previous discussion. The original number of ozone molecules at 30 km is calculated from Equation (11.23) as follows

$$n(30 \text{ km}) = 4.0 \times 10^{12} \times e^{-(30-25)/7} = 2.0 \times 10^{12}$$

Similarly, the original number of ozone molecules at 40 km is calculated as

$$n(40 \text{ km}) = 4.0 \times 10^{12} \times e^{-(40-25)/7} = 4.7 \times 10^{11}$$

These values are used along with the fractions remaining after depletion, taken from Example 11.10, to calculate the final number of ozone molecules at each height. The results are shown in Table 11.3.

We are interested in the overall atmospheric column, so we take the average value of the number density at each height. Essentially, this calculation assumes that the atmosphere is represented by four layers, each 10 km thick, with a uniform number density within each layer. The calculations in Table 11.3 show that the total atmospheric column has been depleted by

$$\frac{(1.87 \times 10^{12}) - (1.70 \times 10^{12})}{1.87 \times 10^{12}} = 0.09, \text{ or } 9\%$$

As in Example 11.10, this calculated depletion is an underestimate because it is based only on CFC-11. Furthermore, the calculation divided the atmospheric column into only four layers; a greater number of layers would provide better accuracy. Note that although the entries in Table 11.3 are given to only two significant figures, we use three significant figures in the calculation because we are calculating a small difference between two large numbers. Using fewer significant figures in the final step would cause a greater error in the overall result.

11.6   SOLUTIONS TO THE CFC PROBLEM: THE MONTREAL PROTOCOL

In 1976 the United States announced that CFCs used as propellant gases in aerosol spray cans would be eliminated. Several other countries, including Canada, Norway, and Sweden, followed suit in the late 1970s. By 1987 the evidence against CFCs was strong enough that a treaty known as the Montreal Protocol was signed by more than 30 countries. By 1998 more than 165 countries had signed. The treaty and its subsequent amendments committed industrialized countries to ending production of CFCs by 1996. Developing countries are committed to a complete elimination of CFC production by 2010. The use of CFCs by consumers is not affected. Consumers may continue to use products they already own that contain CFCs; however, new products such as refrigerators and air conditioners must be designed to avoid the use of CFCs.

Hydrochlorofluorocarbons (HCFCs) have begun to be used as alternatives for CFCs in some applications. HCFCs differ from CFCs in that only some, rather than all, of the hydrogen in the parent hydrocarbon is replaced by chlorine or fluorine. Examples include HCFC-22 (CHF_2Cl), HCFC-123 ($C_2HF_3Cl_2$), and others listed earlier in Table 11.1. The presence of the hydrogen atom decreases the stability of the molecule and makes it more susceptible to destruction in the troposphere; thus many of the HCFC molecules are eliminated before they reach the stratosphere where they can destroy ozone. Although these compounds are still considered a problem because they contain chlorine, most HCFCs destroy only a small percentage of the amount of ozone destroyed by CFCs. HCFC-22, for example, depletes only 3.4 percent as much ozone as CFC-11 (WMO, 1999).

Amendments to the Montreal Protocol in 1992 required that industrialized countries eliminate HCFCs by 2030. Developing countries have yet to agree to any phaseout date for HCFCs. Gases covered by the Montreal Protocol and its amendments are summarized in Table 11.4.

What compounds can replace HCFCs? Researchers are exploring alternative compounds, including hydrofluorocarbons (HFCs) and perfluorocarbons (FCs), none of which contain chlorine. HFCs contain hydrogen, carbon, and fluorine;

Table 11.4 Ozone-depleting gases covered by the Montreal Protocol and its amendments.

Gases Phased Out before 2000	Gases to Be Phased Out by 2030
CFC-11	HCFC-22
CFC-12	HCFC-123
CFC-113	HCFC-124
CFC-114	HCFC-141b
CFC-115	HCFC-142b
Carbon tetrachloride	HCFC-225ca
Methyl chloroform	HCFC-225cb
Halon-1211	
Halon-1301	
Halon-2402	

Source: IPCC, 1996.

examples include HFC-134a ($C_2H_2F_4$), which is now used in place of CFC-12 in new refrigerators (see Chapter 6). Perfluorocarbons do not contain hydrogen and include compounds like FC-116 (C_2F_6) and FC-31-10 (C_4F_{10}). Unfortunately, these compounds are powerful greenhouse gases that contribute to the environmental problem of global warming (see Chapter 12). For this reason, their use as CFC substitutes may be limited. Indeed, many classes of CFC substitutes also are greenhouse gases, although many of them have smaller global warming potential compared to CFCs. Pure hydrocarbons such as propane (C_3H_8) are another alternative to CFCs in some applications such as refrigeration; however, as discussed in Chapter 6, nonenvironmental considerations, such as flammability and safety, may inhibit the widespread use of propane. In short, it will take time to find the best substitutes for some CFC applications.

11.6.1   Ozone Depletion Potential

To compare the amounts of ozone depleted by different compounds, an index called the *ozone depletion potential* (*ODP*) has been developed based on computer models of the chemical reactivities of these compounds. This index is defined as

$$ODP = \frac{\text{Loss in total ozone due to emissions of a unit mass of a compound}}{\text{Loss in total ozone due to emissions of a unit mass of CFC-11}} \quad (11.24)$$

The *unit mass* refers to 1 kg of each substance or any other value that is the same in both numerator and denominator. Note that the loss in total ozone applies to a period long enough to allow the change in ozone to reach steady state. One could also define an ODP based on time-dependent considerations, although doing so is outside the scope of this chapter.

By definition, CFC-11 has an ODP of 1.0. CFC-12 and CFC-113 have ODPs of 0.82 and 0.90, respectively. HCFC-22 has an ODP of 0.034, as mentioned earlier in this section, whereas HCFC-123 has an ODP of 0.012 (WMO, 1999). Chapter 6 (Table 6.1) contains a more extensive tabulation of ODP values for common refrigerants and other halocarbons based on recent modeling results.

Example 11.12

Reduction in ozone depletion with CFC substitutes.   What reduction in stratospheric ozone depletion would occur if all of the CFC-12 in the atmosphere were eliminated and replaced by the same mass of HCFC-22? Assume steady state depletion in both cases.

Solution:

Although the definition of ODP is based on a unit mass, other mass values can be used in Equation (11.24) if the numerator and denominator values are the same. The ODP gives ozone loss relative to CFC-11, so we must use the following calculation to estimate the potency of HCFC-22 relative to CFC-12:

$$\frac{\text{Loss in total ozone due to emissions of a unit mass of HCFC-22}}{\text{Loss in total ozone due to emissions of a unit mass of CFC-12}}$$

$$= \frac{\text{Loss in total ozone due to emissions of a unit mass of HCFC-22}}{\text{Loss in total ozone due to emissions of a unit mass of CFC-11}}$$

$$\times \frac{\text{Loss in total ozone due to emissions of a unit mass of CFC-11}}{\text{Loss in total ozone due to emissions of a unit mass of CFC-12}}$$

$$= (\text{ODP of HCFC-22}) \times \frac{1}{\text{ODP of CFC-12}}$$

$$= 0.034 \times \frac{1}{0.82} = 0.044$$

11.6.2 Potential Environmental Trade-offs

Although the Montreal Protocol and its amendments are considered among the most important environmental regulations yet developed, some potential secondary impacts need to be fully evaluated. One beneficial impact of eliminating CFCs is that their contribution to global warming is also eliminated: CFCs are powerful greenhouse gases as well as depleters of stratospheric ozone. On the other hand, there may also be undesirable impacts. For example, the planned phaseout of HCFCs with very low ozone depletion potential has been criticized for not doing much to attack the problem of ozone loss. In fact, some have argued that phasing out low-ODP gases may actually increase environmental damage. For example, replacing HCFC-123 with other refrigerants by the year 2030, as required by the Montreal Protocol amendments, could increase electricity use because most current alternatives are less efficient. This would mean higher power requirements for cooling, which would likely increase CO_2 emissions from power generation. Thus a very small (or negligible) improvement in ozone depletion may be replaced by increased global warming impacts due to increased CO_2 emissions (Wuebbles and Calm, 1997).

Such trade-offs among different environmental effects are rarely considered in international regulations due to the complexity of these effects; in this case, the Montreal Protocol was developed independently of global warming concerns. The problem of global warming is discussed in detail in Chapter 12, including the contribution of CFCs and other halocarbons. Chapter 15 discusses the role of technological innovation in addressing potential environmental problems such as the trade-offs noted above.

The reduction in CFC production is already beginning to have an impact. The rate of increase in chlorine concentrations was measurably slower in the mid-1990s compared with earlier years. Furthermore, it has been reported that the peak in chlorine may have already occurred. Even if future data show that chlorine levels are now on their way down, it is imperative to continue the search for CFC replacements to enable stratospheric ozone levels to return to normal as quickly as possible. This will take at least several decades, assuming the provisions of the Montreal Protocol are followed.

11.7   CONCLUSION

In this chapter we have considered the problem of chlorofluorocarbons (CFCs) that destroy stratospheric ozone. This naturally occurring ozone is vital for absorbing harmful ultraviolet radiation from the sun. CFCs reduce the amount of stratospheric ozone and thus enable greater amounts of ultraviolet energy to reach the earth's surface.

We found that CFCs are extremely stable and thus can diffuse slowly upward into the stratosphere, where they contribute chlorine atoms that react with the ozone. The problem is found in the midlatitudes, in the range 30–60 degrees north and 30–60 degrees south latitude. Ozone depletion also is especially severe over Antarctica because of special conditions that exist there.

We developed a mass balance model and found that it can roughly approximate the amounts of CFC in the stratosphere. This information can then be used with simple relationships to estimate the amount of ozone depletion at different altitudes. The overall depletion of ozone in the atmospheric column can also be estimated.

The problem of stratospheric ozone depletion was recognized as a major environmental problem in the 1980s, and this led to the development of treaties to reduce CFC production worldwide. Overall, CFCs represent one example of a global environmental problem that developed because of a poor understanding of natural physical and chemical processes. But they also represent an example of how nations around the globe can work together to reach mutually beneficial solutions to environmental problems.

International agreements now in place are expected to bring the problem of stratospheric ozone depletion under control by the middle of the 21st century. However, substantial technical challenges remain to finding acceptable substitutes for the ozone-depleting chemicals still in use. Avoiding new environmental problems—such as contributions to global warming from CFC substitutes—is part of that technical challenge.

11.8   REFERENCES

AFEAS, 1995. "Production, Sales, and Atmospheric Release of Fluorocarbons through 1994," Alternative Fluorocarbon Environmental Acceptability Study, Program Office, U.S. Environmental Protection Agency, Washington, D.C.

Chapman, S., 1930. "A Theory of Upper Atmosphere Ozone." *Roy. Meteor. Soc.,* vol. 3, pp. 103–25.

Elliott, S., and F. S. Rowland, 1987. "Chlorofluorocarbons and Stratospheric Ozone." *J. of Chem. Edu.,* vol. 64, pp. 387–91.

Farman, J. C., B. G. Gardiner, and J. D. Shanklin, 1985. "Large Losses of Total Ozone in Antarctica Reveal Seasonal CLO_x/NO_x Interaction." *Nature,* vol. 315, pp. 207–10.

IPCC, 1996. *Climate Change 1995: The Science of Climate Change.* Intergovernmental Panel on Climate Change, J. T. Houghton, L. G. Meira Fihlo, J. Bruce, H. Lee, B. A. Callander, E. Haites, N. Harris, and K. Marshall, eds., Cambridge University Press, Cambridge, England.

McCormick, M. P., H. M. Steele, P. Hamill, W. P. Chu, and T. J. Swissler, 1982. "Polar Stratospheric Cloud Sightings by SAM II." *J. Atmos. Sci.,* vol. 39, pp. 1387–97.

NAS, 1976. *Halocarbons: Effects on Stratospheric Ozone.* Panel on Atmospheric Chemistry, National Academy Press, Washington, D.C.

NASA, 2000. National Aeronautics and Space Administration, Washington, D.C., http://see.gsfc.nasa.gov/edu/SEES/strat/class/S_class.htm.

Rowland, F. S., 1989. "Chlorofluorocarbons and the Depletion of Stratospheric Ozone." *Amer. Sci.,* vol. 77, pp. 36–45.

Rowland, F. S., 1991. "Stratospheric Ozone Depletion." *Annu. Rev. Phys. Chem.,* vol. 42, pp. 731–68.

WMO, 1999. "Halocarbon Scenarios for the Future Ozone Layer and Related Consequences." S. Madronich and G. J. M. Velders, lead authors. *Scientific Assessment of Ozone Depletion,* World Meteorological Organization, Global Ozone Research and Monitoring Project No. 44.

Wuebbles, D. J., and J. M. Calm, 1997. "An Environmental Rationale for Retention of Endangered Chemicals." *Science,* vol. 278, pp. 1090–91.

11.9 PROBLEMS

11.1 The current federal primary standard for ozone is published as 0.08 ppm or 170 μg/m^3. What temperature has been assumed for this equivalency? Assume the pressure is 1 atmosphere.

11.2 The atmospheric column contains 280 DU of ozone over the eastern United States on a certain day. A ground-level ozone monitor shows a concentration of 0.05 ppm, which is assumed to be uniform up to a temperature inversion height of 700 m. Assume $p = 1$ atmosphere and $T = 300$ K for the region below 700 m. What fraction of the ozone molecules in the total atmospheric column are below the inversion height?

11.3 If the overall global average ozone concentration is 300 DU, calculate the approximate number of ozone molecules in the atmosphere at any given time.

11.4 Sketch the following molecules: CFC-113, CFC-114, HCFC-22, and HCFC-123.

11.5 During 1990–1993 the production of CFC-11 dropped, averaging only 1.9×10^{11} g/year over these four years. By the beginning of 1994, emissions were sufficiently low to be neglected. Write the expression for $C(t)$ over the entire range 1990–2050. For consistency with the notation in Example 11.9, let $C_3(t)$ be the concentration between 1990 and 1993, and let $C_4(t)$ be the concentration after the beginning of 1994. Assume $t = 0$ corresponds to 1960.

11.6 Using the assumptions in Problem 11.5, the airborne concentration of CFC-11 will be a maximum at the beginning of 1994. Determine how long it will take for the concentration to fall to 2 percent of its maximum value. Compare with the number of years it will take for the concentration to reach 98 percent of its steady state value in Example 11.8. Explain your findings when you make this comparison.

11.7 In the previous problems you used approximations for actual emissions of CFC-11 to estimate the rise and fall of CFC-11 airborne concentrations.

 (a) Determine the percentage of ozone depletion in the stratosphere relative to the amount of ozone in 1986 at altitudes of 20, 30, and 40 km at the beginning of 1994. Use the concentration of CFC-11 at the end of 1986 determined in Example 11.10 as the basis for your calculation.

 (b) Estimate the total percentage of ozone depletion in the entire atmospheric column using the depletions you calculated in part (a). Compare your answer with the result of Example 11.11, and discuss reasons for the differences.

11.8 In Example 11.9 you were given the average production rate of CFC-11 as 3.1×10^{11} g/year for the 20 years 1970–1989. The emission rate was assumed to equal the production rate in any given year. The steady state concentration was calculated as 0.624 ppb if the emissions continued at this rate indefinitely. Consider the production of CFC-12, which averaged 3.8×10^{11} g/year for the same 20-year period. Use the same assumption regarding emissions equal to production.

   (a)   Calculate the steady state concentration of CFC-12 if the 1970–1989 emissions continued indefinitely.

   (b)   Which of these two CFCs emits more chlorine atoms to the atmosphere?

   (c)   Calculate the steady state mass of atmospheric chlorine from each of these CFCs. Which of these CFCs would contribute more atmospheric chlorine in steady state if the emissions continued indefinitely?

11.9 In Example 11.8 we computed the steady state mass and concentration of CFC-11 for a constant emission rate of 1.2×10^{11} g/year throughout the 1960s. We assumed that the emissions of this compound were zero before 1960. Data show that CFC-11 production actually averaged 0.2×10^{11} g/year during the 1950s, with negligible production before 1950. Assume all of the 1950–1959 production was emitted to the atmosphere in the same year it was produced.

   (a)   Calculate the CFC-11 concentration at the beginning of 1960.

   (b)   Assume that we account for 1950–1959 CFC-11 emissions as described in part (a). Does this new information influence the calculation of steady state CFC-11 concentration in Example 11.8? Explain.

   (c)   Does this new information influence the calculation of the time it takes for the concentration to reach 98 percent of the steady state value? Determine how long it takes (after 1960) and compare with Example 11.8.

11.10 The production of CFC-11 in a given year resulted in a significant ozone depletion. By what fraction would this depletion be reduced if half the CFC-11 produced in that year were replaced by the equivalent mass production of HCFC-22? Assume steady state ozone depletion in both cases.

11.11 According to data from AFEAS (1995), the total global production of CFC-113 decreased from 2.4×10^{11} g/year at the beginning of 1989 to 3.0×10^{10} g/year at the beginning of 1994, as a result of the Montreal Protocol. Over the same period, total production of HCFC-22 increased from 2.1×10^{11} g/year to 2.4×10^{11} g/year.

   (a)   Assuming linear changes with time between January 1, 1989, and January 1, 1994, determine the month and year in which the rates of production of the two gases were equal.

   (b)   How much more ozone (expressed as a ratio) would eventually be destroyed by all the CFC-113 produced in 1989 compared with the amount of ozone eventually destroyed by all the HCFC-22 produced in 1989?

Global Warming and the Greenhouse Effect

12.1   INTRODUCTION

As you undoubtedly know, a greenhouse is a glass-enclosed structure used to grow plants. When the sun is shining, the temperature inside the greenhouse stays warmer than the outside air temperature. A car sitting in the sun with the windows closed exhibits the same phenomenon—a higher temperature inside than outside. In both cases the warming effect is due partially to the thermal properties of glass, which allows most of the energy in sunlight to pass through and warm the interior surfaces while blocking much of the outward energy flow from the interior. As we shall see later in this chapter, the radiative processes that help to warm a greenhouse also keep the earth at a comfortable average temperature of about 15°C (59°F). Were it not for the greenhouse effect, life on earth as we known it would not exist.

The environmental concern today is over too much of this good thing. As discussed in Chapter 2, increasing emissions of carbon dioxide and other *greenhouse gases* from human activities are believed to be producing a global warming trend whose consequences could be disruptive, and potentially catastrophic, for people and the environment. Thus the key questions that motivate this chapter are

- What is the nature of the greenhouse effect, and how well do we understand it?

- If global warming occurs, how quickly will it happen and how big a change can we expect?

- What measures can be taken to reduce or eliminate the emissions responsible for global warming?

12.1.1   Greenhouse Gas Emissions and Atmospheric Change

The magnitude and sources of the principal greenhouse gas emissions were discussed in Chapter 2 and are summarized in Table 12.1. Figure 12.1 shows the recent increases in atmospheric concentration of several of these gases, and Table 12.2 shows some of their atmospheric characteristics.

Table 12.1   Annual emissions of major greenhouse gases.

	Annual Emissions (Mt/yr)[a]	
Greenhouse Gas	**World**	**U.S.**
Carbon dioxide (CO_2)	29,800	5,300
Methane (CH_4)	375	31
Nitrous oxide (N_2O)	5.7	0.5
CFC-11, -12, -113	0.7	0.1
HCFC-22	0.2	0.1
HFCs, PFCs, SF_6	n/a	0.034

[a]   Mt = millions of metric tons. CO_2 and all U.S. data are for 1996. Other world data are for the early 1990s. n/a = not available.
Source: IPCC, 1996a; USDOE, 1999, 1997; Marland et al., 1999.

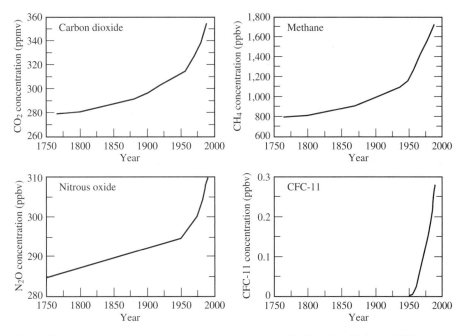

Figure 12.1 Recent trends in atmospheric concentrations of CO_2, CH_4, N_2O, and CFCs. Concentrations of all greenhouse gases have risen sharply in the last century due to human activities. (*Source:* IPCC, 1990)

Table 12.2 Increases in atmospheric concentration of greenhouse gases since 1850.

Quantity	CO_2	CH_4	N_2O	CFC-11	HCFC-22	CF_4
Preindustrial concentration[a]	~280 ppmv	~700 ppbv	~275 ppbv	Zero	Zero	Zero
Concentration in 1994	358 ppmv	1,720 ppbv	312 ppbv	268 pptv	110 pptv	72 pptv
Recent rate of change in concentration	1.5 ppmv/yr (0.4%/yr)	10 ppbv/yr (0.6%/yr)	0.8 ppbv/yr (0.25%/yr)	0 ppbv/yr (0%/yr)	5 pptv/yr (5%/yr)	1.2 pptv/yr (2%/yr)
Atmospheric lifetime (years)[b]	50–200[c]	12	120	50	12	50,000

[a] ppmv = parts per million by volume; ppbv = parts per billion; pptv = parts per trillion.
[b] Time for an initial mass to decay exponentially to $1/e$ = 0.368 of its initial value.
[c] No single lifetime for CO_2 can be defined because of the different rates of uptake by different sink processes.
Source: IPCC, 1996a.

Carbon dioxide (CO$_2$) is the substance emitted in greatest quantity. The main source of CO_2 is the combustion of fossil fuels (oil, coal, and natural gas) that provide energy for transportation, electric power generation, residential and commercial buildings, and industrial processes. The atmospheric concentration of CO_2 has risen steadily for more than a century, with an overall increase of about 30 percent in the past 150 years—a period in which the use of fossil fuels accelerated rapidly (see Chapter 2). The natural sinks for CO_2 emissions include the world's oceans, plus uptake via photosynthesis by trees and other vegetation. But the continued depletion of the world's forests has reduced their natural ability to absorb CO_2, thus indirectly contributing to the increase in atmospheric CO_2 concentration. The burning of trees and vegetation that often accompanies the clear-cutting of forests further increases CO_2 emissions by releasing the carbon previously stored in wood and leaves.

Methane (CH_4) is another gas whose atmospheric concentration has increased in modern times. This growth closely parallels the increase in world population.

Human activities that release CH_4 include energy-related processes such as natural gas production and coal mining. CH_4 also is given off from paddy rice production, animal digestive tracts, decomposing landfill wastes, and natural sources such as wetlands and termites (which also are influenced by human activities).

Nitrous oxide (N_2O) is commonly known as laughing gas, but the increases in atmospheric N_2O concentration are no laughing matter. N_2O emissions are associated mainly with the use of nitrogen-containing chemical fertilizers, which are used worldwide for food production as well as in nonagricultural applications such as lawns and gardens. Combustion also releases small quantities of N_2O. The various magnitudes of N_2O emissions are highly uncertain, although the principal releases to the atmosphere are believed to occur indirectly as a result of chemical processes in soil.

Halocarbons are compounds that combine carbon with chlorine, fluorine, and bromine. Examples include chlorofluorocarbons (CFCs) such as the CFC-11 shown in Figure 12.1. Observed increases in atmospheric concentrations of halocarbons are unquestionably due to human activities because these compounds do not exist in nature. Rather, they have been created since the 1930s for use in refrigeration systems, aerosol cans (until banned in the 1970s), and a host of industrial applications. The subject of halocarbons in the atmosphere was discussed extensively in Chapter 11, which addressed CFCs and the ozone hole. That environmental issue resulted in an international agreement to phase out the production of CFCs and certain other halocarbons by the year 2000. The ban has already stemmed the increase in CFC concentrations in the atmosphere. Nevertheless, many of the substitutes for CFCs, such as HCFC-22 listed in Table 12.2, still contribute to global warming. *Perhalogens* are a class of human-made chemicals that includes compounds such as CF_4 and SF_6. These chemicals are found mainly in industrial processes, but their use has increased since CFC production was banned. Now these chemicals are beginning to appear in measurable concentrations in the atmosphere, as seen in Table 12.2 for CF_4 (a perfluorocarbon, or PFC).

A common characteristic of CO_2, CH_4, N_2O, halocarbons, and perhalogens is that they are all greenhouse gases that can disrupt the temperature balance of the earth. A second characteristic, as seen in Table 12.2, is that they all have long atmospheric lifetimes, typically measured in decades to centuries. This means that once they are in the atmosphere, their global warming effects cannot be easily reversed. In this chapter you will learn how these gases cause the earth to warm and what can be done to mitigate the problem.

12.1.2   The Global Climate System

Before looking at the causes of global warming, it is important to first say a word about the global climate system. Indeed, most of the concerns about global warming really reflect concerns about the potential climate change impacts of warming induced by human activities.

What do we mean by *climate*? Some people think of climate as synonymous with *weather*. However, we all know that weather changes from day to day. Climate is something more stable, commonly defined as the average weather. For example, the climate of the northeastern United States is predictably cold in winter and warm in summer, with an expected annual precipitation of roughly 100 cm of water. In contrast, the climate of southern California is mild to warm year-round, with a lower average rainfall. The climate of a region may vary over time (for example, some winters may be milder or wetter than others), but such fluctuations are usually measured in years or decades—as opposed to changes in the weather, which may occur over hours. Climatologists typically use a 30-year averaging period to characterize the average climate.

The average climate of a region affects the types of trees, plants, and crops that can grow and the types of birds, animals, and insects that can survive. On a more personal level, climate may affect where you choose to live, as well as the type of clothing you wear, the amount of money you spend on home heating and cooling, and the types of recreational activities you can easily enjoy (try finding a ski slope in Florida).

The average atmospheric surface temperature is often the key variable used to describe climate. However, the climate of a region is in fact defined by more variables, including the types and amounts of precipitation, the persistence of cloud cover, average humidity and wind speed, soil moisture content, and the frequency and severity of storms (including extreme events like hurricanes and tornadoes). Climate is determined by the complex interactions of many factors that together constitute the *global climate system*.

The climate system includes not only the atmosphere but also the oceans, sea ice, land ice, snow cover, rivers, lakes, and key features of the land surface, such as vegetation, biomass, human habitats, and ecosystems. All of these factors influence the atmosphere and climate. So do other factors that are usually considered to be outside the climate system, such as the intensity of solar radiation, the slowly changing orbit of the sun–earth system, and geographic features of the earth such as mountain ranges and the topography of the ocean bottom. Although these and other factors are responsible for the natural variations in global climate, the climate system as defined here includes only components that may change as a result of human activities.

12.1.3 Chapter Overview

Because the problem of global warming spans a vast range of scientific and technical disciplines, it is useful to first outline the scope and objectives of this chapter. There are two main objectives. The first is to develop a basic understanding of the processes that give rise to global warming and climate change. The second is to explore potential solutions to the problem, with an emphasis on the role that engineers can play in this complex issue.

Figure 12.2 shows a simple schematic of the basic processes that link emissions of greenhouse gases to global climate change. In terms of environmental science,

Figure 12.2   Basic elements of the global warming problem.

this chapter will focus mainly on the first four elements of this chain. We will begin by looking at the basic physics that triggers global warming when certain gases are added to the atmosphere. A phenomenon known as *radiative forcing* will be introduced as the key link between higher concentrations of greenhouse gases and the resulting change in the earth's average temperature. Understanding and modeling the full dimensions of climate change remain a formidable challenge. We will look briefly at what current models say about the magnitude of future climate change induced by anthropogenic emissions.

The initial link in Figure 12.2 between anthropogenic emissions and atmospheric concentrations of CO_2 and other gases also will be examined later in the chapter in the context of policy objectives and potential solutions to the global warming problem. For now, let us begin by examining the greenhouse effect and its relationship to global temperature. This is an essential first step for understanding the broader problem of global climate change.

12.2   FUNDAMENTALS OF THE GREENHOUSE EFFECT

The basic processes governing the earth's temperature and climate are related to *radiative heat exchange* between the earth and the sun. Recall that heat is a form of energy transferred by virtue of a temperature difference. Radiation is one of three methods by which heat can be transferred from a warmer body to a cooler body. The other two methods are *conduction,* which normally involves heat transfer through a solid substance, and *convection,* which transfers heat from a solid to a liquid or gas. Radiation is unique in that it does not require a physical medium for energy transfer; rather, radiative energy can be transferred across a vacuum (like outer space) in the form of electromagnetic waves or particles known as *photons*.

12.2.1   The Nature of Radiative Energy

Any object with a temperature above absolute zero continuously radiates energy—including this book, the chair you're sitting on, and you while reading this text. The amount radiated depends on the temperature of the material and the nature of its radiating surface. The maximum rate at which energy in the form of heat can be radiated by a body at a given temperature is given by the *Stefan-Boltzmann equation*:

$$\dot{Q}_{max} = \sigma A T^4 \tag{12.1}$$

where $\dot{Q}_{max}$ is the maximum rate of energy radiated (watts), σ is the Stefan-Boltzmann constant (5.67×10^{-8} W/m²-K⁴), A is the surface area of the body (m²), and

T is the absolute temperature of the body (K). An object radiating at this maximum rate is known as a *black body*. Although most real objects at any given temperature emit less than this theoretical maximum, many common substances, including water (which covers most of the earth's surface), radiate at close to the black body amount.

Often it is convenient to express radiative energy in terms of the rate per unit of surface area. This quantity is referred to as a *heat flux* and is denoted here by the symbol $\dot{q}$:

$$\dot{q} = \text{Heat flux} = \frac{\text{Total rate of of heat flow}}{\text{Total surface area}} = \frac{\dot{Q}}{A} \qquad (12.2)$$

For a black body,

$$\dot{q} = \frac{\dot{Q}_{max}}{A} = \sigma T^4 \qquad (12.3)$$

Example 12.1

Radiative heat transfer from a black body. An object with a temperature of 20°C has surface properties approximating those of a black body. Calculate the black body heat flux from this object, and also calculate the total heat flow rate if the surface area is 0.2 m².

Solution:

The black body heat flux is given by Equation (12.3). First convert the surface temperature to Kelvin:

$$T(\text{K}) = 20°\text{C} + 273 = 293 \text{ K}$$

Then

$$\dot{q} = \sigma T^4 = (5.67 \times 10^{-8})(293)^4 = 418 \text{ W/m}^2$$

If the surface area is 0.2 m², the total radiative heat flow rate is

$$\dot{Q} = \dot{q}A = (418 \text{ W/m}^2)(0.2 \text{ m}^2) = 83.6 \text{ W}$$

12.2.2 Solar Energy Reaching Earth

The radiant heat reaching earth from the sun closely approximates the radiation from a black body at a temperature of 5,800 K. At the average distance of the earth from the sun, the amount of solar energy incident on the earth's atmosphere averages about 342 W/m², based on the surface area of the earth.[1] We will call this incident heat flux the *solar input, S_o*. The next section explains why this quantity is so important.

---

1 Incident solar radiation can also be specified as a heat flux passing through a plane just outside the earth's atmosphere. That flux is known as the *solar constant* and has a value of 1,368 W/m². When averaged over the surface of the earth, the resulting heat flux is one-fourth the solar constant, or 342 W/m². The problems at the end of the chapter illustrate this calculation.

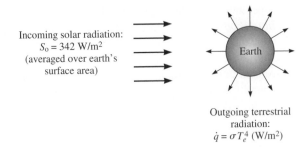

Incoming solar radiation:
$S_0 = 342$ W/m^2
(averaged over earth's
surface area)

Earth

Outgoing terrestrial
radiation:
$\dot{q} = \sigma T_e^4$ (W/m^2)

Figure 12.3   Schematic of incoming solar radiation and outgoing terrestrial radiation. These are the basic energy flows that govern the earth's temperature.

12.2.3   A Simple Earth Energy Balance

To understand the factors that influence the earth's temperature, it is useful to first examine what would happen without the greenhouse effect. Figure 12.3 shows a sketch of the two major energy flows that determine the earth's average surface temperature in this case.

In this simple model, the total solar energy incident on the earth warms the planet to a uniform surface temperature, T_e. In turn, the earth's surface radiates energy back to space, as depicted in Figure 12.3. Eventually, an equilibrium temperature is reached at which the rate of outgoing terrestrial radiation just equals the rate of incoming solar radiation. This stable condition determines the earth's average surface temperature.

Let us assume that the earth's surface approximates a black body. As depicted in Figure 12.3, terrestrial energy is then radiated out to space at a rate per unit surface area of

$$\dot{q}_{earth} = \sigma T_e^4 \qquad (12.4)$$

Although the earth's entire surface continuously radiates energy outward to space, only part of the surface area receives solar radiation at any given time. Averaged over time, however (such as a year), the solar energy input is distributed around the globe. The average rate of solar energy incident on the earth's surface is the solar input, S_0, noted earlier. This means that on average, each square meter of the earth's surface receives 342 W of solar energy.

Not all of this solar energy heats the planet. The nature of electromagnetic waves is such that part of this incoming radiation is reflected back out to space, as illustrated in Figure 12.4. The reflected fraction, *a*, called the *albedo*, depends on the properties of a surface. In general, smooth white surfaces like ice and snow are more reflective, whereas rough dark surfaces are less reflective. Based on measurements from space, the albedo of the earth is currently estimated at about 31 percent. This means that only the remaining 69 percent of incoming solar energy is absorbed and heats the surface. Thus, in general,

Rate of solar energy absorbed (W/m^2) $= S_0 (1 - a) \qquad (12.5)$

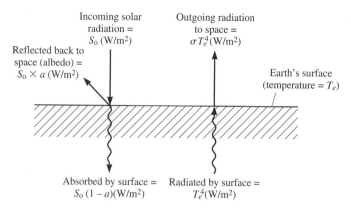

Figure 12.4   A simple energy balance model for the earth, assuming no atmosphere.

Equating this rate of energy absorbed to the rate of terrestrial radiation emitted from Equation (12.4) allows us to find the steady state surface temperature under equilibrium conditions:

$$\text{Rate of energy absorbed} \;=\; \text{Rate of energy emitted} \qquad (12.6)$$

Thus

$$S_o \, (1 \,-\, a) \;=\; \sigma \, T_e^{\,4}$$

Solving for T_e gives us the earth's temperature:

$$T_e \;=\; \left[\frac{S_o \, (1 \,-\, a)}{\sigma} \right]^{\frac{1}{4}} \qquad (12.7)$$

Thus the earth's albedo, a, and the average solar input, S_o, are the two parameters determining the equilibrium surface temperature in this simple model. Substituting the numerical values for these two parameters, along with the Stefan-Boltzmann constant, gives

$$T_e \;=\; \left[\frac{(342)\,(1 \,-\, 0.31)}{5.67 \,\times\, 10^{-8}} \right]^{\frac{1}{4}} = 254 \text{ K} = -19°\text{C}$$

This simple model leaves us with a rather cold and inhospitable earth whose average surface temperature is $-19°$C, or $-12°$F. The actual average surface temperature is a more pleasant (and life-supporting) 15°C. This 34°C difference between the actual average surface temperature and the value predicted by Equation (12.7) is attributable to one major flaw in the simple energy balance model just used: the failure to account for the presence of certain gases in the earth's atmosphere, which affects the overall radiative balance of the planet. It is the presence of these greenhouse gases in the atmosphere that raises the average surface temperature by 34°C. To understand how this process works, it is necessary to examine in more detail the composition of the atmosphere and the nature of radiative heat transfer.

12.2.4   Temperature and the Radiative Spectrum

In the previous simple analysis, the sun and the earth were modeled as black bodies whose radiative energy depended only on their absolute temperatures. Earlier it was noted that radiated energy also can be described either as an electromagnetic wave or as photon particles. The latter characterization was used in Chapters 8 and 11 to quantify the magnitude of solar energy responsible for initiating certain chemical reactions in the atmosphere. The energy of a photon was given as the product of its frequency, v, times Planck's constant, h:

$$E = hv \text{ (joules)} \tag{12.8}$$

This equation shows that photons with the highest frequencies have the greatest energy. Alternatively, when radiation is characterized as a wave traveling at the speed of light, the *wavelength*, λ, is used as the characteristic dimension. Wavelength is usually specified in micrometers (μm) and has an inverse relationship to frequency:

$$\lambda = \frac{c}{v} \tag{12.9}$$

where c is the speed of light (3×10^8 m/s). Substituting Equation (12.9) into Equation (12.8) gives

$$E = \frac{hc}{\lambda} \tag{12.10}$$

Thus the shorter the wavelength, the greater the energy.

But how is the wavelength or frequency of radiated energy related to the temperature of the emitting body? That important relationship can be derived from theoretical physics and is known as *Planck's Law*. It gives the rate of energy radiated at any particular wavelength, λ, by a black body of absolute temperature, T:

$$\dot{E}_\lambda = \frac{C_1}{\lambda^5 \left(e^{C_2/\lambda T} - 1\right)} \tag{12.11}$$

where

$\dot{E}_\lambda$ = rate of energy emitted by a black body per unit area and wavelength (W/m²-μm)

λ = wavelength (μm)

T = absolute temperature of the black body (K)

C_1 = first constant = 3.742×10^8 (W-μm⁴/m²)

C_2 = second constant = 1.44×10^4 (μm-K)

The quantity $\dot{E}_\lambda$ is known as the *emissive power* of a black body. Figure 12.5 illustrates how the emissive power varies as a function of wavelength for two bodies with surface temperatures of 5,800 K and 288 K. These temperatures and spectral distributions approximate the radiation from the sun and earth, respectively.

Notice three important features on this graph.

1. For any given temperature, the black body emissive power peaks at some particular wavelength and falls toward zero at opposite ends of the spectrum.

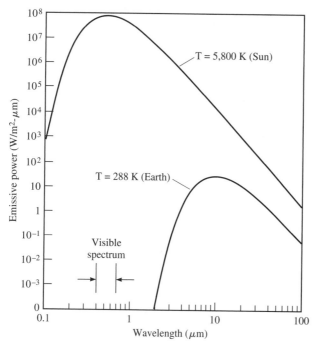

Figure 12.5 Spectral distribution of black body radiation from the sun and earth.

2. The higher the temperature, the shorter the wavelength at which the peak occurs. The location of the peak emissive power can be predicted by a simple equation known as *Wien's displacement law*:

$$\lambda_{max} = \frac{2{,}898}{T} \qquad (12.12)$$

where λ_{max} is the wavelength (μm) of the peak emissive power and T (K) is the absolute temperature of the black body.

3. The higher the temperature, the greater the maximum intensity.

Another useful observation is that the total area under each curve in Figure 12.5 represents the total heat flux (energy flow per unit area) emitted by a black body at the given temperature. Mathematically, that amount can be found by integrating Equation (12.11) over all wavelengths. Earlier we introduced the Stefan-Boltzmann equation, which also gives the total rate of black body radiation per unit area, shown by Equation (12.1). Thus there are two ways of obtaining the same quantity. Setting them equal to each other gives

$$\dot{q}\,(\text{W/m}^2) = \int_0^\infty E_\lambda \, d\lambda = \sigma T^4 \qquad (12.13)$$

To find the radiative heat flux from a body, it is much easier to use the Stefan-Boltzmann equation than to integrate Equation (12.11). The main purpose of introducing Planck's Law was to show that bodies at different temperatures have different spectral distributions. This is extremely important for understanding the greenhouse effect.

Example 12.2

Spectral properties of the earth and the sun.   Assume that the sun radiates with an effective black body temperature of 5,800 K, whereas the earth radiates with a black body temperature of 288 K (15°C). For each body, find the wavelength of maximum emissive power.

Solution:

Use Equation (12.12) (Wien's displacement law) to find the wavelength of maximum emissive power:

For the sun:

$$\lambda_{max} = \frac{2,898}{T} = \frac{2,898}{5,800} = 0.5 \ \mu\text{m}$$

For the earth:

$$\lambda_{max} = \frac{2,898}{T} = \frac{2,898}{288} = 10.1 \ \mu\text{m}$$

Example 12.2 shows that solar radiation is concentrated in the shorter wavelengths, whereas terrestrial radiation occurs at much longer wavelengths.

The visible portion of the solar spectrum lies between 0.4 μm and 0.7 μm and separates the *ultraviolet (UV)* region of shorter wavelengths from the *infrared (IR)* region of longer wavelengths. Most of the solar energy incident on the earth is in the UV (short wavelength) region below 3 μm, whereas outgoing radiation from the earth's surface is almost wholly in the IR (long wavelength) region above 3 μm. This important difference in spectral characteristics sets the stage for the greenhouse warming effect described next.

12.2.5   The Earth's Atmosphere

The atmosphere surrounding the earth is a layer of gases roughly 100 km in height—a very thin layer compared to the earth's radius of 6,370 km. As noted in Chapter 11, atmospheric gases are concentrated primarily in the layer immediately above the earth's surface, known as the *troposphere*. The average height of the troposphere is about 10 km, or one tenth the total height of the atmosphere. But this thin layer contains over 80 percent of the mass of all atmospheric gases.

The present composition of the earth's atmosphere has evolved over the billions of years since the earth was formed; today the atmosphere consists mainly of nitrogen (N_2) and oxygen (O_2) plus trace amounts of additional gases, principally water vapor, argon, and carbon dioxide. Water vapor is unique in that its local concentration in the atmosphere fluctuates substantially—from roughly zero to 3 percent—as water cycles between the atmosphere and the earth's surface due to evaporation and precipitation. As we shall see shortly, water vapor is also a powerful greenhouse gas that plays a major role in atmospheric processes. But because its concentration is so variable, the composition of the atmosphere is usually specified on a "dry" basis, excluding water vapor. This gives us the familiar values of 78 percent N_2 and 21 percent O_2 by volume, with argon accounting for most of the remaining 1 percent. Other naturally occurring trace gases (CO_2, Ne, He, CH_4, Kr, H_2, N_2O, CO, and O_3) account for a total of only 0.04 percent by volume.

Notice that the trace gases of concern in this chapter are measured in units of parts per million (ppm) or parts per billion (ppb) by volume. Some trace gases are measured in units of parts per trillion (ppt), as seen earlier in Table 12.2. These are extremely small quantities.[2] Yet the effect of these small quantities can be substantial because of the radiative properties of these gases.

12.2.6 Radiative Properties of the Atmosphere

Unlike black body surfaces, which emit and absorb radiation over a continuous spectrum of wavelengths, gases can absorb or emit radiation only at certain wavelengths. These wavelengths depend on the molecular structure of the gas. In terms of the earth's energy balance, the key question is which atmospheric gases can absorb incoming solar radiation or outgoing terrestrial radiation, thus altering the simple energy balance described in the previous section. Several atmospheric gases can alter the energy balance, but they are mainly the trace gases like water vapor, CO_2, and methane rather than the dominant atmospheric constituents like nitrogen.

Figure 12.6 shows the absorption bands for the major radiative gases in the atmosphere. An absorptivity, a_λ, of 1.0 means that the gas absorbs all of the radiation at a particular wavelength, while an absorptivity of zero indicates no absorption at that wavelength.

At the short wavelengths typical of incoming solar radiation, the most significant absorption bands are seen for oxygen (O_2) and ozone (O_3). These gases absorb ultraviolet radiation in the stratosphere, which is the atmospheric region above the troposphere. As discussed in Chapter 11, absorption of solar energy by stratospheric O_2 and O_3 protects the earth's surface from the potentially damaging high-energy radiation known as UV-C (wavelengths of 0.20 to 0.28 μm) and UV-B (0.28 to 0.32 μm). About 20 percent of the incoming solar energy is absorbed by the atmosphere; the remainder is either reflected (albedo) or transmitted to the earth's surface, where it is absorbed. Figure 12.6 shows the black body approximation of the incoming solar radiation. This is the same spectrum shown earlier in Figure 12.5 but plotted on a linear scale rather than logarithmic. The shaded area within the solar spectrum corresponds to the total amount of incoming energy absorbed in the upper atmosphere.

---

2 In other terms, one part per million is equivalent to one second in two years, or one cent out of $10,000. One part per billion is equivalent to one cent in $10 million, and one part per trillion is a million times smaller than a part per million.

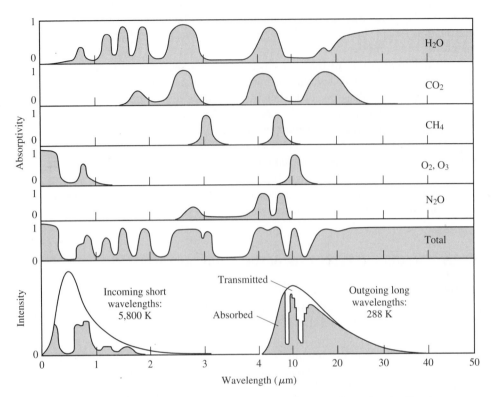

Figure 12.6   Absorptivity of atmospheric gases as a function of wavelength. Also shown are the intensity of incoming solar energy (short wavelengths) and outgoing terrestrial energy (long wavelengths). The shaded area under each of these curves shows the amounts absorbed by atmospheric gases. Note that the wavelength scale changes at 4 μm and that intensities are plotted relative to peak values. (*Source: Masters, G.M., Introduction to Environmental Eng. and Science*, 2nd Ed. © 1998. Reprinted by permission of Prentice-Hall, Inc., Upper Saddle River, NJ.)

At the longer (infrared) wavelengths of radiation outgoing from the earth, the picture is quite different. Several trace gases—especially water vapor and CO_2—have strong, overlapping absorption bands. This means that H_2O and CO_2 capture most of the outgoing surface radiation. There is only one small region of the spectrum, known as the *atmospheric window*, where relatively little absorption occurs. This is the region between about 8 μm and 12 μm, where terrestrial radiation passes directly through the atmosphere to space. Although ozone has a strong absorption band in the middle of this interval, because of its low concentration the overall effect is less pronounced. CFCs (which are not shown in Figure 12.6) also absorb energy within the atmospheric window and are thus important greenhouse gases as well as depleters of stratospheric ozone. Again, however, the low concentration of CFCs limits their overall impact on radiative absorption. So nearly all of the outgoing surface radiation is absorbed by H_2O, CO_2, and other trace gases in the atmosphere, as indicated by the shaded area under the terrestrial spectrum in Figure 12.6.

12.2.7 The Greenhouse Effect Defined

The atmospheric absorption phenomenon just described is referred to as the *greenhouse effect*: Just as in a greenhouse, most of the incoming radiation gets through to warm the earth's surface, and most of the outgoing radiation is blocked (absorbed) by the atmosphere. Gases that absorb infrared radiation are defined as *greenhouse gases*.

The energy absorbed by greenhouse gases does not simply accumulate in the atmosphere. Rather, the energy is reradiated by the greenhouse gas molecules. Part of that reradiated energy escapes to space, but part is directed back to the earth, where it is absorbed by the surface. This additional heat input to the surface warms the earth's temperature above the $-19°C$ calculated earlier in the absence of the greenhouse effect. How much warmer depends on the types and quantities of radiatively active gases that are in the atmosphere. By observation, we can say that the current levels of greenhouse gases in the atmosphere are responsible for a warming of $34°C$—the difference between the actual average surface temperature of $15°C$ and the $-19°C$ temperature calculated earlier. Water vapor is the biggest contributor to this warming, accounting for about 85 percent of the greenhouse effect that makes our planet livable.

12.2.8 Earth Energy Balance Revisited

Let us use this new insight about the role of atmospheric gases to revise the simple earth energy balance model in Section 12.2.3 and see if we can obtain a more realistic result for the earth's average surface temperature. To keep the model simple, we will merely add a thin layer of energy-absorbing material above the earth's surface to represent the atmosphere. We will assume that this layer has a uniform temperature, T_a, and that it radiates as a black body. The top surface radiates to space, whereas the lower surface radiates back to the earth. Because the layer is thin relative to the earth's curvature, we can represent the earth's surface and the atmospheric layer as flat parallel planes with the same surface area, as shown in Figure 12.7. Based on the discussion in the preceding section, let us further assume that this atmospheric layer absorbs all of the outgoing terrestrial radiation (we'll ignore the effect of the atmospheric window) plus a fraction, a_s, of the incoming solar radiation (after accounting for the earth's albedo). A sketch of these energy flows appears in Figure 12.7.

To derive an energy balance, we can again express all energy flows in terms of a heat flux (W/m^2 of surface area) because all surface areas are equal. For outgoing radiation, all surfaces are assumed to radiate as black bodies, so the rate of energy per unit surface area is simply σT^4. For incoming solar radiation, we again have an average flux of $S_o = 342 \ W/m^2$ (based on the earth's surface area), of which a little over 31 percent, or $107 \ W/m^2$, is reflected as albedo. The remaining amount is now absorbed partially by the atmosphere and partially by the earth's surface. As before, the total quantity absorbed, S_a, is

$$S_a = S_o - aS_o = 342 - 107 = 235 \ W/m^2 \qquad (12.14)$$

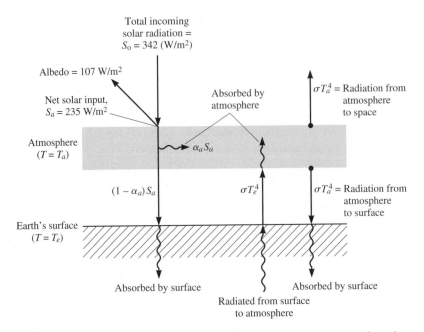

Figure 12.7   A revised energy balance model for the earth, assuming a radiatively active atmosphere.

The energy balance needed to find the new equilibrium temperature of the earth's surface, T_e, is a bit more complicated than for the earlier model with no atmospheric layer, mainly because the equilibrium temperature of that layer, T_a, is also unknown. We now have two unknown quantities (T_e and T_a), so two independent equations are needed to solve the problem. One equation comes from an energy balance on the atmospheric layer, and the other from an energy balance on the earth's surface. Details are left as an exercise for students (see the chapter problems). The result is

For the atmospheric layer:

$$S_a \alpha_a + \sigma T_e^{\,4} = 2\sigma T_a^{\,4} \tag{12.15}$$

For the earth's surface:

$$S_a(1 - \alpha_a) + \sigma T_a^{\,4} = \sigma T_e^{\,4} \tag{12.16}$$

Equations (12.15) and (12.16) must be solved simultaneously to find T_a and T_e. From Equation (12.14), the value of S_a is 235 W/m². Also, in the discussion of Figure 12.6, we saw that the atmosphere absorbed 20 percent of the total incoming solar radiation, S_0. This is equivalent to an absorption of 29 percent of the radiation, S_a, that remains after subtracting out the albedo. Thus the value of α_a is 0.29. We can now solve for T_e:

$$T_e = 290\,\mathrm{K} = 17°\mathrm{C}$$

This result is much closer to the actual earth surface temperature of 15°C compared to the earlier model prediction of −19°C. While still fairly simple, the revised energy balance model adds the key feature of a radiatively active atmospheric layer that increases the heat flux to the earth's surface, thereby increasing its temperature.

12.2.9 Actual Radiative Balance

The previous discussion illustrated how the greenhouse effect can alter the earth's radiative balance. A more accurate accounting of the global energy balance is shown in Figure 12.8, as reported in a 1996 study by the Intergovernmental Panel on Climate Change (IPCC). The IPCC is a group of several hundred prominent scientists, engineers, and policy experts convened from around the world by the World Meteorological Organization (WMO) and the United Nations Environmental Program (UNEP) to assess the problem of global climate change. This chapter draws heavily on their findings.

The left side of Figure 12.8 shows the same average values of 342 W/m² of total incoming solar radiation, and 107 W/m² of total reflected radiation, as used for our previous model. A feature not captured in that model, however, is the reflected radiation (albedo) occurring partially within the atmosphere (especially from clouds) and partially at the earth's surface. We also see an additional 24 W/m² of atmospheric heating provided by thermal convection from the earth's surface, plus 78

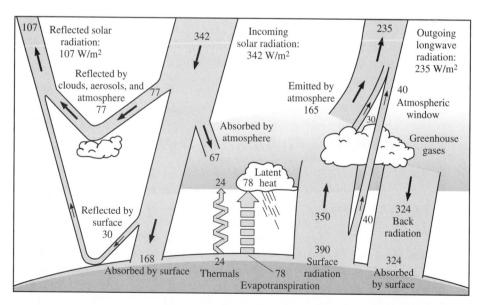

Figure 12.8   The global energy balance for earth. Figures show the average energy flows in W/m² based on the earth's surface area. About 49 percent of the incoming solar radiation is directly absorbed by the surface, but the greenhouse effect adds to the overall energy flow to the surface. (*Source:* IPCC, 1996a)

W/m² of heating when water evaporated at the surface condenses in the atmosphere, releasing the latent heat of evaporation.

More critical to the present discussion is the role of greenhouse gases in the atmosphere. Figure 12.8 shows that the total outgoing surface radiation of 390 W/m² (which corresponds to a black body surface temperature of 288 K or 15°C) is mostly absorbed by the atmosphere, as in our revised model. However, about 10 percent of the surface radiation (40 W/m²) escapes directly to space through the radiative atmospheric window discussed earlier. Together with 195 W/m² of radiation from the atmosphere (including clouds), the overall flow of 235 W/m² outgoing to space exactly balances the 235 W/m² of incoming radiation absorbed by the atmosphere and earth.

Notice, however, that the back-radiation to earth from greenhouse gases in the atmosphere (324 W/m²) is much larger than the energy radiated by the atmosphere out to space. This is a very different picture than our simple atmospheric model in Figure 12.7, which assumed equal amounts of radiation from the top and bottom surfaces of a constant-temperature radiative layer. The fact that the top of the real atmosphere is actually much colder than the lower atmosphere is an important factor in explaining this difference: Less radiative energy is emitted at lower temperatures.

To predict the true amounts of greenhouse gas radiation, an atmospheric model would require much more detail on the physics of radiative and convective heat transfer, the composition and structure of the atmosphere, and the behavior of clouds, particles, and other atmospheric constituents. Sophisticated computer models, validated by data from satellite observations, are now capable of providing good estimates of the radiative heat flows shown in Figure 12.8.

The big question is how well atmospheric models can predict the effects of *changes* in greenhouse gas concentration. For example, as the levels of CO_2 and other radiative gases increase due to anthropogenic emissions, what will be the effects on the global energy balance and on the earth's surface temperature? The following sections provide some preliminary answers to these questions.

12.3   RADIATIVE FORCING OF CLIMATE CHANGE

Thus far we have considered the atmosphere to have a constant composition and to be in an equilibrium or steady state condition, in which incoming radiation from the sun and outgoing radiation from the earth are in balance. Consider now what happens if this system is perturbed by the addition of more greenhouse gases like carbon dioxide, methane, or nitrous oxide. By definition, a greenhouse gas is one that absorbs radiation in the infrared wavelengths, such as emitted by the earth's surface. So if more greenhouse gas is added to the atmosphere, some of the outgoing radiation to space will be absorbed by that gas, resulting in a net decrease of outgoing radiation per unit of area. Let us call this decrease $\Delta \dot{q}_{out}$.

Figure 12.9 depicts the initial radiative equilibrium and the immediate imbalance caused by the addition of a greenhouse gas (GHG) to the atmosphere. Because such gases are generally confined to the troposphere, atmospheric scientists focus on

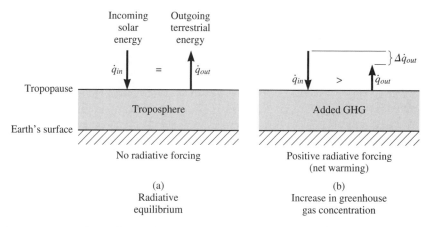

Figure 12.9 Radiative forcing due to a greenhouse gas added to the atmosphere. The initial radiative equilibrium at the top of the troposphere (a) is disturbed by the addition of a greenhouse gas (b), producing a positive radiative forcing.

the average net energy balance at the *tropopause,* which is the upper boundary of the troposphere approximately 10 km above the earth's surface. The length of each arrow in Figure 12.9 represents the magnitude of radiative heat flux (W/m^2) crossing the tropopause.

Initially, as shown in Figure 12.9(a), the rates of incoming and outgoing radiation are equal, resulting in a net energy flux of zero. Figure 12.9(b) shows the temporary imbalance that results from the addition of a greenhouse gas like CO_2. More of the outgoing radiation is now absorbed in the atmosphere, so the incoming solar radiation exceeds the outgoing radiation. Any such change in the average net radiation at the tropopause is referred to as *radiative forcing, ΔF.*

12.3.1 Modes of Radiative Forcing

Radiative forcing may be induced in a variety of ways. In Figure 12.9(b) the change is caused by a decrease in the outgoing radiation due to the addition of a greenhouse gas. The magnitude of radiative forcing in this case is

$$\Delta F(\text{W/m}^2) = \Delta \dot{q}_{out} \qquad (12.17)$$

The term *radiative forcing* is used because any change in the net radiative balance will *force* the climate system to readjust so as to ultimately restore equilibrium. The radiative forcing in Figure 12.9(b) is considered *positive* because there is a *net addition of energy* to the atmosphere. One can anticipate that this additional energy will force the climate system to get warmer.

Figure 12.10 shows two other ways that the earth's energy balance can be perturbed to produce a radiative forcing. One is a change in the solar input. Although we normally regard the rate of solar radiation reaching the earth as constant, in fact the sun's output is known to vary over long periods, as manifest by changes in

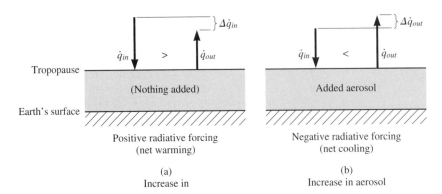

Figure 12.10   Additional modes of radiative forcing. An increase in solar energy input (a) produces a positive forcing, whereas increased aerosol particles in the atmosphere (b) produce a negative forcing.

sunspot activity and other solar phenomena. If the solar input per unit area were to increase by an amount $\Delta \dot{q}_{in}$, as depicted in Figure 12.10(a), a positive radiative forcing would occur, tending to warm the planet. For this case, the magnitude of radiation forcing can be written as

$$\Delta F(\text{W/m}^2) = \Delta \dot{q}_{in} \qquad (12.18)$$

Radiative forcing also can be induced by changes in the earth's albedo. For example, a decrease in the surface albedo of the earth (such as from changes in land use or the melting of reflective ice sheets) would produce positive radiative forcing because less radiation would be reflected back to space. This case would look like Figure 12.9(b), although the cause of the decrease in outgoing energy would be different.

The addition of aerosols to the atmosphere also can change the albedo. An *aerosol* is a suspension of fine particles less than 10 μm in diameter. Although we have not yet talked about aerosols in the atmosphere, their presence can have an important effect on the atmospheric radiative balance. The most important anthropogenic aerosols in the atmosphere are small particles resulting from the combustion of biomass and fossil fuels. Many (but not all) of these aerosols tend to reflect incoming solar radiation back to space. This increases the planetary albedo, resulting in more radiation leaving at the tropopause than coming in. In this case, the radiative forcing is *negative.* That is, rather than adding energy to the climate system, there is a *net decrease in heat flux,* as indicated in Figure 12.10(b). This type of perturbation tends to produce a cooling effect, in contrast to the warming effect of greenhouse gases.

12.3.2   Net Forcing from Atmospheric Changes

As we have just seen, radiative forcing can be induced in several ways. A general expression for radiative forcing combines overall changes in incoming solar radiation with the overall changes in outgoing radiation at the top of the troposphere:

$$\Delta F = \Delta \dot{q}_{out} - \Delta \dot{q}_{in} \qquad (12.19)$$

Distinguishing between positive and negative forcing requires an algebraic convention that gives us a positive forcing when there is a net *increase* in energy *into* the troposphere. These definitions give that result:

$$\Delta \dot{q}_{out} = (W/m^2 \text{ outgoing initially}) - (W/m^2 \text{ outgoing after perturbation}) \quad (12.20)$$

$$\Delta \dot{q}_{in} = (W/m^2 \text{ incoming initially}) - (W/m^2 \text{ incoming after perturbation}) \quad (12.21)$$

The following examples illustrate the use of these equations.

Example 12.3

Radiative forcing from CO_2 addition. An increase in atmospheric CO_2 concentration reduces the outgoing infrared radiation at the tropopause from 235 W/m² to 233 W/m². The incoming solar radiation and albedo do not change. Calculate the radiative forcing.

Solution:

Because there is no change in incoming radiation, we have $\Delta \dot{q}_{in} = 0$.

From Equation (12.20), the net change in outgoing radiation is

$$\Delta \dot{q}_{out} = 235 - 233 = 2 \text{ W/m}^2$$

From Equation (12.19), the net radiation forcing is then

$$\Delta F = \Delta \dot{q}_{out} - \Delta \dot{q}_{in} = 2 - 0 = 2 \text{ W/m}^2$$

This example gives the correct algebraic sign, which is a positive forcing, corresponding to the case of Figure 12.9. The next example illustrates negative forcing.

Example 12.4

Radiative forcing from a sulfate aerosol. An increase in sulfate aerosol from the combustion of coal and oil adds 0.5 W/m² to the global average flux of radiation reflected by the troposphere (albedo). Calculate the radiative forcing.

Solution:

This example corresponds to the case in Figure 12.10(b). We are not given the magnitude of the outgoing radiation initially, but we know that it is 0.5 W/m² larger after addition of the sulfate aerosol. Denote the unknown initial magnitude by x (W/m²). Equation (12.20) gives us

$$\Delta \dot{q}_{out} = x(W/m^2 \text{ initially}) - (x + 0.5)(W/m^2 \text{ after perturbation})$$

$$= -0.5 \text{ W/m}^2$$

Again we have $\Delta \dot{q}_{in} = 0$. Equation (12.19) gives

$$\Delta F = \Delta \dot{q}_{out} - \Delta \dot{q}_{in} = -0.5 - 0 = -0.5 \text{ W/m}^2$$

12.3.3   Quantifying Radiative Forcing

Radiative forcing is a key factor determining how much the climate will change in response to disturbances of the earth's energy balance induced by changes in greenhouse gas concentrations, aerosols, the earth's albedo, and solar input. The larger the radiative forcing (either positive or negative), the larger the anticipated impact. Calculating radiative forcing is thus the first step in predicting quantitative measures of climate change such as changes in average surface temperature, changes in regional temperature, and changes in precipitation.

Given the enormous complexity of the climate system, accurate predictions of climate variables such as temperature and precipitation patterns are still very difficult. Atmospheric scientists are more confident in estimating radiative forcing from greenhouse gases, which is the starting point for climate change predictions.

Radiative forcing itself is a complex quantity to calculate. For example, the forcing caused by an increase in greenhouse gas concentration depends on the radiative absorption characteristics of that particular gas, the initial concentration of that gas in the atmosphere, the thickness and temperature profile of the atmosphere, and the effects of clouds and other radiatively active gases that are present. A correction also may be needed to account for the presence of other energy-absorbing gases and for any changes in the stratosphere that might be indirectly induced by radiative changes in the troposphere.

For most greenhouse gases, however, the physics of direct radiative forcing are now well understood, and computer programs have been developed to calculate the magnitude of radiative forcing as a function of greenhouse gas concentrations with a relatively high degree of confidence. The results of those detailed calculations can be approximated by a simplified set of equations, which are summarized in Table 12.3. Several important features of these equations are discussed next.

12.3.4   Radiative Forcing versus Concentration

The initial concentration of a greenhouse gas strongly affects the magnitude of radiative forcing caused by an additional increment in concentration. The additional radiative forcing is largest when the initial concentration is very low. As the concentration grows, each new increment in concentration produces less additional forcing than the earlier increments. Figure 12.11 illustrates the three different regimes that characterize this effect.

Low-Concentration Regime   When there is little or no greenhouse gas to begin with, there is a linear (proportional) relationship between changes in greenhouse gas concentration and changes in radiative forcing. Because the initial gas concentration is very low, each new molecule added to the atmosphere can absorb the maximum amount of energy permitted by its molecular structure. The radiative forcing thus increases in proportion to the number of molecules, which is reflected by the atmospheric concentration of the gas, C. This type of relationship can be expressed generally as

$$\Delta F = A(C - C_o) \quad \text{(Low concentrations)} \quad (12.22)$$

Table 12.3 Equations for approximation of radiative forcing from greenhouse gases.

Trace Gas	Radiative Forcing, ΔF, W/m²	Units and Range of Concentration
Carbon dioxide	$\Delta F = 6.3 \ln (C/C_o)$	C is CO_2 in ppmv; valid for $C < 1{,}000$ ppmv
Methane[a]	$\Delta F = 0.036(\sqrt{M} - \sqrt{M_o}) - [f(M, N_o) - f(M_o, N_o)]$	M is CH_4 in ppbv, N is N_2O in ppbv; valid for $M < 5$ ppmv
Nitrous oxide[a]	$\Delta F = 0.14(\sqrt{N} - \sqrt{N_o}) - [f(M_o, N) - f(M_o, N_o)]$	M and N as above; valid for $N < 5$ ppmn
Tropospheric ozone	$\Delta F = 0.02 (O - O_o)$	O is ozone in ppbv
CFC-11	$\Delta F = 0.22 (X - X_o)$	X is CFC-11 in ppbv; valid for $X < 2$ppbv
CFC-12	$\Delta F = 0.28 (Y - Y_o)$	Y is CFC-12 in ppbv; valid for $Y < 2$ ppbv

[a] Methane–nitrous oxide overlap term = $f(M, N) = 0.47 \ln [1 + 2.01 \times 10^{-5} (MN)^{0.75} + 5.31 \times 10^{-15} M (MN)^{1.52}]$, where M and N are in ppbv.

Source: IPCC, 1990.

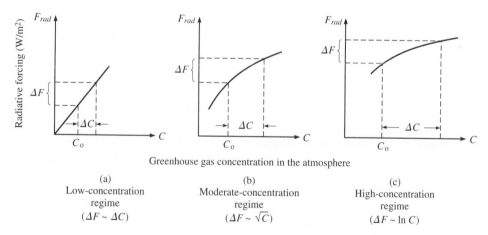

Greenhouse gas concentration in the atmosphere

(a) Low-concentration regime ($\Delta F \sim \Delta C$)

(b) Moderate-concentration regime ($\Delta F \sim \sqrt{C}$)

(c) High-concentration regime ($\Delta F \sim \ln C$)

Figure 12.11 Concentration regimes for radiative forcing. As the greenhouse gas concentration increases from (a) low, to (b) moderate, to (c) high, a larger change in concentration, ΔC, is required to produce the same change in radiative forcing, ΔF.

where C_o is the initial volumetric concentration and A is a proportionality constant.

This type of equation applies to CFCs and other halocarbons, which have relatively low atmospheric concentrations. Indeed, Figure 12.1 earlier showed that the atmospheric concentration of CFCs was zero until they came into use in the 1940s and began to accumulate in the atmosphere. For concentrations less than 2 ppbv (parts per billion by volume), Equation (12.22) can be used to estimate the direct radiative forcing for a variety of halocarbon compounds. The numerical value of A is given in Table 12.3 for two major species, CFC-11 and CFC-12.

Example 12.5

Radiative forcing from CFC-12.   Until its production was banned by the Montreal Protocol and the 1990 Clean Air Act Amendments, CFC-12 was widely used as a refrigerant in household refrigerators and as a chemical for industrial applications. By 1992 the atmospheric concentration of CFC-12 had grown to 500 pptv from an initial value of zero. Calculate the change in direct radiative forcing due to this increased concentration.

Solution:
From Table 12.3, the radiative forcing equation for CFC-12 is

$$\Delta F(\text{W/m}^2) = 0.28(C - C_o) \quad (C, C_o = \text{ppbv}) \tag{12.23}$$

For this problem the initial concentration is $C_o = 0$. To use Equation (12.23) we must first convert the units of C from the given value of 500 pptv (parts per trillion by volume) to the required units of parts per billion:

$$C = 500 \text{ ppt } \times \frac{10^9 \text{ ppb}}{10^{12} \text{ ppt}} = 0.500 \text{ ppb}$$

Then

$$\Delta F = 0.28(0.500 - 0) = 0.14 \text{ W/m}^2$$

Note that this value represents only the *direct* radiative forcing from CFC-12. Later we shall see that additional *indirect* effects offset some of this direct forcing.

Moderate-Concentration Regime   When the concentration of a greenhouse gas increases to a moderate level, the molecules already in the atmosphere have absorbed much of the radiation at the wavelengths where absorption bands are strongest. Thus further absorption occurs increasingly at "off-peak" wavelengths where the absorptivity (α_λ) is lower (refer to Figure 12.6 to see some examples of this). With less energy absorbed per molecule, there is less radiative forcing from each new increment in concentration, as illustrated in Figure 12.11(b). In the moderate-concentration regime ΔF increases approximately in proportion to the square root of concentration:

$$\Delta F = B(\sqrt{C} - \sqrt{C_o}) \quad \text{(Moderate concentrations)} \tag{12.24}$$

This type of relationship applies to methane (CH_4) and nitrous oxide (N_2O). Values of the constant B for those gases are given in Table 12.3, along with the concentration limits that apply.

High-Concentration Regime   Finally, when greenhouse gas concentrations reach a high level, further increases in concentration produce much smaller increases in radiative forcing, as depicted in Figure 12.11(c). This case applies to CO_2. In this regime radiative forcing increases in proportion to the natural logarithm of concentration:

$$\Delta F = k(\ln C - \ln C_o) \quad \text{(High concentrations)} \qquad (12.25)$$

For CO_2, Table 12.3 gives the value of the constant k as 6.3. The difference in logarithms can also be expressed as the log of the concentration ratio. For CO_2 this yields

$$\Delta F_{CO_2} \; (\text{W/m}^2) = 6.3 \ln \left(\frac{C}{C_o} \right) \quad (C, C_o = \text{ppmv}) \qquad (12.26)$$

Notice in Figure 12.11 that as we move from the low-concentration regime to the moderate and high regimes, a larger change in concentration, ΔC, is needed to achieve the same increase in radiative forcing, ΔF, shown in each of the three sketches. This is another way of illustrating the saturation effect that occurs with increasing concentrations of greenhouse gases.

Example 12.6

Radiative forcing from a doubling of CO_2. Estimate the radiative forcing that would result from a doubling of the 1992 CO_2 concentration of 355 ppmv.

Solution:

A doubling of the 1992 CO_2 level would yield a concentration of 710 ppmv. This is within the 1,000 ppmv limit for which Equation (12.26) is valid (see Table 12.3). Thus we can use Equation (12.26) with $C_o = 355$ ppmv and $C = 710$ ppmv:

$$\Delta F_{CO_2} = 6.3 \ln \left(\frac{C}{C_o} \right) = 6.3 \ln \left(\frac{710}{355} \right) = 6.3 \, (\ln 2) = 4.37 \; \text{W/m}^2$$

Effect of Spectral Overlap The preceding discussions assumed that each greenhouse gas acts independently in its ability to induce radiative forcing. This assumption is valid so long as each gas absorbs radiation at different wavelengths. However, if the absorption bands of two or more gases overlap, the radiative forcing functions are no longer independent. Instead, the incremental forcing from a particular gas will depend on the amount of other gases vying for the same portion of the radiative spectrum.

For the greenhouse gases in Table 12.3, there are spectral overlaps for methane and nitrous oxide. As shown in Figure 12.6, both of these gases have absorption bands in the range of 2.5 to 10 μm. The overlap is not strong, however, so Equation (12.24) can still be used independently for each gas to obtain reasonable estimates of forcing versus concentration. For more accurate calculations, Table 12.3 includes a correction term that accounts for the spectral overlap effect on radiative saturation. The problems at the end of this chapter illustrate the use of these correction terms.

12.3.5 Radiative Forcing in the Industrial Age

Before trying to predict future climate change from greenhouse gases, it is instructive to first ask about the levels of radiative forcing and climate change already induced by human influences on the earth's atmosphere. The year 1850 is used by the Intergovernmental Panel on Climate Change (IPCC) to mark the beginning of the industrial era, in which the release of greenhouse gases from fossil fuel combustion and other anthropogenic activities began to accelerate rapidly.

Table 12.4 shows the most recent IPCC estimates of the globally averaged radiative forcing due to changes in atmospheric concentrations of greenhouse gases and aerosols from preindustrial times (1850) to the present day (1990). Also shown for comparison is the estimated forcing due to natural variations in solar intensity over the same period.

Direct Forcing from Greenhouse Gases   The first entry in Table 12.4 shows the direct radiative forcing of individual greenhouse gases, whose atmospheric concentrations since 1850 have increased by the amounts shown. Direct forcings arise from green-

Table 12.4   Global average radiative forcing from 1850 to 1990.

Source of Radiative Forcing	Concentration Increase (ppb)	Radiative Forcing (W/m²)	Uncertainty Estimate[a]	
			Range (W/m²)	Confidence Level
Greenhouse Gases				
Direct forcing from				
Carbon dioxide	78,000	1.56		
Methane	1,014	0.47		
Nitrous oxide	36	0.14		
Halocarbons	0–0.5[b]	0.28		
Total direct forcing		2.45	±15%	High
Indirect forcing from				
Tropospheric ozone		0.4	±50%	Low
Stratospheric ozone		−0.1	Factor of 2	Low
Tropospheric Aerosols				
Direct forcing from				
Sulfate particles		−0.4	Factor of 2	Low
Biomass burning		−0.2	Factor of 3	Very low
Fossil fuel soot		0.1	Factor of 3	Very low
Total direct forcing		−0.5	Factor of 2	Very low
Indirect forcing from				
Cloud formation		?	0 to −1.5	Very low
Solar Variability		0.3	±67%	Very low

[a] This confidence level reflects the judgment of experts that the true radiative forcing is within the uncertainty range shown.

[b] The change in halocarbon concentration since 1850 varies for each halocarbon compound. The upper bound shown is for CFC-12.

Source: Based on IPCC, 1996a.

house gases that are emitted directly into the atmosphere from some identifiable source, such as the CO_2 emitted from the tailpipe of a car or the CFCs emitted from the manufacture of urethane foam.

The greatest contribution (1.56 W/m^2) comes from increased CO_2 levels, which represent 64 percent of the total direct forcing. Other greenhouse gases from human activities raise the total direct forcing from anthropogenic sources to 2.45 W/m^2. The estimated uncertainty in this value is ± 15 percent, i.e., a range of 2.1 to 2.8 W/m^2. The confidence level in Table 12.4 indicates a very good likelihood that the true radiative forcing lies within this estimated uncertainty interval.

Looking back at the global energy balance in Figure 12.8, we see that the total anthropogenic forcing of 2.45 W/m^2 represents a change of about 1 percent in the 235 W/m^2 of longwave radiation leaving the atmosphere under equilibrium conditions. Thus the impact of human activities on the planetary energy balance is beginning to be noticeable.

Indirect Forcing from Greenhouse Gases   Once in the atmosphere, some of the direct emissions may produce other changes in the atmosphere that also affect the earth's radiative balance. These are referred to as *indirect* forcings. We saw in Chapter 8, for example, that tropospheric ozone is the indirect result of emissions of volatile organic compounds (VOCs) and nitrogen oxides (NO_x) which react in the troposphere to produce ozone. Once formed, this ozone adds to the radiative forcing because ozone is also a greenhouse gas with a strong absorption band in the infrared region, as shown earlier in Figure 12.6. Thus Table 12.4 shows a positive indirect radiative forcing of about 0.4 W/m^2 from tropospheric ozone concentrations relative to preindustrial times. Table 12.3 also listed tropospheric ozone as a greenhouse gas in the low-concentration regime based on current atmospheric levels.

Another indirect radiative forcing involves stratospheric ozone, which is the natural layer of "good" ozone that protects the earth's inhabitants from exposure to high-energy ultraviolet radiation. Stratospheric ozone also normally absorbs some of the outgoing infrared radiation from the troposphere. However, stratospheric ozone is destroyed by direct emissions of chlorofluorocarbons (CFCs) and other halocarbons that slowly migrate up to the stratosphere (see Chapter 11). Less ozone in the stratosphere means that more of the outgoing longwave radiation escapes to space, producing cooling. The result is negative radiative forcing, estimated in Table 12.4 at about -0.1 W/m^2. Indirect forcing due to the depletion of stratospheric ozone thus offsets part of the direct forcing from CFCs and other atmospheric halocarbons.

Direct Forcing from Aerosols   As noted earlier, the addition of aerosols to the atmosphere also can disrupt the earth's radiative balance. Table 12.4 shows the current estimates of globally averaged radiative forcings due to changes in anthropogenic aerosol concentrations relative to preindustrial times. Direct effects are listed for three types of aerosols. Sulfate particles arise mainly from the combustion of sulfur-bearing fossil fuels like coal and oil. These tiny particles reflect incoming solar radiation back out to space, resulting in higher planetary albedo as illustrated in Figure 12.12(a)—yielding negative radiative forcing and consequent cooling. Aerosol

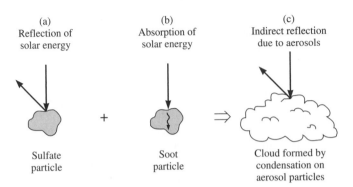

Figure 12.12   Direct and indirect effects of atmospheric aerosols. Most aerosol particles, such as sulfates (a), directly reflect radiation, although some particles, like soot (b), are direct absorbers. Indirect effects are caused by clouds that form as a result of atmospheric aerosols.

particles from the combustion of biomass (mainly wood, paper, and agricultural wastes) exhibit similar behavior and also produce negative radiative forcing.

The third type of aerosol listed in Table 12.4—soot particles from fossil fuel combustion—is different from the other aerosols. The carbon content of these particles absorbs incoming solar radiation that would otherwise be reflected back to space (Figure 12.12b), so the net result here is positive radiative forcing.

Unlike greenhouse gases such as CO_2, CH_4, N_2O, and halocarbons, which are well-mixed throughout the troposphere, anthropogenic aerosols tend to concentrate in areas of the Northern Hemisphere where fossil fuel combustion and related activities are most intense. Nor do aerosols have much time to diffuse and mix throughout the troposphere. Aerosol particles typically remain in the atmosphere for only a few days before they are washed out by precipitation or other atmospheric processes. This is in sharp contrast to the well-mixed greenhouse gases, which have atmospheric lifetimes measured in decades rather than days, as seen earlier in Table 12.2. The spatial variations in aerosol emissions around the globe, combined with uncertainties in regional aerosol concentrations, particle size distributions, and radiative properties, create a large overall uncertainty in the current estimates of radiative forcing from aerosols. The ranges shown in Table 12.4 represent a factor of two uncertainty for sulfates and a factor of three uncertainty for soot particles and biomass burning. For the three aerosol types combined, the IPCC estimate of direct radiative forcing is -0.5 W/m^2 with a factor of two uncertainty.

Indirect Forcing from Aerosols   Even more uncertain are the indirect forcing effects of aerosols. Besides directly scattering or absorbing incoming radiation, microscopic aerosol particles also act as condensation nuclei for cloud formation. In turn, clouds reflect additional incoming radiation, increasing the atmospheric albedo as illustrated in Figure 12.12(c). This indirect cooling adds to the direct cooling effects of aerosols. Indirect processes are not yet understood well enough to predict anything more than a suspected range of 0 to -1.5 W/m^2 for indirect radiative forcing from aerosols.

Forcing Due to Solar Radiation   The final entry in Table 12.4 shows the estimated radiative forcing due to changes in the sun's intensity since 1850. Based on measurements and theoretical studies, the average solar input is believed to have increased slightly, producing a net positive forcing of 0.3 W/m^2 with an uncertainty of ± 67 percent. Perhaps the most important insight from this estimate, however, is that the natural variability of solar input is small relative to the 2.45 W/m^2 of direct greenhouse forcing from human activities over the past 150 years. Thus the consequences of increased radiative forcing cannot be attributed simply to natural fluctuations in solar intensity.

Overall Radiative Forcing   If we were to add up all the positive and negative radiative forcings given in Table 12.4, the globally averaged result would be an overall positive increase in radiative forcing since preindustrial times, indicating a net warming effect. But such an addition is not appropriate because there are significant differences in the geographical distribution of the forcings due to ozone and aerosols, compared to forcings from the well-mixed greenhouse gases (CO_2, CH_4, N_2O, and halocarbons). Thus there could be significant differences in the regional climate changes induced by different sources of radiative forcing. For this reason, combining all of the positive and negative values in Table 12.4 to find the net radiative forcing could be misleading: Regional climate changes could still occur even if the global average sum were zero. Researchers around the world are trying to better understand and predict the overall effect of these combined forcings.

12.3.6   Equivalent CO_2 Concentration

One way of quantifying the combined effect of multiple atmospheric constituents is to focus only on the well-mixed greenhouse gases. In this case, the total radiative forcing from multiple greenhouse gases can be expressed as an equivalent CO_2 concentration that would produce the same overall forcing. This can be done using Equation (12.26). As before, C_o is the initial concentration of CO_2. Now, however, the final concentration, C, is the equivalent CO_2 (C_{equiv}) needed to achieve the overall forcing, ΔF_{total}, produced by multiple greenhouse gases. Thus

$$\Delta F_{total} = 6.3 \ln \left(\frac{C_{equiv}}{C_o} \right) \qquad (12.27)$$

Solving for C_{equiv} gives

$$C_{equiv} \, (\text{ppmv } CO_2) = C_o \exp \left(\frac{\Delta F_{total}}{6.3} \right) \qquad (12.28)$$

The following example gives an application of this equation.

Example 12.7

Equivalent CO_2 increase since preindustrial times.   The preindustrial concentration of CO_2 in the atmosphere was 280 ppmv. Since that time, increases in atmospheric CO_2, CH_4, N_2O, and halocarbons have produced a direct radiative forcing totaling 2.45 W/m^2 (see Table 12.4). Calculate the equivalent CO_2 concentration needed to produce this forcing.

Solution:

Use Equation (12.28) with $C_o = 280$ ppmv:

$$C_{equiv} = C_o \exp\left(\frac{\Delta F_{total}}{6.3}\right) = (280)\exp\left(\frac{2.45}{6.3}\right) = 413 \text{ ppmv CO}_2$$

The equivalent CO_2 concentration in this example represents a 48 percent increase over the initial preindustrial CO_2 level of 280 ppmv. By comparison, the actual CO_2 concentration of 355 ppmv in 1992 represents an increase of 27 percent since preindustrial times. Thus the radiative forcing effect of other greenhouse gases (CH_4, N_2O, and halocarbons) is equivalent to an additional 21 percent increase in CO_2 concentration. As we shall see later, the concept of equivalent CO_2 is also useful in estimating future climate changes due to multiple greenhouse gases.

12.4   TEMPERATURE CHANGES FROM RADIATIVE FORCING

We now (at last) come to one of the key questions that motivated this chapter, namely the relationship between increasing concentrations of greenhouse gases and the resulting impact on global warming. To get to this point we first had to understand the origin of the greenhouse effect and how it alters the radiative balance between the sun and the earth. The presence of atmospheric aerosols also had to be considered because they too affect radiative forcing, especially at the regional level. To now understand how radiative forcing might alter the earth's temperature, let us begin by revisiting the basic energy balance of the planet.

12.4.1   Restoring the Earth's Energy Balance

Figure 12.9 earlier showed how an increase in greenhouse gas concentration caused a positive radiative forcing, ΔF_{rad}, which represented an imbalance between incoming and outgoing radiative heat flow at the top of the troposphere. But such an imbalance cannot persist indefinitely; something must change to restore a new balance. Because a positive forcing means that more energy is coming into the atmosphere than is leaving, the most obvious change is for the earth to warm. As the earth's surface temperature rises, the thermal radiation from the surface increases. Warming of the surface also tends to increase the convective and evaporative heat transfer to the atmosphere, depicted earlier in Figure 12.8.

As before, these additional heat flows from the earth's surface will mostly be absorbed by greenhouse gases in the atmosphere. In turn, the atmosphere will reradiate most of this energy back to the earth's surface, as Figure 12.8 showed earlier. Some of the additional energy also will be radiated out to space. The average surface temperature will therefore increase until it is large enough to force enough additional outgoing radiation to space so as to restore the overall planetary energy balance. At that point the earth's surface temperature will have risen by an amount ΔT_e in response to the original radiative forcing, ΔF_{rad}.

The ratio of the final temperature change, ΔT_e, to the change in radiative forcing, ΔF_{rad}, is known as the *climate sensitivity factor*, γ. Thus

$$\gamma = \frac{\Delta T_e}{\Delta F_{rad}} \qquad (12.29)$$

Climate sensitivity is an important parameter in climate modeling. It relates the net change in radiative forcing caused by greenhouse gases and aerosols to the resulting change in the earth's average surface temperature—exactly the type of relationship we are looking for. Thus the average warming effect can be expressed as

$$\Delta T_e = \gamma(\Delta F_{rad}) \qquad (12.30)$$

This means that for a positive forcing of 1 W/m², the earth's average surface temperature would increase by an amount equal to $\gamma(°C)$.

Quantifying the climate sensitivity factor, however, remains a difficult task—one that has occupied atmospheric scientists for several decades. The following section summarizes some of the results and insights from this research.

12.4.2 Evaluating the Climate Sensitivity Factor

One way to think about the magnitude of the climate sensitivity factor is to ask how much the earth's temperature would change from a 1 W/m² increase in energy absorbed by the surface in the *absence* of the greenhouse effect. If we return to the very simple energy balance model of Equation (12.7), we find that a 1 W/m² increase in the energy absorbed by the surface (keeping the albedo constant) would produce a temperature increase of 0.27°C in the absence of an atmosphere.

In the presence of a radiatively active atmosphere, we might expect the actual temperature increase to be higher than this value because of the additional heat radiated back to the earth's surface from atmospheric greenhouse gases. Higher surface temperatures also could occur for other reasons. For example, as the surface temperature warms, more water from rivers and oceans would evaporate into the atmosphere. We saw earlier that H_2O is a potent greenhouse gas, so the additional water vapor would trap even more outgoing surface radiation. This would be an example of *positive feedback* that adds to the warming effect.

On the other hand, *negative feedback* could occur if higher rates of evaporation produced more clouds to reflect more incoming radiation back to space, causing a cooling effect. So what is the true answer?

12.4.3 Results from Observational Data

Various approaches have been used by atmospheric scientists to quantify the climate sensitivity factor. One method involves correlating surface temperature data at different latitudes with satellite-based measurements of infrared radiation to space

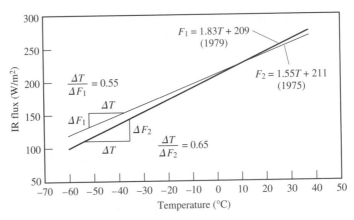

Figure 12.13   Empirical correlations of climate sensitivity. This sketch depicts the best-fit regression lines from two studies that correlated zonally averaged surface air temperature at different latitudes with outgoing radiative flux measured by satellites. (*Source:* Based on North, 1975; Warren and Schneider, 1979)

(North, 1975; Warren and Schneider, 1979). Figure 12.13 shows the results of two such studies, both of which indicate a linear relationship between the locally averaged surface temperature and outgoing infrared radiation. These statistical correlations suggest a climate sensitivity factor ranging from 0.55 to 0.65°C/W-m^{-2}. The average value of 0.6°C/W-m^{-2} is roughly twice the value of 0.27°C obtained from the simple energy balance model. This is an indication that positive feedback effects currently prevail. Although statistical models based on current observations cannot predict future climate changes, they are nonetheless useful for estimating potential temperature increases from climate forcing.

Example 12.8

Temperature change from a doubling of CO_2.   In Example 12.6 the radiative forcing from a doubling of the current CO_2 concentrations was estimated to be 4.37 W/m^2. What is the resulting increase in equilibrium surface temperature if the climate sensitivity factor is 0.6°C/W-m^{-2}?

Solution:

Equation (12.30) gives the temperature increase sought:

$$\Delta T_e = \gamma(\Delta F_{rad}) = (0.6)(4.37) = 2.6°C$$

The main limitation in using observational models to predict future temperature change is that we do not know if the climate sensitivity factor will remain constant as radiative forcing changes from current levels. Various types of positive and negative feedback are among the many factors that could cause the atmosphere to behave differently in the future. Thus an alternative approach to predicting future climate change involves the use of climate models, as described next.

12.4.4   Results from Climate Models

The average surface temperature increase of 2.6°C (about 5°F) in Example 12.8 may seem small, but as we shall see shortly, an increase of this magnitude would dramatically change the earth's climate. For this reason, substantial efforts have been devoted to developing more sophisticated computer models to predict climate change. Such models consider not only the radiative processes emphasized in this chapter, but also atmospheric chemistry and atmospheric dynamics. The latter are especially important in the context of climate change because the general circulation of the atmospheric air mass affects the global distribution of heat, water vapor, and chemical constituents that influence and determine climate.

General circulation models (GCMs) attempt to account for the complex physical and chemical processes that govern the atmosphere. Fundamentally, however, these models are driven by the same basic principles of mass conservation, momentum exchange, and energy conservation that are taught in basic physics and used by engineers to solve a broad range of problems that involve fluids in motion. When applied to the atmosphere, the governing equations are too complex to solve analytically, requiring instead the use of numerical solution methods. Computer solutions usually are obtained at points across a three-dimensional grid that covers the earth horizontally and vertically. Figure 12.14 shows how the world looks on a typical GCM grid. Horizontal resolutions in these models typically range from 300 to 1,000 km, with an additional 2 to 19 layers in the vertical direction (IPCC, 1990).

Detailed computer models also have been developed to simulate the general circulation of the oceans, which are extremely important to the climate system. The most ambitious climate modeling efforts to date have coupled the atmospheric and ocean GCMs, together with models of the cryosphere (ice layers) and land-surface processes. Because these coupled models are still extremely time-consuming and cumbersome to run, a variety of simpler models are frequently used in climate studies to simulate the role of oceans and land surfaces on the climate system.

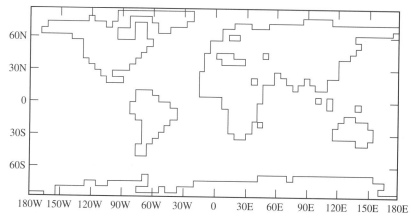

Figure 12.14   A typical GCM grid of the earth's surface. (*Source:* IPCC, 1990)

Despite the considerable advances embodied in current climate models, much is still not known about key climate system processes, especially the radiative properties of clouds, heat exchange between the atmosphere and the oceans, and the cycling of water between land surfaces and the atmosphere. One reflection of these knowledge gaps is the broad range of results that are predicted by different climate models based on different assumptions about these underlying processes.

The 1996 IPCC study used climate models to estimate the surface temperature increase resulting from a doubling of the atmospheric CO_2 concentration. This temperature increase represented the equilibrium value after a radiative balance was restored. Results ranged from 1.5°C to 4.5°C, with a best-estimate value of 2.5°C.

This equilibrium temperature change from a doubling of CO_2 is another measure of climate sensitivity that is frequently used in comparing climate models. The resulting temperature change is called $\Delta T_{2\times}$, where "2×" refers to a doubling of CO_2. This measure of climate sensitivity is related to the earlier climate sensitivity factor γ by

$$\Delta T_{2\times} = 4.37\gamma \qquad (12.31)$$

where the value 4.37 is the radiative forcing (W/m²) for a doubling of CO_2 as calculated in Example 12.6.

Example 12.9

Range of the climate sensitivity factor.   Use the IPCC values of 1.5°C, 2.5°C, and 4.5°C for the low, best, and high estimates of equilibrium temperature change from a CO_2 doubling to calculate the implied values of the climate sensitivity factor, γ.

Solution:

Equation (12.31) relates the two sensitivity measures. In this case we are given the values of $\Delta T_{2\times}$. Therefore

$$\gamma = \frac{\Delta T_{2\times}}{4.37}$$

For the three given values of $\Delta T_{2\times}$, the results are as follows:

IPCC Estimate	$\Delta T_{2\times}$ (°C)	γ (°C/W-m^{-2})
Low	1.5	0.34
Best	2.5	0.57
High	4.5	1.03

We see that the best-estimate value of γ derived from climate models is similar to the average value of 0.6°C/W-m^{-2} found earlier from observational models.

The preceding results also show a factor of three range in the climate sensitivity factor used in current climate models. This indicates a high level of uncertainty as to how climate sensitivity might vary in response to sizable changes in greenhouse gas concentrations.

12.4.5   Time Lags and Temperature Commitment

The average temperature changes discussed thus far represent the equilibrium temperature adjustment of the earth's surface in response to a change in radiative forcing induced by greenhouse gases and aerosols in the atmosphere. However, we have not yet talked about how long it takes to achieve this new equilibrium. In general, there will be a time lag between the initial disruption of the radiative balance and the reestablishment of radiative equilibrium. When we consider that approximately 70 percent of the earth's surface is covered by ocean water, we see that the thermal inertia of the oceans governs the speed at which the average surface temperature will change.

Figure 12.15 illustrates the time lag phenomenon. This case shows the surface temperature response to a step change in radiative forcing, such as from an increase in greenhouse gas concentration at time $t = 0$. The earth's surface temperature responds by gradually rising from its initial equilibrium temperature T_e to a new equilibrium temperature, $T_e + \Delta T_e$. At any time, t, the time lag can be expressed as

$$\Delta t_{lag} = t_{eq} - t \qquad (12.32)$$

where Δt_{lag} is the time until the equilibrium temperature t_{eq} is reached. Notice also that at any given time there is a difference between the actual (realized) temperature and the higher equilibrium temperature. This difference is referred to as the *temperature commitment*, ΔT_{commit}:

$$\Delta T_{commit} = T_{eq}(t) - T_e(t) \qquad (12.33)$$

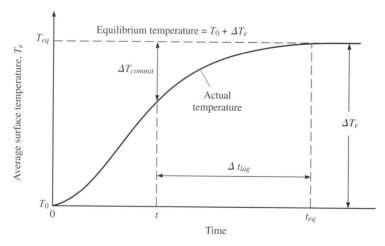

Figure 12.15   Time lag for global temperature change in response to a step change in radiative forcing at $t = 0$.

where $T_e(t)$ is the actual surface temperature at any time t, and $T_{eq}(t)$ is the equilibrium surface temperature at that time. The significance of the temperature commitment is that the earth's temperature will continue to rise even if further increases in radiative forcing immediately cease. Thus, just as a large ship continues to move forward even after the engines are stopped, the inertia of a changing climate system pushes the earth's temperature upward until the equilibrium value is reached.

The magnitude of time lags and temperature commitments are commonly estimated from climate models.

Example 12.10

Temperature commitment after a CO_2 doubling.   A climate model assumes a 1 percent/year increase in CO_2 concentration, doubling CO_2 after 70 years. At that time the realized increase in surface temperature is 62 percent of the equilibrium value. The assumed climate sensitivity for a CO_2 doubling $(\Delta T_{2\times})$ is 2.5°C. What is the actual increase in surface temperature after 70 years, and how large is the temperature commitment at that time?

Solution:

By definition, the climate sensitivity for a CO_2 doubling $(\Delta T_{2\times})$ is the value of the *equilibrium* change, which is given as 2.5°C. So the actual (realized) temperature change after the CO_2 has doubled is 62 percent of this value:

$$\Delta T_e = (0.62)(2.5°C) = 1.6°C$$

From Equation (12.33) the temperature commitment at that time is

$$\Delta T_{commit} = T_{eq} - T_e = 2.5 - 1.6 = 0.9°C$$

Thus, if further increases in CO_2 were to cease after 70 years, the average surface temperature would still continue to warm by another 0.9°C.

12.5 CLIMATE CHANGE PREDICTIONS

The ability of current climate models to accurately predict future changes in climate is an actively debated topic that will not be resolved definitively for many years to come. As already noted, there are still many gaps in our knowledge of fundamental processes that can affect how climate responds to changes in greenhouse gas concentrations and aerosols. Nonetheless, current climate models—when properly calibrated—have been able to correctly simulate large-scale features of the current global climate such as temperature, pressure, wind, and precipitation in both summer and winter. Confidence in these models has been enhanced by their ability to also reproduce climate features of the last ice age and other historical periods. Recent years have seen progress in the development of coupled climate models, which are expected to improve the accuracy of climate simulations. Let us see what these models say about the climate impacts of human activities.

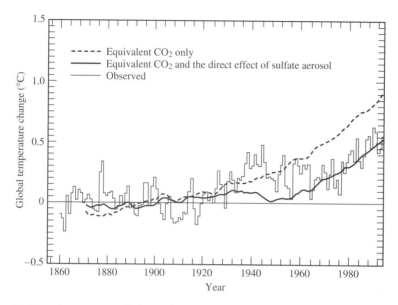

Figure 12.16   Comparisons of observed and predicted global warming since 1860. Adding the effects of aerosols substantially improves the agreement with observed data. (*Source:* IPCC, 1996a)

12.5.1   Temperature Change since Preindustrial Times

Table 12.4 showed the radiative forcings from greenhouse gases and anthropogenic aerosols since the beginning of the industrial era. It is useful to first see how well climate models predict the average global temperature change from these historical forcings compared to observed temperature changes over that period.

Figure 12.16 shows one such comparison, which is typical of the results from several studies. Surface temperature measurements since the mid-19th century show the average global temperature has increased by about $0.5°C$ ($1°F$) over the past century. Climate models that include only CO_2 and other greenhouse gases (as equivalent CO_2) typically predict a temperature change of about $1°C$ using best estimates of climate sensitivity. As seen in Figure 12.16, much better agreement is obtained when the cooling effect of atmospheric aerosols is included in the simulation. These results suggest that anthropogenic aerosols might be masking a warming effect from greenhouse gases that is greater than the observed temperature increases to date.

12.5.2   Global Warming in the 21st Century

Various modeling studies have explored the possible consequences of increased greenhouse gas emissions over the next century and beyond. Recent projections by the IPCC illustrate the magnitude of global warming estimates and their uncertainties.

The IPCC estimates come from six scenarios for global emissions of CO_2, CH_4, N_2O, halocarbons, and sulfate aerosols from 1990 to 2100, assuming no intervention measures to curb global warming. The dominant factor in these scenarios is the assumed growth in CO_2 emissions from energy use. Figure 12.17(a) shows the range of global CO_2 emissions for the six cases. Emissions are expressed in units of giga-tonnes of carbon (Gt C) per year.[3] Each scenario reflects many assumptions, the most critical of which relate to future changes in world population, rates of economic growth, and rates of technological change in different regions of the world. Later in this chapter we will discuss emission scenarios in greater detail.

For the moment, the most important observations are that the scenarios in Figure 12.17(a) span nearly an eightfold range in projected CO_2 emissions. By 2100 annual emissions in the reference scenario (labeled IS92a) are nearly three times greater than in 1990. In the highest scenario (IS92c) they are nearly five times greater. Only one case (IS92c) shows a gradual decline in emission rates over the next century.

The accumulation of CO_2 in the atmosphere raises the atmospheric CO_2 levels in these scenarios to between 500 and 1,200 ppmv. Figure 12.17(b) shows a value of 700 ppmv for the reference scenario, representing a doubling of the 1990 concentration.

Figure 12.17(c) shows the increases in radiative forcing for the high, low, and reference scenarios. These results include the effect of other greenhouse gases and aerosols in addition to CO_2 (although these results are similar to CO_2 alone, according to the IPCC). By 2100 the radiative forcing has increased to between 3.9 and 8.5 W/m^2 across the range of scenarios.

Finally, Figure 12.17(d) shows the average temperature increases predicted for these scenarios. The central or reference case (IS92a) shows a warming of 2.0°C to 2.4°C, depending on whether anthropogenic aerosols are assumed to be constant or changing. The temperature range across all scenarios is an increase of 0.8°C to 4.5°C. This range reflects not only different emission assumptions but also a range of climate sensitivity assumptions ($\Delta T_{2\times}$ values) in the climate models used for this analysis.

Climate models also paint a picture of how the temperature and other climate variables like precipitation would vary regionally for a given scenario. Figure 12.18 shows an example for the case of a doubling of the current CO_2 concentration. The map shows the regional distribution of annual average temperature increases after equilibrium is established. The largest temperature increases occur at the high latitudes over North America, Europe, and Asia. In some regions the magnitude of these increases is several times greater than the global average warming.

Given the limitations of current climate models, as well as the uncertainty in forecasting future emission levels, any projection of climate change far into the future must be viewed with a large grain of salt. Nonetheless, the scenarios in Figure 12.17 broadly indicate the magnitude of possible global warming over the next century based on current knowledge. Because of the time lags discussed earlier, the global average temperature in these scenarios would continue to rise for roughly another century, even if *no* greenhouse gases were emitted after 2100.

---

3 A gigatonne (Gt) is 10^9 metric tons (which are also called tonnes, to distinguish them from the short ton of 2,000 lbs). A Gt also is equivalent to a petagram (1 Pg = 10^{15} g). To convert Gt of carbon (C) to Gt of carbon dioxide (CO_2), multiply by 3.67 (that is, by the ratio of molecular weights, 44/12).

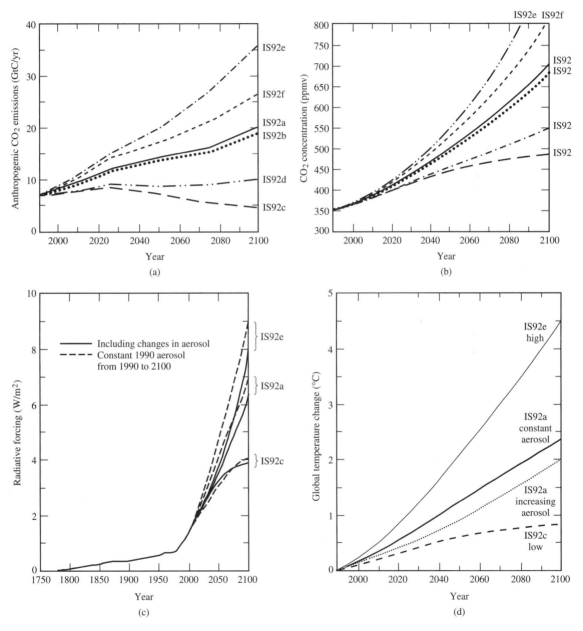

Figure 12.17 IPCC scenarios for global warming 1990–2100. (a) CO_2 emissions; (b) atmospheric CO_2 concentrations; (c) total radiative forcing; (d) global average temperature change. (*Source:* IPCC, 1996a)

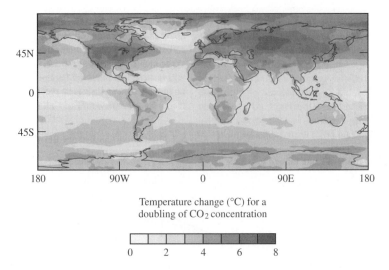

Temperature change (°C) for a
doubling of CO_2 concentration

0	2	4	6	8

Figure 12.18   Regional temperature changes for an equilibrium doubling of current CO_2 concentration. Climate models predict that the greatest temperature increases will occur at middle and high latitudes, although the effect of atmospheric aerosols will offset the effects of CO_2 alone. (*Source:* IPCC, 1996a)

How significant are the temperature changes predicted in these scenarios? After all, most people are accustomed to seeing the temperature change by 10–20°C during a single day, with even wider fluctuations over a year. So a warming of only a few degrees over a hundred years may not seem like a big deal. To gain some perspective, one must look back in time at the magnitude of global temperature changes over the past 10,000 years or more.

12.6   HISTORICAL TEMPERATURE CHANGES

Because direct measurements of the earth's temperature are available for only the past century or two, the science of paleoclimatology has employed a variety of proxy measures to deduce past climate trends going back millions of years. Pollen remains, ocean sediments, and polar ice cores are among the many different data sources used to infer the characteristics of past climates. The relative abundance of temperature-sensitive isotopes of oxygen and other gases trapped in ice cores and sediments is one way that temperature records have been constructed. Ice core data also directly indicate past atmospheric concentrations from air bubbles trapped in the ice. The deeper the ice core, the further back in time we can peer.

For most of the past million years, the earth has been much colder than today, characterized by ice ages with brief (geologically speaking) intermittent warming periods. Since the end of the last ice age about 10,000 years ago, the temperature of the earth has been relatively stable, with average temperature fluctuations generally

Scientists extracting an ice core section at the Vostok, Antarctica test station. Analysis of the full (3.3 km deep) core provides dramatic evidence of climate trends over the past 420,000 years.

within 1°C of present temperatures (IPCC, 1990). Over the past century the earth's average temperature has risen by approximately 0.5°C relative to preindustrial times (see Figure 12.16).

There is still uncertainty about the causes of past temperature fluctuations. Variations in the earth's orbit about the sun, plus changes in the earth's tilt and rotation about its axis, are among the factors that may explain past ice ages. Changes in the circulation of ocean currents that distribute heat around the globe also have been implicated in past climate change.

What about the relationship between temperature and greenhouse gas concentrations? One piece of evidence is a remarkable record from a 3.3-km deep ice core from Vostok, Antarctica, which allows us to examine trends over the past 420,000 years (Petit et al., 1999). The data in Figure 12.19 show part of that record, revealing temperature trends similar to those from other sources. The data also show a striking correlation between changes in atmospheric CO_2 and methane concentrations, as measured in trapped air bubbles, and corresponding temperature variations. Cause–effect relationships cannot be clearly established by these data, but calculations show that changes in greenhouse gas concentrations were at least partially responsible for past climate changes (IPCC, 1990). Historical concentrations of CO_2 and methane also have been well below present levels, as Figure 12.19 illustrates.

The fundamental message from the analysis of past climate changes is that a doubling of the atmospheric CO_2 concentrations to 700 ppm, coupled with global warming of several degrees centigrade over the next century—the scenario suggested by current climate models—would indeed be a big deal. Both the magnitude and rate of average temperature change would be unparalleled in the earth's recent history.

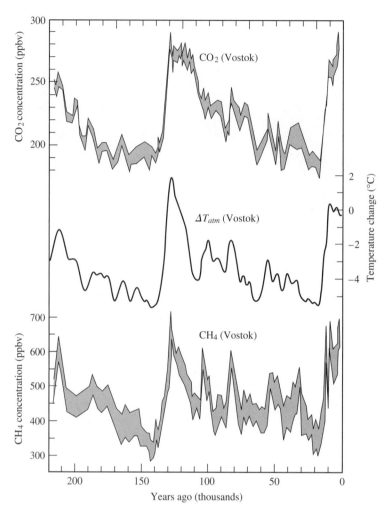

Figure 12.19   Temperature and greenhouse gas trends from the Vostok, Antarctica, ice core. This record shows that historical changes in atmospheric carbon dioxide (CO_2) and methane (CH_4) concentrations correlate well with changes in Antarctic air temperature throughout the record. (*Source:* Adapted from IPCC, 1996a)

12.7   STABILIZING ATMOSPHERIC CONCENTRATIONS

The potential consequences of global warming were described earlier in Chapter 2. They include a relentless rise in sea level, which could threaten or destroy many coastal and low-lying regions; an increase in the severity of storms, droughts, and floods in some parts of the globe; the more rapid spread of infectious diseases; and adverse impacts on the viability of forests, vegetation, and wildlife in regions of rapidly changing climate. On-going assessments are attempting to better elucidate these impacts at the regional level (USGCRP, 2000).

In light of these concerns, the United Nations Framework Convention on Climate Change was adopted in 1992 by more than 150 countries, including the United States, at the Earth Summit conference held in Rio de Janeiro. The stated goal of this historic accord is to

> . . . achieve stabilization of greenhouse gas concentrations in the atmosphere at a level that would prevent dangerous anthropogenic interference with the climate system. Such a level should be achieved within a time-frame sufficient to allow ecosystems to adapt naturally to climate change, to ensure that food production is not threatened, and to enable economic development to proceed in a sustainable manner.

Although the specific timetable and concentration levels judged to be dangerous remain to be determined, the general goal of stabilizing atmospheric concentrations of greenhouse gases is indeed ambitious. How might such a goal be achieved? And how long would it take? We briefly examine the science needed to address such questions, then summarize the results of recent analyses that provide some preliminary answers.

12.7.1   Atmospheric Lifetime of Greenhouse Gases

To understand how atmospheric concentrations may be stabilized, we must first understand the fate of gases emitted to the atmosphere. The longer the atmospheric lifetime of a substance, the more difficult the stabilization objective becomes.

Early in this chapter, Table 12.2 showed the atmospheric lifetime of different greenhouse gases. However, we did not rigorously define that term; nor have we yet discussed the relationship between greenhouse gas emissions and resulting atmospheric concentrations over time. To understand these relationships, consider what happens when a greenhouse gas is released into the atmosphere.

One possibility is for natural processes to physically remove or absorb the substance from the atmosphere and return it to the earth's surface. Indeed, the most powerful greenhouse gas in the atmosphere—water vapor—is also released to the atmosphere from fossil fuel combustion, along with CO_2. But the natural hydrological cycle of precipitation and evaporation limits the accumulation of H_2O in the atmosphere. Precipitation also is effective in washing out other atmospheric constituents such as aerosol particles. The time scale of the natural precipitation cycle is a matter of days or weeks, implying a very short atmospheric lifetime for anthropogenic aerosol emissions. For atmospheric gases like carbon dioxide, however, the time scale for removal by natural processes is very long, as we shall see shortly.

Another possibility is for greenhouse gases to be removed by chemical reactions in the atmosphere. For example, methane (CH_4) is slowly oxidized to CO_2 and H_2O by reactions with other atmospheric species. The amount of methane remaining in the atmosphere thus gradually diminishes over time. All of the major greenhouse gases except CO_2 undergo some type of chemical transformation in the atmosphere. However, the speed of these reactions varies considerably.

In general, it has been found that the rate at which a substance is depleted is proportional to the amount present at any given time. Mathematically, this can be expressed as

$$\frac{dm}{dt} = -\beta m \qquad\qquad (12.34)$$

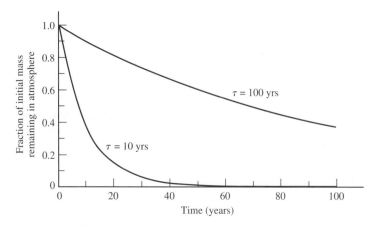

Figure 12.20   The exponential decay of a chemically reacting gas for two values of atmospheric lifetime, τ.

where dm/dt is the instantaneous rate of change, m is the mass remaining at any time t, β is a proportionality constant, and the minus sign means the mass is decreasing as a result of atmospheric removal processes. The solution to Equation (12.16) is the well-known exponential decay function:

$$m = m_o e^{-\beta t} \tag{12.35}$$

where m_o is the initial mass of the chemical at time $t = 0$. Because the proportionality constant β has the units of (time)$^{-1}$, it is more useful to define a new constant, τ, as

$$\tau = \frac{1}{\beta} \tag{12.36}$$

The constant τ thus has the units of time, typically measured in years. This time constant τ is defined as the *atmospheric lifetime*. Equation (12.35) can then be rewritten as

$$m = m_o e^{-t/\tau} \tag{12.37}$$

Figure 12.20 plots this function for two values of τ. As you can see, the longer the atmospheric lifetime τ, the slower and more gradual the depletion. The average concentration of a substance in the atmosphere follows the same declining curve because the volume of any gas is proportional to its mass. The numerical value of the atmospheric lifetime for any substance must be determined from experimental data.

Example 12.11

Atmospheric decay of a greenhouse gas.   Table 12.2 lists the atmospheric lifetime of methane (CH_4) as 12 years. If 1 kg of CH_4 is added to the atmosphere today, how much will remain after one year?

Solution:

To find the amount remaining after one year, use Equation (12.37) with $m_o = 1$ kg, $\tau = 12$ years, and $t = 1$ year.

$$m = m_o \, e^{-t/\tau} = 1.0 \exp\left(-\frac{1}{12}\right) = 0.92 \text{ kg}$$

So after one year, 8 percent of the initial amount has been removed by atmospheric reactions, and 92 percent remains.

Notice that an exponential decay rate means that the mass remaining in the atmosphere after one atmospheric lifetime is $e^{-1} = 0.368$, or 36.8 percent of the initial amount. After two lifetimes ($t = 2\tau$) 13.5 percent remains, and after three lifetimes only 5 percent is left. Figure 12.20 illustrates the fraction remaining for two values of atmospheric lifetime.

To stabilize the atmospheric concentration of a greenhouse gas that undergoes exponential decay, the amount of gas added to the atmosphere each year must not exceed the amount lost via chemical reactions. Thus, if 8 percent of the atmospheric methane is lost each year as calculated in Example 12.11, then no more than 8 percent of the original amount can be added each year, including natural as well as anthropogenic emission sources. For gases like halocarbons and perhalogens with extremely long atmospheric lifetimes (often measured in centuries), stabilization can be achieved only by completely eliminating anthropogenic emissions to the atmosphere.

12.7.2 The Carbon Cycle

The fate of carbon dioxide emissions in the atmosphere is more complicated than that of the chemically reacting gases described above. CO_2 is a very stable, nonreacting chemical. Carbon is a basic constituent of all life forms on earth and is found throughout the natural environment. Carbon in various chemical forms, including CO_2, is exchanged between the land, the atmosphere, and the oceans in a cycle linking biological and physical processes, including human activities.

Figure 12.21 depicts the global carbon cycle, showing the magnitudes of reservoirs and average exchange rates during the 1980s. All values are expressed in gigatonnes of carbon rather than CO_2. The carbon cycle is a dynamic system in which some processes occur relatively quickly, such as the carbon uptake by vegetation during photosynthesis, while other processes occur very slowly, such as the accumulation of carbon in deep ocean sediments. Over thousands of years, the global carbon cycle achieved a relatively stable balance until human activities disrupted the system by extracting carbon from the earth's crust (mainly in the form of fossil fuels) and injecting it into the atmosphere as CO_2 formed during combustion. Additional anthropogenic disruptions came from industrial processes such as cement production, which gives off CO_2 from carbonate minerals, and from land use practices such as clear-cutting forests, which reduces the biomass uptake of CO_2.

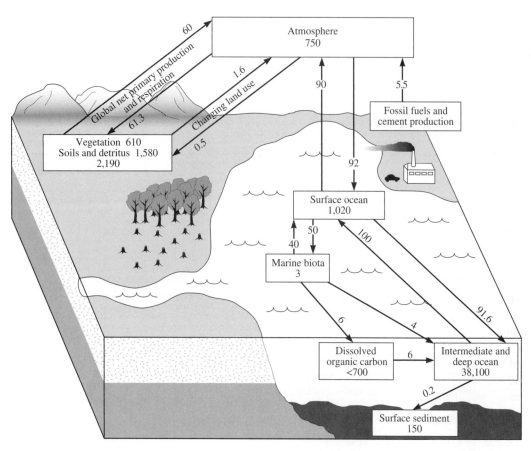

Figure 12.21   The global carbon cycle, showing the major reserves (in GtC) and annual average fluxes (in GtC/yr) over 1980–1989. (*Source*: IPCC, 1996a)

Because different parts of the climate system exchange carbon with the atmosphere at very different rates, no single atmospheric lifetime can be ascribed to CO_2. Rather, carbon cycle models reflect different uptake and release processes that govern the concentration of CO_2 in the atmosphere. For this reason Table 12.2 shows a range of 50–200 years for the atmospheric lifetime of CO_2. The shorter periods typically reflect uptake by terrestrial sinks such as forests, whereas longer time constants reflect the much slower ocean exchanges.

A simple carbon cycle model can be represented by a combination of two exponential decay rates (Enting and Newsam, 1990), as shown in Equation (12.38). This equation gives the amount of carbon remaining in the atmosphere after an initial mass is injected at time $t = 0$:

$$\frac{m_C}{m_{C,o}} = 0.375 \exp\left(-\frac{t}{10.43}\right) + 0.625 \exp\left(-\frac{t}{291.5}\right) \qquad (12.38)$$

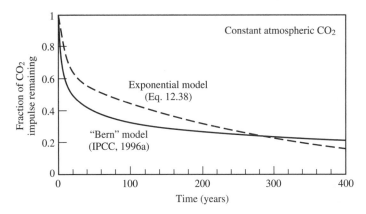

Figure 12.22 Atmospheric response to a CO_2 impulse using a simple carbon cycle model. The figure also shows the response predicted by the Bern carbon cycle model used by IPCC (1996a). These responses assume that the CO_2 impulse is only a small addition to a constant atmospheric CO_2 concentration.

Here $m_{C,o}$ is the initial carbon mass, and m_C is the amount remaining after t years. This model applies only to CO_2 emissions from fossil fuel combustion; it assumes that other carbon exchanges in the biosphere remain in balance. Figure 12.22 plots the CO_2 mass remaining in the atmosphere after an impulse of 1 kg is added at $t = 0$. More complex models show a somewhat faster initial decay rate, as seen in Figure 12.22 for the Bern carbon cycle model used in the 1996 IPCC assessment. The next example illustrates the use of a simple carbon cycle model.

Example 12.12

An application of a carbon cycle model. From Table 12.1, U.S. CO_2 emissions in 1996 totaled 5.30 Gt (equivalent to 1.45 Gt C). How much of the CO_2 emitted in 1996 will still be in the atmosphere in the year 2100 according to the carbon cycle model in Equation (12.38), assuming no further CO_2 additions? How does this compare to the Bern cycle estimate from Figure 12.22?

Solution:

First use Equation (12.38) with $m_{C,o} = 1.45$ Gt C and $t = 104$ years (from 1996 to 2100):

$$\frac{m_C}{m_{C,o}} = 0.375 \exp\left(-104/10.43\right) + 0.625 \exp\left(-104/291.5\right)$$

$$= (1.8 \times 10^{-5}) + 0.437 = 0.437$$

$$m_C = 0.437\, m_{C,o} = (0.437)(1.45) = 0.63 \text{ Gt C}$$

So after more than a century nearly 44 percent of the CO_2 will still remain in the atmosphere according to this model. From Figure 12.22, the more detailed Bern cycle model estimates that about 30 percent will remain after 100 years. For longer time horizons, the difference between the two estimates diminishes. Again, these results assume a single pulse with no further additions of CO_2.

Table 12.5   Average annual CO_2 sources and sinks for 1980–1989.

CO_2 Sources and Sinks	GtC/yr
CO_2 sources:	
(1) Emissions from fossil fuel combustion and cement production	5.5 ± 0.5
(2) Net emissions from changes in tropical land use	1.6 ± 1.0
(3) Total anthropogenic emissions = (1) + (2)	7.1 ± 1.1
Partitioning among reservoirs:	
(4) Storage in the atmosphere	3.3 ± 0.2
(5) Ocean uptake	2.0 ± 0.8
(6) Uptake by Northern Hemisphere forest regrowth	0.5 ± 0.56
(7) Other terrestrial sinks = (3) − [(4) + (5) + (6)]	1.3 ± 1.5

Source: IPCC, 1995.

The carbon cycle model is a critical element of climate change predictions because it links anthropogenic emissions of CO_2 to changes in atmospheric CO_2 concentration. Quantifying the fate of atmospheric carbon is not an easy task, however. Table 12.5 shows the estimated average annual flows of CO_2 during the 1980s, along with the associated uncertainties. Of the 7.1 Gt/year of total anthropogenic emissions, roughly half (47 percent) ultimately remains in the atmosphere, while the remaining half is partitioned almost equally between the oceans and various terrestrial sinks.

Because CO_2 is being added to the atmosphere at an unprecedented rate, the fraction of CO_2 remaining in the atmosphere is likely to increase relative to the estimate for the 1980s. More sophisticated carbon cycle models such as the Bern model indicate that roughly *half* of the CO_2 added to the atmosphere will remain there, even if future anthropogenic emissions of CO_2 were to completely cease. The policy goal of stabilizing atmospheric CO_2 concentrations in the face of continual and increasing anthropogenic emissions is thus extremely formidable.

12.7.3   Stabilization Scenarios

Figure 12.23 shows several scenarios developed by the Intergovernmental Panel on Climate Change (IPCC) to show how atmospheric CO_2 concentrations might be stabilized at levels from 450 to 1,000 ppmv CO_2. Stabilization is assumed to be achieved over the next 75 to 400 years, depending on the scenario.

Achieving the stabilized concentrations in Figure 12.23 requires substantial reductions in anthropogenic CO_2 emissions. However, various pathways can achieve the same final outcome. Figure 12.23(b) shows two possible trajectories for CO_2 emission reductions (the solid lines and the dotted lines) that lead to the same stabilized concentration values. The solid lines assume that global CO_2 reduction measures

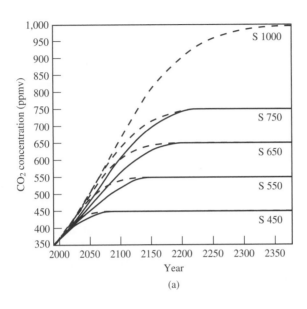

(a)

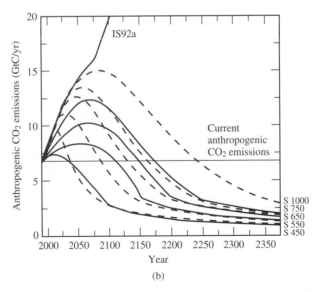

(b)

Figure 12.23   Scenarios for stabilizing atmospheric CO_2 concentrations at levels of 450 to 1,000 ppmv (a) require the reductions in CO_2 emissions shown in (b). The dashed lines and solid lines represent different timetables for emission reductions. (*Source:* IPCC, 1996a)

begin in 2000, whereas the dashed lines assume that emission reductions begin some time later. (Neither case refers to actual or planned actions.)

The most important insight from the scenarios in Figure 12.23 is that *the international goal of stabilizing atmospheric CO_2 levels will require anthropogenic emissions to decline to roughly 60 to 80 percent below 1990 emission rates.* As a fraction of future CO_2 emissions (such as the 20 Gt C/yr by 2100 in the IPCC reference scenario), the required percentage reductions are in excess of 90 percent. Achieving such reductions would require a major change in the world's use of energy. The following section explores this issue in greater detail.

12.8   CO$_2$ EMISSIONS AND ENERGY USE

Most of the energy used throughout the world is in the form of fossil fuels consisting of oil, coal, and natural gas. Chapter 5 discussed the chemical nature of these fuels and their use for electric power generation. Fossil fuels also supply energy for transportation, industrial processes, and residential and commercial buildings. Figure 12.24 shows the magnitude of U.S. energy consumption, along with primary energy sources. Approximately 85 percent of U.S. energy is supplied by fossil fuels, predominantly oil (petroleum products). Globally the picture is similar, with fossil

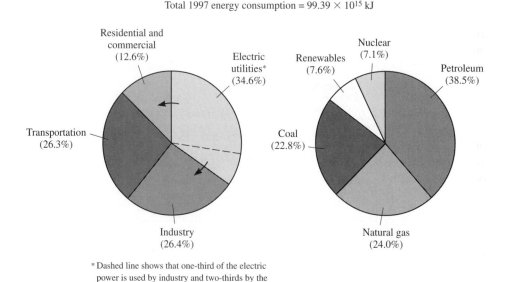

Total 1997 energy consumption = 99.39 × 10^{15} kJ

Residential and commercial (12.6%)
Electric utilities* (34.6%)
Transportation (26.3%)
Industry (26.4%)

Nuclear (7.1%)
Renewables (7.6%)
Petroleum (38.5%)
Coal (22.8%)
Natural gas (24.0%)

* Dashed line shows that one-third of the electric power is used by industry and two-thirds by the residential and commercial sector.

Figure 12.24   U.S. energy consumption for 1997. (*Source:* Based on USDOE, 1998)

Table 12.6 Typical carbon and energy content of fossil fuels.

Energy Source	Carbon Content (Weight %C)	Heating Value (kJ/g)[a]	Carbon Intensity (g C/MJ)[a]
Natural gas	74	54.4	13.7
Crude oil	85	44.3	19.2
Coal	59[b]	24.2	24.4

[a] Based on higher heating value (HHV). The lower heating value (LHV) is approximately 10% lower for natural gas and 5% lower for coal and crude oil. Use of the LHV would *increase* the carbon intensity value of each fuel by these percentages.
[b] Values vary by about ±30% for different coal types.
Source: USDOE, 1998; IPCC, 1996b.

fuels accounting for 75 percent of the world's total annual energy consumption (which includes biomass fuels) and 85 percent of commercial energy.

12.8.1 Carbon Content of Fuels

As discussed in Chapter 5, fossil fuels contain varying proportions of carbon and hydrogen, depending on whether the fuel is a liquid (oil), gas (natural gas), or solid (coal). When burned, the carbon in fuel is converted to carbon dioxide:

$$C + O_2 \rightarrow CO_2 \qquad (12.39)$$

The atomic weight of carbon is 12 and the molecular weight of CO_2 is 44. Thus every 12 g of carbon in fuel produces 44 g of CO_2. The ratio of these two numbers provides a convenient way of expressing CO_2 emissions in terms of equivalent carbon, which is how many organizations (including the IPCC) report CO_2 emissions:

$$1 \text{ mass unit of C} = \frac{44}{12} = 3.667 \text{ mass units of } CO_2 \qquad (12.40)$$

To determine the total mass of CO_2 emitted each year, we must know the amount of each fossil fuel consumed and the mass fraction of carbon in each fuel type. Table 12.6 gives typical values of the carbon content of fossil fuels. Assuming all of the carbon is converted to CO_2, the total emission from each fuel type is simply

$$\text{Mass of carbon emitted} = \frac{\text{wt\%C}}{100} \times \text{Mass of fuel burned} \qquad (12.41)$$

If there are multiple fuels with different carbon contents, the total carbon emission rate $(\dot{m}_C)$ is simply the sum over all fuels.

Example 12.13

CO_2 emissions from coal consumption. In 1996 approximately 4,700 million metric tons of coal were burned throughout the world. Estimate the resulting carbon emission rate in units of Gt C/yr and also Gt CO_2/yr.

Solution:

The coal mass of 4,700 million metric tons = 4.700×10^9 tonnes = 4.7 Gt. Based on the data in Table 12.6, approximately 59 percent of the coal mass is carbon. So the mass flow of CO_2 emissions from coal burning, expressed as equivalent carbon, is

$$\dot{m}_C = \text{mass of carbon emitted/yr} = \left(4.7 \frac{\text{Gt coal}}{\text{yr}} \right)\left(0.59 \frac{\text{Gt C}}{\text{Gt coal}} \right) = 2.8 \text{ Gt C/yr}$$

To express this result in terms of CO_2, multiply by 3.667: $\dot{m}_{CO_2} = 10 \text{ Gt } CO_2/\text{yr}$

Biomass fuels such as wood also contain carbon and release CO_2 when burned. In this case, however, the CO_2 released equals the CO_2 absorbed by the biomass via photosynthesis during the growth cycle. Thus there is no *net* addition of CO_2 to the atmosphere, although the absorption and release of CO_2 occur at different times, often decades apart.

12.8.2　Energy Content of Fuels

An alternative method for calculating CO_2 emissions (or equivalent carbon) is based on fuel energy content rather than fuel mass. The energy delivered by a fossil fuel is determined mainly by its content of carbon and hydrogen, which release chemical energy when the fuel is burned. The quantity of energy per unit of fuel mass is known as the *heating value* and is typically expressed in units of kJ/kg or kJ/mol. Numerical values are determined from standard laboratory measurement methods.

Given the importance of this measure, one might think there would be universal agreement on how to report the heating value of a given fuel. But in fact, the heating value may be specified in one of two ways: the *higher* heating value (HHV) and the *lower* heating value (LHV). These two measures reflect different assumptions about whether energy can be recovered from water vapor formed during combustion. If the vapor is cooled and condensed into a liquid, additional useful energy (equal to the heat of vaporization) can be extracted. This is the basis for the HHV, which is the prevailing measure used in the United States, the United Kingdom, Japan, and elsewhere. In most of Europe and many other parts of the world, however, the LHV is used to quantify fuel energy content. The LHV reflects the fact that existing combustion systems do not usually recover the heat of vaporization because it is not economical to do so with current technology. Accordingly, the fuel is assumed to have less energy available for use. Thus LHV makes fuel-using technologies appear more efficient.

The numerical difference between HHV and LHV ranges from about 2 percent to over 10 percent, depending on the specific fuel. Typical values are shown in Table 12.6. The difference is greatest for natural gas, whose combustion produces more water vapor per unit of fuel than oil or coal. Manufacturers of gas-using equipment (such as gas turbines) thus favor the use of LHV in reporting thermal efficiency. Care must be exercised when using energy data (or efficiency values) because many sources do not clearly state whether HHV or LHV is being used. This book uses HHV except where explicitly noted.

12.8.3 Carbon Intensity of Fuels

Expressing the carbon content of a fuel in relation to the fuel energy content gives a direct link between energy consumption and carbon emissions. The ratio of a fuel's carbon content to its energy content defines the *carbon intensity*:

$$\text{Carbon intensity} = \frac{\text{Fuel carbon mass}}{\text{Fuel energy}} \qquad (12.42)$$

Once the fuel heating value is determined (based either on HHV or LHV), the corresponding carbon intensity is calculated as

$$\text{Carbon intensity} = \frac{\text{Fraction of C in fuel}}{\text{Fuel heating value}} \qquad (12.43)$$

The fraction of carbon in fuel may be specified on either a mass basis or a molar basis, consistent with the units used for heating value. Table 12.6 gives representative values of the carbon intensity of fossil fuels.

Example 12.14

Carbon intensity of methane gas.   Methane (CH_4) has a higher heating value (HHV) of 55.64 kJ/g. Calculate the carbon intensity of methane and compare it to the value for natural gas in Table 12.6.

Solution:

The carbon intensity is defined by Equation (12.43). Because the heating value is given on a mass basis, we must first find the mass fraction of carbon in methane. This is

$$\frac{\text{Mass of C}}{\text{Mass of } CH_4} = \frac{\text{Atomic wt of C}}{\text{Molecular wt of } CH_4} = \frac{12}{16} = 0.75 \, \frac{\text{g C}}{\text{g } CH_4}$$

Thus 75 percent of methane is carbon by weight. The carbon intensity is then

$$CH_4 \text{ carbon intensity} = \frac{0.75 \text{ g C/g } CH_4}{55.64 \text{ kJ/g } CH_4} = 0.0135 \text{ g C/kJ} = 13.5 \text{ g C/MJ}$$

Notice that the carbon intensity for natural gas in Table 12.6 (13.7 g C/MJ) is slightly higher than that of pure methane because natural gas includes small amounts of ethane, propane, and butane, which contain more carbon per unit mass than methane.

Once we know the carbon intensity of an energy source, we can calculate total carbon emissions by multiplying the carbon intensity by the total energy use:

$$\text{Mass of carbon emitted} = \text{Energy use} \times \text{Carbon intensity} \qquad (12.44)$$

So long as a consistent system of units is used to express fuel heating value and fuel carbon intensity, total carbon emissions will be calculated correctly. For example, the U.S. energy consumption data in Figure 12.24 are based on the higher heating value of all fuels. Hence the carbon intensity values in Table 12.6 are appropriate for calculating total carbon emissions, as illustrated next.

Example 12.15

Carbon emissions from natural gas use.   Estimate the total carbon emissions from U.S. natural gas usage in 1997 based on the energy data in Figure 12.24.

Solution:

From Figure 12.24, natural gas accounted for 24 percent of total 1997 U.S. energy use of 99.39 EJ. Thus

$$\text{Energy from natural gas} = (0.24)(99.39) = 23.85 \text{ EJ}$$

U.S. energy data are based on HHV, so multiply by the carbon intensity of natural gas from Table 12.6 to estimate the total carbon release:

$$\text{Carbon emitted} = (23.85 \text{ EJ})\left(13.7\frac{\text{g C}}{\text{MJ}} \right)\left(\frac{10^{18} \text{ J/EJ}}{10^6 \text{ J/MJ}} \right)$$

$$= 3.3 \times 10^{14} \text{ g C}$$

$$= 0.33 \text{ Gt C}$$

Note that had the lower heating value (LHV) convention been used, the reported U.S. energy consumption of natural gas would have been about 10 percent lower, but the carbon intensity value would be 10 percent higher. The resulting calculation of carbon (or CO_2) emissions would therefore be unchanged.

12.8.4   Regional Sources of CO_2 Emissions

World energy production and consumption in 1997 resulted in CO_2 emissions of 6.2 GtC/yr, expressed in terms of equivalent carbon. Figure 12.25(a) shows the breakdown by country for the 10 largest emitters. The United States accounted for the largest fraction (24 percent) of energy-related carbon emissions, followed by China (13 percent) and Russia (7 percent). On a per capita basis, Figure 12.25(b) shows the United States again is the largest emitter, followed by Canada and Russia. The less industrialized countries of China and India emit substantially less CO_2 per person than the industrialized nations even though their total emissions are relatively high because of their large population.

As we look to the future, scenarios developed by the IPCC and others project a changing picture of where CO_2 emissions will originate over the next century. These projections show increasing CO_2 emissions from developing regions of the world that today emit relatively little CO_2. Indeed, carbon emissions from most of these regions, especially Asia and Africa, will overtake those of the United States within the next several decades, according to these scenarios. Rapid growth in energy use is the key reason for the projected growth in CO_2 emissions.

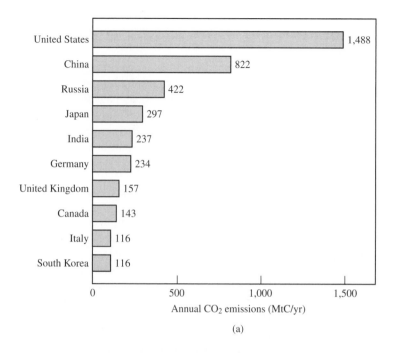

(a)

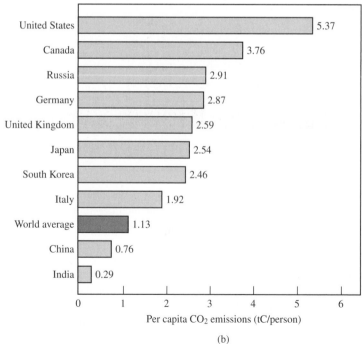

(b)

Figure 12.25　CO_2 emissions from the 10 largest emitting countries in 1997 showing (a) total annual emissions and (b) emissions per capita for each country. (*Source:* Marland et al., 1999)

12.9   REDUCING GREENHOUSE GAS EMISSIONS

Options and methods for reducing emissions of CO_2 and other greenhouse gases are far-reaching topics involving a host of technical, economic, and societal issues. The limited objective of this chapter is to provide a basic understanding of the factors that contribute to the projected growth in anthropogenic emissions. Some of the methods available to curtail those emissions also are discussed. The primary emphasis is on CO_2 emissions from energy use, which are the major contributor to global warming.

12.9.1   Factors Affecting CO_2 Emissions Growth

Example 12.15 showed how annual CO_2 emissions from energy use can be calculated as the product of annual energy consumption times the carbon intensity of the energy source. But what factors determine the total annual energy consumption?

There are various ways to decompose the problem in order to address this question. One common method is to relate energy use to population and economic well-being. In general, the greater the population and the higher the standard of living, the greater the energy consumption. A common measure of economic well-being is the gross domestic product (GDP) of a country expressed on a per capita basis. The total energy consumption can then be written as the product of three terms:

$$\text{Energy use} = \text{Population} \times \frac{\text{GDP}}{\text{per capita}} \times \frac{\text{Energy use}}{\text{per GDP}} \tag{12.45}$$

We can multiply this equation by the average carbon intensity to calculate the annual carbon (or CO_2) emission rate:

$$\frac{CO_2 \text{ emissions}}{\text{per year}} = \left(\frac{\text{Population}}{\text{per year}}\right) \times \left(\frac{\text{GDP}}{\text{per capita}}\right) \times \left(\frac{\text{Energy use}}{\text{per GDP}}\right)$$
$$\times \left(\frac{CO_2 \text{ emissions}}{\text{per unit energy}}\right) \tag{12.46}$$

The four factors on the right side are the key determinants of CO_2 emissions at the national level. Assumptions about these four factors—their magnitude and rate of change—form the basis for the emission scenarios that drive global climate change projections.

Population Growth   The first term in Equation (12.46) reflects the size of the population and effect of population growth. Increasing population generates greater demand for food, clothing, shelter, and other human needs. Meeting these increased demands results in additional greenhouse gas emissions from the use of energy and other activities. At the end of the 20th century, the world's population was just over 6 billion people. The United Nations projects nearly a doubling to more than 10 billion people by the end of the 21st century. This population growth fuels much of the projected growth in regional energy use and resulting carbon emissions.

GDP per Capita   The second term in Equation (12.46), GDP per capita, measures average affluence. As this term grows, an individual's demand for goods and services—such as improved education, improved health care, and more consumer products—also grows. As standards of living increase, so does the potential to emit greater quantities of greenhouse gases. Questions arise, however, as to whether changes in GDP per capita are linked to rates of population growth, or whether these two factors are indeed independent of one another as Equation (12.46) suggests. There are varying opinions and conflicting studies on this subject. This lack of understanding is one of the limitations in forecasting future CO_2 emissions.

Energy Intensity   The last two terms in Equation (12.46) are most closely related to technology and technological change. The ratio of primary energy consumption to GDP shows how much energy a society uses to generate a unit of economic activity. This ratio is often referred to as the *energy intensity* of an economy. Figure 12.26 compares the energy intensity of several countries based on recent data. Among the seven most industrial nations, the energy intensity varies by more than a factor of two, with Canada and the United States having the highest values and Japan the lowest. Energy use for transportation is a key factor accounting for these differences. For the less industrialized countries the energy intensity values are much higher. Russia, for example, uses 11 times more energy than Japan per dollar of GDP.

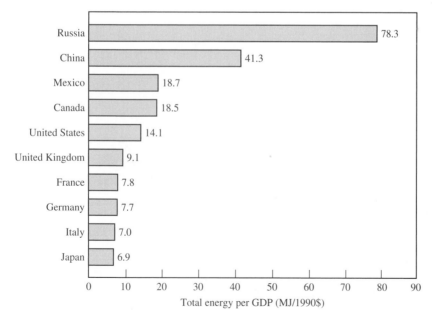

Figure 12.26   Energy intensity of selected countries expressed as total energy consumption per dollar of gross domestic product. All data are for 1998, with GDP expressed in constant 1990 dollars. (*Source:* Based on USDOE, 2000)

The question again arises of whether energy intensity is correlated with other factors in Equation (12.46). Although the IPCC scenarios treat these four factors as independent, historical studies suggest that energy intensity depends on the structure of an economy and the level of economic development. These factors are discussed more fully in Chapter 15, which addresses environmental forecasting.

Carbon Intensity   The final term in Equation (12.46) is carbon intensity, which was discussed earlier. Figure 12.27 shows the magnitudes and trends of carbon intensity for five countries. The three industrialized nations (United States, Japan, and France) have lower carbon intensity than the two less developed countries (China and India), which rely predominantly on coal as an energy source. In all cases carbon intensity has trended downward, reflecting a greater diversification of energy supplies. The case of France is especially dramatic. The decline there is due mainly to the increased use of nuclear energy for electricity, displacing coal and other carbon-intensive fuels. Worldwide, average carbon intensity has decreased by about 32 percent over the past century (IPCC, 1996b).

Reducing CO_2 Emissions   The four parameters in Equation (12.46) reflect the major factors governing CO_2 emissions growth. If future CO_2 emissions are to be reduced, long-term changes in world population growth, economic growth, energy intensity,

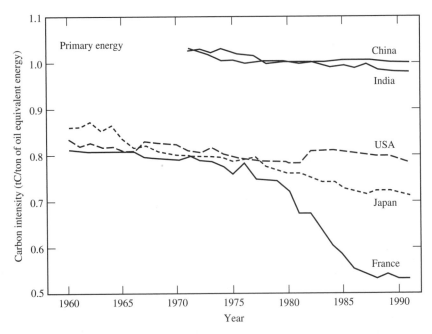

Figure 12.27   Carbon intensity of primary energy supplies for selected countries, showing different rates of decline in carbon emissions per unit of energy over the past several decades. (*Source:* PCAST, 1997)

and carbon intensity must be reconciled. The first two parameters are influenced strongly by the cultural and political forces of individual countries. Chapter 15 provides some additional insights into the factors affecting population and economic growth. The remainder of this chapter focuses on the two more technology-related parameters—energy intensity and carbon intensity.

12.9.2 Reducing Energy Intensity

Figure 12.28 tells an interesting story. It shows that from 1970 to the mid-1980s the energy intensity of the United States economy fell significantly even as the GDP per capita continued to grow. In fact, total U.S. energy consumption actually *decreased* during the early 1980s, with economic growth decoupled from energy growth.

How did this reduction in energy intensity come about? Two major things happened. One was a change in the makeup of industries that contribute to the national economy. This period saw a decline in energy-intensive industries like steel manufacturing and a rise in light industries like semiconductor assembly, which require less energy to produce a dollar's worth of goods and services. These types of *structural changes* in the economy helped reduce the overall energy intensity in the United States.

Efficiency improvements were the other factor behind the decline in U.S. energy intensity in the 1970s and 1980s. During that period, manufacturing industries found new ways to substantially reduce energy use, as did many homeowners, shopping centers, and office buildings. In response to new regulations, the average fuel economy of U.S. automobiles also began to improve. The stimulus for most of these efficiency gains was the dramatic increase in oil prices in the 1970s following the Arab

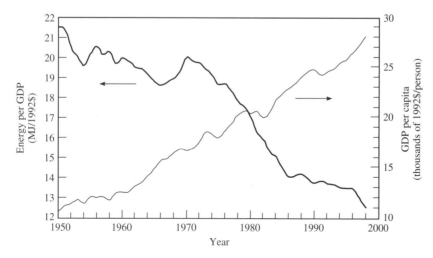

Figure 12.28 Trends in U.S. energy intensity and per capita GDP, 1950–1998. Energy intensity fell sharply from about 1970 to 1985, even as the overall economy continued to grow. (*Source:* Based on USDOE, 1999)

oil embargo and (five years later) the change of government in Iran. With energy much more expensive, and sometimes in short supply, new ways were found to reduce energy consumption in all sectors of the economy.

By the late 1980s world energy prices had fallen dramatically, eliminating many incentives to improve energy efficiency. From an engineering perspective, however, opportunities still abound to substantially reduce energy consumption through efficiency improvements that also yield economic savings (for example, see Rubin et al., 1992; NAS, 1992). Table 12.7 lists some of the areas that have been studied in detail. Other chapters of this book also examine how improved engineering design can reduce the energy consumption of modern technologies. Chapters 3–6 analyze automobiles, battery-operated appliances, electric power plants, and household refrigerators. Chapter 7 shows how life cycle assessments can help save energy across the economy. And Chapter 13 shows the economic analysis for some of these options. In the context of global climate change, technology innovations and efficiency improvements that reduce energy consumption are an essential component of any long-term strategy to reduce greenhouse gas emissions.

12.9.3   Reducing Carbon Intensity

Along with reductions in energy intensity, reductions in the carbon intensity of world energy use will be essential if the U.N. Framework Convention goal of stabilizing atmospheric CO_2 concentration is to be realized. (Recall that the stabilization scenarios required CO_2 reductions of 60–80 percent below 1990 levels.) Table 12.6 showed representative values of the carbon intensity of fossil fuel energy sources. In contrast, nuclear energy and most renewables emit no net carbon. The amounts of each energy source in national and global energy budgets determine the average

Table 12.7   Some areas for improving the energy efficiency of current technology. Significant reductions in energy use are possible in all of these areas using currently available measures.

Energy Demand Sectors and Technologies		
Residential/Commercial	*Industrial Processes*	*Transportation Systems*
Lighting	Electricity use (motors, drives, etc.)	Light-duty vehicles
Water heating	Process heat	Heavy-duty vehicles
Cooking	Fuel use	Off-road vehicles
Refrigeration	Cogeneration systems	Locomotives
Space heating	New process technology	Aircraft
Air conditioning		Marine vessels
Ventilation		Transportation management
Appliances		
Building shells		

value of overall carbon intensity. The historical data shown in Figure 12.27 reveal that the average carbon intensity of many countries has in fact been declining, as is global carbon intensity. Many analysts predict that this trend will continue into the future. Additional perspectives on these forecasts can be found in Chapter 15, which discusses environmental forecasting.

Example 12.16

Reducing the carbon intensity of U.S. energy use.   Calculate the average carbon intensity of U.S. energy use for 1997. If half the 1997 coal use were replaced by natural gas, by how much would the carbon intensity decrease?

Solution:

Figure 12.24 gives the percentages of each energy source used in 1997, and Table 12.6 gives the carbon intensity of each fossil energy source. For nuclear and renewables (most of which is hydroelectric power), the energy intensity is zero because these energy sources emit no carbon. The average carbon intensity can thus be calculated by multiplying the energy fraction and carbon intensity, then summing over all energy sources, as shown in the following table:

Energy Source	Fraction of 1997 Energy Use	Carbon Intensity (g C/MJ)	Weighted Average (g C/MJ)
Petroleum	0.385	19.2	7.39
Natural gas	0.240	13.7	3.29
Coal	0.228	24.4	5.56
Nuclear	0.071	0	0
Renewables	0.076	0	0
Total	1.000		16.24

The average 1997 carbon intensity for the United States is thus found to be 16.2 g C/MJ. If half the coal were replaced by natural gas, the new fraction of coal use would be 0.114, and the new fraction for gas would be 0.354. The total weighted average would then fall to 15.0 g C/MJ. This would be a decrease of 7.5 percent in average carbon intensity.

As Example 12.16 illustrates, reducing the average carbon intensity of energy use in general implies a shift away from high-carbon fuels like coal and oil toward lower-carbon or zero-carbon fuels like natural gas and renewables. But this is easier said than done. The following sections outline some of the opportunities for and barriers to carbon intensity reductions.

Alternative Energy Sources   For electric power generation, one of the most attractive options today is the use of natural gas in a combined-cycle power plant (see Chapter 5 for details). This option combines the lower carbon intensity of natural gas with a higher efficiency of power generation. As a result, CO_2 emissions per kilowatt-hour of electricity generated are about 80 percent lower than from a conventional coal-fired power plant. In the United States and many other parts of the world, gas-fired combined-cycle plants are also the most economical option for new power-generating facilities. The use of natural gas is thus expected to grow substantially over the next few decades, which will reduce the carbon intensity of energy use.

The key question is how long it will be until natural gas supplies run out or become too expensive relative to other energy sources. The same question has been asked about oil supplies for the past hundred years. Estimates of remaining oil and gas reserves in the past have been notoriously wrong, as new discoveries have continued to add to the worldwide reserve base faster than the growth in fuel consumption. Table 12.8 shows one current estimate of the fossil fuel supplies that are potentially available under different degrees of certainty (and cost of recovery). Coal supplies are most abundant, but proven oil and gas reserves also are substantial. Eventually, of course, these nonrenewable resources will be depleted, and their prices will increase as scarcity sets in. But for at least the next decade or two, the supplies of low-carbon natural gas appear plentiful and relatively low in cost in many parts of the world. The cost and availability of gas will nonetheless vary substantially from one region to another, so other fuels, like coal, will remain more economical in many cases. Political events also may affect the cost and availability of low-carbon fuels.

Achieving zero carbon emissions poses much greater difficulty. One option is nuclear power, which is technologically capable of supplying much of the world's growing demand for electricity while emitting zero CO_2. But high capital costs, coupled with societal and environmental concerns over nuclear safety, nuclear proliferation, and radioactive waste disposal (see Chapter 2), limit or preclude nuclear plants

Table 12.8   Estimated worldwide reserves of fossil fuels; all figures in 10^9 metric tons of oil equivalent (Gtoe).[a]

Energy Source	Consumption		Proven Reserves	Probable Resources[b]	Resource Base[c]	Additional Occurrences
	1850–1990	1990				
Oil						
Conventional	90	3.2	150	145	295	n/a
Unconventional	—	—	193	332	525	1,900
Natural gas						
Conventional[d]	41	1.7	141	279	420	n/a
Unconventional	—	—	192	258	450	400
Hydrates	—	—	n/a	n/a	n/a	18,700
Coal	125	2.2	606	2,794	3,400	3,000
Total[e]	256	7.0	1,282	3,808	5,090	24,000

[a] 1 Gtoe = 41.87 × 10^{18} J based on lower heating value (LHV); dash = negligible amounts; n/a = data not available.

[b] Resources to be discovered or developed to reserves.

[c] Sum of reserves and resources.

[d] Includes natural gas liquids.

[e] All totals have been rounded.

Source: Nakicenovic et al., 1998.

in many parts of the world, especially in industrialized nations. In the United States, nuclear power generation is projected to decline as existing plants retire.

Renewable energy technologies such as wind, solar, and biomass are viewed more favorably as zero-carbon options for electric power generation. In the case of biomass, zero *net* carbon emission is achieved by offsetting the carbon released in power generation by the carbon stored during biomass growth. However, many renewable energy technologies are not yet economically competitive with fossil fuel technologies at today's energy prices. Policy measures such as financial incentives, or a limit on CO_2 emissions, would be needed to stimulate a large-scale shift to renewable energy forms over several decades. Other types of environmental impacts would result from such a shift, as discussed in Chapter 5.

Alternative energy sources for the transportation sector represent the greatest technological challenge. The world's mobility today depends almost exclusively on carbon-based petroleum products such as gasoline, diesel fuel, and jet fuel. Electric cars powered by batteries or fuel cells are among the possible paths for reducing the carbon intensity of transportation services. Liquid fuels derived from biomass are another potential route to lower carbon intensity. This requires fuels to be produced in a manner that yields no net CO_2 emissions, considering the carbon uptake by biomass, the CO_2 released in combustion, and the energy required for processing. Chapter 3 discusses options for alternative transportation fuels in greater detail.

Carbon Sequestration   The natural ability of biomass to absorb CO_2 from the atmosphere represents a well-known method of limiting or offsetting anthropogenic CO_2 emissions from fossil fuel combustion. Measures such as reforestation can indeed help slow the growth in atmospheric CO_2 levels. But indirect sequestration in biomass cannot, by itself, achieve the sizable long-term reductions needed to stabilize atmospheric concentrations in the face of global energy demands.

Another possible approach to reducing the carbon intensity of energy use is to capture and sequester CO_2 before it enters the atmosphere. The technological capability to remove CO_2 from the gas streams of industrial processes already exists. Such techniques could be applied to sources such as power plants, although the current cost would be extremely high (roughly doubling the cost of electricity). Disposing of the captured CO_2 then remains as the major problem. Several projects have demonstrated the ability to store CO_2 in geologic formations such as abandoned oil wells, coal mines, and deep saline reservoirs. Ocean storage also appears feasible, at least technically. The ability to permanently sequester CO_2 on a large scale, however, remains highly uncertain and is the subject of ongoing research. If this research succeeds in developing safe, publically acceptable, and low-cost sequestration options, the world's fossil energy reserves could continue to be used without releasing CO_2 to the atmosphere.

In summary, the economic cost of reducing carbon intensity poses the greater overall barrier to lowering global CO_2 emissions. This is true not only in the industrialized world but also in less industrialized countries like China and India, which rely heavily on cheap, abundant domestic coal supplies. Thus the long-term technological challenge is to develop more affordable means of supplying the world's energy needs in a less carbon-intensive manner.

12.9.4   Reducing Non-CO$_2$ Emissions

Any efforts to minimize future greenhouse warming must address other greenhouse gases in addition to CO$_2$. Atmospheric aerosols also play a role in greenhouse warming.

Greenhouse Gases   A number of measures are available to lower anthropogenic emissions of methane, nitrous oxide, and halocarbons, which contribute to radiative forcing. Some of these measures already are being implemented to some degree in the United States and elsewhere. One example is the capture of methane gas leaked from landfills and underground coal seams. This not only prevents methane releases to the atmosphere but also provides a supplementary source of natural gas. Methane leakage from natural gas pipelines also is being reduced by improved technology for leak detection and pipeline inspection.

The primary source of anthropogenic N$_2$O is the use of nitrogenous fertilizers for agriculture and related activities. Combustion sources also produce some N$_2$O. Reducing fertilizer use is a challenging task, given the strong growth in fertilizer applications shown earlier in Chapter 2 (Figure 2.13). Nonetheless, some reductions are possible without a loss of agricultural productivity (NAS, 1992).

Emissions of CFCs and certain other halocarbons that contribute to global warming already are being reduced or eliminated as a result of national and international actions to prevent stratospheric ozone depletion (see Chapter 11). However, many of these halocarbon compounds will remain in use for several decades to come. Nor does the Montreal Protocol banning CFC production extend to industrial chemicals such as perfluorocarbons (PFCs), sulfur hexafluoride (SF$_6$), and other perhalogens, which are powerful greenhouse gases with long atmospheric lifetimes (see Table 12.2). Only by eliminating the production of these chemicals can we ultimately stabilize their atmospheric concentrations.

Atmospheric Aerosols   Earlier we saw that anthropogenic aerosols significantly affect radiative forcing and subsequent climate change. These particles stem mainly from the combustion of biomass and sulfur-bearing fossil fuels (predominantly coal), so we can reduce anthropogenic aerosols by using less of these fuels. Indeed, the same strategies that reduce carbon intensity—such as substituting renewables and natural gas for coal and oil—also reduce anthropogenic aerosols. Sulfate aerosols and soot particles can be reduced by technological means as well. For example, coal-fired power plants can be equipped with air pollution control systems that capture nearly all the soot particles and remove most of the sulfur dioxide responsible for the formation of sulfate aerosols in the atmosphere (see Chapter 5 for details). More extensive use of such technology can therefore reduce anthropogenic aerosols considerably.

Ironically, reducing anthropogenic aerosols is likely to cause *increased* warming because most aerosols provide a negative radiative forcing that cools rather than warms regions of the planet (refer to Figure 12.12). Because aerosol particles have a very short atmospheric lifetime of only a few days, we would see this warming effect relatively quickly if aerosols were suddenly eliminated. As a practical matter, however, it is unlikely that aerosol-related emissions can be reduced rapidly in regions of the world that rely heavily on coal or biomass as an energy source. Especially in less industrialized countries the cost of modern emission controls also is prohibitive.

12.9.5 Evaluating Emission Reduction Strategies

We have seen that the problem of global warming arises from a broad range of human activities that involve the use of energy, the use of land, and the use of chemicals for food production, industrial processes, and personal convenience. Thus no single "magic bullet" can solve the problem. Rather, reducing greenhouse gas emissions will require actions across a broad spectrum of activities.

Because different greenhouse gases with different atmospheric lifetimes all contribute to radiative forcing, it is not immediately obvious how to evaluate emission reduction strategies that involve multiple gases. For example, if we reduce carbon intensity and CO_2 emissions by substituting natural gas for coal at electric power plants, to what extent will increased leakage of methane gas from pipelines offset the benefits of lower CO_2 emissions? Similarly, what are the global warming benefits of reducing nitrous oxide emissions from fertilizer use compared to eliminating an equal mass of tetrafluoromethane (CF_4, a perfluorocarbon) from aluminum manufacture?

A measure called the *Global Warming Potential* (*GWP*) has been devised to help answer such questions. The GWP combines the two principal attributes of a gas molecule that determine its contribution to global warming: radiative forcing and atmospheric lifetime. A gas molecule that efficiently absorbs infrared (longwave) radiation and persists in the atmosphere for a long time has a greater influence on warming than one that absorbs less heat and has a short atmospheric lifetime.

The GWP is defined for a unit mass (1 kg) of a substance relative to a unit mass of CO_2. This provides a convenient way to compare different emissions in terms of an equivalent mass of CO_2. Mathematically, the GWP is defined as

$$\text{Global Warming Potential (GWP)} = \frac{\int_0^Y \Delta F_{GHG} f_{GHG}(t)\, dt}{\int_0^Y \Delta F_{CO_2} f_{CO_2}(t)\, dt} \tag{12.47}$$

The numerator of this equation expresses the total radiative forcing of a greenhouse gas, ΔF_{GHG}, over a time, Y (years), in the atmosphere. The denominator does the same for CO_2. The term ΔF_{GHG} is the radiative forcing due to 1 kg of the gas introduced at time $t = 0$. The term $f_{GHG}(t)$ is the fraction of the gas remaining in the atmosphere at any time t. The numerical value of GWP thus depends on the time frame of interest. It also depends on the current concentrations of gases in the atmosphere, which determine the radiative forcing regimes (shown in Figure 12.11) and the extent of spectral overlap between gases.

Table 12.9 gives the GWP values for several greenhouse gases computed for 20-, 100-, and 500-year time horizons. Note that the GWP values for halocarbons and perhalogens are thousands of times greater than for CO_2, which gives increased importance to those substances. However, except for methane, the GWP values in Table 12.9 do not include indirect effects, which can either augment or offset direct heating effects (such as the indirect cooling effects of halocarbons caused by depletion of the stratospheric ozone layer). As discussed earlier in Section 12.3.5, indirect effects are more difficult to quantify. The next example illustrates the use of GWP.

Table 12.9   Global warming potential of selected greenhouse gases.

Substance	Chemical Formula	Atmospheric Life (Years)	Global Warming Potential (Time Horizon)		
			20 Years	100 Years	500 Years
CO_2	CO_2	50–200	1	1	1
Methane[a]	CH_4	12.2	56	21	6.5
Nitrous oxide	N_2O	120	280	310	170
Methylchloroform	$C_2H_3Cl_3$	5.4	360	110	35
CFC-11	$CFCl_3$	50	5,000	4,000	1,400
CFC-12	CF_2Cl_2	102	7,900	8,500	4,200
CFC-113	$C_2F_3Cl_3$	85	5,000	5,000	2,300
HCFC-22	CF_2HCl	13.3	4,300	1,700	520
HCFC-141b	$C_2FH_3Cl_2$	9.4	1,800	630	200
HCFC-142b	$C_2F_2H_3Cl$	19.5	4,200	2,000	630
HFC-125	C_2HF_5	32.6	4,600	2,800	920
HFC-134a	$C_2H_2F_4$	14.6	3,400	1,300	420
HFC-152a	$C_2H_4F_2$	1.5	460	140	42
Perfluoromethane	CF_4	50,000	4,400	6,500	10,000
Sulfur hexafluoride	SF_6	3200	16,300	23,900	34,900

[a] The GWP for methane includes indirect effects of tropospheric ozone production and stratospheric water vapor production.
Source: IPCC, 1996a.

Example 12.17

Calculation of equivalent CO_2 emissions.   Table 12.1 showed the worldwide emission rates of CO_2, CH_4, N_2O, and several halocarbons. Calculate the equivalent CO_2 emission rates based on a 100-year GWP. Assume that CFCs are divided equally among the three compounds listed. How much of the total CO_2-equivalent emissions are contributed by CO_2 alone?

Solution:

Multiply the actual emission rate of each substance by the 100-year GWP values from Table 12.9, as summarized in the following table:

Greenhouse Gas	Emissions (Mt/yr)	100-Year GWP	Equiv. CO_2 (Mt/yr)
CO_2	29,800	1	29,800
CH_4	375	21	7,875
N_2O	5.7	310	1,767
CFC-11	0.23	4,000	920
CFC-12	0.23	8,500	1,955
CFC-113	0.23	5,000	1,150
HCFC-22	0.2	1,700	340
Total			43,860

The total equivalent CO_2 is thus 43.9 Gt CO_2/yr, or 47 percent more than the actual CO_2 mass emissions. Expressed in terms of carbon rather than CO_2 (divide by 3.667), the equivalent emission rate is 12.0 Gt C/yr. CO_2 alone contributes 68 percent (about two-thirds) of the total CO_2-equivalent emissions. Note, however, that this calculation does not take into account the indirect cooling effects of the halocarbon emissions, which partially offset the direct warming effects reflected by the GWP.

The GWP also can be useful in evaluating the overall implications of a greenhouse gas mitigation plan that involves different greenhouse gases. The next example illustrates this application.

Example 12.18

Effects of methane pipeline leakage. It has been found that CO_2 emissions at an electric power plant can be reduced from 1.40 Mt/yr to 0.70 Mt/yr by substituting natural gas for coal. However, 1 percent of the gas supplied—equal to 3,000 t/yr (assumed to be pure methane)—leaks directly into the atmosphere from the supply pipeline. What is the net reduction in equivalent CO_2 emissions based on the 20-year GWP values?

Solution:

$$\text{Original coal plant emissions} = 1.40 \text{ Mt } CO_2/\text{yr}$$
$$\text{New gas plant emissions} = 0.70 \text{ Mt } CO_2/\text{yr} + 0.003 \text{ Mt } CH_4/\text{yr}$$

To evaluate the net effect, calculate the equivalent CO_2 emissions before and after the new measure, using the 20-year GWP values from Table 12.9: $CO_2 = 1$, $CH_4 = 56$. Thus

Coal plant:

$$\text{Equivalent } CO_2 = (1.40)(1.0) = 1.40 \text{ Mt } CO_2/\text{yr}$$

Gas plant:

$$\text{Equivalent } CO_2 = (0.70)(1.0) + (0.003)(56) = 0.87 \text{ Mt } CO_2/\text{yr}$$

The net reduction in equivalent CO_2 is $1.40 - 0.87 = 0.53$ Mt CO_2/yr. On a percentage basis, this is a 38 percent reduction in equivalent CO_2, compared to a 50 percent reduction without the methane leakage. Improved technology to reduce pipeline leakage clearly would be beneficial.

In the preceding example, use of the 20-year GWP maximized the impact of methane leakage, reflecting a near-term perspective. Over a longer time horizon, methane emitted today would be less significant because methane has a relatively short atmospheric lifetime of 12 years (see Table 12.2). The smaller values of the 100-year and 500-year GWP values for methane reflect this difference in perspective. On the other hand, perhalogens and many halocarbons are more important from a longer-term perspective because of their long atmospheric lifetimes. Accordingly, the GWP values for these compounds are greatest for the longer averaging times.

12.10   FUTURE OUTLOOK

Since the 1992 Earth Summit in Rio de Janeiro, worldwide awareness of the global warming problem has grown. Efforts to address the problem also have accelerated, especially among the highly industrialized countries. In addition, research programs are under way to better understand and predict global climate change and its consequences, especially at the regional level. Adaptation strategies also are being investigated in anticipation that some degree of warming may be inevitable in the decades ahead. Some nations, including the United States, have undertaken voluntary programs to improve the efficiency of energy use as part of a "no regrets" strategy that can save money as well as reduce CO_2 emissions.

Five years after the U.N. Framework Convention on Climate Change established the goal of stabilizing atmospheric greenhouse gas concentrations, the first step was taken toward binding targets and timetables for reducing greenhouse gas emissions. This agreement was forged at a December 1997 conference held in Kyoto, Japan.

12.10.1   The Kyoto Protocol

In the 1997 accord, known as the *Kyoto Protocol,* the leaders of major industrialized countries agreed to reduce their overall emissions of CO_2, CH_4, and N_2O to an average of 5.2 percent below 1990 levels by the period 2008 to 2012. In the same time frame, overall emissions of perfluorocarbons (PFCs), hydrofluorocarbons (HFCs), and sulfur hexafluoride (SF_6) are to be reduced to 5.2 percent below 1995 levels. Different countries have different emission limitations relative to the 5.2 percent average. These limits range from an 8 percent decrease to a 10 percent increase in emissions, as summarized in Table 12.10.

Table 12.10   Greenhouse gas emission limitations under the Kyoto protocol.

Countries	Emission Limit for 2008–2012[a]
Bulgaria, Czech Republic, Estonia, European Union (15 countries), Latvia, Liechtenstein, Lithuania, Monaco, Romania, Slovakia, Slovenia, Switzerland	8% reduction
United States	7% reduction
Canada, Hungary, Japan, Poland	6% reduction
Croatia	5% reduction
New Zealand, Russian Federation, Ukraine	no change
Norway	1% increase
Australia	8% increase
Iceland	10% increase

[a] Relative to 1990 emissions for CO_2, CH_4, and N_2O, and to 1995 for PFCs, HFCs, and SF_6.

Many critical details remain to be resolved, however, before this treaty can be implemented. The limitations agreed to in Kyoto do not become legally binding until ratified by individual country governments according to a process laid out in the agreement. For the United States, this will require ratification by Congress. As of 2000, that approval appears unlikely because of political opposition to the terms of the treaty and concerns about its economic impact. It therefore remains to be seen whether the Kyoto Protocol will ever gain the force of law. Nonetheless, it marks the first time that leaders of the world's industrialized nations agreed to binding reductions of greenhouse gas emissions within a specified time. Ongoing negotiations can be expected to build on this development, even if the pace is slowed. The Kyoto targets are still far short of the emission reductions needed to stabilize atmospheric concentrations, but the experience gained from efforts to implement these initial measures will set the stage for future efforts to reduce greenhouse gas emissions.

12.10.2 Beyond Kyoto

The Kyoto Protocol does not include the participation of economically developing countries like China and India, whose greenhouse gas emissions are projected to grow dramatically in coming decades. Without the participation of all nations, substantial worldwide reductions in future greenhouse gas emissions will be impossible. Major challenges thus remain in developing a global strategy to address global warming. Proposed steps in that direction include a variety of incentives and mechanisms to reduce emissions in the most cost-effective manner. Among these is CO_2 emissions trading, a proposed market-based system of emission allowances that could be bought or sold to minimize the cost of CO_2 reductions. Students interested in the details of global climate policy issues and proposals will find a wealth of additional information in the references at the end of this chapter, as well as in other current reports, journals, and climate-related Web sites.

12.11 CONCLUSION

Over the next decade, we can expect significant improvements in our ability to forecast changes in global climate and the consequences of those changes. This will reduce many of the uncertainties that currently cloud our understanding of the impacts of greenhouse gas emissions on human welfare and environment. It remains to be seen whether that improved understanding will lead to actions that reduce future climate change. The development of more cost-effective energy and environmental technologies will play a critical role in determining the pace of future actions. Engineering innovations are thus an essential ingredient of a workable solution to this most challenging of environmental problems.

12.12   REFERENCES

Enting, I. G., and G. N. Newsam, 1990. "Atmospheric Constituent Inversion Problems: Implications for Baseline Monitoring." *Atmos. Chem.,* vol. 11, pp. 69–87.

IPCC, 1990. *Climate Change: The IPCC Scientific Assessment.* Editors J. T. Houghton, G. J. Jenkins, and J. J. Ephraums, Intergovernmental Panel on Climate Change, Cambridge University Press.

IPCC, 1995. *Climate Change 1994: Radiative Forcing of Climate Change.* Editors J. T. Houghton et al., Intergovernmental Panel on Climate Change, Cambridge University Press.

IPCC, 1996a. *Climate Change 1995: The Science of Climate Change.* Editors J. T. Houghton et al., Intergovernmental Panel on Climate Change, Cambridge University Press.

IPCC, 1996b. *Climate Change 1995: Impacts, Adaptations, and Mitigation of Climate Change.* Editors R. T. Watson, M. C. Zinyowera, and R. H. Moss, Intergovernmental Panel on Climate Change, Cambridge University Press.

Marland, G., R. J. Andres, T. A. Boden, C. Johnston, and A. Brenkert, 1999. "Global, Regional, and National CO_2 Emission Estimates from Fossil Fuel Burning, Cement Production, and Gas Flaring: 1751–1996 (revised March 1999)." Carbon Dioxide Information Analysis Center, Oak Ridge National Laboratory, Oak Ridge, TN. http://cdiac.esd.ornl.gov/ftp/ndp030/.

Masters, G. M., 1998. *Introduction to Environmental Engineering and Science,* 2nd Edition, Prentice Hall, Upper Saddle River, NJ.

Nakicenovic, N., A. Grubler, and A. McDonald, 1998. *Global Energy Perspectives.* International Institute for Applied Systems Analysis, Cambridge University Press.

NAS, 1992. *Policy Implications of Greenhouse Warming.* National Academy of Sciences, National Academy Press, Washington, DC.

North, G. R., 1975. "Theory of Energy-Balance Climate Models." *J. of the Atmospheric Sciences,* vol. 32, no. 11, p. 2033.

PCAST, 1997. *Report to the President on Federal Energy Research and Development for the Challenges of the 21st Century.* President's Committee of Advisors on Science and Technology, Executive Office of the President, Washington, DC.

Petit, J. R. et al., 1999. "Climate and Atmospheric History of the Past 420,000 Years from the Vostok Ice Core, Antarctica." *Nature,* vol. 399, p. 429, June.

Rubin, E. S. et al., 1992. "Realistic Mitigation Options for Global Warming." *Science,* vol. 256, pp. 148–149, 261–266.

USDOE, 1997. *Emissions of Greenhouse Gases in the United States 1996.* DOE/EIA-0573(96), Energy Information Administration, U.S. Department of Energy, Washington, DC.

USDOE, 1998. *Annual Energy Review 1997.* DOE/EIA-0384(97), Energy Information Administration, U.S. Department of Energy, Washington, DC.

USDOE, 1999. *Annual Energy Review 1998.* DOE/EIA-0384(98), Energy Information Administration, U.S. Department of Energy, Washington, DC.

USDOE, 2000. International Database, http://www.eia.doe.gov/emeu/international. Energy Information Administration, U.S. Department of Energy, Washington, DC.

USGCRP, 2000. U.S. Global Change Research Program, Washington, DC. www.usgcrp.gov.

Warren, S. G., and S. H. Schneider, 1979. "Seasonal Simulation as a Test for Uncertainties in the Parameterizations of a Budyko–Sellars Zonal Climate Model." *J. of the Atmospheric Sciences,* vol. 36, p. 1377, August.

12.13 PROBLEMS

12.2 The solar constant, S_o, was defined as the average solar energy input *per unit area of the earth's surface,* with a numerical value of 342 W/m². An alternative definition in common use is based on the amount of incident solar energy per unit area of a circular plane perpendicular to the solar rays a short distance away from the earth, as sketched below. Let us denote this value as S_o*. Assume the earth is a sphere of radius R. Show that the average solar flux, S_o, based on the earth's surface area is related to the flux S_o* by $S_o = S_o*/4$.

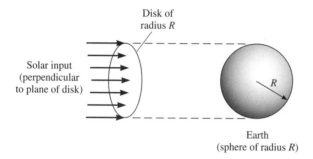

Disk of
radius R

Solar input
(perpendicular
to plane of disk)

R

Earth
(sphere of radius R)

12.2 The planet Venus is closer to the sun than earth and has a thick atmosphere of mostly CO_2. Because of the greenhouse effect, the average surface temperature is about 475°C. What would the average temperature be if there were no atmosphere? The solar constant, S_o, for Venus is 655 W/m² (based on surface area), and the planetary albedo is about 76 percent.

12.3 Equations (12.15) and (12.16) summarized the energy balance for a simple model of the earth surrounded by a radiatively active atmosphere. Derive these two equations with the aid of Figure 12.7 and solve them to find the earth's surface temperature. Use the numerical values of solar input, albedo, and atmospheric absorptivity given in the text.

12.4 Estimate the increase in radiative forcing from methane if its atmospheric concentration were to increase from 1,720 ppb to 1,800 ppb over the next 50 years.

(a) Ignore the methane–nitrous oxide overlap term.

(b) Include the overlap term in your calculation based on a constant N_2O level of 312 ppb. How much of a difference does this make? Give your answer both in absolute terms (W/m²) and as a percentage difference from your estimate in part (a).

12.5 Repeat Problem 12.4 for the case where the nitrous oxide concentration increases from 312 ppb to 370 ppb while the methane concentration remains constant at 1,720 ppbv.

12.6 The third column of Table 12.4 gave an estimate of the uncertainty in radiative forcing since 1850 from each of the factors listed in the table. Convert those measures into a numerical range (in W/m²) for each of the sources listed. Then use "error bars" to display on a graph (or histogram) the nominal value and uncertainty range for

(a) Total direct forcing from greenhouse gases.

(b) Total direct forcing from tropospheric aerosols.

(c) Indirect forcing from each of the three sources listed.

(d) Direct forcing from solar variability.

Comment on the usefulness of a graphical presentation of these data.

12.7 Suppose the earth's current albedo were to decrease by just 1 percent due to a decrease in surface ice cover caused by warmer temperatures.

(a) What would be the resulting increase in radiative forcing?

(b) How does this increase compare to the estimated radiative forcing from CO_2 since preindustrial times?

12.8 Assume that anthropogenic CO_2 emissions increase rapidly over the next 100 years, leading to an exponential increase in atmospheric CO_2 concentration given by $C = C_o e^{0.01t}$, where t is the time (in years) and C_o is the initial CO_2 concentration at $t = 0$, equal to 360 ppm.

(a) Plot the CO_2 concentration from $t = 0$ to 100 years.

(b) Estimate and plot the resulting increase in radiative forcing over the same period.

12.9 (a) Use the radiative forcing results of the previous problem to estimate and plot the increase over time in equilibrium average surface temperature, assuming the climate sensitivity factor remains constant at the best-estimate value of $0.57°C/W\text{-}m^{-2}$.

(b) Plot the equilibrium temperature changes for the low and high values of climate sensitivity estimated in the text—that is, 0.34 and $1.03°C/W\text{-}m^{-2}$, respectively.

(c) Based on your results, what is the best estimate and uncertainty range for equilibrium warming after 100 years for this CO_2 scenario?

12.10 Assume that the equilibrium surface temperature of the earth is projected to increase at a constant rate in response to an increase in radiative forcing beginning at time $t = 0$. Sketch such an increase on a graph of ΔT_{eq} versus time (no numbers needed, just a qualitative sketch). On the same graph, sketch the actual (or realized) temperature change versus time. Pick any point in time away from the origin and show on your graph the magnitude of the temperature commitment at that time.

12.11 In Example 12.10, the actual surface temperature increase after 70 years was 62 percent of the equilibrium value, resulting in a temperature commitment of 0.9°C. How long would it take to realize that commitment if the earth continued to warm at the same average rate as the first 70 years?

12.12 Equation (12.38) presented a simple model of the fractional decay in a unit mass of CO_2 injected into the atmosphere. Other researchers (IPCC, 1996a) have developed a more complex model based on the sum of three terms rather than two:

$$\frac{m_{CO_2}(t)}{m_o} = 0.30036 \exp\left(-\frac{t}{6.993}\right) + 0.34278 \exp\left(-\frac{t}{71.10}\right) + 0.35686 \exp\left(-\frac{t}{815.727}\right)$$

Plot this equation as a function of time from $t = 0$ to $t = 400$ years. On the same graph plot Equation (12.38).

(a) In general, what do you conclude about the similarity and difference between these two models?

(b) What is the difference between these two models in the predicted fraction of carbon remaining at $t = 10$ years? 50 years? 100 years? 400 years?

(c) Which of the two models compares more favorably to the detailed Bern model shown in Figure 12.22?

12.13 Equation (12.46) presented four factors that determine annual CO_2 emissions when multiplied together. Suppose you were responsible for reducing the annual CO_2 emissions of the United States to half the current value before the end of this century.

(a) Devise a strategy you would pursue to achieve that objective.

(b) Define your approach in terms of the percentage increase or decrease in each of the four parameters you would seek to achieve.

(c) Briefly describe the kinds of measures or changes that would be needed.

12.14 Suppose all of the world's proven reserves of fossil fuels (see Table 12.8) were extracted and burned to supply the world's energy needs.

(a) What is the approximate total mass of carbon that would be emitted to the atmosphere?

(b) How much of this would come from each major fuel type (oil, gas, coal)?

(c) If roughly half of this additional carbon remained in the atmosphere, what would be the percentage increase in the current atmospheric mass of carbon, which is estimated to be 750 GtC?

(d) What would be the resulting atmospheric concentration of CO_2?

(e) Compare your answer from part (d) to the IPCC scenario results in Figure 12.17. Would the magnitude of potential global warming be greater or smaller relative to the IPCC results for 2100?

12.15 Use the 20-year Global Warming Potential (GWP) values in Table 12.9 to calculate an equivalent CO_2 emission rate for worldwide greenhouse gas emissions as given in Table 12.1. Assume that total CFCs are divided equally among the three compounds listed. What is the percentage contribution of actual CO_2 emissions to the total equivalent CO_2? What is the next most important greenhouse gas emission based on this analysis? How do these results compare to those using the 100-year GWP in Example 12.17?

12.16 Repeat the previous problem using the 500-year GWP. Discuss the similarities and differences in perspective provided by the GWPs for the three different time horizons.

12.17 A new climate change assessment by the Intergovernmental Panel on Climate Change (IPCC) will be completed in late 2000 or early 2001. Search the Internet for the new IPCC study. In particular, look for the latest scenarios of future global CO_2 emissions and average temperature increases over the next century. Compare these new results with those of the 1995 assessment shown in Figure 12.17. Prepare a brief report summarizing your findings.

12.18 Figure 12.25 displayed the differences in current levels of per capita CO_2 emissions in several countries. These values reflect different levels of energy use and economic development. Suppose all nations of the world emitted CO_2 at the same per capita rate as the United States.

(a) Based on the current world population of 6 billion people, what would be the total annual emission of CO_2? How does this compare to the current emission rate shown in Table 12.1?

(b) Assuming half of this CO_2 remained in the atmosphere, what would be the new CO_2 concentration after 50 years? Assume the atmosphere currently has 750 GtC and a CO_2 level of 360 ppm.

(c) What would be the resulting increase in radiative forcing? How does this compare to the current CO_2 forcing since preindustrial times (see Table 12.4)?

12.19 (a) What global energy demand for fossil fuels is implied by the scenario in the previous problem? Base your estimate on the weighted average carbon intensity of proven energy reserves (Table 12.8) using data from Table 12.6.

(b) What fraction of the world's total proven reserves of fossil fuels would be needed to supply this annual energy demand?

(c) If this total demand for fossil fuels remained constant, with no change in population, how long would it take to deplete the proven reserve base?

(d) Comment on the sustainability of a scenario in which the entire world population emitted CO_2 at the same per capita rate as the United States.

PART
4

TOPICS IN ENVIRONMENTAL POLICY ANALYSIS

13

Economics and the Environment

13.1 INTRODUCTION

Cost is a major factor in solving environmental problems. This chapter looks at the economics of environmental control, focusing on methods to evaluate the cost of proposed solutions. Two general topics are addressed. The first is the cost of a particular project or technology, a subject known as *engineering economics.* The second topic, *cost–benefit analysis,* is concerned with broader questions involving the balance between money spent on environmental control and the resulting benefits from such expenditures.

This chapter briefly introduces these two topics, emphasizing the fundamentals needed to address economic issues. Topics from earlier chapters illustrate the application of economic principles to environmental problem solving. In Chapter 15 the subject of *macroeconomics* is introduced to examine how economic activity in general—and economic growth in particular—affect environmental quality.

13.2 FUNDAMENTALS OF ENGINEERING ECONOMICS

Let us begin by looking at the cost of a specific project or technology. Rather than an environmental system, let us start with something more familiar. For example, what is the cost of your family automobile? If you own or drive a car, how would you answer this question?

Actually, there could be several different answers because the term *cost* has not yet been clearly defined. Does it mean only the cost of purchasing the car? Or does it also include the cost of gasoline, insurance, repairs, and other expenses?

13.2.1 Categories of Cost

In general, two broad categories of costs are associated with any project or technology. One is the *capital cost,* also known as the *first cost* or *initial cost.* This is the amount of money required to purchase, build, or otherwise acquire the item.

The second category is *operating and maintenance (O&M) costs.* This is the amount of money required each year to operate and maintain the technology. In the case of an automobile, the costs of gasoline, insurance, and repair would fall into this category. The O&M category can be further subdivided into *variable costs* and *fixed costs.* Variable costs depend on how much or how often the technology is used, whereas fixed costs are independent of use. The annual cost of gasoline for a car is an example of a variable cost (it depends on how much you drive). In contrast, an annual license or registration fee is an example of a fixed cost (it does not depend on how often the car is used). If you borrowed money to purchase the car, your monthly or annual car payment is another example of a fixed cost.

13.2.2 Cash Flow Diagrams

These two general cost categories—capital costs and annual O&M costs—constitute the total cost of a project or technology. A useful method for visualizing and keeping

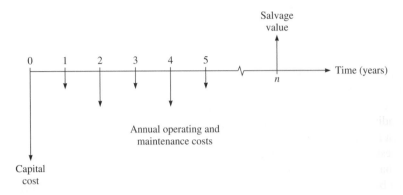

Figure 13.1   An illustrative cash flow diagram. Downward (negative) arrows represent expenditures and upward (positive) arrows represent income.

track of these different cost elements is the *cash flow diagram.* As illustrated in Figure 13.1, this is simply a time line, typically incremented in years, with a vertical arrow to indicate the magnitude of expenditures or income in each year. An upward-pointing arrow (positive direction) represents income, and a downward-pointing arrow (negative direction) indicates an expenditure. Multiple expenditures in the same year can be shown by arrows placed end to end.

The illustration in Figure 13.1 could represent the cash flow diagram for an automobile or any other project or technology, such as an environmental control system. The large downward arrow at time $t = 0$ represents the capital cost of the item, and the smaller downward arrows at the end of each year represent the annual operating and maintenance costs. At the end of the time line, the positive (upward) arrow represents income from selling the item at the end of its useful lifetime. This income is referred to as the *salvage value* or *scrap value.*

To evaluate the overall cost of a particular technology, or to compare the costs of alternative options, a method is required to combine costs that occur at different points in time. Unfortunately, this is not as simple or straightforward as summing all of the positive and negative costs on a cash flow diagram to get the overall total. The reason, as we discuss next, is the time value of money.

13.3   THE TIME VALUE OF MONEY

Suppose you bought a car for $15,000 and kept it for seven years, spending $1,000 each year on operating and maintenance expenses. After seven years you sell it for $2,000. What has been the total cost of owning and operating the car? If your answer is $20,000 (from summing $15,000 + $7,000 − $2,000), you have neglected to account for the time value of money.

Simply put, a dollar today is not worth the same as a dollar in the future. The main reason is that money (as a surrogate for what it can buy) is always in short supply. Do you know of anyone (including yourself) who has enough money and couldn't use more?

Because there is never enough money in the world to satisfy everyone's needs and desires, those who possess funds are in a position to profit by lending some of their money to others. This profit comes in the form of *interest* paid on the loan. Thus, in addition to paying back the amount borrowed, an additional amount is paid to the lender as the cost of borrowing money for some period of time. In general, the longer the period, the greater the total amount of interest that must be paid.

Time is therefore an important parameter in any calculation involving money. If an individual or organization wants to borrow money for some purpose (such as to buy a new car or to purchase an environmental control technology), the amount of interest paid over the period of the loan must be factored into the overall cost of the purchase. Even if you use your own money to purchase an item, that expenditure must be weighed against other opportunities to use those funds for the period under consideration.

To formalize a method of accounting for the time value of money, we introduce five terms that will form the basis for engineering economic calculations. These terms, and their representation on a cash flow diagram, are summarized in Table 13.1 and discussed more fully in the next sections.

13.3.1 Present and Future Amounts

The amount P shown in Table 13.1 represents an amount of money at the present time. In engineering economics problems, this is the amount of money needed today to initiate a project or to make a purchase (that is, the capital cost). In investment decisions, this is the present amount of money available to lend or invest.

Table 13.1 Engineering economic cost parameters.

Parameter	Definition	Diagram
P	Present value (also present amount or present worth)	
F	Future amount	
U	Uniform series amount	
n	Number of time periods	
i	Interest rate per time period	

The term F refers to a single *future* amount. This could be a future amount of income (upward-pointing arrow) or an anticipated future expense (downward-pointing arrow).

The amounts F and P are related by the time value of money. One common example is the case of depositing an amount, P, in a bank account today and asking what its future value, F, will be n years from now. The answer depends not only on the time period, n, but also on the *interest rate, i*. The interest rate is the fraction of P paid at the end of each time interval. Interest rates are commonly specified as an annual percentage, expressed as a decimal for calculations (for example, $6\% = 0.06$).

If the time period, n, is measured in years, and i is the annual interest rate payable at the end of each year, then after one year an initial amount, P, will grow to

End of first period:

$$F_1 = P(1 + i)$$

Similarly, if the amount F_1 is left in the bank for another year, we have

End of second period:

$$F_2 = F_1(1 + i) = [P(1 + i)](1 + i)$$

$$= P(1 + i)^2$$

In general, at the end of any time period n we have

$$F = P(1 + i)^n \tag{13.1}$$

Equation (13.1) is commonly known as the *compound interest* formula. The value of P grows exponentially because the interest paid after each period earns additional interest in subsequent periods (a process known as *compounding*). On a cash flow diagram, the growth of an initial amount, P, into a future amount, F, is sketched in Figure 13.2. A simple numerical example illustrates the use of Equation (13.1).

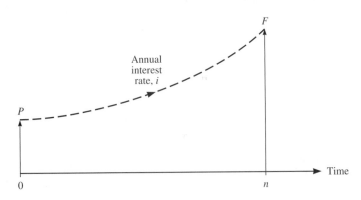

Figure 13.2   Relationship between present and future values of a single amount. The two values are related by the time interval, n, and the annual interest rate (or discount rate), i.

Example 13.1

Growth of a single present amount.   You deposit $10,000 in a bank savings account paying interest at 4 percent/year compounded annually. What is the value of this amount after five years?

Solution:

Use Equation (13.1) with $P = 10,000$, $i = 0.04$, and $n = 5$:

$$F = P(1 + i)^n = (10,000)(1.04)^5 = \$12,167$$

So the future value of $10,000 today is $12,167 for this case. The difference of $2,167 represents the total interest earned over the five years.

The relationship between present and future values embodied by Equation (13.1) is one of the fundamental relationships in engineering economics. Equally useful is the form of that equation that solves for P in terms of F (as well as n and i):

$$P = \frac{F}{(1 + i)^n} = F(1 + i)^{-n} \tag{13.2}$$

The term $(1 + i)^{-n}$ is referred to as the *present value factor* or the *present worth factor.* It directly measures how much a future amount is worth today, as illustrated in the next example.

Example 13.2

Present value of a future amount.   Four years from now you plan to buy a new car that will cost $25,000. In order to have that amount, how much money must you deposit today in a bank certificate of deposit (CD) paying an annual interest rate of 6 percent/year?

Solution:

Use Equation (13.2) with $F = 25,000$, $n = 4$, and $i = 0.06$:

$$P = F(1 + i)^{-n} = (25,000)(1.06)^{-4} = \$19,802$$

Thus you would have to deposit $19,802 today to have $25,000 four years from now.

Notice that the present value of a future amount is always smaller than the future amount as long as the interest rate is greater than zero. This means that one dollar earned or spent some time in the future is worth less than a dollar today because of the time value of money.

13.3.2   Uniform Series Amounts

Another useful set of relationships in engineering economics involves calculating or comparing costs in terms of a uniform series of equal payments (or income) in each time period. Common examples include monthly or annual car payments or home mortgage payments. In these examples, an initial amount, P, is borrowed to purchase the car or home, and then a uniform amount is paid back to the lender at regular intervals over time. The amount, U, paid back at each interval can be calculated based on the amount of the initial loan (P), the number of payment periods (n), and the interest rate per period on the unpaid balance (i). It can be shown that the uniform payment amount is

$$U = P \left[\frac{i(1 + i)^n}{(1 + i)^n - 1} \right]$$

This can be written more compactly as

$$U = P \left[\frac{i}{1 - (1 + i)^{-n}} \right] \tag{13.3}$$

The quantity in the brackets is known as the *capital recovery factor* or the *fixed charge factor*. This important parameter is commonly used in engineering economics to find the uniform annual amount needed to repay a loan or investment with interest.

Example 13.3

Annual payments on a loan.   A company borrows $100,000 from a bank to purchase an environmental control technology needed to comply with air pollution laws. The repayment period is 20 years at an annual interest rate of 8 percent/year on the unpaid balance. What is the uniform annual payment over the term of the loan?

Solution:

Use Equation (13.3) with $P = 100,000$, $n = 20$, and $i = 0.08$:

$$U = P \left[\frac{i}{1 - (1 + i)^{-n}} \right] = (100,000) \left[\frac{0.08}{1 - (1.08)^{-20}} \right] = \$10,190/\text{yr}$$

The annual payment is thus $10,190/yr. After 20 years the company will have paid a total of $10,190 \times 20 = \$203,800$, which is more than twice the original loan amount. The total interest paid is thus $103,800—larger than the original loan amount! This is common for long-term loans.

Suppose we want to calculate the amount of a *monthly* series of uniform payments instead of an annual series of payments. Over 20 years the number of time periods would then be $20 \times 12 = 240$. Recall that i is defined as the interest rate *per period*. For an 8 percent annual rate, the interest rate per month would be 8 percent/ $12 = 0.667$ percent/month. Repeating the same calculation as in Example 13.3 with

$n = 240$ and $i = 0.00667$ yields a uniform monthly payment of $836.69. Because a monthly series pays back the loan slightly faster than annual payments, the total interest paid over the life of the loan (240 payments) is also slightly lower.

Equation (13.3) also can be used to find the present value of a given series of annual payments. A general expression can be obtained simply by rewriting Equation (13.3) to solve for P:

$$P = U\left[\frac{1 - (1 + i)^{-n}}{i}\right] \qquad (13.4)$$

The following example illustrates an application of this equation.

Example 13.4

Present value of a uniform annual series of payments. A car owner anticipates spending $1,000 each year for operating and maintenance (O&M) expenses. She expects to own the car for seven years and plans to pay the O&M expenses by annually withdrawing money from a savings account that earns 4 percent/year in simple interest. How much money must she place in the savings account today to cover the anticipated O&M expenses over the life of the car?

Solution:

Assume for simplicity that the annual O&M expense of $1,000/year is paid at the end of each year. The cash flow diagram is depicted in Figure 13.3.

We wish to find the single amount, P, invested at the present ($t = 0$) that will provide $1,000/yr for seven years while earning 4 percent/year interest on the remaining balance. Therefore, use Equation (13.4) to solve for P, given $U = 1,000$, $n = 7$, and $i = 0.04$:

$$P = U\left[\frac{1 - (1 + i)^{-n}}{i}\right] = (1,000)\left[\frac{1 - (1.04)^{-7}}{0.04}\right] = \$6,002$$

Thus $6,002 invested today at 4 percent/year interest will cover the $7,000 in total annual O&M expenses over the life of the car.

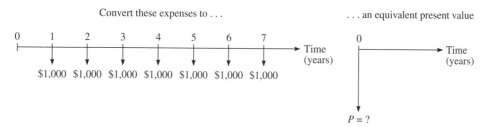

Figure 13.3 Cash flow diagram for Example 13.4.

In some types of economic analyses, a uniform series amount extends over a very long period. For example, the annual economic benefits from pollution control may extend indefinitely. A useful form of Equation (13.4) is the limiting case where n approaches infinity. For this special case, Equation (13.4) reduces to

$$P = \frac{U}{i} \quad (n \to \infty) \tag{13.5}$$

Example 13.5

Present value of an infinite uniform series.   The annual economic benefit derived from a water pollution control project has been estimated at $10,000/year. What is the total present value of this benefit, assuming that it continues indefinitely? Assume the annual interest rate is 5 percent.

Solution:

The cash flow diagram for this case would be a uniform series of $10,000/year extending out to $n \to \infty$. Therefore, use Equation (13.5) to calculate the present value for $i = 0.05$:

$$P = \frac{U}{i} = \frac{\$10,000}{0.05} = \$200,000$$

By way of comparison, if we had used Equation (13.4) for a finite time of, say, $n = 100$ years, we would get a present value of $P = \$198,479$. This is within 0.8 percent of the result for $n \to \infty$.

Yet another useful equation that can be derived in engineering economics gives the uniform series amount, U, needed to obtain a single future amount, F:

$$U = F\left[\frac{i}{(1 + i)^n - 1}\right] \tag{13.6}$$

The expression in brackets is known as the *sinking fund factor*. Again, n is the number of time periods in the uniform series and i is the interest rate per period.

Example 13.6

Saving annually for a future expense.   You plan to buy a new car four years from now by putting a certain amount each year into a savings account paying 6 percent interest. The car will cost $25,000. How much must you save each year?

Solution:

Use Equation (13.6) with $F = 25,000$, $n = 4$, and $i = 0.06$:

$$U = F\left[\frac{i}{(1 + i)^n - 1}\right] = (25,000)\left[\frac{0.06}{(1.06)^4 - 1}\right] = \$5,715/\text{yr}$$

Notice that the uniform annual amount of $5,715/year in the preceding example is smaller than the amount you would calculate by simply dividing the $25,000 purchase price by 4 ($25,000 ÷ 4 = $6,250/year). This difference reflects the time value of money. Money put into an interest-bearing account earns additional funds that contribute to the future goal. Dividing the future amount by 4 in this example is equivalent to saving the money under a mattress, where it earns no interest. Because money is always in short supply, any funds that are not needed immediately can always grow in value by earning interest in a "safe" investment like a federally insured savings account.[1]

Finally, we write an equation to calculate the future amount of a uniform series. This is obtained simply by solving Equation (13.6) for F:

$$F = U\left[\frac{(1 + i)^n - 1}{i}\right] \tag{13.7}$$

This expression is analogous to Equation (13.1) in that the future amount includes the compounding effect of interest paid over time. In this case the equation is more complicated because the interest is applied to a series of payments rather than to a single amount.

Example 13.7

Future value of a uniform series of payments. Each year, beginning at age 25, you put $10,000 into a retirement fund that earns 9 percent/year in tax-free income. How much will you have accumulated at age 65 when you plan to retire?

Solution:

At age 65 you will have contributed $10,000/year for 40 years. To find the total future amount, use Equation (13.7) with $U = 10,000$, $n = 40$, and $i = 0.09$:

$$F = U\left[\frac{(1 + i)^n - 1}{i}\right] = (10,000)\left[\frac{(1.09)^{40} - 1}{0.09}\right] = \$3,378,824$$

Notice again how the time value of money makes a considerable difference, especially over a long time. If you had saved your money in a cookie jar (or under your mattress) for 40 years, you would have only $400,000 instead of $3.38 million!

13.3.3 Summary of Key Equations

The preceding examples used six equations relating three key economic variables: P, F, and U. Table 13.2 summarizes these relationships. These six equations allow

---

1 In this case you are lending your money to the bank, which pays you interest in order to attract your funds. The bank, in turn, lends your money (and that of fellow depositors) to other individuals or businesses. The bank earns a profit by charging a higher rate of interest on its loans than the rate offered to you on your savings account. Of course, the bank bears the risk of losing money through loans that are not repaid. By insuring your deposits, the federal government bears the risk that the bank will fail.

Table 13.2 Summary of cost factors for engineering economic analysis.

Cost Factor			
Symbol	**Name**	**Converts**	**Equation**
P/F	Present value	F to P	$\dfrac{1}{(1 + i)^n}$
P/U	Uniform series present value	U to P	$\dfrac{1 - (1 + i)^{-n}}{i}$
F/P	Single payment compound amount	P to F	$(1 + i)^n$
F/U	Compound amount	U to F	$\dfrac{(1 + i)^n - 1}{i}$
U/P	Capital recovery	P to U	$\dfrac{i}{1 - (1 + i)^{-n}}$
U/F	Sinking fund	F to U	$\dfrac{i}{(1 + i)^{-n} - 1}$

monetary amounts to be converted from one time basis to another. For example, the first two factors in Table 13.2 convert future and uniform amounts to their equivalent present values. *These six equations relating present, future, and uniform series amounts form the heart of engineering economic analysis,* including applications to environmental issues. They are also widely used in business accounting and other economic applications.

Notice that each factor listed in Table 13.2 involves the two parameters i and n. The numerical value of each factor thus can be tabulated for different values of i and n to provide a convenient reference for cost calculations. Table 13.3 gives such a tabulation for several interest rates and time periods. The six cost factors listed are *P/F, P/U, F/P, F/U, U/P,* and *U/F.* Modern computer spreadsheet programs also have these functions embedded for easy use.

To use the cost factor tables, simply select the interest rate and number of time periods, then find the value of the cost factor needed. To select the appropriate cost factor, make the numerator the quantity desired and the denominator the quantity known. For example, if a future amount is known and its present value is desired, the factor needed is *P/F.* Multiply the *P/F* value by the known future amount to find the quantity desired:

$$F \times (P/F) = P$$

The value of any factor depends on i and n, so the convention used to fully identify the value of *P/F* is (*P/F, i%, n*). The next example illustrates the use of these tabulated factors.

Table 13.3   Illustrative cost factor values for selected interest rates.

	_i = 4%					
	Find P		Find F		Find U	
n	P/F	P/U	F/P	F/U	U/P	U/F
1	0.9615	0.9615	1.0400	1.0000	1.0400	1.0000
2	0.9246	1.8861	1.0816	2.0400	0.5302	0.4902
3	0.8890	2.7751	1.1249	3.1216	0.3603	0.3203
4	0.8548	3.6299	1.1699	4.2465	0.2755	0.2355
5	0.8219	4.4518	1.2167	5.4163	0.2246	0.1846
6	0.7903	5.2421	1.2653	6.6330	0.1908	0.1508
7	0.7599	6.0021	1.3159	7.8983	0.1666	0.1266
8	0.7307	6.7327	1.3686	9.2142	0.1485	0.1085
9	0.7026	7.4353	1.4233	10.5828	0.1345	0.0945
10	0.6756	8.1109	1.4802	12.0061	0.1233	0.0833
20	0.4564	13.5903	2.1911	29.7781	0.0736	0.0336
30	0.3083	17.2920	3.2434	56.0849	0.0578	0.0178
40	0.2083	19.7928	4.8010	95.0255	0.0505	0.0105
50	0.1407	21.4822	7.1067	152.6671	0.0466	0.0066
100	0.0198	24.5050	50.5049	1,237.6237	0.0408	0.0008

	i = 6%					
	Find P		Find F		Find U	
n	P/F	P/U	F/P	F/U	U/P	U/F
1	0.9615	0.9615	1.0400	1.0000	1.0400	1.0000
1	0.9434	.9434	1.0600	1.0000	1.0600	1.0000
2	0.8900	1.8334	1.1236	2.0600	0.5454	0.4854
3	0.8396	2.6730	1.1910	3.1836	0.3741	0.3141
4	0.7921	3.4651	1.2625	4.3746	0.2886	0.2286
5	0.7473	4.2124	1.3382	5.6371	0.2374	0.1774
6	0.7050	4.9173	1.4185	6.9753	0.2034	0.1434
7	0.6651	5.5824	1.5036	8.3938	0.1791	0.1191
8	0.6274	6.2098	1.5938	9.8975	0.1610	0.1010
9	0.5919	6.8017	1.6895	11.4913	0.1470	0.0870
10	0.5584	7.3601	1.7908	13.1808	0.1359	0.0759
20	0.3118	11.4699	3.2071	36.7856	0.0872	0.0272
30	0.1741	13.3648	5.7435	79.0582	0.0726	0.0126
40	0.0972	15.0463	10.2857	154.7620	0.0665	0.0065
50	0.0543	15.7619	18.4202	290.3359	0.0634	0.0034
100	0.0029	16.6175	339.3021	5,638.3681	0.0602	1.0002

(continued)

Table 13.3 *(continued)*

$i = 10\%$

n	Find P		Find F		Find U	
	P/F	*P/U*	*F/P*	*F/U*	*U/P*	*U/F*
1	0.9091	0.9091	1.1000	1.0000	1.1000	1.0000
2	0.8264	1.7355	1.2100	2.1000	0.5762	0.4762
3	0.7513	2.4869	1.3310	3.3100	0.4021	0.3021
4	0.6830	3.1699	1.4641	4.6410	0.3155	0.2155
5	0.6209	3.7908	1.6105	6.1051	0.2638	0.1638
6	0.5645	4.3553	1.7716	7.7156	0.2296	0.1296
7	0.5132	4.8684	1.9487	9.4872	0.2054	0.1054
8	0.4665	5.3349	2.1436	11.4359	0.1874	0.0874
9	0.4241	5.7590	2.3579	13.5795	0.1736	0.0736
10	0.3855	6.1446	2.5937	15.9374	0.1627	0.0627
20	0.1486	8.5136	6.7275	57.2750	0.1175	0.0175
30	0.0573	9.4269	17.4494	164.4940	0.1061	0.0061
40	0.0221	9.7791	45.2593	442.5926	0.1023	0.0023
50	0.0085	9.9148	117.3909	1,163.9085	0.1009	0.0009
100	0.0001	9.9993	13,780.6123	137,796.1234	0.1000	0.0000

Example 13.8

Use of the cost factor tables.   Use Table 13.3 to find the present value of a $1,000 expense five years from now for an interest rate of 6 percent/year.

Solution:

The desired quantity is P and the known quantity is F, so the factor needed is P/F for $i = 6\%$ and $n = 5$. From Table 13.3, the value of $(P/F, 6\%, 5)$ is 0.7473. Thus

$$P = F \times (P/F, 6\%, 5)$$

$$= (\$1{,}000)(0.7473) = \$747.30$$

This result can be verified by using Equation (13.2) directly.

13.4   EVALUATING TOTAL LIFE CYCLE COST

The relationships in Table 13.2 allow costs that occur at different points in time to be converted to a common basis. This allows us to evaluate the total cost of a project or technology over its useful lifetime. This total cost is often referred to as the *life cycle cost.* There are several ways in which the total (life cycle) cost can be evaluated; two of the most common methods are described next.

13.4.1   Net Present Value

The net present value (NPV) method of evaluating the total cost of a project is often used in engineering economics. The cash flows at different points in time are all converted to their present value amounts using either Equation (13.2) (for single amounts) or Equation (13.4) (for uniform series amounts). Because the present value amounts all occur at the same point in time ($t = 0$ on a cash flow diagram), positive and negative amounts—corresponding to income and expenses—can be combined to determine the *net present value* (*NPV*) of the project. This value represents the amount of money that would be needed today to cover all expenses for the project over its lifetime, including the capital cost and all annual O&M costs. The next example illustrates the use of this method for a consumer purchase.

Example 13.9

Net present value of a pollution control system.   The owner of an automotive repair shop has just purchased a pollution control system to capture emissions of volatile organic compounds (VOCs). The equipment was purchased for $15,000 using money from the owner's savings account. He plans to keep the equipment for seven years and expects to pay $1,000/year in annual operating and maintenance expenses. After seven years he anticipates selling the equipment for $2,000. What is the total present value of owning and operating the system? Assume the interest rate is 4 percent/year, corresponding to the bank savings account rate.

Solution:

A cash flow diagram for this project is shown in Figure 13.4(a).

The capital cost of the system ($P_{cap} = \$15,000$) is already expressed as a present value amount at $t = 0$. The annual O&M expense of $1,000/year represents a uniform series

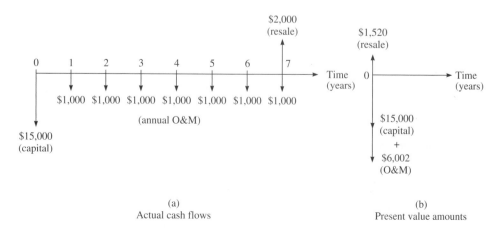

(a)
Actual cash flows

(b)
Present value amounts

Figure 13.4   Cash flow diagram for Example 13.9.

amount (U). The identical amount was converted to an equivalent present value amount in Example 13.4. That amount was

$$P_{O\&M} = \$6{,}002$$

Finally, the resale value of $2,000 after seven years represents a single future amount whose present value can be calculated from Equation (13.2) or Table 13.2. Using Table 13.2 gives $(P/F, 4\%, 7) = 0.7599$. Thus

$$P_{resale} = F \times (P/F, 4\%, 7)$$

$$= (\$2{,}000)(0.7599) = \$1{,}520$$

The equivalent cash flow diagram expressed as present values looks like Figure 13.4(b).

To compute the net present value, use the algebraic convention of positive values for income and negative values for expenditures. Thus

$$NPV = P_{resale} - P_{cap} - P_{O\&M}$$

$$= \$1{,}520 - \$15{,}000 - \$6{,}002$$

$$= -\$19{,}482$$

The negative sign indicates a net cost rather than income (which of course we know). The value of $19,482 represents the total amount of money required today to purchase and operate the pollution control system over its expected life, based on an interest rate of 4 percent.

Note that this example assumed that annual O&M costs remained constant each year in order to simplify the calculations. More complex assumptions, in which O&M costs change over time, would require the present value of each future amount to be calculated individually using Equation (13.2) or the P/F values from Table 13.3.

Notice too that the choice of interest rate plays an important role in the result of this calculation. Had the interest rate been higher, the net present value would have been smaller. Later in this chapter we will discuss the choice of an appropriate interest rate for different situations. First let us look at another common method of combining costs that occur at different points in time.

13.4.2 Levelized Annual Cost

It is often useful to express the total cost of a technology in terms of a uniform annual amount. This can be done by converting the total net present value (NPV) to a uniform series amount over the life of the project. For example, the net present value of buying and operating the pollution control system in Example 13.9 was $19,482. Equation (13.3) can convert this present value to a uniform annual amount over the seven-year life. The result is a total annual cost of $3,246/year, based on the 4 percent interest rate in Example 13.9.

The total annual cost calculated in this manner is known as the *levelized annual cost*. It combines the annual operating and maintenance costs with the annualized "mortgage payment" on the capital cost (adjusted for any salvage value income streams). This is an alternative way of expressing the overall cost of a project that may be more meaningful, especially for people working with annual budgets. The following example illustrates an application to environmental control technology.

Example 13.10

Levelized annual cost of sulfur dioxide control. A 500 megawatt (MW) electric power plant is being equipped with a flue gas desulfurization (FGD) system to control emissions of sulfur dioxide (SO_2). The capital cost of the system is $140/kW of plant capacity. The annual O&M costs include several types of fixed costs and variable costs, as given in the following table. These costs are assumed to remain constant over the life of the plant.

Annual O&M Costs for FGD System

O&M Cost Item	Fixed O&M ($/kW-yr)	Variable O&M ($/MW-hr)
Labor and personnel	7.47	
Maintenance materials	3.58	
Chemical reagent		0.290
Electric power		0.853
Steam and water		0.374
Waste disposal	____	0.263
Total	11.05	1.780

The power plant operates at a capacity factor of 65 percent, meaning that it operates for the equivalent of 5,694 hours/year at full capacity (0.65 × 365 da/yr × 24 hr/da). The FGD system has an expected lifetime of 30 years with no salvage value. The capital cost of the system must be repaid at an interest rate of 10 percent/year. What is the levelized annual cost of this project?

Solution:

First convert the given cost measures (which are expressed in terms of plant size) to total dollars based on the plant size of 500 MW:

(a) Total capital cost = $140/kW × 500 MW × 1,000 kW/MW

= $70,000,000

(b) Fixed O&M costs = $11.05/kW-yr × 500 MW × 1,000 kW/MW

= $5,525,000/yr

(c) Variable O&M costs = $1.780/MW-hr × 500 MW × 5,694 hrs/yr

= $5,068,000/yr

(d) Total O&M cost = Fixed O&M + Variable O&M

= 5,525,000 + 5,068,000 = $10,593,000/yr

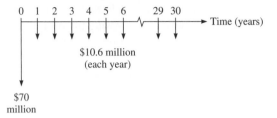

Figure 13.5   Cash flow diagram for Example 13.10.

The cash flow diagram for the FGD system cost is shown in Figure 13.5. In this case, because the salvage cost is zero, the simplest way to obtain the total levelized annual cost is to convert the $70 million capital cost to a uniform annual amount (as in Example 13.3), then add that annualized value to the annual O&M cost. Thus

$$U_{cap} = (\$70,000,000)(U/P, 10\%, 30)$$

$$= (70,000,000)(0.1061) = \$7,427,000/\text{yr}$$

This amount is analogous to an annual mortgage payment. The total levelized annual cost is then

$$U_{total} = U_{cap} + U_{O\&M}$$

$$= \$7,427,000/\text{yr} + \$10,593,000/\text{yr}$$

$$= \$18,020,000/\text{yr}$$

13.4.3   Cost per Unit of Product

The levelized annual cost just described can be used to derive another useful measure of total cost, namely the cost per unit of product or service delivered by a project or technology. In general, the total cost per unit of product can be expressed simply as

$$\left[\begin{array}{c} \text{Cost per unit} \\ \text{of product} \end{array}\right] = \frac{\text{Levelized annual cost}}{\text{Annual quantity of product}} \qquad (13.8)$$

In more complex analyses, engineers and economists may investigate how the unit cost of a technology changes in response to different levels of output or different assumptions about cost items. For the electric power plant in Example 13.10, the useful service provided is electricity (kilowatt-hours), which is the product sold to customers. Environmental control technologies that are added to reduce pollution typically increase the cost of production, as illustrated in the next example.

Example 13.11

Cost per kilowatt-hour for SO₂ control.   Use the data in Example 13.10 to calculate how much the SO_2 control system adds to the cost per kilowatt-hour (kW-hr) of generating electricity at that power plant. Compare this added cost to the total generating cost of 4.50¢/kW-hr before the FGD installation.

Solution:

First find the total kilowatt-hours generated by the power plant each year. From the data in Example 13.10, this is

$$\text{Annual generation} = (500 \text{ MW} \times 1{,}000 \text{ kW/MW}) \times 5{,}694 \text{ hrs/yr}$$

$$= 2.847 \times 10^9 \text{ kW-hr/yr}$$

From Example 13.10, the levelized annual cost of SO_2 control is $18.02 million/year. Use Equation (13.8) to find the added cost per kW-hr of electricity generated by the plant:

$$SO_2 \text{ cost per kW-hr} = \frac{\text{Levelized annual } SO_2 \text{ control cost}}{\text{Annual electricity generation}}$$

$$= \frac{18.02 \times 10^6 \text{ \$/yr}}{2.847 \times 10^9 \text{ kW-hr/yr}}$$

$$= \$0.0063/\text{kW-hr (or } 0.63¢/\text{kW-hr)}$$

Compared to the base plant cost of 4.5¢/kW-hr without SO_2 control, the FGD cost represents a 14 percent increase in the cost of generating electricity from this plant.

13.4.4 Average Cost-Effectiveness

Another type of unit cost that is commonly employed for environmental control technologies is the average cost per unit mass of pollutant removed. This ratio, called the *average cost-effectiveness,* can be calculated as

$$\left[\begin{array}{c} \text{Average} \\ \text{cost-effectiveness} \end{array}\right] = \frac{\text{Levelized annual cost}}{\text{Annual mass of pollutant removed}} \qquad (13.9)$$

The smaller the cost-effectiveness ratio, the less money is needed to reduce emissions of an environmental pollutant (a desirable attribute for an environmental technology). The following example illustrates this calculation.

Example 13.12

Cost-effectiveness of SO_2 control. The SO_2 control system installed on the power plant in Example 13.10 removes 2.765×10^7 kg of SO_2 each year. What is the average cost per metric ton of SO_2 removed?

Solution:

Use Equation (13.9) with the levelized annual cost of $18.02 million from Example 13.10. Recall that a metric ton (also known as a *tonne*) is defined as 1,000 kg. Thus

$$[\text{Average cost-effectiveness}] = \frac{\text{Levelized annual cost}}{\text{Annual } SO_2 \text{ removed}} = \frac{\$18{,}020{,}000/\text{yr}}{27{,}650 \text{ tonnes/yr}}$$

$$= \$652/\text{tonne } SO_2 \text{ removed}$$

Although measures like average cost-effectiveness often express the overall cost of a technology, a more common application is in comparing different options for doing a job. The following section discusses this topic in greater depth.

13.5   COMPARING TECHNOLOGY OPTIONS

When we choose among different options for accomplishing an objective, the cost of each alternative is usually an important consideration. Often, however, the competing options have different capital costs, operating costs, salvage value, and perhaps even different lifetimes. In such cases it is neither easy nor straightforward to know which option has the lowest overall cost.

There are several techniques for making systematic comparisons. A rigorous approach is to consider the full life cycle cost of each option based on either net present value or levelized annual cost. As we saw earlier, both of these methods solve the "apples and oranges" problem of how to combine costs that occur at different points in time.

13.5.1   Comparisons Based on Net Present Value

In general, the easiest way to compare the total cost of different options that achieve the same objective is to compare their net present value. This method is often preferred because it represents the total amount of money needed today to cover all costs over the life of a project or technology. The key requirement, however, is that *the options being compared must have the same useful lifetime.* Other factors such as the level of environmental control also must be identical. If all lifetimes and other important attributes are the same, the option with the highest (most positive) net present value is the economically preferred choice. The following example illustrates this type of comparison.

Example 13.13

Comparison of two SO$_2$ removal options.   The power plant flue gas desulfurization (FGD) system described in Example 13.10 (call it Technology A) creates solid waste as a consequence of removing SO$_2$. The annual O&M cost thus includes a cost of waste disposal. An alternative, more environmentally friendly technology removes the same amount of SO$_2$ but creates a useful by-product (gypsum) that can be sold for use in wallboard manufacturing. The capital cost of this technology (call it Technology B) is $160/kW, compared to $140/kW for Technology A. However, the annual O&M cost now includes an income from by-product sales instead of a waste disposal cost. This reduces the variable O&M cost for the plant. Annual maintenance costs, however, increase slightly because of additional equipment needs; this increases the fixed O&M costs. Overall, the costs of Technology B are estimated to be

Total capital cost $=$ $80.000 million

Fixed O&M cost $=$ $5.802 million/yr

Variable O&M cost $=$ $4.014 million/yr

Technologies A and B both have a 30-year lifetime with zero salvage value. The interest rate for financing both systems remains at 10 percent. Which option has the lowest overall life cycle cost?

Solution:

The following table summarizes the cost data for each technology:

Cost Data for Two FGD Systems

Cost Item	Technology A	Technology B
Total capital cost ($M)	70.000	80.000
Annual O&M costs ($M/yr)		
Fixed	5.525	5.802
Variable	5.068	4.014
Total O&M	10.593	9.816

Use the net present value (NPV) method to compare the two options. The capital cost already represents a present value amount; so it is only necessary to convert the uniform series of annual O&M costs to an equivalent present value using Table 13.3:

$$P_{O\&M} = U_{O\&M}(P/U, 10\%, 30) = U_{O\&M}(9.4269)$$

where $U_{O\&M}$ is the total annual O&M cost ($M/yr) of each technology shown in the previous table.

The following table summarizes the present value cost amounts. Recall that expenditures are reported as negative values:

Present Value of Alternative Options (Millions of Dollars)

Cost Item	Technology A	Technology B
Total capital cost	−70.000	−80.000
Total O&M cost	−99.859	−92.534
Net present value	−169.859	−172.534

Both technologies have negative NPV amounts because they reflect expenditures, not income. However, Technology A has the more positive (less negative) value and is thus the lower cost option. The life cycle cost of Technology B (the by-product recovery option) is $2.7 million more expensive than the system producing a solid waste. The analysis shows, however, that if the capital cost of Technology B could be reduced by about 4 percent or more it would be the lower cost system.

In general, if the lifetime of competing options is not the same, we cannot use the net present value method to compare the options unless we adjust to achieve identical lifetimes. We can do this by replicating a technology at the end of its life. For example, to compare the total cost of Option A with a 10-year life to that of Option B with a 5-year life, assume that a new Option B is installed at the end of 5 years. The equivalent cash flow diagrams for the capital cost component (only) would look like Figure 13.6. Of course, any annual O&M costs, plus any income or revenue streams, also would be added to the capital costs shown in Figure 13.6. In this way we can compare the total costs of both options over equivalent periods.

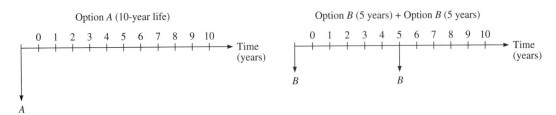

Figure 13.6   Equalizing the lifetimes of two competing options. In this illustration Option B is repeated at the end of its 5-year life to obtain a 10-year time horizon equivalent to the lifetime of Option A.

13.5.2   Comparisons Based on Levelized Annual Cost

An easier way of comparing options with different lifetimes is to express the total cost on a levelized annual basis. The least-cost option is thus the one with the lowest levelized cost (or greatest positive value, assuming that a "cost" means a negative quantity).

Comparisons based on this method also implicitly assume that a shorter-lived option is replaced by an equivalent system at the end of its useful life. Here, however, the differences in lifetime are automatically accounted for when costs are annualized, as the next example illustrates.

Example 13.14

A comparison based on levelized annual cost.   A series of 150-watt electric lamps provides outdoor lighting for an office building. Each lamp operates for 3,000 hours/year and must be replaced every two years at a cost of $4.50. A new, energy-efficient, 60-watt lamp costing $14.50 offers a similar amount of lighting but lasts for eight years (at 3,000 hours/year). The cost of electricity is 7.0¢/kW-hr, and the cost of capital (interest rate) for the office complex is 8 percent. Which lamp has the lower life cycle cost?

Solution:

Because the lifetimes of the two lamps are different, compare their total life cycle costs using the levelized annual cost method. The two relevant cost items are the initial (capital) cost and the annual operating cost, which is the cost of electricity. (This example does not include any additional operating costs for maintenance labor to replace the dead lamps.)

Current lamp:

$$\text{Annual energy use} = 150 \text{ W} \times 3,000 \text{ hrs/yr} \times \frac{1 \text{ kW}}{1,000 \text{ W}} = 45 \text{ kW-hr}$$

$$\text{Annual operating cost} = 45 \text{ kW-hr} \times 0.070 \text{ \$/kW-hr} = \$3.15/\text{yr}$$

$$\text{Annualized capital cost} = \$4.50(U/P, 8\%, 2) = (4.50)(0.5608) = \$2.52/\text{yr}$$

$$\text{Total levelized cost} = \$3.15/\text{yr} + \$2.52/\text{yr} = \$5.67/\text{yr}$$

Energy-efficient lamp:

$$\text{Annual energy use} = 60 \text{ W} \times 3,000 \text{ hrs/yr} \times \frac{1 \text{ kW}}{1,000 \text{ W}} = 18 \text{ kW-hr}$$

$$\text{Annual operating cost} = 18 \text{ kW-hr} \times 0.070 \text{ \$/kW-hr} = \$1.26/\text{yr}$$

$$\text{Annualized capital cost} = \$14.50(U/P, 8\%, 8) = (14.50)(0.1740) = \$2.52/\text{yr}$$

$$\text{Total levelized cost} = \$1.26/\text{yr} + \$2.52/\text{yr} = \$3.78/\text{yr}$$

The energy-efficient lamp is cheaper by $1.89/year. On a life cycle basis, the energy-efficient lamp is thus more economical, costing 33 percent less than the standard lamp. The total amount of money saved each year will depend on the number of lamps used at the facility.

13.5.3   Comparisons Based on Payback Period

Another concept used to compare and evaluate the economics of different technologies is the period of time for an initial capital cost to be recovered or offset from savings in annual O&M costs. This is often referred to as the *payback period*. A common example in the environmental domain is an investment in a technology that reduces the cost of energy use. In Chapter 6, for instance, advanced refrigerator designs were able to dramatically cut annual electricity consumption, but they required new insulation materials and other technological improvements that added to the capital cost of the unit. How long would it take for the annual savings in electricity to offset the higher initial cost of the new technology?

To answer such questions, we can use Equation (13.4) or Equation (13.2) to solve for the time period, n, at which *the present value of future savings equals the capital cost needed to achieve those savings.* By definition, this is the payback period, n_{pb}. For the case where an incremental capital cost, ΔP, yields a uniform periodic (such as annual) savings, ΔU, Equation (13.4) yields

$$\Delta P = \Delta U \left[\frac{1 - (1 + i)^{-n_{pb}}}{i} \right] \tag{13.10}$$

Solving this equation for n_{pb} gives

$$n_{pb} = \frac{\log \left[1 - \left(\dfrac{\Delta P}{\Delta U} \right) i \right]^{-1}}{\log (1 + i)} \tag{13.11}$$

Often we simply divide the additional capital cost, ΔP, by the annual cost savings, ΔU, to obtain the *simple payback period, n_{spb}*:

$$n_{spb} = \frac{\Delta P}{\Delta U} \tag{13.12}$$

This simple payback period is shorter than n_{pb} because it ignores the time value of money. An example illustrates.

Example 13.15

Payback period for an energy-efficient refrigerator.   Table 6.7 of Chapter 6 showed two advanced refrigerator designs. Unit B saved 190 kW-hr/yr in electricity relative to the base design (Unit A) but cost the consumer $107 more initially. Unit C saved 274 kW-hr/yr but cost $269 more initially. Assume the average residential price of electricity is 8.5¢/kW-hr. Based on an annual interest rate of 6 percent, what are the payback periods for the two energy-efficient designs? How do these values compare to the simple payback periods?

Solution:

For Unit B, the annual savings in electricity cost are

$$\text{Unit B annual savings} = (190 \text{ kW-hr/yr}) (\$0.085/\text{kW-hr}) = \$16.15/\text{yr}$$

The simple payback period is thus

$$n_{spb} = \frac{\Delta P}{\Delta U} = \frac{\$107}{\$16.15/\text{yr}} = 6.6 \text{ yr}$$

This was the "advertised" value reported in Table 6.7 (Chapter 6). However, for a nonzero interest rate, the actual payback period is calculated from Equation (13.10), assuming a constant energy savings each year. Thus, for Unit B,

$$\frac{\Delta P}{\Delta U} = \frac{1 - (1 + i)^{-n_{pb}}}{i}$$

$$\frac{107}{16.15} = \frac{1 - (1.06)^{-n_{pb}}}{0.06}$$

$$n_{pb} = \frac{\log 1.66}{\log 1.06} = 8.7 \text{ yr}$$

Considering the time value of money thus adds two years to the payback period. However, because the typical lifetime of a refrigerator is 15–20 years, the cost of the Unit B efficiency measures is easily recouped over the lifetime of the unit.

Repeating the same calculations for Unit C yields a simple payback period of 11.4 years but a much longer actual payback period of 20.2 years based on a 6 percent interest rate. In this case the savings in energy do not compensate for the higher capital cost of the Unit C technology, considering the time value of money over a 15- to 20-year refrigerator life.

13.5.4   Comparisons Based on Average Cost-Effectiveness

When analyzing options to reduce environmental emissions, the *average cost-effectiveness* of competing measures is another way we can compare the available options. To illustrate, the two power plant SO_2 control technologies in Example 13.13 both removed 27,650 tonnes/yr of SO_2. In Example 13.12, the average cost-effectiveness of Technology A was found to be \$652/tonne SO_2 removed. A similar calculation for Technology B (which is left as an exercise for the student) yields an average cost-effectiveness of \$662/tonne SO_2 removed. Hence Technology A (which has the lower value) is more cost-effective than Technology B.

In some cases the cost-effectiveness of an environmental control option is calculated based on *indirect* reductions in emissions. For instance, in Example 13.14, the reduction in electricity consumption from using more energy-efficient lamps at an office complex also indirectly reduces environmental emissions at the power plants that generate the electricity. One of these emissions is carbon dioxide (CO_2), which contributes to global warming. U.S. power plants currently emit an average of 0.61 kg CO_2 per kW-hr of electricity generated (see Chapter 5 for details).

The energy-efficient technology in Example 13.14 reduced the use of electricity by 27 kW-hr/yr for each lamp. This translates into an average annual reduction of $27 \times 0.61 = 16.5$ kg CO_2 from U.S. power plants. We can calculate the cost-effectiveness of this measure and compare it to other strategies for reducing CO_2 emissions by the same amount. In Example 13.14, the reduction in CO_2 was actually achieved at a net cost *savings* of \$1.89/year over the life of the energy-efficient lamp. In mathematical terms, the levelized cost of this measure is therefore negative (a "negative cost" means a savings). Using Equation (13.9) to calculate the average cost-effectiveness for this case gives

$$\text{Average cost-effectiveness} = \frac{-\$1.89/\text{yr}}{16.5 \text{ kg } CO_2/\text{yr}} = -\$0.115/\text{kg } CO_2 \text{ removed}$$

$$= -\$115/\text{tonne } CO_2 \text{ removed}$$

A negative cost-effectiveness value represents a win–win situation because environmental impacts can be reduced at a net *cost savings*. All such measures are economically worthwhile when compared to more costly alternatives that achieve the same emissions reduction.

13.6 MARGINAL COST ANALYSIS

Often the environmental control technologies that are most cost-effective have only a limited ability to reduce the overall level of emissions. As higher levels of emissions reduction are sought, the total cost and average cost-effectiveness tend to increase. The following example illustrates this point.

Example 13.16

Average cost-effectiveness of increasing emission reductions. Consider again the power plant described in Examples 13.10 through 13.13, in which an FGD system was installed to remove 27,650 tonnes/year of SO_2. Two other technological options available to reduce SO_2 emissions are (1) switching to a lower-sulfur coal without an FGD system and (2) converting the plant to burn natural gas instead of coal. The coal-switching option would reduce SO_2 emissions by

14,550 tonnes SO_2/year (a 50 percent reduction). The added levelized cost is estimated to be $2.50 million/year. Converting to natural gas would eliminate all SO_2 emissions (29,100 tonnes/year). Because natural gas is more expensive than coal, the levelized annual cost of this option is estimated at $40.0 million/year. Calculate the average cost-effectiveness of all three SO_2 control options, and plot the results on a graph of average cost-effectiveness (*y*-axis) versus SO_2 emission reduction (*x*-axis).

Solution:

From Example 13.12, installation of the flue gas desulfurization (FGD) technology would remove 27,650 tonnes SO_2/year at a cost of $18 million/year, giving an average cost-effectiveness of $652/tonne SO_2 removed. The average cost-effectiveness of the other two options is

Switch to low-sulfur coal:

$$\text{Average cost-effectiveness} = \frac{\$2.50 \times 10^6/\text{yr}}{14,550 \text{ tonnes/yr}} = \$172/\text{tonne } SO_2 \text{ removed}$$

Convert to natural gas:

$$\text{Average cost-effectiveness} = \frac{\$40.0 \times 10^6/\text{yr}}{29,100 \text{ tonnes/yr}} = \$1,375/\text{tonne } SO_2 \text{ removed}$$

Figure 13.7 plots these three values as requested. Coal switching is the most cost-effective option but is limited to a reduction of 50 percent (14,550 tonnes SO_2/year). If greater reductions are required, FGD technology can remove 27,650 tonnes/year at a higher average cost per tonne removed. Converting to natural gas can achieve a slightly greater SO_2 reduction but has the highest average cost per tonne removed.

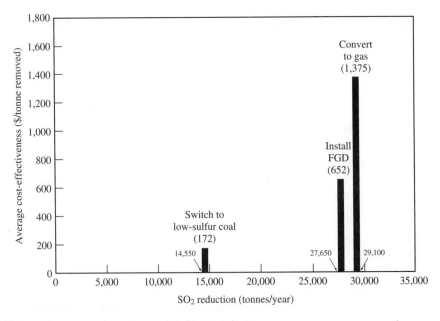

Figure 13.7   Cost-effectiveness graph for Example 13.16.

The preceding example illustrates a common situation in which the average cost-effectiveness of environmental control measures increases with the overall level of emission reduction. A more general case can be envisioned where there are many options for reducing the emissions of a given pollutant. If we plot the levelized annual cost of each option as a function of the emission reduction achieved, we get a series of "steps" as each new measure is added to achieve a higher level of emission reduction (at a higher overall cost). This is illustrated in Figure 13.8(a).

We can further idealize this relationship into a smooth curve of cost versus emission reduction, as sketched in Figure 13.8(b). The average cost-effectiveness at any Point A on this curve is simply the total cost divided by the total emission reduction.

13.6.1   Marginal Cost-Effectiveness

Another important concept in economics, however, is the incremental cost—more commonly called the marginal cost—of an incremental reduction in emissions, say from Point *A* to Point *B* in Figure 13.8(b). As seen in the sketch, the *marginal*

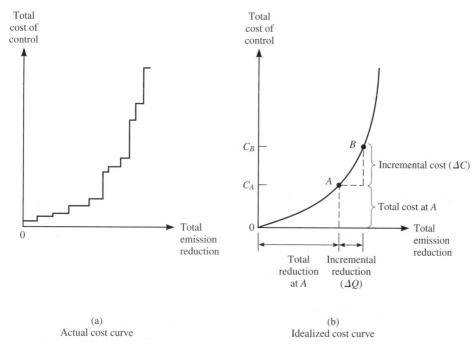

(a)
Actual cost curve

(b)
Idealized cost curve

Figure 13.8   Total cost versus emission reduction. On the actual cost curve (a) each step is a discrete measure or control technology. The idealized cost curve (b) represents this as a smooth, continuous function. The slope of the idealized curve at any point (such as Point A) represents the marginal cost of control for an additional increment of emission reduction.

cost-effectiveness can be defined as the ratio of the incremental cost, ΔC, for an increase in the quantity of emissions reduction, ΔQ:

$$\left[\begin{array}{c} \text{Marginal} \\ \text{cost-effectiveness} \end{array} \right] = \frac{\text{Incremental total cost}}{\text{Incremental emissions reduction}} \qquad (13.13)$$

The next example illustrates this calculation.

Example 13.17

Marginal cost of SO_2 control.   In Example 13.16, converting the power plant to natural gas would reduce SO_2 emissions by an additional 1,450 tonnes/year beyond the FGD system capability. What is the marginal cost-effectiveness of this reduction?

Solution:

From Example 13.16, the natural gas option would cost \$40 million/year, whereas the FGD option costs \$18 million/year. The incremental cost is thus \$40 − \$18 = \$22 million/year. The incremental SO_2 reduction is 29,100 − 27,650 = 1,450 tonnes/year. The marginal cost-effectiveness is thus

$$\left[\begin{array}{c} \text{Marginal cost-effectiveness} \\ \text{of converting to gas} \end{array} \right] = \frac{\text{Incremental total cost}}{\text{Incremental emissions reduction}}$$

$$= \frac{\$22 \times 10^6/\text{yr}}{1{,}450 \text{ tonnes/yr}} = \$15{,}200/\text{tonne } SO_2 \text{ removed}$$

This means that the cost of each additional tonne removed is about 25 times greater than the average cost of the FGD option (which was \$652/tonne).

Example 13.17 and Figure 13.8 show that the cost of eliminating the last increment of a pollutant is often very high. An important relationship between marginal cost and average cost (or cost-effectiveness) is that *marginal costs are always greater than average costs if the average cost is rising.* A marginal cost analysis can quantify these incremental costs relative to their effectiveness in reducing environmental emissions.

13.6.2   Application to Market-Based Solutions

The preceding examples all focused on the cost of reducing emissions at a single facility by the application of various technologies. But the same ideas and cost curves derived for a single facility can also apply to a *collection* of facilities. Suppose, for example, we wanted to reduce the environmental impacts of acid rain by reducing SO_2 emissions from U.S. power plants by 50 percent. Environmental studies have shown that the principal need is to reduce total SO_2 emissions across a broad geographic area; the specific location and magnitude of individual SO_2 sources is relatively unimportant so long as total emissions are reduced. In this type of situation, the least-cost solution usually calls for different power plants to reduce emissions by different amounts while still achieving the overall goal (that is, the 50 percent reduction). Imposing identical reduction requirements on *each* plant will be more costly overall because some power plants are able to reduce emissions more cheaply than others due to differences in plant size, plant age, fuel costs, and other factors.

When a collection of facilities is required to achieve an overall emissions reduction, the least-cost solution resembles the marginal cost curve in Figure 13.8(a), with each step on the curve corresponding to a particular facility and control measure. But how does one implement such a scheme in a practical and equitable manner to achieve the overall emission reduction goal?

The modern answer to this question is to let the market figure it out. That is, allow all emitters (of SO_2 in this case) to efficiently buy or sell the right to emit one unit of SO_2 per year, but limit the total number of emission "rights" or *allowances* that are allocated (for instance, to half the current level of emissions if a 50 percent reduction is required). This type of *market-based approach,* adopted in the United States in 1990 to address the problem of acid rain, is credited with having achieved significant overall cost reductions relative to the traditional "command and control" approach of uniform emission limits for each source. Problem 13.15 at the end of this chapter illustrates how such a market-based approach—also known as *emissions trading*—can reduce the overall cost of environmental compliance.

As noted earlier, this approach is viable only in cases where environmental impacts depend on the aggregate level of emissions from many dispersed sources, and not on specific source–receptor relationships. There must also be a sufficient number of sources to constitute a competitive market for the trading of emission allowances. Where these conditions are met, market-based approaches offer an opportunity to reduce overall emissions most cost-effectively.

13.7   CHOOSING AN INTEREST RATE

In all time value of money calculations, the choice of an interest rate has a big effect on the result. For example, if the interest rate is zero, a dollar at any time in the future has the same value as today. In contrast, as the interest rate grows larger, a future dollar has less worth today.

What determines the value or choice of an interest rate? You should be aware of several important factors when conducting an economic analysis.

13.7.1   Effect of Inflation

You may have heard stories that your parents told about the unbelievably low price of a movie ticket when they were kids or the 10-cent candy bar that costs 50 cents today. In most societies price levels tend to increase over time, a phenomenon known as *inflation.* Figure 13.9 shows the annual U.S. inflation rate from 1960 to 1998 as measured by a government index of price levels known as the *implicit price deflator.*

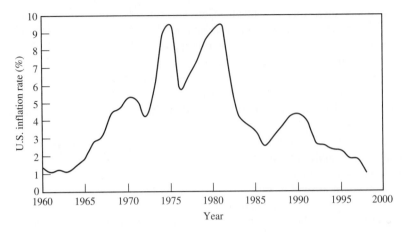

Figure 13.9   U.S. inflation rates, 1960 to 1998. (*Source:* Based on USDOE, 1999)

According to this broadly based index, inflation rates have fluctuated widely over the past several decades, from over 10 percent per year in the early 1980s to under 2 percent per year in the early 1960s and 1990s.

13.7.2   Constant versus Current Dollars

Inflation increases the price of a given item over time. In quoting a cost or price, therefore, it is important to state the year in which those costs are based. For instance, the capital cost of an environmental control technology might be specified as the cost in 1998 dollars.

Table 13.4 gives the values of the implicit price deflator index for the United States, which allows us to adjust prices for the effects of inflation. The annual percentage change in this index gives the annual inflation rate shown in Figure 13.9. For example, the price deflator index of 109.7 for 1996 means that \$109.70 in 1996 had approximately the same purchasing power as \$100.0 in 1992. Thus, to adjust price levels from one year to another, use the implicit price deflator index as follows:

$$\text{Price in year } X = (\text{Price in year } Y)\left(\frac{\text{Year } X \text{ deflator index}}{\text{Year } Y \text{ deflator index}}\right) \quad (13.14)$$

Amounts that are adjusted for inflation in this fashion are said to represent *constant dollar* amounts. Costs that are not adjusted for inflation are called *current dollar* or *nominal dollar* amounts. These are the amounts we use daily. If you are paying off a loan, for instance, the amount of your actual payment is in nominal (current) dollars, uncorrected for inflation.

For many types of economic analyses, constant dollars provide a more useful picture of cost trends and a more systematic basis for cost comparisons. Figure 13.10 illustrates the different perspective obtained using constant dollar values versus current dollar amounts for the price of gasoline in the United States. Although the nominal

Table 13.4   Implicit price deflator index for the United States.

Year	Price Deflator (1992=100)	Annual Percentage Change	Year	Price Deflator (1992=100)	Annual Percentage Change
1960	23.3	1.39	1980	60.4	9.23
1961	23.6	1.16	1981	65.9	9.43
1962	23.9	1.27	1982	70.1	6.30
1963	24.2	1.17	1983	73.1	4.26
1964	24.5	1.49	1984	75.9	3.78
1965	25.0	1.96	1985	78.4	3.42
1966	25.7	2.84	1986	80.6	2.61
1967	26.5	3.19	1987	83.1	3.06
1968	27.7	4.38	1988	86.1	3.65
1969	29.0	4.70	1989	89.7	4.22
1970	30.6	5.32	1990	93.6	4.32
1971	32.2	5.18	1991	97.3	3.95
1972	33.5	4.24	1992	100.0	2.74
1973	35.4	5.62	1993	102.6	2.63
1974	38.5	8.98	1994	104.9	2.39
1975	42.2	9.41	1995	107.6	2.29
1976	44.6	5.87	1996	109.7	1.87
1977	47.4	6.46	1997	111.5	1.84
1978	51.0	7.29	1998	112.8	1.17
1979	55.3	8.52	1999	114.4	1.45

Source: USDOE, 1999; USDOC, 2000.

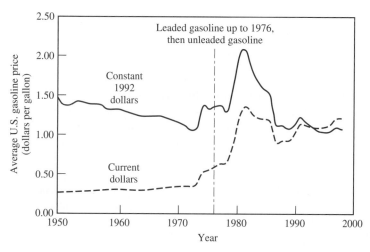

Figure 13.10   Nominal versus real price trends for gasoline in the United States, 1950 to 1997. The curve labeled "current dollars" is the nominal price seen by consumers in that year. The curve labeled "constant 1992 dollars" adjusts the current dollar prices for inflation to obtain a real cost based on 1992 price levels. (*Source:* USDOE, 1998)

573

(current dollar) price has increased considerably since 1950, the constant dollar price has actually fallen, making gasoline cheaper (not more expensive) relative to general price levels for other goods and services.

13.7.3   Real versus Nominal Interest Rates

Interest rates at any point in time vary with the inflation rate. When inflation rates are low, interest rates fall; but when inflation rates rise, interest rates also rise to compensate for the higher nominal price levels expected in the future. In the early 1980s, for example, when inflation rates were high, interest rates for home mortgage loans were often at 15 percent, compared to half that rate in the low-inflation period of the mid-1990s.

Just as with prices, in many types of economic analysis it is useful to evaluate interest rates in "real" terms that exclude the effects of inflation. An inflation-adjusted interest rate is especially useful for comparing fundamental economic trends at different points in time. An analysis using *real* (inflation-adjusted) interest rates utilizes constant dollar prices or costs that are also adjusted for inflation. On the other hand, interest rates expressed in nominal terms—which are the rates most people are familiar with—are used in conjunction with nominal prices and costs that are not adjusted for inflation.

It can be shown that the *real* interest rate, adjusted for inflation, is related to the *nominal* interest rate (which includes the effect of inflation) by this equation:

$$1 + i_{real} = \frac{1 + i_{nom}}{1 + r_{infl}} \qquad (13.15)$$

where

i_{real} = real interest rate (without inflation)

i_{nom} = nominal interest rate (including inflation)

r_{infl} = annual rate of inflation

The real interest rate calculated from this equation is slightly lower than the value that would be obtained by simply subtracting the inflation rate from the nominal interest rate. The following example illustrates this calculation.

Example 13.18

Calculation of real interest rate.   A local savings bank advertises a 6 percent annual interest rate on deposits of $10,000 or more held for one year. If the current inflation rate is 2 percent/year, what is the real interest rate earned on these deposits?

Solution:

The 6 percent interest rate represents the nominal value. Thus, use Equation (13.15) with i_{nom} = 0.06 and r_{infl} = 0.02:

$$1 + i_{real} = \frac{1 + i_{nom}}{1 + r_{infl}} = \frac{1.06}{1.02} = 1.039$$

$$i_{real} = 0.039 \text{ or } 3.9\%$$

Notice that this result is slightly less than the value of 4.0 percent obtained by simply subtracting the inflation rate from the nominal rate (which is a perhaps more intuitive but unfortunately incorrect procedure).

The real interest rate provides a useful measure of the true income from an investment, or the true cost of a loan, independent of the inflation rate. It also forms the basis for a constant dollar cost analysis, as described earlier.

13.7.4   The Analysis Perspective

Whether expressed in nominal or real terms, the value of the interest rate used for engineering economic calculations must reflect the perspective of the person or organization performing the analysis. A few basic questions help guide the choice of an interest rate: Whose money is it? Is the money being borrowed or lent? For how long? What is the purpose of the project or investment? And, when money is lent, how risky is the venture (that is, what is the chance of losing all or some of the money)?

A typical situation in engineering economics is the need for funds to initiate a project or to purchase equipment. This capital cost is often obtained by borrowing the money from others, thus incurring *debt*. Alternatively, the money can come out of one's own pocket. Money you already have is called *equity*.

If funds are borrowed, the interest rate paid often depends on whose money it is. A general distinction is between *public* and *private* funds. Government organizations often lend public funds for "worthwhile" projects—such as environmental control— at interest rates lower than those charged by private institutions such as banks.

When very large sums of money are required (like hundreds of millions of dollars), private corporations as well as public agencies often borrow from the general public by issuing *bonds*. Government and corporate bonds are promises to repay the money borrowed at a specified interest rate over a period of time. These interest rates typically vary with the duration of the loan. In the late 1990s, 30-year government bonds paid roughly 5–6 percent interest, whereas long-term corporate bonds paid roughly 6–8 percent. In all cases the interest rate must be high enough to attract investors who are willing to lend their money for the specified period.

The reputation or creditworthiness of the borrower is another key factor in establishing the interest rate: The riskier the investment, the higher the interest rate. An important function of the financial markets in any country is to evaluate the creditworthiness of potential borrowers and share information on the interest rates available from different sources.

Because it is always in short supply, those who possess equity capital, whether public or private, may invest it in many different ways. Thus, when spending equity capital (money from your own pocket), the choice of an interest rate is the rate that would have been earned by an alternative investment. For an individual, that alternative might be to put the money into a secure bank savings account (earning perhaps

3–6 percent interest in the late 1990s). For a private corporation, the expenditure of equity funds is expected to yield profits for shareholders (owners). Thus corporations often assume fairly high interest rates (such as 20–30 percent) in evaluating the engineering economics of a project using equity capital. In contrast, expenditures of public funds by government organizations historically earn lower rates of return of roughly 3–10 percent in real terms.

Capital investments for large projects often are made with a combination of debt and equity capital. In this case the appropriate interest rate is the *weighted cost of capital* obtained by prorating the rates for each source of funds, as illustrated below.

Example 13.19

Weighted average cost of capital.   An electric power company plans to purchase the SO_2 control technology in Example 13.10 with 60 percent debt and 40 percent equity capital. The debt portion will come from issuing a corporate bond paying 9 percent annual interest. The equity portion will come from issuing common stock offering a return to stockholders of 11.5 percent. What is the weighted cost of capital for this case?

Solution:

Use the information provided to calculate the weighted cost of capital:

$$60\% \text{ debt} = 0.60 \times 9\% = 5.4\%$$

$$40\% \text{ equity} = 0.40 \times 11.5\% = \underline{4.6\%}$$

$$\text{Weighted cost of capital} = 10.0\%$$

This weighted average interest rate of 10 percent is the value used in Example 13.10.

The message from this discussion of interest rates is that *there is no single answer as to what interest rate should be used to evaluate the cost of a particular project.* As in most engineering analyses, there is ample room for judgment and alternative assumptions within the guidelines discussed. The analytical framework presented in this chapter allows different interest rate assumptions to be evaluated in the context of a specific analysis.

13.7.5   Taxes and Depreciation

Finally, a detailed economic analysis also should consider the effects of taxes and depreciation allowances on the bottom-line cost of a project or technology. Very briefly, tax considerations can alter the cash flow picture of a project because certain investment-related expenses can be deducted on federal and state income taxes in the United States, thus lowering the tax bill. Many types of capital investments also earn tax credits that further lower the bill. Government rules for depreciating capital equipment similarly can contribute to tax deductions.

Any savings in income taxes effectively lowers the overall (present value) cost of a project. This savings can also be viewed as lowering the effective interest rate (also known as the *after-tax rate of return*). Detailed calculations involving taxes and depreciation can be tedious and require up-to-date knowledge of current tax law. A more complete discussion of this subject is thus left to more advanced courses and texts in engineering economics (such as Au and Au, 1992; White et al., 1989).

13.8   COST–BENEFIT ANALYSIS

So far we have focused only on the cost of a particular project or technology. In the context of environmental issues, the larger question that often arises is whether money spent on environmental controls is worth the resulting benefits to public health and the environment.

Cost–benefit analysis is an economic analysis method designed to help answer these broader types of questions. As its name suggests, the objective is to compare the economic benefits of an activity to the total cost of the undertaking. If the benefits exceed the costs, the endeavor can be said to have merit from an overall economic point of view.

In its simplest form, cost–benefit analysis is a direct extension of the net present value method for evaluating alternative projects. As illustrated in Figure 13.11, a stream of economic benefits (B) can be depicted by positive values on a cash flow diagram; costs (C) again are shown by negative values depicted by downward-pointing arrows. The net present value (NPV) is then equal to the total present value of all benefits minus the total present value of all costs. Mathematically, this can be written as

$$\text{NPV} = \sum_{k=0}^{n} \frac{B_k}{(1 + i)^k} - \sum_{k=0}^{n} \frac{C_k}{(1 + i)^k} \qquad (13.16)$$

The term within each summation is simply the present value factor shown earlier in Equation (13.2) and Table 13.2. The symbol k refers to any future year from 1 to n, and i is again the applicable interest rate (which is often called the *discount rate* because it discounts future values to the present). The costs C_k in Equation (13.16) include the same types of engineering economic costs discussed earlier. For some projects the costs may also involve a variety of nonengineering costs (such as the

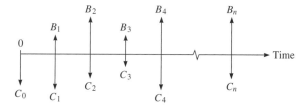

Figure 13.11   An arbitrary stream of costs and benefits. Benefits are represented as positive arrows (equivalent to income); costs are negative, representing expenditures.

cost of resettling people affected by the construction of a new dam). Valuing the benefits, however, is often more difficult.

13.8.1   The Nature of Economic Benefits

The concept of economic benefits in a cost–benefit analysis is similar to a positive income stream in an engineering economic analysis. In the latter case, however, the income is usually a tangible amount of money received from a project or technology (like income from the sale of widgets produced at a widget plant). In a cost–benefit analysis the economic benefits often are less tangible. This is especially true in the case of environmental problems.

Money spent on environmental controls usually does not generate income for a company as would expenditures to build a new widget factory. Rather, the benefits of environmental expenditures are manifest in improved public health and a cleaner, safer environment. Economists can place values on many (though not all) of these benefits, such as reductions in medical expenses and time lost from work due to pollution-related illness. For the most part, these economic benefits accrue indirectly to the public at large rather than to the companies that spend money on pollution control. Nonetheless, they are an important part of the overall economic equation. A cost–benefit analysis weighs these economic benefits relative to the cost of achieving them. If the benefit–cost ratio exceeds 1.0, then the overall economic benefits exceed the overall costs. The following example illustrates the general approach.

Example 13.20

Cost–benefit analysis of a proposed regulation.   A new environmental regulation has been proposed to reduce industrial discharges of toxic chemicals to rivers and lakes. Companies would be given three years to comply. The cost of reducing these discharges is estimated (in constant 1999 dollars) to be $100 million in Year 1, $150 million in Year 2, and $100 million in Year 3. After Year 3, the economic benefits resulting from this measure are estimated at $50 million/year for each of the next 12 years. Calculate the net present value for this proposed regulation, and determine the benefit–cost ratio. Assume a real discount rate of 6 percent/year, with all costs and benefits occurring at the end of each year.

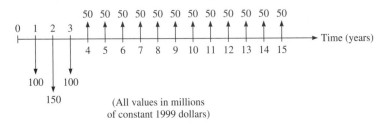

Figure 13.12   Cash flow diagram for Example 13.20.

Solution:

The cash flow diagram for this problem is shown in Figure 13.12. Use Equation (13.16) and Table 13.3 to find the present value of each future amount for a discount rate (interest rate) of 6 percent:

Calculation of Present Value Amounts

n (Year)	$(P/F, 6\%, n) \times$ Future Amount ($M) =			Present Value ($M)	
		Costs	Benefits	Costs	Benefits
1	0.9434	100		94.34	
2	0.8900	150		133.50	
3	0.8396	100		83.96	
4	0.7921		50		39.61
5	0.7473		50		37.37
6	0.7050		50		35.25
7	0.6651		50		33.26
8	0.6274		50		31.37
9	0.5919		50		29.60
10	0.5584		50		27.92
11	0.5268		50		26.34
12	0.4970		50		24.85
13	0.4680		50		23.44
14	0.4423		50		22.12
15	0.4173		50		20.87
Total				311.80	351.97

The net present value (in millions of constant 1999 dollars) is thus

$$\text{NPV} = \sum_{n=4}^{15} B_n(P/F, 6\%, n) - \sum_{n=1}^{3} C_n(P/F, 6\%, n)$$

$$= 351.97 - 311.80 = 40.17$$

The NPV thus has a positive value of $40.2 million, meaning that benefits exceed costs by that amount. This indicates that the cost of the proposed regulation would be warranted by the overall benefits that accrue. The benefit–cost ratio is

$$\frac{\text{Benefits}}{\text{Costs}} = \frac{351.97}{311.80} = 1.13$$

This ratio is greater than 1.0 and shows that the present value of benefits is 13 percent greater than costs. Note that this is not a very large difference considering the likely uncertainties in the various cost and benefit estimates. A more detailed analysis would attempt to quantify those uncertainties and their effect on the overall result.

Estimating the economic benefits of environmental control is discussed in more specialized textbooks (such as Zerbe and Dively, 1994; Hanley and Spash, 1993). Environmental impacts must first be quantified in physical terms and then valued economically. For human health impacts, such methods rely to a large extent on laboratory and epidemiological studies coupled with advanced statistical techniques. These methods attempt to isolate the effects of environmental pollutants on the number and severity of illnesses and premature deaths attributed to pollution. Economic measures then can be applied to value these impacts in monetary terms. Procedures also have been developed to value other types of environmental damages besides human health (see Chapter 2 for a discussion of environmental effects). From these estimates of economic damages, the economic benefits of reducing pollution can then be determined. The following section presents a general framework for such an analysis.

13.8.2   A General Cost Optimization Framework

The costs and benefits of an environmental control program or policy can be described qualitatively using a general framework that is best illustrated by a series of graphs. First we consider the adverse effects of environmental pollution. As discussed in the preceding section, we can ascribe (at least theoretically) an economic cost to the damages to human health and the environment caused by a given level of pollution. In general, we anticipate that damages will increase with higher levels of pollution. Figure 13.13 sketches this type of relationship, with the cost of environmental damages shown on the y-axis and the level of pollution shown on the x-axis. As used here, *pollution* is simply a general term for emissions and other impacts.

In this framework, the economic benefits from environmental control correspond to a *reduction in damage costs*. To illustrate, assume the current level of pol-

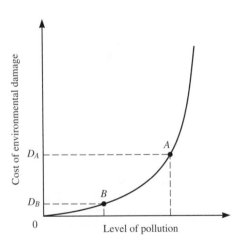

Figure 13.13   Cost of environmental damages versus pollution level. Higher levels of pollution cause greater environmental damage.

lution is initially at Point A in Figure 13.13, with economic damage costs to society at the level D_A. If the pollution level is now reduced to Point B, the economic damage cost falls to D_B. The difference between these two damage costs represents the economic benefit of reducing the pollution level from A to B. Thus, as illustrated in Figure 13.13,

$$\text{Economic benefits} = \text{Reduction in damage cost} = D_A - D_B \qquad (13.17)$$

A similar graph can be constructed for the cost of reducing pollution. In general, control costs start low but tend to increase as the level of pollution emitted approaches zero. Figure 13.14 sketches such a relationship. This is the same general curve shown earlier in Figure 13.8, except now the x-axis shows the level of pollution emitted rather than the level of emissions reduced. The extra cost of reducing emissions from level A to level B is illustrated on this graph as the cost difference $C_B - C_A$. Although the direct cost of pollution control is borne by the polluting industries, these costs usually are passed on to consumers and the public in higher prices for goods and services. Thus, in a general framework, these costs too are borne by society as a whole. In some cases the effect of price increases due to pollution controls may reverberate throughout the economy, resulting in additional indirect costs that should be included in the overall cost–benefit framework. Chapter 15 discusses this type of macroeconomic perspective.

Once pollution control costs are evaluated, they can be combined with the cost of environmental damages to obtain the total cost associated with different levels of environmental pollution. That result is sketched in Figure 13.15.

The curve labeled "total cost to society" effectively combines the costs and benefits into one picture. That is, for any given level of environmental pollution, Figure 13.15 shows that there is a cost of controlling to that level, plus a damage cost from the pollution emissions that still remain. At some point (labeled M in Figure 13.15)

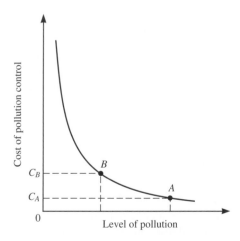

Figure 13.14 Cost of environmental controls versus pollution level. Reducing the level of pollution becomes increasingly expensive as emissions approach zero.

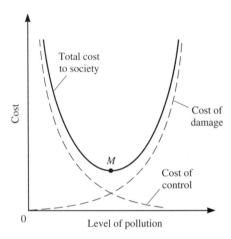

Figure 13.15   Total societal cost of environmental pollution. The sum of damage costs plus control costs yields the total cost to society. This curve has a minimum value at Point M.

the sum of these two costs is a minimum. That minimum point defines the level of pollution control that can be considered optimal in terms of overall cost. Moving to either lower levels of pollution (more control) or higher levels of pollution (less control) results in a greater total cost to society. This means that the economic benefits of such moves are outweighed by the change in costs.

13.8.3   Limitations of Cost–Benefit Analysis

The theoretical framework illustrated by Figure 13.15 is a useful way to display and account for the economic trade-offs involved in problems of environmental control. In practice, however, it is usually difficult to quantify all of the costs and benefits needed to perform a rigorous cost–benefit analysis or to identify an optimal economic solution.

As mentioned earlier, it is especially difficult to quantify the economic damages due to environmental pollution and the economic benefits from reducing those damages. Although progress has been made in quantifying some types of damages (especially from air pollutants like sulfur dioxide and particulate matter), there are still very large uncertainties in current estimates of damage costs and the economic benefits of environmental control measures.

Even the costs of pollution control may not be readily quantified in all cases. Aside from the direct cost of pollution control equipment, there may be additional indirect costs to an industry or to the overall economy, as mentioned earlier. For example, regulation that requires a large reduction in emissions from manufacturing plants across the country could result in job losses and higher product prices for consumers. In turn, consumers would have less money to spend on other goods and services, thus slowing the overall economy. These indirect effects represent additional economic costs attributable to the new regulation. Yet quantifying such costs may be difficult; often they may not even be recognized. One of the limitations of cost–benefit analysis, therefore, is that all of the appropriate costs and benefits might not be included or quantified in economic terms.

Another important limitation is that cost–benefit analysis does not consider issues related to *equity*. Thus total costs are compared to total benefits without regard as to *who* is paying the costs and who is deriving the benefits. When one group of people pays a disproportionate share of the costs while others derive the benefits, issues of fairness and equity must be analyzed by other means (although more advanced treatments of cost–benefit analysis provide some guidance). Despite these important limitations, cost–benefit analysis remains a useful and often powerful tool for the economic analysis of environmental issues.

13.9 CONCLUSION

This chapter presented a brief overview of methods that can be used to analyze environmental issues from an engineering economics perspective. The cost of environmental control is often a key factor in solving the environmental problems of our day. Thus it is important that engineers understand not only the technical issues related to environmental solutions, but the economic dimensions as well. Toward that end, the basic analysis tools presented in this chapter represent a first step in gaining knowledge of this important subject. Chapter 15 provides additional perspectives in the context of environmental forecasting. More advanced textbooks are available for students desiring greater depth in the study of engineering and environmental economics. A few of these titles are listed in the references for this chapter.

13.10 REFERENCES

Au, T., and T. P. Au, 1992. *Engineering Economics for Capital Investment Analysis,* 2nd ed. Prentice-Hall, Upper Saddle River, NJ.

Hanley, N., and C. L. Spash, 1993. *Cost–Benefit Analysis and the Environment.* Edward Elgar Publishing Ltd, Brookfield, VT.

USDOC, 2000. http://www.bea.doc.gov/bea. U.S. Department of Commerce, Bureau of Economic Analysis, Washington, DC.

USDOE, 1998. *Annual Energy Review 1997.* DOE/EIA-0384(97), Energy Information Administration, U.S. Department of Energy, Washington, DC.

USDOE, 1999. *Annual Energy Review 1998.* DOE/EIA-0384(98), Energy Information Administration, U.S. Department of Energy, Washington, DC.

White, J. A., M. H. Agee, and K. E. Case, 1989. *Principles of Engineering Economic Analysis.* John Wiley & Sons, New York, NY.

Zerbe, R. O., and D. D. Dively, 1994. *Benefit–Cost Analysis in Theory and Practice.* HarperCollins College Publishers, New York, NY.

13.11   PROBLEMS

13.1   A wastewater treatment plant adds $600,000 to the cost of a computer chip manufacturing facility.

(a)   What is the annualized capital cost of this plant if its lifetime is 15 years, the interest rate is 10 percent, and the end-of-life scrap value is $30,000?

(b)   If the facility manufactures 1 million chips per year, what is the added cost per chip due to capital expenditure for wastewater treatment?

13.2   An emission control system for a new car costs $700 and reduces NO_x emissions from 2.0 g/mi to 0.4 g/mi. Assume the car is driven 10,000 miles per year, and the life of the emission control system is 10 years with zero scrap value.

(a)   What is the annualized cost of this system in dollars per year if the discount rate (interest rate) is 6 percent/year?

(b)   What is the cost per 100 miles driven?

(c)   What is the cost per metric ton of NO_x removed? (1 metric ton $= 1,000$ kg $= 10^6$ g).

13.3   The owner of a widget factory plans to convert the plant to a new "clean production" system based on the principles of green design when the current facility becomes obsolete eight years from now. The new plant will cost $5.85 million. How much money must be put aside at the end of each year, beginning now, to accumulate the required capital? Assume these funds will be put into an account earning 6 percent interest, and the scrap value of the current plant will be $750,000 when it retires.

13.4   (a)   Repeat Problem 13.3 for interest rates of 3 percent, 9 percent, 12 percent, and 15 percent to reflect a range of potential investment options (with different degrees of risk). Plot your results for the annual amount (on the y-axis) as a function of interest rate (on the x-axis).

(b)   There is also uncertainty about the future salvage value of the current plant. Repeat the calculations in part (a) for salvage values of zero and $1.5 million. Plot these results on the same graph as part (a). How might a graph like this be useful to the factory owner?

13.5   New instrumentation is needed to monitor the level of environmental emissions at a large manufacturing plant. Two available systems give identical performance. One system costs $100,000 and will last for 10 years; the other costs $60,000 and will last for 6 years. The scrap value is negligible in both cases. Both systems also have an annual maintenance cost of $1,000/year. Is one of these systems preferable to the other based on cost? Assume the company's interest rate is 10 percent.

13.6   Two remediation options are being considered for a contaminated land area formerly used for industrial operations. Option 1 involves removing all of the contaminated soil over a two-year period at a cost of $2.2 million per year. Option 2 is to leave the soil in place but treat it with a bioremediation agent at a cost of $960,000/year over a three-year period. Subsequently, the soil would be sampled each year for the next five years to ensure the effectiveness of the treatment system. The cost of the sampling program would be $250,000 the first year and $100,000/year for the remaining four years.

(a)   Draw a cash flow diagram for each of the two options.

(b) Calculate the net present value of each option based on a discount rate of 6 percent/year.

(c) Which option has the lowest overall cost? What is the difference in total cost between the two options based on NPV? Give your answer both in dollars and as a percentage difference.

13.7 Assume the power plant described in Examples 13.10 through 13.13 has been operating for 30 years so that the capital cost of the SO_2 control system has been fully paid off. The plant expects to continue operating for another 15 years, but at a slightly lower capacity factor of 60 percent. Assume that the annual O&M costs given in Example 13.10 remain unchanged.

(a) What is the new value of total levelized annual cost for this plant during years 31 to 45?

(b) What is the new cost of SO_2 control per kW-hr?

(c) What is the new average cost per tonne of SO_2 removed?

13.8 Suppose the power plant in Problem 13.7 receives an offer from a wallboard manufacturing company. The company will build a new wallboard manufacturing plant adjacent to the power plant if the electric plant agrees to give it (at no cost) all of the solid waste produced by the SO_2 removal system. This waste will be the raw material used to produce wallboard.

(a) How would this change the economics of SO_2 removal at the power plant during years 31 to 45, as calculated in Problem 13.7?

(b) Why would this industrial ecology arrangement be beneficial to the wallboard company?

13.9 An environmental life cycle assessment reveals a number of opportunities where a company can reduce its energy consumption and eliminate waste materials that harm the environment. Several of these measures require capital expenditure to realize the environmental and operating cost benefits. It is estimated that a total capital investment of $25 million can reduce annual O&M costs by $4.5 million/year.

(a) What is the simple payback period for this investment?

(b) What is the payback period if the company's "huddle rate" (the effective interest rate applied to a capital expenditure) is 20 percent/year?

13.10 A manufacturer of consumer goods spends $10 million in one year to eliminate the toxic wastewater discharges reported to EPA's Toxics Release Inventory. That same year it spends another $2 million to publicize the cleanup and its corporate commitment to "green" manufacturing. As a result of favorable customer reaction, its sales increase by $5 million/year over each of the next three years. What is the effective rate of return earned by this overall expenditure? (Note: This is known as the *internal rate of return,* a measure often used to gauge the profitability of an investment.)

13.11 As part of a negotiation process, representatives of an industrial association and a government environmental agency are separately analyzing the total life cycle cost of a proposed new cleanup requirement for hazardous wastes.

(a) If you were the industrial representative, would you be more likely to use a real interest rate or a nominal interest rate in preparing your cost estimate? Explain.

(b) What if you were the environmental representative? Would your answer be the same? Explain.

13.12 Draw a graph showing the present value, P (on the y-axis), of $1,000 of environmental benefits realized at some future time, t (on the x-axis), where t varies from zero to 40 years. Plot this curve for annual interest rates (discount rates) of 20 percent, 10 percent, 5 percent, and 0 percent. Show all curves on a single graph. Based on your graphical results, answer the following questions:

(a) How much would you be willing to spend today to obtain $1,000 of environmental benefits for your children 30 years from now if you expect a 20 percent return on your investment?

(b) What if you expect only a 5 percent return?

(c) What general conclusion do you draw regarding the effect of discount rate on investments in future environmental quality?

(d) What general conclusion do you draw regarding the effect of the time horizon on investments in future environmental quality?

13.13 Battery-powered electric vehicles (EVs) have been introduced in some parts of the country to reduce urban air pollution. Assume the electrical energy required to recharge the batteries is 0.5 kW-hr/mile, and the car is driven 10,000 miles per year.

(a) If the capital cost of the EV is $30,000, compare its total life cycle cost over seven years to that of a gasoline-powered car costing $20,000. Assume both vehicles are financed at an interest rate of 8 percent, and other annual expenses are as shown in this table:

Cost Item	Conventional Vehicle	Electric Vehicle
Purchase price	$20,000	$30,000
Fuel consumption	28 mi/gal	0.5 kW-hr/mi
Fuel cost	$1.40/gal	8.3¢/kW-hr
Annual travel	10,000 mi	10,000 mi
Insurance	$1,000/yr	$1,000/yr
License and fees	$50/yr	$50/yr
Maintenance repairs	$150/yr	$150/yr
Parking and tolls	$300/yr	$300/yr
Resale value	$2,000	$3,000

(b) If you found in part (a) that the EV was more costly than the gasoline-powered car, how low would its capital cost have to be in order to compete over a seven-year life cycle? If you found in part (a) that the EV was less costly, how much more could the auto company charge and still compete?

13.14 You have been asked to evaluate a proposed energy efficiency standard for desktop computers. Assume the average energy consumption of a desktop computer is 2,000 kW-hr of electricity per year. The proposed efficiency standard would cut the energy consumption to half the current value. The technology for doing this is available, but computer manufacturers claim it would add $200 to the purchase price of a computer now costing $1,500.

(a) From a life cycle perspective, would the energy-efficient model be economically attractive to consumers? Assume the average lifetime of the computer is five years, after which it is sold for $200. The energy-efficient model would have the same lifetime and resale value. The residential price of electricity is 8.3¢/kW-hr, and the interest rate on a consumer's money market savings accounts is 5 percent. Be sure to show all work and discuss any assumptions you have made.

(b) Behavioral research shows that most people do not actually invest in energy-efficient equipment unless the return on investment (effective interest rate) is at least 25 percent. Would this criterion alter your conclusion in part (a)? Explain.

13.15 Power plants A and B emit 100,000 and 60,000 tons/year of sulfur dioxide, respectively. Under a new market-based acid rain control program, both plants receive allowances to emit no more than 40,000 tons/year each. Plant A estimates it would cost $400/ton to reduce its emissions to the required level. However, Plant B determines that it can reduce its emissions to as low as 30,000 tons/year at a cost of $150/ton. This would allow it to sell the extra 10,000 tons/year in allowances.

(a) Assuming that each plant reduces its own emissions to the required level (40,000 tons/year) at the indicated costs, what is the overall cost of control?

(b) Assume that Plant B reduces its emissions to 30,000 tons/year (that is, overcontrol) and sells its excess allowances to Plant A at the market price of $200/ton. This allows Plant A to emit 50,000 tons/year instead of 40,000 tons/year. What is the new overall cost of control? Has there been any change in the overall level of emissions reduction? Has there been any reduction in the overall cost?

13.16 You are an engineer/policy analyst working on a cost–benefit analysis of a proposed emission reduction regulation to achieve ambient ozone standards. You have been asked to provide a five-year perspective. You estimate the proposed measure would cost (in millions of dollars) $500 in Year 1 and $100 in Year 2. The estimated benefits of the measure (also in millions) are $100 in Year 3, $300 in Year 4, and $300 in Year 5.

(a) Draw a cash flow diagram for this analysis.

(b) Calculate the net present value of the proposed measure based on an interest rate of 6 percent/year.

(c) From a cost–benefit perspective, does this policy measure seem justified over a five-year time horizon? Explain.

(d) Suppose there were additional benefits of $15 million in Year 6. Would this affect your conclusion in part (c)? Explain.

13.17 The EPA estimates that emissions of toxic pollutants from certain manufacturing plants are currently causing $750 million per year worth of environmental damages. They want your help in identifying an optimal level of control. Your analysis of options for reducing emissions concludes that the annual cost of control (in millions of dollars per year) is given by

$$\text{Cost of control ($ million/yr)} = \frac{R^3}{1,000}$$

where R = percentage emissions reduction from the current level ($0 < R < 100$). You also find that the annual environmental damage cost decreases linearly with higher emission reductions:

$$\text{Cost of damages ($ million/yr)} = 750 - 7.5R$$

(a) Prepare a graph showing the emission reduction level (R) on the x-axis and the annual cost ($ million/year) on the y-axis. Then plot the (1) cost of emission control, (2) cost of environmental damages, and (3) total societal cost.

(b) Based on your analysis, what level of emissions reduction would be optimal for this case? Briefly discuss and justify your conclusion.

(c) What is the value of annual benefits (relative to no control) for the optimal emission reduction level?

(d) What is the benefit–cost ratio for this emission reduction level?

chapter
14

Risk Assessment and
Decision Analysis

14.1   INTRODUCTION

People take risks all the time. Every time you leave your house you face some risk of being injured in an accident whether you drive a car, take a bus, or cross a street. If you smoke cigarettes the risk to your health goes up considerably, and if you breathe polluted air the risk to your health also increases.

Many risks are taken voluntarily, but others are not under our immediate control. To protect ourselves from involuntary harm, we frequently turn to government agencies to establish and enforce "safe" levels for such things as drinking water quality, air quality, food quality, and other types of known or potential risks to human health and safety.

This chapter examines the concept of risk as applied to environmental pollutants. In particular, we will learn how government agencies assess the level of risk to human health from exposure to hazardous or toxic chemicals in the environment. We also will examine some of the methods used to decide what action, if any, to take in response to a given risk. Risk assessment and decision analysis methods are among the policy analysis tools that are now widely used to address difficult or complex environmental problems. Although a comprehensive treatment of these topics is beyond the scope of this text, the fundamentals covered in this chapter can provide a starting point for more detailed studies.

14.2   DEFINING ENVIRONMENTAL RISKS

Risk involves a chance of injury or loss. If the loss is something well understood—such as death—risk is sometimes defined simply as the probability of that result. In this case

$$\text{Risk} = \text{Probability of a } \textit{specific} \text{ undesired consequence} \qquad (14.1)$$

Table 14.1 lists various causes of death along with the odds that when the average American dies, it will be from one of these specific causes. This probability is sometimes defined as *risk* (of dying from a particular cause). For example, the lifetime risk of dying from cancer is 0.23, meaning a 23 percent chance that death will be from this cause, based on national statistics. On the other hand, the odds of dying from botulism are less than one in a million. Risk levels thus vary by many orders of magnitude.

When several different kinds or magnitudes of injury or loss may occur, risk is sometimes defined as a product of the probability and the size of the loss:

$$\text{Risk} = \left(\begin{array}{c} \text{Probability of an} \\ \text{undesired consequence} \end{array} \right) \times (\text{Size of the loss}) \qquad (14.2)$$

For example, the risk of a nuclear power plant accident might be viewed differently for a plant located in a sparsely populated rural area as compared to one near a major urban center. Similarly, an investment in the stock market might be considered riskier if the amount invested were $100,000 as compared to $1,000.

Table 14.1 Risk of death for Americans from various activities.

Cause of Death	Approximate Number of U.S. Deaths Each Year from This Cause[a]	Approximate Odds of Death from This Cause	Lifetime Risk of Death from This Cause
Disease (all kinds)	2,000,000	1 in 1.1	9.1×10^{-1}
Heart disease	770,000	1 in 2.7	3.7×10^{-1}
Cancer (all kinds)	480,000	1 in 4.4	2.3×10^{-1}
Accidents (all kinds)	95,000	1 in 22	4.5×10^{-2}
Auto accidents	48,000	1 in 44	2.3×10^{-2}
Diabetes	37,000	1 in 57	1.8×10^{-2}
Suicide	31,000	1 in 68	1.5×10^{-2}
Homicide	21,000	1 in 100	1.0×10^{-2}
Drowning	5,900	1 in 360	2.7×10^{-3}
Fire	4,800	1 in 440	2.3×10^{-3}
Asthma	4,000	1 in 530	1.9×10^{-3}
Firearm accident	1,500	1 in 1,400	7.1×10^{-4}
Viral hepatitis	1,000	1 in 2,100	4.8×10^{-4}
Electrocution	850	1 in 2,500	4.0×10^{-4}
Car–train accident	570	1 in 3,700	2.7×10^{-4}
Appendicitis	510	1 in 4,100	2.4×10^{-4}
Pregnancy and related	470	1 in 4,500	2.2×10^{-4}
Lightning	78	1 in 27,000	3.7×10^{-5}
Flood	58	1 in 36,000	2.8×10^{-5}
Tornado	58	1 in 36,000	2.8×10^{-5}
Fireworks	8	1 in 260,000	3.8×10^{-6}
Botulism	2	1 in 1,100,000	9.1×10^{-7}

[a] The number of deaths from all causes is 2.1 million, out of a total U.S. population of 245 million in 1988.
Source: Based on USHHS, 1988.

Equations (14.1) and (14.2) are simple, but in the real world things are not this straightforward. When psychologists study how risky people consider different hazards, they find that peoples' judgment about risk depends on factors other than the probability and size of the loss. These other factors include how well the risk is understood, how well people can control their exposure to the risk, how equitably the exposure is distributed across the population at risk, and a number of other factors.

From Equation (14.2) it is clear that one way to reduce a risk is to reduce the size of the loss (for example, expose fewer people to the hazard or replace deaths and serious injuries with minor injuries). Another way to reduce risk is to reduce the probability that the loss will occur.

For simplicity, in most discussions in this chapter we will consider a well-specified loss, such as death, and we will define risk in terms of just the probability of occurrence. That is, we will use the definition of Equation (14.1).

As discussed in Chapters 1 and 2, human activities release into the environment a variety of chemicals that can adversely affect human health and welfare. In the United States, the public policy response has sought to identify safe levels of the most common contaminants in the air we breathe and the water we drink. This has resulted in the promulgation of national health-based environmental quality standards for a number of common air and drinking water pollutants. Contaminant concentrations below government standards are generally considered to be free of known health risks (although this is not always the case).

But what of the hundreds of chemicals in the environment that are not as ubiquitous or well-studied as the common pollutants in air and drinking water? Many such chemicals find their way to human receptors via the air we breathe, the water we drink, or the food we eat. Are there any health risks of concern from exposure to such chemicals?

Many chemicals have been designated as hazardous or toxic by the U.S. Environmental Protection Agency (EPA). However, as discussed in Chapter 2 (Section 2.5), such designations often are based on criteria other than health impacts. Furthermore, the presence of chemicals in the environment is often highly localized in nature. Such releases commonly originate from industrial plants in a community or from leaking storage tanks or waste containers buried in the ground. But if we have detected a hazardous chemical in the air, soil, or water, is there a risk to public health that warrants attention? If so, what actions are feasible, and which option is best in a particular case?

It was not until the early 1980s that questions about risk from chemical exposure gained prominent attention in the environmental regulatory community. Incidents such as the 1984 toxic chemical spill in Bhopal, India, and the 1978 discovery of soil contamination at Love Canal, New York, focused public concern on the potentially severe health effects of hazardous chemicals in the environment (see Section 2.5). One result was the passage of national legislation to identify and remediate the many toxic waste dumps across the country. But the sheer number of chemicals of concern, coupled with the vast number of contaminated waste disposal sites (tens of thousands across the country), forced a new approach to dealing with hazardous chemicals in the environment. Thus did the science (and art) of risk assessment and risk management enter as a way of helping to prioritize problems and identify solutions in the face of limited resources and capability to respond to a potentially massive problem.

The primary focus has been on assessing and reducing risks to human health. Two categories of health risk are commonly analyzed in environmental risk assessments. One is the risk of contracting cancer from exposure to a chemical. Such *carcinogenic* chemicals are generally of highest concern. The second category of health risk covers all other human illnesses and disease induced by chemicals in the environment. These effects often are associated with damage to a particular organ or tissue, such as the liver, kidney, or nerves. If the risk of either carcinogenic or noncarcinogenic effects from exposure to a particular chemical is unacceptable, measures to reduce or eliminate the risk must be considered. But before getting to that stage, we must first address the question of what levels of risk are unacceptable.

14.3 HOW SAFE IS SAFE?

Is it possible to eliminate risk completely? Based on Equation (14.2), zero risk from environmental contaminants implies either zero consequences from exposure to a pollutant, or zero probability of that exposure. But zero is a very small number! Eliminating exposure to a chemical altogether, such as by shutting down a factory, removing all contaminated soil, or containing a chemical after it is released, may be technically difficult or prohibitively expensive. On the other hand, exposure to only very low levels of a chemical might be free of harmful consequences or might pose risks that are so small as to be acceptable to society. In a few cases, such as with fluoride in drinking water, such exposure might actually confer health benefits.

Researchers who study risk and the statistics that underlie its quantification provide us with some interesting insights about acceptable levels of risk. Studies have found that most people are willing to accept a higher level of risk in activities over which they have direct control (like skiing) or from which they derive a direct benefit (like riding in cars, buses, or airplanes), compared to risks that are imposed involuntarily or that confer no direct benefit (like having a chemical waste dump built near your home).

The U.S. Environmental Protection Agency (EPA) has endorsed certain numerical criteria to assess whether a particular exposure to chemicals in the environment poses an acceptable or unacceptable risk to public health. As elaborated later in the chapter, determinations of acceptable risk levels involve a substantial amount of judgment (in contrast to "hard" scientific facts). Nonetheless, these EPA criteria provide the basis for most current risk assessments for chemicals that are designated as toxic or hazardous.

The criteria for acceptability are different for carcinogenic and noncarcinogenic chemicals. For carcinogens, for example, a one in a million chance (10^{-6} probability) of an additional human cancer over a 70-year lifetime is the level of risk considered generally acceptable. Even higher risks can be accepted depending on circumstances. To better understand these criteria and their application, the following sections elaborate on current risk assessment methods.

14.4 RISK ASSESSMENT METHODOLOGY

A four-step process for risk assessment was defined in 1983 by the National Research Council (NRC, 1983), and these steps have been widely adopted by government agencies and others involved in risk assessments. As illustrated in Figure 14.1, the four steps include the assessment of hazards, the development of dose–response relationships, exposure assessment, and risk characterization.

14.4.1 Hazard Assessment

The first step is to determine whether there is any *potential* problem from exposure to a given chemical. This step is called *hazard assessment*. Its purpose is to determine whether exposure to a given chemical can cause an observable increase in

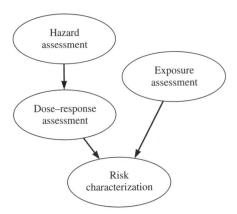

Figure 14.1   Elements of a risk assessment.

some illness or health condition. This step draws on all available data from laboratory studies, animal studies, and epidemiological studies (based on observations of people in the real world). The outcome is an evaluation and description of the nature and severity of any effects that might be caused by exposure to a particular chemical, including carcinogenic and noncarcinogenic effects.

14.4.2   Dose–Response Assessment

If the hazard assessment establishes that a chemical can cause some type of health effect, the next step is to quantify the relationship between the dose of that chemical—the mass of chemical ingested or received—and the resulting response or adverse effect. This step is referred to as a *dose–response assessment.* It is one of the most difficult and controversial steps in a risk assessment because there are usually insufficient data to characterize a dose–response relationship over the full range of interest. Ideally, dose–response relationships are sought both for carcinogenic and noncarcinogenic effects.

Carcinogenic Effects   The evaluation of chemical carcinogens relies heavily on animal studies in which laboratory animals such as rats and mice are given measured doses of a particular chemical. If significantly more tumors develop in the exposed animals, the chemical is clearly a carcinogen for those animals. But is it also carcinogenic for humans, especially at the lower doses typical of human exposure? This question cannot be answered scientifically by conducting experiments on people, so other approaches are needed. In some cases information on human exposure to suspected carcinogens may be available from epidemiological studies of accidents or of workers at chemical plants and other facilities where chemicals are prevalent. In most cases, however, only animal data are available, and even those data may be sparse for many chemicals.

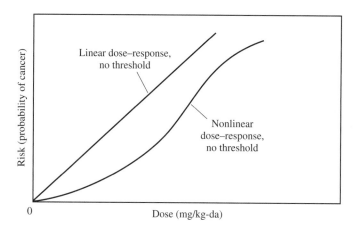

Figure 14.2 Dose–response assessment relationships for a carcinogenic chemical typically assume a linear relationship with no threshold (any dose above zero results in some risk). In some cases nonlinear relationships have been proposed.

The available data for a chemical provide the basis for a dose–response function that quantifies the incremental risk of cancer above the normal or *background* level. Risk is typically plotted on the *y*-axis as a function of the average daily dose of toxicant on the *x*-axis. Examples of such relationships are shown in Figure 14.2.

The dose of a chemical is commonly normalized on the body weight of the test animal or subject and expressed as milligrams per day of chemical per kilogram of body weight (mg/kg-da). The average daily dose provides a measure of the long-term or *chronic* exposure, in contrast to short-term or *acute* exposure from a single incident or event.

A key assumption in dose–response relationships for carcinogens is that *any* exposure to a carcinogenic substance is considered to increase the lifetime risk of cancer, regardless of the dosage. This conservative assumption implies that the risk of cancer can never be zero unless the lifetime exposure to a chemical is also zero. This also means that the dose–response curve must pass through the origin of the dose–response graph. Typically, a linear dose–response relationship is assumed, as indicated by the solid line in Figure 14.2. As Example 14.1 illustrates, it is not generally feasible to experimentally verify assumptions about cancers induced by very small doses.

Example 14.1

Data requirements for low dosage response studies. In a population of 100 mice, 2 animals developed tumors (a 2 percent cancer rate, or a risk of 0.02) when exposed to a daily average dose of 1 mg/kg-da of a particular chemical. The dosage for typical human exposure to this chemical is approximately 1×10^{-5} mg/kg-da. Assuming a linear dose–response relationship and the same incidence rate in both mice and people, roughly how many animals would have to be tested to detect the expected cancer rate at the lower dose?

Solution:

The lower dose value is 10^{-5} times smaller than the test dose administered to the mice. Based on a linear dose–response assessment relationship, the expected rate of tumors in mice thus would be 10^{-5} times smaller than the 2 percent rate observed, or 2×10^{-5} percent. This corresponds to a fraction of 2×10^{-7}, or 2 animals out of 10 million. Because some animals can be expected to develop tumors due to other causes (that is, with no exposure to the test chemical), more than 10 million animals would have to be tested to observe the expected effect. Clearly, a routine test of this magnitude would not be feasible.

Noncarcinogenic Chemical Effects   Chemicals that do not induce tumors in test animals even at high dosages are judged noncarcinogenic. However, other types of adverse health effects that might be identified in a hazard assessment include kidney or liver damage, nerve damage, and diseases such as diabetes. Animal studies again are often the major source of data for dose–response relationships.

For noncarcinogenic effects, a different form of dose–response relationship is commonly employed for risk assessments. A key assumption for noncarcinogens is the existence of a dosage *threshold,* below which there is no increase in adverse effects above natural background rates. This type of dose–response assessment relationship is sketched in Figure 14.3. The intersection of the dose–response curve with the *x*-axis indicates the threshold dosage. Below that level there is no observable adverse effect. Above the threshold level are risks of injury, disease, or other adverse effects.

The concept of a threshold dosage is supported by medical science, which has found that humans and other living organisms often can tolerate some degree of exposure to various chemicals before they become toxic or cause ill effects. Indeed, many trace metals that are toxic in high concentrations are essential nutrients in very low quantities. The existence of a threshold level also has been proposed for some

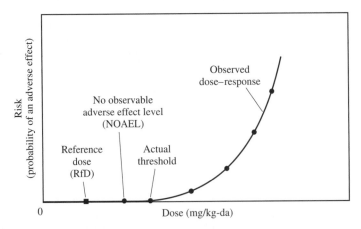

Figure 14.3   Dose–response assessment relationship for a noncarcinogenic chemical. These relationships typically have a threshold value above which adverse effects are observed.

known or suspected carcinogens. However, because such levels are difficult or impossible to verify, the more conservative assumption of a zero threshold is commonly used for carcinogens, whereas a nonzero threshold is employed for other types of health effects.

14.4.3 Exposure Assessment

The dose–response assessment just discussed determines the probability of an individual being adversely affected by a given chemical dose. The next step, called the *exposure assessment,* is to quantify the dose actually received in a particular situation. The purpose is to measure or estimate the frequency, intensity, and duration of human exposure to a chemical agent in the environment. From this information the total exposure can be determined.

As depicted in Figure 14.4, there are various pathways by which people can be exposed to chemicals in the environment. The most significant route is often inhalation. Breathing polluted air carries chemical contaminants into the human respiratory tract. From there, contaminants can subsequently enter the bloodstream via the microscopic air sacs of lung tissue. Chapter 10 described this process in the context of human exposure to lead in the environment.

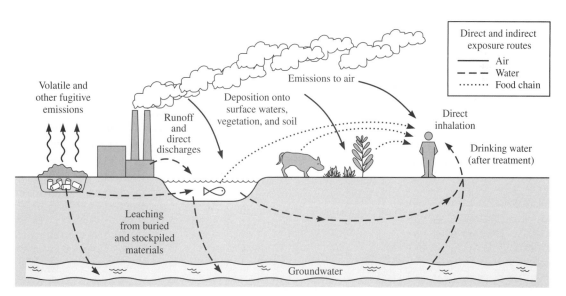

Figure 14.4   Exposure pathways for risk assessments. Pollutants enter the environment via the air, water, and land and can be dispersed both near and far from the source. Contaminant deposits on ground surfaces may be carried by local drainage and runoff into lakes, rivers, and groundwater systems. Some substances, such as PCBs and mercury, bioaccumulate in fish. Human exposure may occur via inhalation of polluted air, dermal contact (especially for children), and ingestion of contaminated water, soil, or food.

Just as in the case of lead, additional exposure routes include oral ingestion (swallowing) and dermal contact (touching the skin). The latter can occur when chemicals in contaminated soils, sediments, or water are absorbed into the body through pores in the skin when people swim in polluted water or handle contaminated soil (especially children at play). Direct oral ingestion of contaminated soil or waterborne pollutants can occur either accidentally or from eating contaminated foods. Many foods, especially vegetation and fish, can accumulate and store certain chemicals found in contaminated water and soil—a process known as *bioaccumulation*. These food products then have higher contaminant concentrations than found in the surrounding environment.

The process of exposure assessment involves either measuring or estimating the concentrations of chemicals in the environment where people are likely to be exposed. The mathematical models and calculation methods described in Chapter 8 for urban air pollutants and Chapter 9 for waterborne organic chemicals (PCBs) are among the techniques used to estimate environmental concentrations in the absence of actual measurements (which often are quite limited). Additional assumptions and models are then used to estimate the dosage of chemicals that an individual has received from different exposure pathways. Chapter 10 gave an example for the case of human exposure to lead in the air, water, and soil. Data required for exposure assessments include the quantities of air, water, and soil ingested by adults and children. Table 14.2 summarizes the typical values of these parameters used to estimate chemical dosage. The following example illustrates how the data in Table 14.2 can be used to estimate human exposure to a hazardous chemical.

Table 14.2   Standard default exposure factors for environmental risk assessments. These EPA guidelines are used for calculating reasonable maximum exposure at a contaminated site.

Land Use at or Near Site	Exposure Pathway	Daily Intake	Exposure Frequency (Days/Year)	Exposure Duration (Years)[a]	Body Weight (kg)
Residential and agricultural	Ingestion of potable water	2 liters	350	30	70
	Ingestion of soil and dust	200 mg (child) 100 mg (adult)	350	6 24	15 (child) 70 (adult)
	Inhalation of contaminants	20 m^3 (total) 15 m^3 (indoors)	350	30	70
Commercial/industrial	Ingestion of potable water	1 liter	250	25	70
	Ingestion of soil and dust	50 mg	250	25	70
	Inhalation of contaminants	20 m^3 per workday	250	25	70
Agricultural	Consumption of homegrown produce	42 g (fruit) 80 g (vegetables)	350	30	70
Recreational	Consumption of locally caught fish	54 g	350	30	70

[a] These values are only for noncarcinogenic chemicals. For carcinogens the value is 70 years.
Source: USEPA, 1991.

Example 14.2

Inhalation exposure to a hazardous chemical. The average ambient air concentration of formaldehyde in an urban region is estimated to be 4.6 $\mu g/m^3$ of air. What is the average daily dose (in mg/kg-da) received by an average adult, assuming that all the inhaled material is taken up by the body?

Solution:

From Table 14.2, an average adult is assumed to weigh 70 kg and breathe 20 m^3 air/day. The total mass of formaldehyde inhaled by this adult during one day is therefore

$$(4.6 \ \mu g/m^3)(20 \ m^3/da) = 92 \ \mu g/da$$

The average chemical dose in units of milligrams per day per kilogram of body weight is thus

$$\frac{\left(92\dfrac{\mu g}{da}\right)\left(\dfrac{1 \ mg}{1{,}000 \ \mu g}\right)}{70 \ kg \ body \ wt} = 1.3 \times 10^{-3} \ \frac{mg \ formaldehyde}{kg\text{-}da}$$

Similar calculations can be used to estimate the dose of chemical exposures from other pathways such as oral ingestion. The contributions of all exposure pathways can be summed to determine the total dosage received.

14.4.4 Risk Characterization

The fourth and final step of the risk assessment process is to combine the results of the exposure assessment with the dose–response function for each chemical of concern. This *risk characterization* then yields the expected incidence of the adverse health effects identified earlier in the hazard assessment. For carcinogenic chemicals, the risk is expressed as the probability of a chemically induced cancer during the lifetime of an exposed individual. The total number of additional cancers in a given population also can be estimated in this way. For noncarcinogenic chemicals, the risk of an adverse health effect is expressed by other parameters that are defined later in this chapter.

The following sections elaborate on the mathematical procedures used to assess carcinogenic and noncarcinogenic risks. Because the scientific basis for risk assessments is still largely in its infancy, there are great uncertainties in the quantitative values that emerge from these methods. Later in the chapter we will discuss the sources of some of these uncertainties and their implications.

14.5 ASSESSING RISK FOR CARCINOGENS

As noted earlier, risk assessments have become an important element of the federal, state, and local response to problems of hazardous chemicals in the environment, especially for problems of hazardous waste disposal. Inevitably, a set of acronyms and formulas has emerged to help standardize and facilitate key calculations.

14.5.1   Chronic Daily Intake

For carcinogenic chemicals, a parameter called the *chronic daily intake* (*CDI*) is defined. This is the average daily dose of a chemical over the lifetime of an individual, normalized on his or her body weight. Thus

$$\text{CDI (mg/kg-da)} = \text{Chronic daily intake}$$
$$= \frac{\text{Average daily dose (mg/da)}}{\text{Body weight (kg)}} \qquad (14.3)$$

Notice that this parameter is identical to the quantity calculated earlier in Example 14.2. The average daily dose represents an average over a lifetime of exposure to a carcinogen. To evaluate the average daily dose, the conservative assumption typically used by the EPA and other government agencies is a lifetime of 70 years of continual exposure to the maximum concentration of a carcinogen from a particular source. A typical adult body weight of 70 kg is then assumed to calculate the CDI.

14.5.2   Potency Factor

To calculate the resulting risk of cancer from a given exposure, the CDI is multiplied by a parameter called the *potency factor* (*PF*). This factor is based on the dose–response curve for the chemical and exposure pathway of concern. As illustrated in Figure 14.5, the potency factor represents the incremental lifetime cancer risk corresponding to a chronic daily intake of 1 mg/kg-da of a particular chemical:

$$\text{PF (mg/kg-da)}^{-1} = \text{Potency factor}$$
$$= \text{Incremental cancer risk} \qquad (14.4)$$
$$\text{for a chronic daily intake (CDI) of 1 mg/kg-da}$$

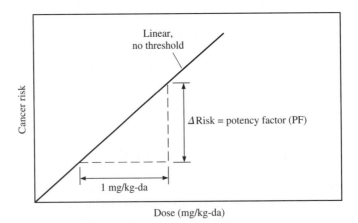

Figure 14.5   Illustration of the potency factor for a carcinogenic chemical.

Table 14.3 Potency factor values for selected chemicals.

Substance	Oral PF $(mg/kg\text{-}da)^{-1}$	Inhaled PF $(mg/kg\text{-}da)^{-1}$
Metals and Inorganics		
Arsenic	1.5	50.0
Beryllium	4.3	8.4
Cadmium	—	6.3
Chromium VI	—	42
Organic Compounds		
Benzene	5.5×10^{-2}	2.9×10^{-2}
Benzol(a)pyrene (BaP)	7.3	3.10
Chloroform	6.1×10^{-3}	8.1×10^{-2}
1,4 Dioxane	1.1×10^{-2}	—
Formaldehyde	—	4.5×10^{-2}
Methylene chloride	7.5×10^{-3}	1.65×10^{-3}
Polychlorinated biphenyls (PCBs)	0.4–2.0	0.4–2.0
Tetrachloroethylene	5.2×10^{-2}	2.0×10^{-3}
Trichloroethylene (TCE)	1.1×10^{-2}	6.0×10^{-3}

Source: ORNL, 2000; USEPA, 1999.

Because the dose–response curve for a carcinogen is assumed to be linear as in Figure 14.5, the potency factor is equivalent to the slope of the dose–response line (where dose is normalized on body weight). For this reason it is also sometimes referred to as the *slope factor.* Table 14.3 lists the value of the potency factor for several known or suspected carcinogens.

14.5.3 Incremental Risk

Based on a linear dose–response model, the total incremental cancer risk over a lifetime is then equal to

$$\text{Incremental lifetime cancer risk} = \text{CDI} \times \text{PF} \qquad (14.5)$$

An example illustrates how this risk is calculated.

Example 14.3

Cancer risk from inhalation exposure. An industrial facility in a community emits formaldehyde into the atmosphere with a peak concentration in the surrounding community of 4.6 $\mu g/m^3$. What is the lifetime cancer risk to a maximally exposed individual?

Solution:

Assume an average adult weighing 70 kg and breathing 20 m^3 air/day, as in the preceding example. Further assume the person is exposed to the peak concentration for an entire lifetime (this is a maximally exposed individual). The chronic daily intake of formaldehyde is then identical to the value calculated in Example 14.2:

$$CDI = \frac{(4.6\ \mu g/m^3)(20\ m^3/da)}{(70\ kg)(1,000\ \mu g/mg)} = 1.3 \times 10^{-3}\ mg\ formaldehyde/kg\text{-}da$$

From Table 14.3, the potency factor for formaldehyde is

$$PF = 4.5 \times 10^{-2}\ (mg/kg\text{-}da)^{-1}$$

Thus the incremental lifetime cancer risk is

$$\begin{aligned}
Incremental\ lifetime\ risk &= CDI \times PF \\
&= (1.3 \times 10^{-3})(4.5 \times 10^{-2}) \\
&= 5.8 \times 10^{-5}
\end{aligned}$$

The risk value calculated from Equation (14.5) is the *added risk* (probability) of developing a cancer *over and above* whatever background risk existed in the absence of the chemical exposure being studied. By comparing this incremental risk to a level of acceptable risk, we can judge the severity of a particular situation.

14.5.4   Levels of Acceptable Risk

What level of risk is acceptable for a known or suspected carcinogen? The EPA has concluded that a *lifetime* risk level of 10^{-6} (one chance in a million) or less can generally be regarded as acceptable or inconsequential, whereas a lifetime risk of 10^{-3} (one in a thousand) or greater is considered serious and is a high priority for attention. Risk levels between 10^{-6} and 10^{-4} (one in 10,000) also may be acceptable according to EPA criteria, although this range is more of a gray area that may require a case-specific judgment as to the acceptability of a particular risk. For instance, in Example 14.3 the cancer risk of 5.8×10^{-5} was higher than the one-in-a-million risk level, but still within the potentially acceptable range of 10^{-6} to 10^{-4} endorsed by the EPA for chemical carcinogens. One way of judging the acceptability of this risk is to compare it to background cancer rates, as in the next example.

Example 14.4

Additional cancers from community exposure.   For the community in Example 14.3, estimate the total number of expected additional cancers per year from inhalation of formaldehyde based on a population of 30,000 people. Compare the result to the average risk of cancer deaths in the United States inferred from Table 14.1.

Solution:

A conservative (and common) approach is to assume that the entire population of 30,000 people is exposed to the peak formaldehyde concentration for their entire lifetime. The total number of additional lifetime cancers for this community is then

$$Lifetime\ community\ risk = \left(5.8 \times 10^{-5}\ \frac{cancers}{lifetime}\right)(30,000\ people)$$

$$= 1.7\ additional\ lifetime\ cancers$$

Assuming an average lifetime of 70 years, the increase in annual cancer *rate* would be

$$\text{Annual community risk} = \frac{1.7 \text{ additional lifetime cancers}}{70 \text{ years per lifetime}}$$
$$= 0.024 \text{ cancers/yr}$$

By comparison, the average (background) U.S. cancer death rate derived from Table 14.1 is

$$\text{U.S. cancer rate} = \frac{480,000 \text{ cancers/yr}}{245 \times 10^6 \text{ people}}$$
$$= 196 \text{ cancers/yr per 100,000 people}$$

Applying this rate to the local community gives

$$\text{Background cancer rate} = \left(\frac{196 \text{ cancers/yr}}{100,000 \text{ people}}\right)(30,000 \text{ people})$$
$$= 59 \text{ cancers/yr}$$

Against this background, the potential for an additional 0.024 cancers/year from formaldehyde inhalation would not be detectable.

14.5.5 Application to Contaminated Sites

Many U.S. risk assessments involve contaminated waste disposal sites or leaking underground storage tanks. In these situations exposure to toxic chemicals can occur via ingestion of contaminated drinking water or contaminated soil, as well as via inhalation of vapors given off by soil or water. Table 14.2 showed typical maximum ingestion rates via different environmental media for adults and children.

In many cases exposure to a pollutant occurs during only a portion of an individual's lifetime, such as during childhood. Accordingly, the chronic daily intake (CDI) can be adjusted to account for the actual exposure via each environmental medium or exposure route. An example best illustrates this calculation.

Example 14.5

Risk assessment of a proposed playground. Adjacent to a residential neighborhood is an undeveloped parcel of land where quantities of benzene from an old factory were dumped many years ago. The concentration of benzene in the soil was recently measured and found to be at or below 0.9 mg benzene/kg of soil (0.9 ppmw). No other exposure route is evident. The site is now being considered for a children's playground. Is the cancer risk low enough to allow children to play here? For risk estimation, assume that a child would use the playground four hours a day, 350 days per year for 10 years, and that this is the only source of lifetime exposure to benzene.

Solution:

From Table 14.2, assume a child weighs 15 kg and ingests about 200 mg of soil per day (equal to 2×10^{-4} kg/da). First calculate the chronic daily intake over the fraction of a child's life spent at play in the proposed playground:

$$CDI = \frac{\text{Concentration in soil} \times \text{Soil intake rate}}{\text{Body weight}} \times \frac{\text{Total exposure time}}{\text{Total lifetime}}$$

$$= \frac{\left(0.9 \, \dfrac{\text{mg benzene}}{\text{kg soil}}\right)\left(2 \times 10^{-4} \, \dfrac{\text{kg soil}}{\text{da}}\right)\left(4 \, \dfrac{\text{hr}}{\text{da}} \times 350 \, \dfrac{\text{da}}{\text{yr}} \times 10 \text{ yrs}\right)}{(15 \text{ kg})\left(70 \, \dfrac{\text{yr}}{\text{life}} \times 365 \, \dfrac{\text{da}}{\text{yr}} \times 24 \, \dfrac{\text{hr}}{\text{da}}\right)}$$

$$= 2.7 \times 10^{-7} \, \frac{\text{mg benzene}}{\text{kg-da}}$$

From Table 14.3, the potency factor for benzene (oral ingestion) is 5.5×10^{-2} $(\text{mg/kg-da})^{-1}$. Thus, the incremental lifetime cancer risk is:

$$\text{Incremental cancer risk} = CDI \times PF$$
$$= (2.7 \times 10^{-7})(5.5 \times 10^{-2}) = 1.5 \times 10^{-8}$$

Thus, even after incorporating a series of conservative assumptions regarding exposure, this level of risk is 67 times smaller than the one-in-a-million chance endorsed by the EPA as an acceptable lifetime risk. Even if we had assumed the child was exposed to this dose for an entire lifetime, the incremental risk would be well below 10^{-6}.

The two preceding examples illustrate another—sometimes controversial—feature of risk assessments: namely, that estimated risks often are at or below levels that are considered acceptable according to EPA criteria. As a result, the contaminant is more likely to be left in place, compared to the more traditional environmental engineering approach in which *any* level of contamination requires a site to be remediated to a pristine (precontaminated) level. We will return to this point a little later in the chapter when summarizing the strengths and weaknesses of environmental risk assessments. First we continue with a look at how risks are evaluated for noncarcinogenic effects.

14.6   ASSESSING RISK FOR NONCARCINOGENS

As noted earlier, a key feature in evaluating noncarcinogenic health risks is the incorporation of a threshold dose for the health effect of concern. As a result, several new acronyms emerge in the risk assessment procedure. The dosage of a chemical below which no adverse health effects are observed is called (not surprisingly) the *no observable adverse effects level,* commonly abbreviated as *NOAEL.* Doses of a chemical at or below this level (in units of mg/kg-da, as before) are nominally considered safe. But many uncertainties can make it difficult to determine the NOAEL dose for a particular chemical, so the procedures for assessing noncarcinogenic health risks are a bit more involved.

14.6.1   Reference Dose

To account for the uncertainties in defining a threshold value of dose for no adverse effects on humans, several new factors are introduced to arrive at a parameter called the *reference dose (RfD)*. *The reference dose is a key parameter used in risk assessments to characterize the safe dose of a noncarcinogenic chemical.* The reference dose is obtained by adjusting the available NOAEL data from animal studies and other sources:

$$RfD \text{ (mg/kg-da)} = \frac{NOAEL}{UF \times MF} \tag{14.6}$$

Unlike equations in physics and chemistry that are based on laws of nature, this equation is simply a convenient definition based on subjective human judgment. In this equation *UF* refers to an uncertainty factor. Typically *UF* increases by a factor of 10 for each of the following conditions that apply:

1.   Extrapolating NOAEL data from animals to humans (interspecies effects).

2.   Extrapolating data from subchronic exposures (ranging from weeks to a few years) to chronic exposures over a lifetime.

3.   Variable responses in the affected population (intraspecies effects, such as different responses for males, females, elderly, pregnant women, asthmatics, and other specific groups).

4.   Lack of NOAEL data, with reliance instead on the *lowest* dose with observed adverse effects (LOAEL).

If all four of these conditions applied, the calculated reference dose (RfD value) would be 10,000 times *smaller* than the lowest dose causing an observed adverse effect. More commonly, however, the uncertainty factor is in the range of 10–1,000.

The factor *MF* in Equation (14.6) refers to a *modifying factor* that represents an additional factor reflecting different professional judgments about the quality and scientific uncertainties of the data used to estimate a particular reference dose. This factor allows different experts or organizations to modify the RfD value for a particular chemical or exposure pathway. Nominally, however, the value of *MF* is 1.0.

Examples of reference dose values for selected chemicals are summarized in Table 14.4. These values form the basis for health risk assessments, as discussed next.

Example 14.6

Calculation of a reference dose.   Animal tests involving exposure to acetone show no observable adverse effects (NOAEL) at or below a dose of 100 mg/kg-da. The data reflect subchronic exposure and indicate some intraspecies variability among the test animals. Determine the reference dose (RfD) for this case, assuming a modifying factor (*MF*) value of 1.0.

Solution:

The uncertainty factor (UF) for this case should reflect extrapolations for interspecies effects (animals to humans), intraspecies effects (among the test animals), plus subchronic to chronic exposure. According to Equation (14.6), each item brings a factor of 10 uncertainty. Thus

$$UF = 10 \times 10 \times 10 = 1,000$$

Then, from Equation (14.6),

$$RfD = \frac{NOAEL}{UF \times MF} = \frac{100 \text{ mg/kg-da}}{(1,000)(1.0)} = 0.1 \text{ mg/kg-da}$$

Table 14.4   Reference dose values for selected chemicals.

Substance	Oral RfD (mg/kg-da)	Inhaled RfD (mg/kg-da)
Metals and Inorganics		
Ammonia	—	2.86×10^{-2}
Arsenic	3×10^{-4}	—
Beryllium	2×10^{-3}	5.7×10^{-6}
Cadmium	1.0×10^{-3} (diet)	6.1
	5.0×10^{-4} (water)	6.1
Chlorine	0.1	5.7×10^{-5}
Chromium VI	3×10^{-3}	2.86×10^{-5}
Hydrogen chloride	—	5.71×10^{-3}
Manganese	1.4×10^{-1} (diet)	1.43×10^{-5}
	4.60×10^{-2} (water)	1.43×10^{-5}
Mercury	—	8.57×10^{-5}
Selenium	5.0×10^{-3}	—
Zinc	3×10^{-1}	—
Organic Compounds		
Carbon disulfide	0.1	0.2
Chloroform	1×10^{-2}	—
Formaldehyde	0.2	—
n-Hexane	6.00×10^{-2}	5.71×10^{-2}
Methanol	5×10^{-1}	—
Methyl ethyl ketone	0.6	0.286
Methylene chloride	6.0×10^{-2}	0.857
Styrene	0.2	0.286
Tetrachloroethylene	1.0×10^{-2}	0.171
Toluene	0.2	0.114
Xylene	2.0	—

Source: ORNL, 2000; USEPA, 1999.

14.6.2 Hazard Quotient

We're almost there. The final step in risk assessments for noncarcinogenic chemicals is to compare the actual or estimated daily intake of the chemical to the reference dose (RfD value) for the applicable exposure pathway. Because the RfD is a conservative estimate of the dosage that produces no observable adverse effect, any dose equal to or less than the RfD does not pose a known health concern. Doses above the RfD may require actions to reduce or eliminate the health risk.

The metric used in risk assessments to compare an actual dose of a chemical to the reference dose is a parameter known as the *hazard quotient* (*HQ*). This is defined as the ratio of the *average daily dose* (*ADD*) of a chemical divided by the reference dose:

$$\text{Hazard quotient (HQ)} = \frac{\left[\begin{array}{c}\text{Average daily dose (ADD)}\\ \text{during exposure period (mg/kg-da)}\end{array}\right]}{\text{Reference dose (RfD)(mg/kg-da)}} \tag{14.7}$$

The average daily dose is similar to the chronic daily intake (CDI) used for carcinogenic risk assessments. The principal difference is that for noncarcinogens the average dose is usually evaluated only for the duration of the exposure, when toxicity is most important, rather than over the 70-year lifetime assumed for cancer risks. The default exposure values in Table 14.2 are typically 25–30 years.

A hazard quotient less than or equal to 1.0 means there are no known adverse effects from the exposure. In other words, the actual exposure is less than the reference dose. Thus, according to EPA guidelines,

$$\text{Acceptable risk for a noncarcinogen} = \text{HQ} \le 1.0 \tag{14.8}$$

In cases where several chemicals are present at the same time, the hazard quotients (HQ values) for the individual chemicals or exposure routes are summed to yield a *hazard index* (*HI*):

$$\begin{aligned}\text{Hazard index (HI)} &= \text{Sum of hazard quotients}\\ &= \sum_{i=1}^{N}(\text{HQ})_i\end{aligned} \tag{14.9}$$

where the index *i* refers to each chemical or exposure route. In this case an acceptable risk for noncarcinogens is a hazard index less than 1.0. If the hazard index is greater than 1.0, there is a potential for a noncarcinogenic effect.

Example 14.7

Health risk from exposure to trace chemicals.   Recent measurements in the well water serving a small community found trace concentrations of arsenic (0.2 μg/L), chloroform (70 μg/L), and toluene (2.5 mg/L). Would these levels trigger a concern about noncarcinogenic health effects?

Solution:

Because the exposure route is via ingestion of water, first find the oral RfD value for each chemical from Table 14.4:

$$\text{Arsenic RfD} = 3 \times 10^{-4} \text{ mg/kg-da}$$

$$\text{Chloroform RfD} = 1 \times 10^{-2} \text{ mg/kg-da}$$

$$\text{Toluene RfD} = 0.2 \text{ mg/kg-da}$$

Next calculate the average daily dose (ADD) of each chemical, assuming a typical water consumption of 2.0 L/da for a 70 kg adult (from Table 14.2). The RfD values for arsenic and chloroform are in units of micrograms, so a conversion factor is needed to obtain milligrams:

$$\text{ADD (arsenic)} = \frac{(0.2 \ \mu g/L)(1 \text{ mg}/1,000 \ \mu g)(2 \text{ L/da})}{70 \text{ kg}} = 5.7 \times 10^{-6} \text{ mg/kg-da}$$

$$\text{ADD (chloroform)} = \frac{(70 \ \mu g/L)(1 \text{ mg}/1,000 \ \mu g)(2 \text{ L/da})}{70 \text{ kg}} = 2.0 \times 10^{-3} \text{ mg/kg-da}$$

$$\text{ADD (toluene)} = \frac{(2.5 \text{ mg/L})(2 \text{ L/da})}{70 \text{ kg}} = 7.1 \times 10^{-2} \text{ mg/kg-da}$$

Divide the ADD value for each chemical by its oral RfD value to determine the hazard quotient for each chemical:

$$\text{HQ (arsenic)} = \frac{5.6 \times 10^{-6} \text{ mg/kg-da}}{3.0 \times 10^{-4} \text{ mg/kg-da}} = 0.019$$

$$\text{HQ (chloroform)} = \frac{2.0 \times 10^{-3} \text{ mg/kg-da}}{1.0 \times 10^{-2} \text{ mg/kg-da}} = 0.20$$

$$\text{HQ (toluene)} = \frac{7.1 \times 10^{-2} \text{ mg/kg-da}}{0.2 \text{ mg/kg-da}} = 0.36$$

None of the individual HQ values exceeds 1.0. However, because multiple chemicals are present, it is necessary to sum the HQ values to determine the overall hazard index:

$$\text{Hazard index} = \sum_{i=1}^{3} (\text{HQ})_i = 0.019 + 0.20 + 0.36 = 0.57$$

The overall hazard index also is less than 1.0, so there is no obvious health concern from these exposure levels.

14.7   LIMITATIONS OF RISK ASSESSMENTS

As we have seen, the principal strength of an environmental risk assessment is that it is a systematic and scientifically based process for characterizing risks from contaminants in the environment. A risk assessment thus can help identify and prioritize the problems of greatest concern. In turn, this helps direct economic and human resources to where they will do the greatest good. A number of standardized procedures have been developed to guide the risk assessment process (for example, USEPA, 1989, 1991, 1997; ASTM, 1998).

Nonetheless, risk assessment procedures invariably require some degree of professional judgment in the use and interpretation of the data and mathematical models that underlie the assessment process. The limitations of risk assessments are reflected in the many uncertainties that characterize each step of the process. The practice of risk assessment thus is far from the "cookbook" procedure it can sometimes appear to be at first glance.

14.7.1 Sources of Uncertainty

Table 14.5 lists some of the sources of uncertainty for the case of waste disposal sites with contaminated soil and/or groundwater (NRC, 1999). These general characteristics also apply to other sources of environmental risk. For waste disposal sites, there is seldom a good record of the wastes that have been buried or spilled at a particular location. Thus uncertainties exist first with regard to the types, amounts, and location of contaminants in place. Even time-consuming and costly measurement programs cannot fully resolve these problems. Sampling programs often are beset by measurement errors and an inability to completely characterize large areas and different types of terrain that might contain toxic chemicals beneath the surface.

Further uncertainties are introduced in modeling the relationship between contaminant sources and the amounts of chemicals that ultimately reach people or eco-

Table 14.5   Sources of uncertainty in risk assessment.

Sources	Contaminant Pathways	Receptors
Lack of information on source location(s)	Unknown pattern of subsurface heterogeneity	Limitations of the dose–response models:
Poorly known history of contaminant releases	Complexities due to natural and anthropogenic stresses	• Extrapolation of hazard and toxicity data
Unknown variability in mass or concentration distributions of contaminants	Inability to define and characterize physical, chemical, and biological fate and transport processes	• Insufficient data to identify hazards or dose–response assessment relationship
		• Model selection
Complexity in the chemical composition of contaminants	Limitations of models of contaminant fate and transport processes	• Parameter estimation for dose–response model
	Difficulties in estimating parameters for contaminant fate and transport models	Problems characterizing exposure and outcome:
		• Identification of toxicants
		• Identification of target population over time
		• Variability in receptors

Source: NRC, 1999.

logical receptors. This *source–receptor relationship* requires an understanding of the physical process by which contaminants are transported through soils, water, and air, as well as the chemical and biological processes by which they are transformed into other species and other chemical phases (such as chemical liquids that vaporize to create airborne pollutants). Although there is a substantial body of scientific and engineering knowledge about these transport and transformation processes, we often need site-specific data to assess a particular situation. Even when such data are available, considerable uncertainties in the source–receptor relationship may remain for complex situations involving multiple exposure pathways and multiple pollutants.

Finally, there are both uncertainty and variability from one individual to another in the way different contaminants affect people and ecological systems. This was mentioned earlier in the discussion of dose–response assessments. For humans, there is uncertainty in the actual exposure to a contaminant and in the actual dose received. For instance, as discussed in Chapter 10, not all contaminants are absorbed by the human body after they are ingested, so the actual dose received may be less than that predicted from simple exposure estimates. There are further uncertainties in the health outcome from a particular dose or exposure. An individual's health, age, genetic makeup, and other factors can affect this outcome. And finally, there are uncertainties in the dose–response relationship for different doses and different time periods ranging from acute to chronic exposure. Especially uncertain is the toxic potential of chemicals that have not been well studied, for which metrics like potency factor and reference dose may not exist.

All of these uncertainties apply as well to ecological receptors, which encompass the full spectrum of biological species. Indeed, that enormous diversity makes the process of ecological risk assessment far more complex than for humans. Although still in its infancy, some progress in ecological risk assessment is being made through site-specific studies. Still, relatively little is currently known about the response of different types of organisms to different chemical exposures, or about the impact of contaminants on the viability and integrity of whole ecosystems.

14.7.2   Dealing with Uncertainty

The practice of environmental risk assessment is beginning to include uncertainty as an explicit component of the analytical framework. Traditionally, risk analysis has relied on various "safety factors" and the use of conservative (worst-case) assumptions to account for known or suspected uncertainties. Examples include the uncertainty factor (*UF*) and modifying factor (*MF*) parameters used in Equation (14.6). This approach nonetheless results in a deterministic (single-valued) estimate of risk, such as calculated in Example 14.3. One limitation of this approach is that it conveys little or no information about the *confidence* in the resulting risk estimate, or the likelihood of a higher or lower risk based on current knowledge and data.

More advanced analytical methods include *probabilistic risk assessments* that present risk results as a probability distribution rather than a single value. An example of a probability distribution related to health risks was described in Chapter 10 for the case of lead poisoning from environmental exposure.

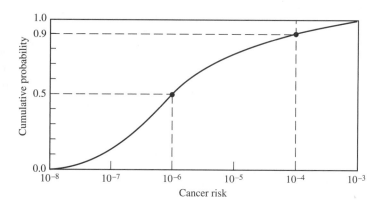

Figure 14.6   Example of a probabilistic risk result showing the cumulative probability (on the y-axis) that the cancer risk will be equal to or less than the value indicated on the x-axis.

An advantage of a probabilistic analysis is that it conveys a quantitative measure of the confidence or conservatism in a risk estimate. Figure 14.6 illustrates such a case, in which risk is expressed as a curve known as a *cumulative distribution function*. In Figure 14.6, the estimate of risk corresponding to the 50th percentile value means that there is a 50–50 chance that the risk may be higher or lower than the corresponding value shown on the x-axis (which in this case is a risk of 10^{-6}). However, Figure 14.6 shows that there is also a 10 percent probability that the risk could exceed 10^{-4}, which would exceed the EPA's range of acceptable risk for a chemical carcinogen. This type of probabilistic information can be especially useful to analysts and decision makers who must judge the implications of risk analysis results in the context of a specific situation. In this case, they would have to determine whether a 10 percent probability of a higher risk would justify further study or remedial actions to reduce the risk under these particular circumstances.

Probabilistic procedures also can identify which of many factors contribute most to the overall uncertainty. This information can help target ways of reducing key uncertainties, such as by conducting specific measurement and monitoring programs. A disadvantage of probabilistic analysis is that it is more time-consuming and data-intensive than the traditional deterministic approach. But depending on the stakes involved in a particular case (in terms of community health risks, potential cleanup costs, and other factors), the additional effort may nonetheless be warranted. Recent trends suggest that probabilistic analysis is becoming a more common component of environmental risk assessments (NRC, 1999).

14.8   APPROACHES TO RISK MANAGEMENT

What happens when the level of risk revealed by a risk assessment exceeds an acceptable level? The term *risk management* is used to describe the process for defining an acceptable risk in the context of a particular situation, and for deciding on the appropriate action to reduce, control, or eliminate an unacceptable risk.

The options for dealing with unacceptable risks fall into four general categories, as illustrated in Figure 14.7:

1. The source of the risk can be reduced or eliminated, such as by removing contaminated soil, closing a facility, or installing environmental control technology to reduce emissions.

2. The exposure pathway can be modified or avoided, such as by installing an engineered barrier that prevents contaminant migration through the soil, or a tall chimney that disperses pollutants beyond the local community.

3. Human exposure to the contaminants can be reduced or eliminated, such as by relocating the affected population or prohibiting access to a contaminated area.

4. In the least desirable option, effects can be treated or compensated for after they occur, such as by medical treatment or monetary payments from parties responsible for the contamination.

14.8.1   Defining Goals and Procedures

To date, the greatest body of experience in environmental risk management lies in dealing with the tens of thousands of hazardous waste disposal sites across the United States, including many leaking underground storage tanks containing fuels and chemicals. The most severely contaminated of these sites are listed by the EPA

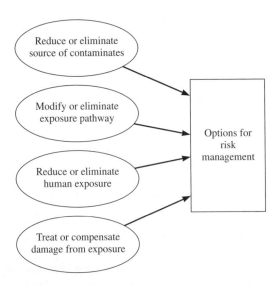

Figure 14.7   Schematic of alternative risk management options.

as the highest priority for cleanup under the federal program known as Superfund. An army of environmental consulting and remediation firms has been tackling this problem since the mid-1980s in response to federal and state cleanup requirements (see Chapter 2).

In many cases, however, the cleanup goals initially adopted by the EPA for contaminated groundwater and soils have proved unworkable. Those goals often required reducing the concentration of waterborne species to their maximum contaminant levels (MCLs, as defined in Table 2.5) and cleaning the soil to a level suitable for unrestricted use. In some cases the technological capability to achieve such goals simply does not exist. In other cases the cost of cleanup measures has been extremely high and the rate of progress very slow. Some Superfund sites have taken more than a decade to remediate. And in some cases an environmental cleanup measure might actually cause more damage than it avoids. Dredging contaminated river sediments, for example, may disperse the contaminants more widely. For this reason, the state of New York has been cautious about dredging PCBs from sediments in the Hudson River (see Chapter 9).

As with most environmental problems, we inevitably must balance the need to protect public health and the environment against the capabilities of economic and technological resources. The introduction of risk-based methods for environmental management is widely viewed as an important step in trying to achieve that balance. Elaborate procedures to guide the risk management process at sites with contaminated soil or groundwater have been developed by the EPA, state and local agencies, and professional societies such as the American Society for Testing and Materials (ASTM). Figure 14.8 shows the sequential steps and procedures for dealing with a contaminated site, as defined by the EPA.

In general, the goals of risk management are to identify the level of risk reduction afforded by different remedies and to choose the preferred alternative in light of the cost and feasibility of different solutions. The actual or intended use of a particular site is an important factor in this process because it can affect the criteria for acceptable risk or cleanup goals. But it is not just environmental professionals who are involved in these decisions. Risk management also must involve interactions with the affected communities and citizens who are likely to be most affected by decisions that are taken. In this regard, a variety of outreach activities, risk communication activities, and the formation of citizens' advisory panels also are important elements of sound risk management practices.

14.8.2 Finding Workable Solutions

A general criticism of risk-based approaches to environmental management is that they are more likely to result in a decision to leave contaminants in place (at some low level of acceptable risk) compared to the more conservative approach (historically preferred by the EPA) of treating and/or removing *any* contaminants found at a contaminated site. Furthermore, risk-based remedies such as engineered containments, or "institutional controls" such as a fenced-off area to prevent access to a contaminated site, may not necessarily insure against future exposure. Nor do such

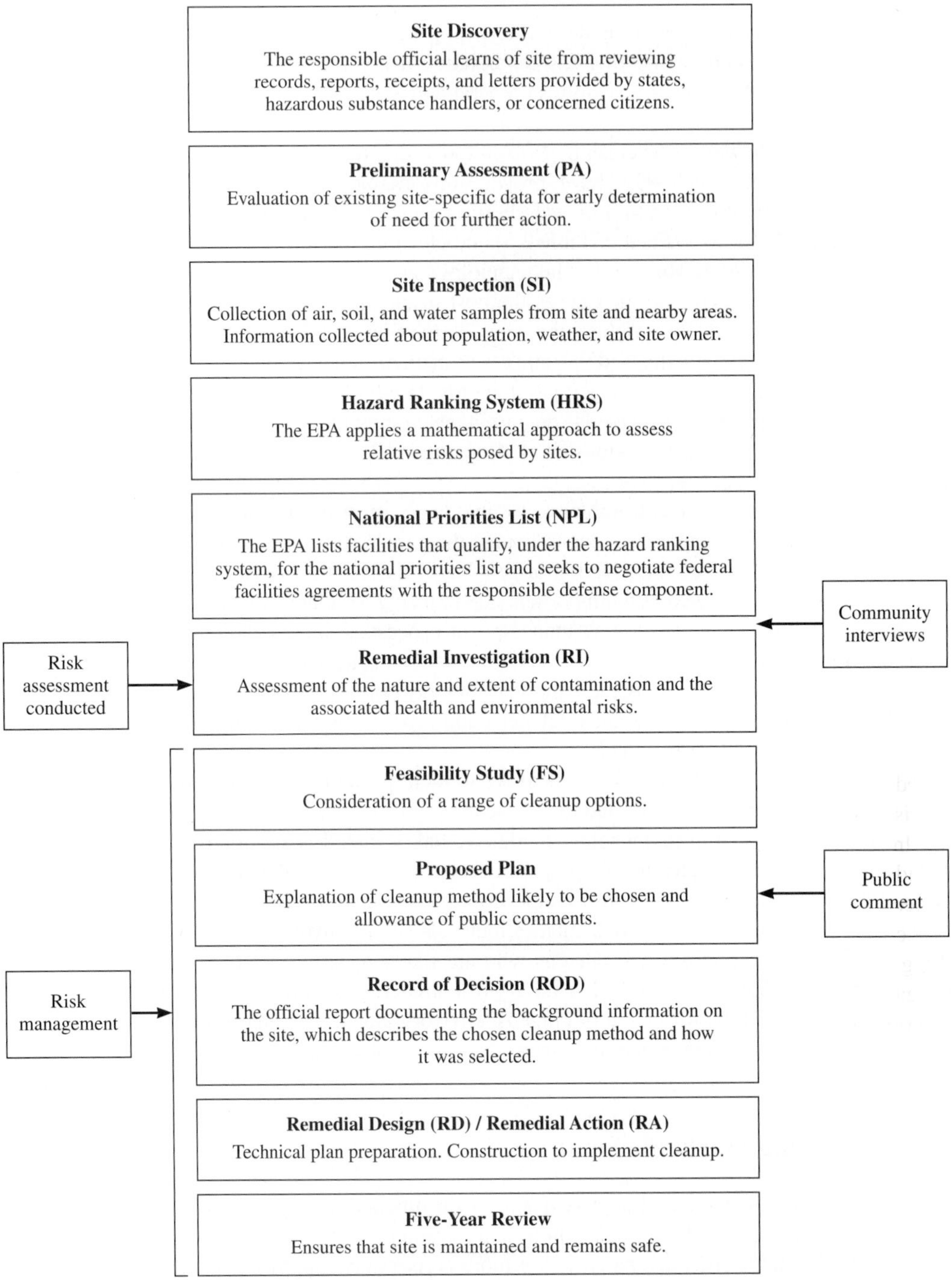

Site Discovery

The responsible official learns of site from reviewing records, reports, receipts, and letters provided by states, hazardous substance handlers, or concerned citizens.

Preliminary Assessment (PA)

Evaluation of existing site-specific data for early determination of need for further action.

Site Inspection (SI)

Collection of air, soil, and water samples from site and nearby areas. Information collected about population, weather, and site owner.

Hazard Ranking System (HRS)

The EPA applies a mathematical approach to assess relative risks posed by sites.

National Priorities List (NPL)

The EPA lists facilities that qualify, under the hazard ranking system, for the national priorities list and seeks to negotiate federal facilities agreements with the responsible defense component.

Community interviews

Risk assessment conducted

Remedial Investigation (RI)

Assessment of the nature and extent of contamination and the associated health and environmental risks.

Feasibility Study (FS)

Consideration of a range of cleanup options.

Proposed Plan

Explanation of cleanup method likely to be chosen and allowance of public comments.

Public comment

Risk management

Record of Decision (ROD)

The official report documenting the background information on the site, which describes the chosen cleanup method and how it was selected.

Remedial Design (RD) / Remedial Action (RA)

Technical plan preparation. Construction to implement cleanup.

Five-Year Review

Ensures that site is maintained and remains safe.

Figure 14.8 Steps in the risk assessment and risk management process for a contaminated site under the Comprehensive Environmental Response, Compensation, and Liability Act (CERCLA), commonly known as Superfund. (*Source:* USEPA, 1992).

Soil sampling at a toxic waste dump.

remedies insure against unforeseen risks from a substance that was overlooked during risk assessment.

In response to such criticism, risk management procedures have been refined to include more extensive follow-up monitoring requirements, as well as more sophisticated analysis methods—such as uncertainty analysis—to better anticipate unforeseen consequences. Such methods are still evolving as governmental agencies, citizen groups, and private organizations seek more effective ways to prioritize and mitigate the risks of chemicals in the environment. The "brownfield" development program described in Chapter 2 (Section 2.5) is one recent effort to improve on this process. Here a more flexible approach is being sought for the cleanup and development of former industrial sites. Other references listed at the end of this chapter provide more detailed discussion of modern risk management methods and their application to environmental problem solving.

14.9 INTRODUCTION TO DECISION ANALYSIS

Engineering students are sometimes surprised to learn that there is a field of study called *decision analysis.* After all, people make decisions all the time, so what's the big deal?

In many areas of human endeavor—including environmental management— deciding what is the best or most appropriate course of action can be extremely difficult. In the case of a contaminated waste disposal site there may be over a dozen different options that could be pursued. Such options could range from doing nothing to a variety of technological and/or institutional responses. Which of these actions should be selected?

Even more complex decisions are faced by environmental policymakers. For instance, what governmental actions, if any, should be taken to avoid the potential future threats of global warming? Or of nuclear waste disposal? Within the private sector, should Company A invest its resources in developing a new control technology for hazardous waste disposal? Or should Company B attempt to develop a more environmentally friendly automobile that exceeds current requirements?

Such decisions traditionally are made by corporate executives and government officials who weigh the apparent pros and cons to ultimately reach a decision. But in many cases the process of merely identifying all the factors that affect a particular decision can be difficult. In complex situations, sorting out the options and the consequences of alternative decisions requires a careful, systematic approach if decisions are to be based on the best available information.

This is where decision analysis comes in. It is a tool not only for helping to *make* decisions, but also for *structuring and identifying information* relevant to the decision at hand. This chapter briefly introduces this topic to illustrate its application to environmental problem solving. In particular, two techniques from the field of decision analysis are highlighted: influence diagrams and decision trees.

14.10   INFLUENCE DIAGRAMS

There is an old saying that a picture is worth a thousand words. Thus our ancestors knew what modern social science research has confirmed—namely, that a graphic or pictorial representation of a situation is often easier to comprehend than a written or verbal description. The graphic interface on your computer screen today is but one modern manifestation of this type of communication tool.

An *influence diagram* is a way of visualizing the important connections among different elements of a problem. In its simplest form, the diagram consists of a series of ovals representing the key variables that can influence a decision or some outcome of interest. For example, you might be interested in your final grade for a course you're taking. Your instructor has announced that the grade will be based on your scores on class exams, homework problems, and the final exam, plus your class attendance and participation. Thus a simple influence diagram might look like Figure 14.9(a).

More detail can be added by recognizing that your class attendance is likely to influence your understanding of the course material and hence your performance on homework and exams. This leads to the diagram in Figure 14.9(b). Stepping back even further, as in Figure 14.9(c), your class attendance and participation could depend on how much (or how little) sleep you get, as well as your general interest in the subject. The latter, in turn, may be influenced by the quality of the lectures and textbook, as well as other factors.

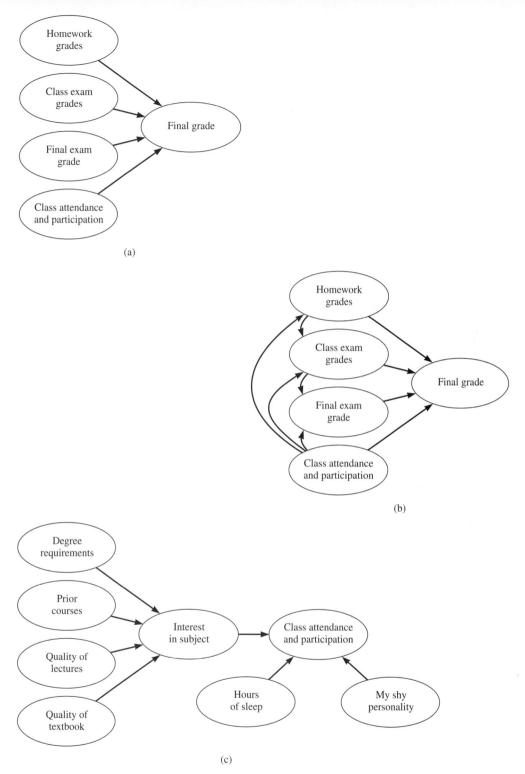

(a)

(b)

(c)

Figure 14.9   An influence diagram for your final grade in a course. The sketch shows increasing levels of refinement, including (a) identification of major factors, (b) their interactions, and (c) examples of other influences on one of the key factors.

14.10.1   Symbols and Conventions

As the problem is *decomposed* into more and more elements like this, one begins to see how different factors interact and how different actions, assumptions, or decisions can affect the outcome of interest.

In the formal development of an influence diagram, symbols with different shapes represent different types of elements. Figure 14.10 shows the four common shapes that are used. *Ovals,* such as those shown in Figure 14.9, represent *chance events,* which are factors that are variable or whose value is uncertain (such as your scores on homework and exams). *Rectangles* represent *decision nodes.* Decisions typically influence—and are influenced by—other factors, as illustrated in Figure 14.11.

Influence diagrams also can include nodes that represent *math calculations or constant values.* The symbol used here is a *rectangle with rounded corners.* Figure 14.12 illustrates a calculated quantity (the average grade) that is determined by the scores on three separate class exams. Although the value of each exam score is uncertain prior to the exam (and so is shown by an oval), the average value is calculated by the known formula of summing the three scores (whatever they turn out to be) and dividing by 3. Another symbol sometimes used in influence diagrams is a *hexagon.* This is used to display an *objective* that one is trying to achieve or minimize or maximize, such as minimum cost, minimum risk, or maximum benefits.

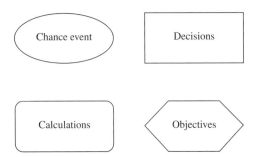

Figure 14.10   Symbolic shapes used in influence diagrams.

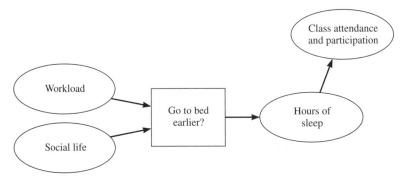

Figure 14.11   Example of an influence diagram decision node.

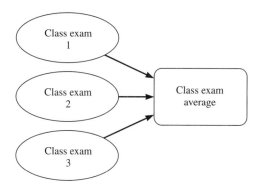

Figure 14.12 Example of an influence diagram calculation node.

For the purposes of this chapter, the detailed procedures and symbols used to draw an influence diagram are less important than the fact that such diagrams can be extremely useful for understanding environmental problems and the factors that influence environmental control decisions. The following example illustrates such applications.

14.10.2 An Environmental Example

Figure 14.13 shows an example of a simplified influence diagram related to the problem of global warming (the subject of Chapter 12). At a *qualitative* level, the influence diagram reveals the linkages among major factors that determine the atmospheric concentration of carbon dioxide (CO_2), the main greenhouse gas accumulating in the atmosphere. In this case, the diagram displays the link between human activities that require energy (like transportation, buildings, and industrial processes) and the resulting buildup of CO_2 in the atmosphere from the burning of fossil fuels to supply energy. The diagram also demonstrates the connection between land use activities like deforestation and the reduced uptake of CO_2 for photosynthesis.

Influence diagrams can further be used as the basis for *quantitative* analysis and model building. Thus the quantity of CO_2 produced each year can be calculated by specifying the types and amounts of fossil fuels that are used to supply the energy needs depicted in Figure 14.13. Other quantitative relationships can then be used to calculate the buildup of CO_2 in the atmosphere (see Chapter 12 for details). These relationships are indicated by calculation nodes.

Decision nodes also could be added to display and analyze various intervention options. For instance, a policy decision to change the fuels used by power plants and automobiles would directly influence the magnitude of CO_2 emissions and the resulting CO_2 concentration in the atmosphere. Indirectly, such a policy decision also could influence economic activity as fuels and technology are changed. A more elaborate influence diagram would depict these interactions.

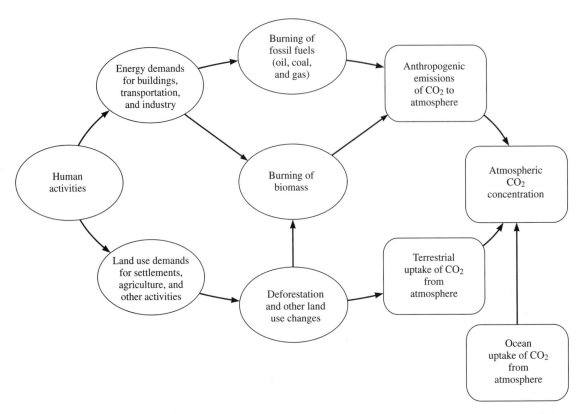

Figure 14.13   A simplified influence diagram for the buildup of CO_2 in the atmosphere (a major contributor to global warming). The climate change resulting from increased atmospheric CO_2 concentrations will feed back to, and influence, a number of the processes shown in this diagram.

A number of commercial software products have been developed to support decision analysis and quantitative model building. Figure 14.14 shows an example using a program that displays the influence diagram that underlies the calculations of a particular model. A hypertext format allows a user to explore the model structure at different levels (or layers) of detail by simply clicking on an element of the influence diagram. This capability can be extremely helpful in understanding the structure of complex environmental problems and decisions.

14.10.3   Further Applications

As you can see from the preceding examples, there is no single right answer as to the correct or most appropriate influence diagram for a particular problem. Rather, influence diagrams are a generic tool that can help structure, explain, and analyze complex problems and decisions. The principal challenge is to *recognize* and include

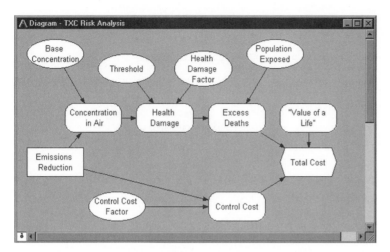

Figure 14.14   An example of a computer-generated influence diagram. This illustration shows a cost–benefit analysis of reducing emissions of a toxic air pollutant. A software package called Analytica displays the interactions among all variables of a model to facilitate understanding and communication.

all key variables and their relationships; failure to do so can erroneously represent a problem and lead to incorrect conclusions.

Influence diagrams are sufficiently intuitive that we will leave it to readers (and their instructors) to further explore their use in describing environmental problems and analyzing proposed solutions. The problems at the end of this chapter also include examples requiring influence diagrams. More detailed treatments of this subject, including advanced concepts and specific rules for constructing influence diagrams, can be found in more advanced references (such as Clemen, 1996; Golub, 1997).

14.11   DECISION TREES

Another useful graphic tool for decision analysis is the *decision tree.* This tool is designed to highlight the ramifications of alternative decisions and uncertain events. Like a tree, it continues to branch out as it grows. As a tool for gaining insight into a problem, a decision tree can be used both qualitatively and quantitatively, just like an influence diagram. For a quantitative analysis, the tree can be "solved" to iden-tify the optimal decision (or series of decisions) for a particular problem.

14.11.1   Building a Decision Tree

The building blocks of a decision tree are based on two types of nodes: a *decision node* represented by a small square and a *chance node* represented by a small circle.

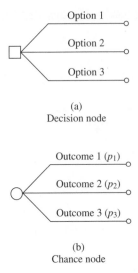

(a)
Decision node

(b)
Chance node

Figure 14.15   Decision tree nodes showing (a) a decision node with three options and (b) a chance node with three possible outcomes. The probability of each outcome is shown as p_1, p_2, and p_3, respectively.

At a decision node we must choose which branch to follow. Each branch from the decision node represents a discrete choice, as illustrated in Figure 14.15(a).

At a chance node there is uncertainty about the outcome of a future event. In this case each branch from the node represents one of several possible outcomes covering the complete spectrum of possibilities. A probability (usually expressed as a fraction) is assigned to each of these possible outcomes, as illustrated in Figure 14.15(b).

A decision tree commonly begins with a node that has two or more branches, such as a decision between doing this or that, or a choice of yes and no. The purpose of the decision tree is to explore the consequences of choosing either branch. Typically, each branch results in some uncertainty about what will happen next. That uncertainty is represented by a chance node. Drawing the decision tree forces you to first think through the problem systematically; this process alone is often very helpful. In a quantitative analysis the tree can be used to identify the best decision based on some specified measure or metric, such as achieving the lowest risk, the minimum cost, or the maximum profit. An example best illustrates the development and use of a decision tree.

Example 14.8

Decision tree for an industrial site selection.   Your company wants to buy and develop a site to build a new assembly plant for its product. Two available properties would serve equally well. The main difference is cost, with Site A costing $1,000,000 whereas Site B costs $600,000. Site A is known to be uncontaminated. However, Site B may have some environmental con-

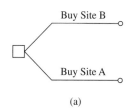

(a)

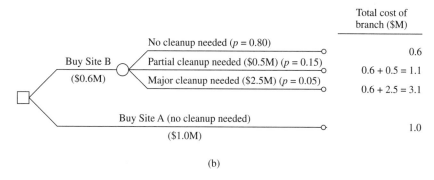

(b)

Figure 14.16   Decision tree for Example 14.8, showing (a) the basic decision node and (b) the expanded decision tree.

taminants that would require cleanup. You are assigned to further analyze the problem and recommend which property to buy *based purely on the lowest expected cost.*

Background research revealed that an old factory at the less expensive location (Site B) produced leaded paint at that site up until 50 years ago, when the factory was torn down. The property has been vacant ever since. Your company hired a well-known environmental consulting firm to do a preliminary assessment of whether the site might be contaminated or pose any health risk to workers, visitors, or nearby residents. Based on their data and experience at other sites, the consulting firm developed a mathematical model to predict the likelihood of environmental risks and potential cleanup costs. On the basis of this mathematical model, the firm concluded that there is an 80 percent chance the property is clean and free of any risk. However, there is a 15 percent chance the land is partially contaminated and would require a cleanup effort costing $500,000. There is also a 5 percent chance the property is highly contaminated and would require a $2,500,000 cleanup program.

Draw a decision tree for this problem based on the information given.

Solution:

The basic decision tree for this problem is shown in Figure 14.16(a). The decision involves choosing between Site A and Site B, with minimum expected cost as the only criterion for selection. Thus there is one decision node with two branches. Based on the information provided, a decision to buy Site B could incur costs beyond the initial purchase price if the site proves contaminated. These additional factors are shown in Figure 14.16(b). A chance node shows the three possible outcomes corresponding to the consulting firm's preliminary assessment. Site A in this example is free of any environmental concerns.

Notice that the decision tree shows the probability for each branch of the chance node, with the sum of all branches totaling 1.0. The dollar value or cost associated with each outcome also is shown at the far right of the figure. For Site A this is just the $1 million purchase price. For Site B the value includes the initial site purchase price of $600,000, plus the cost of any cleanup measures that may be required according to the preliminary estimate.

14.11.2   Solving a Decision Tree

Now that we have a decision tree, how does it help solve the problem of what decision to make? To use or "solve" a decision tree, we work from right to left, focusing first on the chance nodes of the problem. The idea is to collapse the branches of each chance node into a single branch that represents the *expected value* of the uncertain outcomes. This is a process known as *folding back the tree*.

The concept of an expected value comes from statistics and probability theory. In essence, it reflects the average or best-guess outcome that would be expected if an experiment (such as a lottery) could be repeated many times with the given probabilities for each branch of the chance node. Thus, if we have assigned (or estimated) the probability (p_i) for each branch of the node and know the economic cost or value associated with each branch (V_i), the expected value (EV) at any chance node is simply the weighted average across all branches of the node, with the weights being the probability assigned to each branch. If the probability of each branch is expressed as a fraction, we have

$$\begin{bmatrix} \text{Expected value} \\ \text{at a chance node} \end{bmatrix} = \text{Sum of} \left[\begin{pmatrix} \text{Probability of} \\ \text{each branch} \end{pmatrix} \times \begin{pmatrix} \text{Value of} \\ \text{each branch} \end{pmatrix} \right]$$

or

$$EV = \sum_i p_i V_i \qquad (14.10)$$

The single expected value is then assigned to the chance node in the decision tree. To illustrate such a calculation, we return to the previous example.

Example 14.9

Expected value at a chance node.   Calculate the expected value of the chance node associated with purchasing Site B in Example 14.8. Then redraw the folded decision tree.

Solution:

For this problem there are three branches with the probabilities and costs shown in Figure 14.16(b). The expected value at the chance node is therefore

$$EV = p_1 V_1 + p_2 V_2 + p_3 V_3$$
$$= (0.80)(\$0) + (0.15)(\$500,000) + (0.05)(\$2,500,000)$$
$$= \$200,000$$

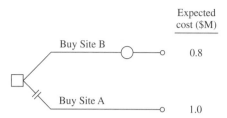

Figure 14.17   Example 14.9 decision tree after calculating the expected value of the chance node.

This is the expected value of environmental cleanup costs for Site B. It gives a measure of what the average cost would be if the purchasing decision could be made over and over again with the given probabilities of having to actually pay nothing, $0.5 million, or $2.5 million for cleanup.

Figure 14.17 shows the folded back decision tree incorporating the expected value calculation for Site B. Based on that calculation, Site B has a total expected cost of $800,000 ($600,000 purchase plus $200,000 expected cleanup cost) compared to $1 million for Site A.

In Example 14.9, a decision based solely on the lowest expected cost would reject Site A and select Site B. A convention used in solving decision trees is to draw double-slashed lines through an option not chosen, as shown in Figure 14.17. Note that the use of expected value as the decision criterion represents "going with the odds" as to the likely environmental cleanup costs at Site B. A little later in the chapter we will consider the added dimension of risk aversion, which also can affect a decision.

14.11.3   Adding Complexity

The previous example involved only one decision node and one chance node. Decision problems often are more complex. In such cases the decision tree will include additional nodes and branches, such as depicted in the next example.

Example 14.10

The site selection problem revisited.   Suppose the environmental consulting firm that did the preliminary evaluation of Site B in the previous example offers you the option of a more detailed site investigation. For an additional $100,000 they will survey the site and sample the soil at several points around the property. Their experts say this procedure will yield an 85 percent chance of knowing conclusively whether the site is safe or in need of either partial or major cleanup. However, there remains a 15 percent chance that the deeper investigation will be no more conclusive than the preliminary study based on the consultant's computer model. Draw a decision tree for this revised problem and calculate the probabilities and costs associated with the new chance nodes. Use information from Example 14.8 for any additional assumptions that may be needed.

Solution:

The option of a more detailed study of Site B adds a new branch to the initial decision node, as depicted in Figure 14.18(a).

A decision to further investigate Site B will incur a certain cost of $100,000. The outcome will be an 85 percent chance of knowing definitively whether the site requires partial or major cleanup. Based on the preliminary investigation (Example 14.8) there was a 20 percent chance the site would require some level of cleanup and an 80 percent chance it was safe with no cleanup required. Thus the more detailed study is likely to produce the following results:

(85% conclusive) × (20% chance of cleanup) = 17% cleanup definitely needed

(85% conclusive) × (80% chance of no cleanup) = 68% definitely no cleanup

In addition, there remains a 15 percent chance that the results of the deeper investigation will still be inconclusive. After incorporating the results of the preceding calculations, the new chance node on the decision tree looks like Figure 14.18(b).

If cleanup of Site B is definitely needed, the preliminary study indicated that partial cleanup was three times more likely than major cleanup. So the probabilities for partial versus major cleanup are 75 percent and 25 percent, respectively. But in this case, if *any* cleanup is needed, that would eliminate Site B from consideration because the minimum cleanup cost of $500,000, in addition to the $600,000 purchase price, would make Site B more expensive than Site A.

On the other hand, if there is definitely no cleanup needed at Site B, the choice between Sites A and B clearly favors Site B. Its overall cost would now be $700,000 (the purchase price plus the $100,000 study cost) compared to $1 million for Site A.

What happens if the results of further study are inconclusive? We are then faced with the original decision posed in Example 14.8—that is, a choice between Sites A and B based only on results of the preliminary assessment. Thus another decision node is added as shown in Figure 14.18(c). At this node, a decision to buy Site B faces the same uncertainties depicted previously in Example 14.8. A new chance node is therefore added, as shown in Figure 14.18(c). This completes the additions to the previous decision tree.

Figure 14.19 shows the final decision tree for Example 14.10 in its full glory. Included are three final decision nodes omitted from Figure 14.18(c) for clarity. At the far right is the total cost associated with each branch of the tree. To find the site selection decision that gives the lowest expected cost, we would again solve the tree by working from right to left, replacing each *chance* node with its expected value, as in Example 14.9. Then we fold back the tree by choosing the branch of each decision node with the lowest expected cost. As we eliminate branches from consideration, we draw double lines through them, as depicted earlier in Figure 14.17. When the tree is completely folded back, the final decision node reveals the option having the lowest expected cost overall.

We leave the final solution of this particular problem as an exercise for students in the problems section at the end of the chapter. But if you absolutely must know the answer now, take a peek at the footnote.[1]

---

1 The decision with the lowest expected cost is to forgo the more detailed study of Site B and purchase Site B based on the consultant's initial assessment. In other words, given the particular odds and the cost of a more detailed study, the decision maker is likely to end up better off by just planning to spend money on cleanup if it turns out to be necessary, rather than trying to predict whether the expense will be required.

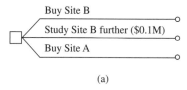

(a)

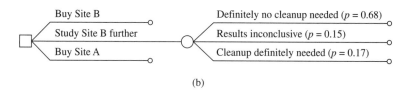

(b)

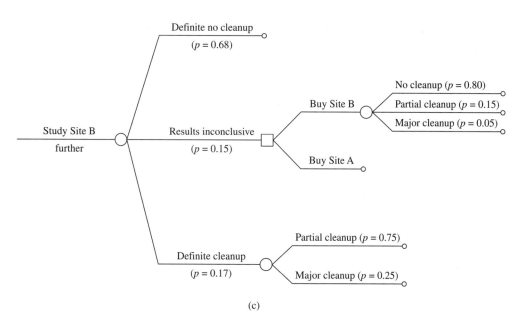

(c)

Figure 14.18   Revised decision tree for Example 14.10, showing (a) the basic decision node, (b) the additional chance node for the "study Site B" option, and (c) the expanded tree for the "study" option.

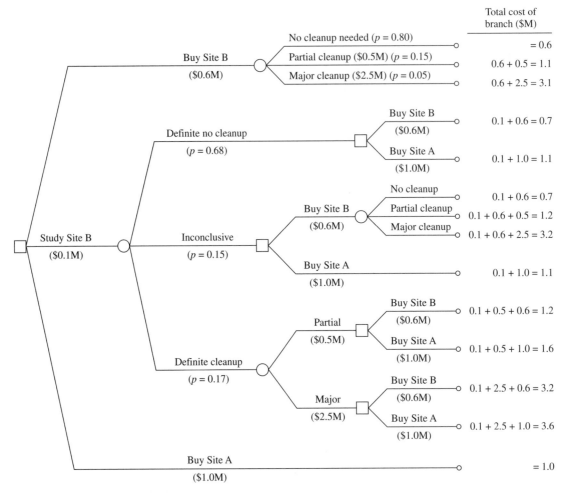

Figure 14.19   Full decision tree for Example 14.10, showing all decision nodes and chance nodes, along with the total cost of each branch.

14.12   CONCLUSION

The goal of this chapter was to briefly introduce the topics of risk assessment and decision analysis as tools for environmental problem solving. As with any new subject, there is much more depth that one can pursue. As environmental issues become increasingly complex, skill in dealing with the subjects of risk assessment and decision analysis become more important.

The treatment of risk assessment was limited here to a brief overview of the methods commonly used to quantify human health risks from exposure to hazardous chemicals. However, the details of those risk assessment methods and the science behind them remain subjects for further study, as are the topics of uncertainty analysis and probabilistic risk assessment. More advanced treatments of these subjects are found in books such as Morgan and Henrion (1990).

In addition, the topic of *risk perception* can be important in dealing with environmental issues, especially as it relates to the development of public policy. As we noted at the beginning of this chapter, research has shown that people often care about factors besides the probability and consequences associated with a given risk. In such cases peoples' subjective assessments of how risky something is do not correspond with the measured probability or with the product of actual probability and consequence. In addition, people often have incorrect impressions about the true probabilities and consequences of different risks. For both these reasons, public demands for action sometimes do not correspond to what one would expect based on the actual data for probability and consequences. Understanding the role of public perception, and the factors that influence it, can therefore be extremely helpful in developing policies and solutions that are most appropriate and that make the best use of available resources.

The introduction to decision analysis similarly skimmed the surface of a broad field of study. More advanced treatments of decision analysis would include two important topics that were not addressed in this chapter. One is the question of how to analyze decisions that are based on more than one criterion or *attribute*. For instance, in the site selection problem used to illustrate decision trees, it was assumed that the decision between Site A and Site B was based only on the expected cost of each site. All other factors were assumed to be equal and therefore not relevant to the decision. In practice this is seldom the case. Other site attributes such as proximity to labor markets, transportation routes, and regional cost of living are likely to differ from one location to another, and these factors too will affect the final choice of a site. Attributes that are not easily translated into monetary terms also can be important for a decision. Often these attributes reflect the personal values or preferences of the decision maker. The president of the company, for example, might prefer to live in the urban community near Site A rather than the suburban neighborhoods of Site B.

Frequently trade-offs must be made among these multiple attributes of a decision problem. Analytical methods for *multiattribute decision making* have been developed to help people structure, think through, and resolve these types of complex multifaceted decisions. In the environmental domain, such methods have been applied to problems such as facility siting decisions (like where to build a new airport, factory, or power plant) and the selection of remedies for hazardous waste management.

As you know from your own experience, some people are more cautious than others when it comes to making important decisions. People who are *risk averse* are less willing to take chances even when the odds are in their favor. On the other hand, some people who are *risk seekers* are very comfortable taking substantial risks. The perspective reflected in the expected value calculations we used to solve decision trees can be characterized as *risk neutral*. This implies a willingness to accept the most probable outcome of an uncertain event.

More advanced methods of decision analysis, however, are able to identify and incorporate the actual risk profile of an individual decision maker faced with a complex decision. Discussions of such methods and their application can be found in several references listed at the end of this chapter. Findings indicate that most companies are willing to be risk neutral when the risks or potential losses are small compared with their resources. However, they become more risk averse when the loss could involve "betting the company." The same has been found to be true for nations and individuals.

14.13   REFERENCES

ASTM, 1998. *Standard Provisional Guide for Risk-Based Corrective Action,* PS 104-98, American Society for Testing and Materials, West Conshohocken, PA.

Clemen, R. T., 1996. *Making Hard Decisions: An Introduction to Decision Analysis,* 2nd ed., PWS-Kent Publishing Co., Boston, MA.

Golub, A. L., 1997. *Decision Analysis: An Integrated Approach.* John Wiley & Son, New York, NY.

Morgan, M. G., and M. Henrion, 1990. *Uncertainty: A Guide to Dealing with Uncertainty in Quantitative Policy Analysis,* 3rd printing. Cambridge University Press, Cambridge.

NRC, 1983. *Environmental Risk Assessment in the Federal Government: Managing the Process.* National Academy Press, Washington, DC.

NRC, 1999. *Cleanup at Navy Facilities: Risk-Based Methods.* National Research Council, National Academy Press, Washington, DC.

ORNL, 2000. *Risk Assessment Information System.* Toxicity and Chemical-Specific Data Base, Oak Ridge National Laboratory, Oak Ridge, TN. http://risk.lsd.ornl.gov/cgi-bin/tox/TOX_9801.

USEPA, 1989. *Risk Assessment Guidance for Superfund, Volume 1, Human Health Evaluation Manual (Part A).* EPA/540/1-89/002, U.S. Environmental Protection Agency, Washington, DC.

USEPA, 1991. "Human Health Evaluation Manual." OSWER Directive 9285.6-03, Office of Solid Waste and Emergency Response, U.S. Environmental Protection Agency, Washington, DC.

USEPA, 1992. *Understanding Superfund Risk Assessment.* USEPA 9285.7-06FS, U.S. Environmental Protection Agency, Washington, DC.

USEPA, 1997. *Ecological Risk Assessment Guidance for Superfund: Process for Designing and Conducting Ecological Risk Assessments.* EPA/540-R-97-006, U.S. Environmental Protection Agency, Washington, DC.

USEPA, 1999. *Integrated Risk Information System (IRIS) Database.* U.S. Environmental Protection Agency, Washington, DC. http://www.epa.gov/iris.

USHHS, 1988. *Vital Statistics for the United States, 1986.* PHS88-1122, U.S. Department of Health and Human Services, Washington, DC.

14.14   PROBLEMS

14.1   Table 14.1 showed a lifetime risk of death from cancer of 0.23.

(a)   Based on those data, what cause of death would be 10 times less likely? 100 times less likely? 1,000 times less likely?

(b)   What is the lifetime risk of dying from any cause?

14.2 Conduct a small experiment in risk perception. List the causes of death in Table 14.1 in alphabetical order. At the top of the list write the following statement: "This list shows various causes of death in the United States listed in alphabetical order. Circle any items for which you think there is less than a one in a million chance of dying from that cause." Obtain responses from at least 10 people (excluding other students in this class). All responses should be anonymous. Compare the responses to the values in Table 14.1 and prepare a brief report summarizing the results of your survey.

14.3 A family consists of an adult male weighing 70 kg, an adult female weighing 50 kg, and a child weighing 15 kg. The air they breathe contains 1 μg/m³ of a listed carcinogenic chemical, and the water they drink contains 1 μg/L of the same chemical.

(a) Use the EPA default parameters in Table 14.2 to estimate the total chemical dose per kg of body weight for each family member. Assume the body retains 80 percent of the air pollutant and 100 percent of the water pollutant.

(b) Which person receives the highest average dose?

(c) Which exposure route is most important for this chemical?

(d) Comment as to whether you found the information in Table 14.2 adequate for this problem. If not, explain why.

14.4 Suppose the family members in the previous problem also ingest the amounts of soil listed in Table 14.2. If the soil contains 10 parts per million (0.001 percent by weight) of the carcinogenic chemical, what are the new values of the total daily dose per kg of body weight of the chemical for the adult family members? What are the new values for the relative importance of each exposure route?

14.5 Assume the chronic daily intake of a carcinogenic chemical is equal to the value found in Problem 14.4 for the adult male weighing 70 kg. Using the potency factor values from Table 14.3, find his incremental lifetime cancer risk if the chemical of concern is

(a) Arsenic.

(b) Trichloroethylene (TCE).

How do these risks compare to the EPA guideline of 1×10^{-6}? Give your answer as a ratio or multiple (such as 150 times greater, or smaller, than 10^{-6}).

14.6 Chapter 9 discussed the problem of polychlorinated biphenyls (PCBs) in U.S. surface waters. Suppose the drinking water supplied to a region of 10 million people had a PCB level equal to the EPA's maximum contaminant level (MCL) of 0.5 μg/L for PCBs.

(a) Based on the EPA exposure parameters for an average adult, what is the highest incremental lifetime cancer risk from drinking water with 0.5 μg PCB/L? Is this risk level within the EPA guidelines for acceptable risk?

(b) Based on your answer to part (a), how many additional cancers per year are expected in the total population of this region?

(c) How does this value compare to the number of cancers per year expected in this region based on the average (background) cancer rate for the United States derived from Table 14.1? Express your answer as the ratio of PCB-related cancers to the total number of cancers expected in the region.

14.7 Based on the results of Problem 14.6, what is the maximum contaminant level (MCL) for PCBs in drinking water needed to achieve an incremental lifetime cancer risk of one in a million for the average adult? How does this value compare to the current MCL of 0.5 μg/L for PCBs?

14.8 In the movie *Erin Brockovich,* the water supply of a small California town was contaminated with hexavalent chromium (chromium VI) from a nearby industrial plant, causing severe illness and disease. Suppose the level of chromium VI in drinking water had been 0.6 mg/L. Would there be any reason to suspect a potential health problem? Explain your reasoning and provide supporting calculations.

14.9 According to the EPA's Toxics Release Inventory (TRI) for 1998, the environmental chemical released in greatest quantity was hydrochloric acid (HCl), emitted to the atmosphere in gaseous form mainly by coal-burning electric power plants. Studies show that the resulting maximum atmospheric concentration of HCl in the communities near these sources is typically 0.7 μg/m^3, with a range of about 0.1 to 1,000 μg/m^3. Calculate the hazard quotient (HQ) for this range of exposure. Do your results support or refute the claim by electric utility companies that these levels of HCl in the air pose no significant health concern? Explain.

14.10 Name three major sources of uncertainty in an environmental risk assessment. Briefly discuss the measures or programs you would propose to reduce each of these uncertainties.

14.11 A probabilistic risk assessment presents results in the form of a distribution function rather than a single value of risk. Use the distribution function in Figure 14.6 of this chapter to identify the 90 percent confidence interval for the risk in that case. This interval (or range) is defined by the risk levels having cumulative probabilities of 5 percent and 95 percent. Thus there is a 90 percent probability that the true risk lies somewhere in this interval.

14.12 Identify and briefly discuss four risk management options that might be used to address the problem of contaminated drinking water outlined in Problem 14.8. Assume the town's water supply comes from a local groundwater source contaminated with chromium VI. What are some of the pros and cons of each risk management option?

14.13 Assume you own and drive an automobile and are concerned about the problem of global warming. You want to calculate your personal contribution to annual CO_2 emissions and how this might be reduced. Draw an influence diagram to help you think through the problem. You know that the mass of CO_2 emitted is proportional to the amount of fuel burned each year. Identify the main factors that influence the annual mass emissions of CO_2 and sketch their interactions using the symbols outlined in this chapter. Try to identify as many factors as possible that can influence the final result.

14.14 You are responsible for conducting a risk assessment of a former industrial site adjacent to a new suburban housing development. Trace amounts of a carcinogenic organic chemical have been detected in the groundwater, the air, and the soil around the site. You have been asked to recommend whether any action is warranted to protect the health of the community. Draw an influence diagram displaying your approach to the problem and the factors that would affect your recommendation.

14.15 A leak detection system for an underground storage tank consists of several monitoring wells located both upstream and downstream of the tank. If there is a leak of chemical from the tank, the downstream monitors should show a higher reading than the upstream monitors, which measure the normal background concentrations of the chemical. However, the measuring instruments are accurate only to within about 20 percent of the indicated value, and there are also fluctuations in the background chemical concentration around the tanks. Thus a higher downstream reading may not always be due to a leak from the tank. By the same token, there is some chance that a leak could occur without being detected by this system. If a leak does occur, immediate remedial action can prevent any serious consequences. On the other hand, if no action is taken, a leak could cause wide-spread damage that would be very expensive to remediate.

(a) Draw a decision tree that displays all of the decision options for this situation. Begin with a yes/no chance node for whether the measured downstream concentration exceeds the normal background level. If yes, a decision must be made about whether to remediate immediately or take no action (in the belief that this is not a true leak). If no, the same options apply (you may still want to remediate just to be safe because the tank is rather old). In either case there is a probability, p_1, that a high measurement signifies a true leak, and a probability, p_2, that there is a leak even when the monitors indicate no difference.

(b) Assume that for this monitoring system, $p_1 = 0.8$ and $p_2 = 0.1$. Assume further that the cost of remediating immediately is $100,000, whereas the cost of dealing with a leak not detected is $1,000,000. Show on your decision tree the expected cost of each possible option.

14.16 Revisit Example 14.8 and consider a case where the odds of Site B needing either a partial or major cleanup are interchanged: a 5 percent chance of a cleanup costing $500,000 and a 15 percent chance of a cleanup costing $2.5 million. Would this affect the choice between Site A and Site B? Explain.

14.17 Solve the decision tree in Figure 14.19. Show each step in folding back the tree to arrive at your final answer.

14.18 Prepare a brief report discussing the strengths and limitations (as you see them) of the decision analysis methods presented in this chapter.

chapter

15

Environmental Forecasting

15.1 INTRODUCTION

People always have been interested in the future. Fortune-tellers and astrologers earn their living that way. So do weather forecasters. And so do many engineers, economists, mathematicians, and other professionals whose job is to provide insights on what the future might bring in various domains.

Questions about the future invariably arise in the analysis of environmental issues. For instance, what actions are required to achieve and maintain national air quality standards into the future? What measures are needed to prevent future environmental degradation of waterways and ecosystems? What are the environmental consequences of continuing our current "business-as-usual" practices?

Answers to such questions are critical for developing sound public policies that address environmental concerns. Of course, attempting to predict the future is a risky business. History is full of projections and forecasts that never came true. On the other hand, many environmental trends and changes have been successfully anticipated by thoughtful, forward-looking analysis. Thus the better we understand the factors that influence future environmental quality, the more able we will be to set and achieve appropriate environmental policy goals.

This chapter discusses several aspects of environmental forecasting from an interdisciplinary perspective that links environmental engineering studies with subjects in the social sciences domain. In particular, we examine how mathematical models can lend insight about the future to help guide actions and policy decisions about environmental laws, regulations, guidelines, and standards. Because the subject area of this chapter is very broad, only a brief introduction to key topics is possible. The main objective is to stimulate thinking about the role of quantitative analysis in addressing questions about future environmental issues.

15.2 FRAMING THE QUESTION

Perhaps the best place to begin is by asking what the objectives are for an environmental forecast. Indeed, there are many possible objectives. Consider, for instance, the range of topics presented in earlier chapters of this book. Chapter 8 discussed the problem of urban air pollution, in particular the complex problem of ozone produced by chemical reactions in the atmosphere. One objective of an environmental forecast could be to estimate the future regional emissions of substances that contribute to ozone formation. Two major sources of such emissions are automobiles (discussed in Chapter 3) and electric power plants (Chapter 5). What level of emissions can we anticipate from cars and power plants 5 or 10 years from now? Such projections are essential for determining whether air quality standards for ozone can be achieved and maintained in the future.

Similar questions can be asked about the future levels of other environmental contaminants like trace metals and toxic chemicals. For global environmental concerns like stratospheric ozone depletion and global climate change, the need to look ahead is especially imperative because of the long time horizons inherent in these problems.

15.2.1   Environmental Attributes of Concern

What exactly would we like to predict? The answer depends on the scope and nature of the questions posed. For the types of problems just mentioned, the general need is to estimate future levels of environmental quality. Usually this is expressed in terms of a pollutant concentration in the water, soil, or atmosphere. Estimates of future pollutant concentration may be used to anticipate problems that require action now, or to assess future compliance with current environmental regulations and standards. Assessments of future impacts on human health and ecological systems also may stem from these projections.

Other attributes of environmental quality include local, regional, and global patterns of land use. Future changes in the land area devoted to human settlements, agriculture, forests, wetlands, and other ecosystems can profoundly affect human health and welfare. Understanding and predicting how such patterns may change over time is thus another area of potential interest.

15.2.2   Forecasts versus Scenarios

There are some important distinctions in the words commonly used to talk about the future. The term *forecast* is generally used when one purports to know what will actually happen in the future. A common example is a weather forecast, which tells us what the forecaster thinks will happen tomorrow, next week, or next month. The terms *projection* and *prediction* often have the same connotation as a forecast—that is, trying to determine as accurately as possible an outcome at some future time.

In contrast to forecasts, a *scenario* attempts to describe what *would* happen *given some specified set of circumstances.* Rather than trying to predict what *will* happen in the future, scenarios typically describe a *range* of possible outcomes that would logically follow from different *assumptions* about key factors or events that can affect the outcome. Thus a scenario begins with a set of "what if" questions and then elaborates on the future consequences of those assumptions.

For instance, what if all automobiles in Los Angeles 20 years from now were powered by electric batteries instead of gasoline? How would future ozone levels change? And what would be the impacts of manufacturing and disposing of or recycling all the additional batteries that would be needed? Such a scenario might be used to analyze the impacts of a policy proposal to promote the use of electric vehicles. However, there is no claim or forecast that such a scenario is actually expected to occur. Rather, scenarios explore a range of future environmental outcomes that depend on specific assumptions, such as a policy decision taken at some specified time. Scenarios are especially useful as the time horizon becomes more distant and the reliability of environmental forecasts is inevitably diminished.

Despite the title of this chapter, we will purposely blur the distinction between environmental forecasts and environmental scenarios because the quantitative methods that we will present are useful for both purposes. More fundamentally, the main thrust of this chapter is to explore ways of thinking about the future that can be translated into mathematical terms. Although still an imperfect tool, analytical models let us peer into the future in ways that can often be helpful for environmental problem solving and decision making.

15.2.3 Time Period of Concern

An important characteristic of any environmental forecast or scenario is the time frame of concern. Are we trying to anticipate environmental impacts a year from now? Ten years from now? A hundred years from now? Often the phrases *near-term, midterm,* and *long-term* are used to denote different periods of interest. The specific meaning of these terms depends on the context of the problem at hand. In general, near-term usually means within the next few years, perhaps even out to a decade. Midterm might look ahead perhaps 10 to 30 years, whereas long-term could range from several decades to several centuries from now.

The choice of time frame is important because it dictates the types of models and methods that are appropriate for a given analysis. Detailed models and data usually are most applicable for near-term and midterm horizons, whereas simpler, more transparent approaches are often best suited for long-term analyses when uncertainties are large. Many other factors also affect the level of modeling complexity, including the types of questions posed, the availability of models and data, and the time and resources required for the analysis.

15.2.4 Spatial Scale of Concern

Different geographic scales may be of interest for different types of environmental problems. Analysis at the *local* scale includes urban areas with a geographic extent of perhaps 50 km or less. Problems such as water pollution and urban air pollution are important at this scale. Problems at the *regional* scale encompass a larger area with characteristic distances on the order of hundreds of kilometers. Acid rain and the land use impacts of human activities are examples of regional issues. Beyond that are geographic scales at the *national* and *global* levels. Global climate change and stratospheric ozone depletion are important at this scale.

The choice of geographic scale affects the feasibility and accuracy of environmental analysis, especially when looking toward the future. This is because the ability to model or predict environmental consequences at different geographic scales is often limited by the availability of necessary data, as well as by the fundamental understanding of environmental processes. Assumptions and scenarios can sometimes be helpful, but a lack of scientific knowledge is often the most limiting factor.

15.3 MODELING THE FUTURE

Try to imagine the world 30 years from now, when perhaps your own children will be of your age today. What will have changed during this time to affect future environmental quality?

There are many ways to answer such a question, but for discussion let us focus on three key drivers of environmental change.

15.3.1   Drivers of Environmental Change

The first major factor affecting environmental change is *population*. As noted in Chapter 1, the environmental impacts of human activities depend strongly on the number of people that inhabit the planet or a particular piece of it. So projections of future population trends clearly are important for environmental forecasts or scenarios. The further into the future one tries to peer, the more important this factor becomes.

The second key factor is the *standard of living* or the level of affluence of the population. This is usually measured in economic terms such as the gross domestic product (GDP) per capita. The more affluent the population, the more goods and services that are demanded, and the greater the resulting environmental impacts per person. Thus changes in economic conditions and standard of living over time are another dimension of environmental forecasting.

A third critical factor is *technology*—the vehicle for delivering the goods and services that people demand. Broadly defined, technology includes the methods used to provide food, shelter, comfort, transportation, consumer products, and a wide array of other services such as education and health care. The design and deployment of modern technology gives rise to the environmental issues discussed in Chapter 2 and elaborated on throughout this text. Expectations for technological change over time (including the development of environmental control technologies) are thus the third element needed to describe an environmental future. Among the many measures of technological change are the changes in environmental emissions, energy requirements, and natural resource requirements of a particular technology, as well as changes in the nature or types of technologies in use at any future time.

These three factors—population growth, economic growth, and technological change—are the primary focus of this chapter. Although the three are closely interrelated, we will examine them individually to facilitate the discussion. In particular, we will look at ways of describing these phenomena in mathematical terms for use in environmental forecasts or scenarios.

15.3.2   Modeling Environmental Processes

Of course, these three factors alone do not tell the whole story of environmental futures. Rather, they are the principal drivers that determine future land use patterns, natural resource requirements, and pollutant emissions to air, water, and land.

To assess the resulting environmental impacts (such as changes in pollutant concentrations, effects on human health, and changes in ecological systems), additional mathematical models and data are required. Chapters 8–12 of this text described some of the mathematical models available to translate emissions from human activity into measures of environmental change. Those types of mathematical models

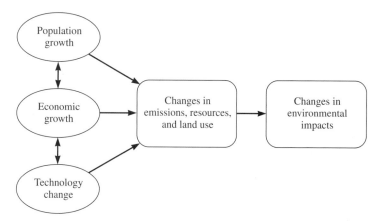

Figure 15.1 Basic elements of environmental forecasting, illustrating the three main "drivers" of population growth, economic activity, and technological change. Future changes in environmental impacts also can feed back to and affect these processes.

reflect our current understanding of how the world works based on principles of physics, biology, and chemistry. Many of these science-based models of environmental processes also are dynamic, meaning they can predict how factors like pollutant concentrations will change over time in response to a specified input or stimulus such as an increase or decrease in emissions from human activities.

The study of environmental engineering and science focuses mainly on developing the science-based process models just described. Such models are essential for predicting the environmental consequences of changes in anthropogenic emissions, both now and in the future. In contrast, the study of population growth, economic activity, and technological change lies primarily in fields of the social sciences, where mathematical models also are used for prediction. Figure 15.1 illustrates some of the links between these social science models (reflecting human behavior) and the physical science models of environmental processes. Both types of models are important for environmental forecasting. The social science models emphasized in this chapter provide a broader perspective on the factors affecting environmental futures.

15.4 POPULATION GROWTH MODELS

According to the United Nations, the world's population reached 6 billion people on October 12, 1999. The trend in world population growth through the end of the 20th century A.D. is shown in Figure 15.2. Over the past 100 years, the world population has quadrupled. Although it took thousands of years for the population to grow to 1 billion, the last billion people arrived in only 13 years! Several billion new neighbors are expected to join us on the planet over the next several decades. Because population growth is a major determinant of environmental impacts, we look first at how such growth can be expressed in mathematical terms and used for environmental modeling.

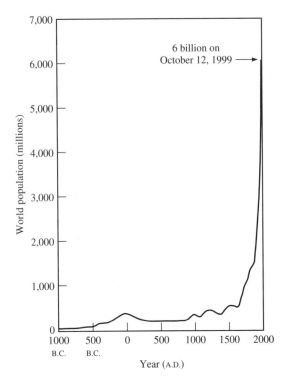

Figure 15.2   World population growth, 1000 B.C. to 2000
A.D. (*Source:* Based on USDOC, 1999a)

Depending on the time frame and geographic region of concern, one could envision a broad array of population trajectories, as illustrated in Figure 15.3. These include populations that grow quickly or slowly, as well as populations that stabilize, or that decline over time. The scope and purpose of an environmental analysis play a large role in defining the importance and type of population projections needed.

Each curve or trajectory in Figure 15.3 can be represented by an equation or mathematical model. Indeed, virtually any path that can be envisioned can be represented numerically in an environmental forecast or scenario. The next sections present a set of population growth models that span a range of complexity. The parameters governing the behavior of each model represent the variables of an environmental forecast or scenario. The value of key variables is often guided by analysis of historical data, but it is up to the analyst to specify how these parameters might change in the future.

15.4.1   Annual Growth Rate Model

One of the simplest and most common ways to quantify the growth of a population is to assume a *constant annual growth rate, r,* expressed either as a percentage or as a fraction. For example, if a population grows at a rate of 2 percent per year, next year's population will be 1.02 times greater than this year's population, and the following year it will be 1.02 times greater than that (an overall increase of 2.04 per-

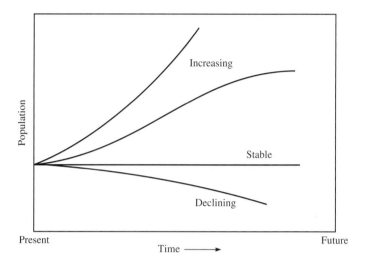

Figure 15.3   Possible trajectories of population change.

cent). If the annual growth rate, r, is expressed as a fraction (such as 0.02), a general expression for the total population, P, after t years is

$$P = P_o (1 + r)^t \qquad (15.1)$$

where P_o is the initial population at the time $t = 0$.

This equation has the identical form as the compound annual interest equation used in engineering economics to calculate monetary growth (see Chapter 13). The key characteristic of this equation is a nonlinear increase in the total quantity over time (be it population, money, or any other quantity that grows at a constant annual rate). Figure 15.4 illustrates this trend for three different rates of annual population growth. The higher the annual growth rate, the more dramatic the rise in population over time.

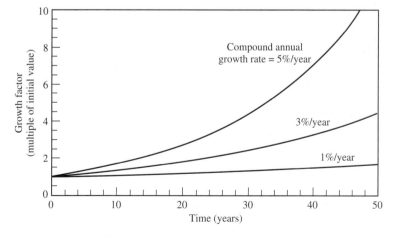

Figure 15.4   Population increase for three annual growth rates based on a compound annual growth model.

Example 15.1

Population growth of an urban area.   The current population of an urban area is 1 million people. The region has experienced rapid growth at an annual rate of 7 percent/year. City planners and environmental officials anticipate that this annual growth rate will continue for the next 10 years. If so, what would the population be 10 years from now?

Solution:

Assuming a constant annual growth rate, use Equation (15.1) with P_o = 1 million, r = 0.07, and t = 10 years:

$$P = P_o(1 + r)^t = (1 \times 10^6)(1.07)^{10} = 2 \times 10^6$$

Thus the population would double in 10 years at a 7 percent/year rate of growth.

The implication of a compound annual growth model is an ever-rising population. This type of model is frequently used in environmental forecasts or scenarios to estimate future environmental emissions from human activity. The simplest types of projections use population figures together with per capita measures of environmental impact, as in the next example.

Example 15.2

Projected growth in municipal solid waste.   Pleasantville is a city of 100,000 people that currently collects 8×10^7 kg (80,000 metric tons) of municipal solid waste (MSW) each year. The waste is disposed of in a sanitary landfill the city owns. Based on recent trends, the city's population is projected to grow at a rate of 3 percent/year over the next 15 years. Assuming that per capita waste production remains constant over this period, how much *additional* waste will the city have to collect and dispose of annually 15 years from now?

Solution:

First use Equation (15.1) to calculate the future population of the city 15 years from now, based on the 3 percent/year annual growth rate:

$$P = 100,000(1.03)^{15} = 155,800 \text{ people}$$

The number of additional people is therefore

$$\text{Added population} = 155,800 - 100,000 = 55,800 \text{ people}$$

The current annual waste generation per capita is

$$\text{MSW/person} = \frac{8 \times 10^7 \text{ kg}}{100,000 \text{ people}} = 800 \text{ kg/person-yr}$$

Assuming this rate remains constant, the total additional waste generated 15 years from now would be

$$\text{Additional waste} = (55,800 \text{ people}) \times (800 \text{ kg/person-yr})$$

$$= 4.5 \times 10^7 \text{ kg/yr}$$

$$= 45,000 \text{ metric tons}$$

Estimates of this sort may be used to anticipate the magnitude of future environmental problems, such as the need for additional landfill area or alternative methods of waste disposal. Because the future is always uncertain, a good analysis also employs a range of assumptions for key parameters that affect the outcomes of interest. In this case, both the annual population growth rate and the amount of waste generated per person should be treated as uncertain. Similarly, different growth rates might apply to different time periods, yielding a *multiperiod growth model*. Problems at the end of the chapter include examples that illustrate these types of projections.

15.4.2 Exponential Growth Model

The annual growth models just described assume that population increases occur in annual spurts at the end of each year, a process known as compound annual growth. This works well for compound annual interest added to a bank account, but a more realistic model for population increases would be *continuous*. This can be modeled by shortening the time period for compound growth from annual to continuous. The result is an alternative model of pure *exponential growth:*

$$P = P_0\, e^{rt} \qquad\qquad (15.2)$$

This equation is based on the assumption that at any point in time the rate of change in population is proportional to the total population at that moment. Mathematically, this can be written as

$$\frac{dP}{dt} = rP \qquad\qquad (15.3)$$

where the proportionality constant, r, is the growth rate expressed as a fraction of the current population. The solution to this differential equation is Equation (15.2), where P_0 is the original population at time $t = 0$.

Figure 15.5 compares the exponential growth model of Equation (15.2) with the compound annual growth model of Equation (15.1). As you can see, the two models give very similar results for low values of growth rates and time periods. But as r and t increase, there is greater divergence, with the exponential model growing more rapidly.

Example 15.3

Exponential versus compound annual growth.   In Example 15.2 the population of Pleasantville was assumed to grow at a compound annual rate of 3 percent/year for 15 years. Suppose instead that an exponential growth model had been used based on the same 3 percent growth rate. How would this change the estimate of Pleasantville's population 15 years from now?

Solution:

Use Equation (15.2) with $P_0 = 100{,}000$ people, $r = 0.03$/yr, and $t = 15$ years:

$$P = P_0\, e^{rt} = (100{,}000)\, e^{(0.03)(15)} = 100{,}000 e^{0.45} = 156{,}800 \text{ people}$$

This compares to 155,800 people using the compound annual growth model. The difference in this case is less than 1 percent, or an additional 1,000 people, assuming exponential growth.

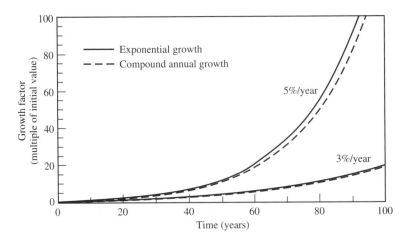

Figure 15.5   Comparison of two simple growth models.

Increases in the world's population over the past 5,000 years resemble an exponential growth function.[1] Over the past 100 years the rate of increase has been approximately 1.3 percent/year. We know, however, that exponential growth cannot continue indefinitely. In an environment with finite space and finite resources to support the needs of a population, growth eventually is curtailed. The next section presents a mathematical model that exhibits such characteristics.

15.4.3   Logistic Growth Model

Biologists have found that the population growth of many living organisms tends to follow an S-shaped curve like the one sketched in Figure 15.6, a shape known as *sigmoidal*. Initially the population begins to grow exponentially, but over time the growth rate gradually slows until it finally reaches zero. At that point the population stabilizes at the limit labeled P_{max} in Figure 15.6. This limit is known as the *carrying capacity* of the environment. It defines an equilibrium condition in which the total demands of the population for food, water, waste disposal, and natural resources are in balance with the capability of the environment to supply those needs. That balance defines a stable level of population with no further growth. The result is known as a *logistic growth curve*, represented by the sketch in Figure 15.6.

The leveling-off phenomenon in a logistic growth model represents a resistance to further growth as the population nears the carrying capacity of the environment. Mathematically, this can be represented by adding an "environmental resistance" term to the simple exponential growth model of Equation (15.3):

$$\frac{dP}{dt} = rP\left(1 - \frac{P}{P_{max}}\right) \tag{15.4}$$

---

1 Of course, over shorter periods and on smaller geographic scales, the population varies more erratically, especially due to catastrophes such as wars or famine.

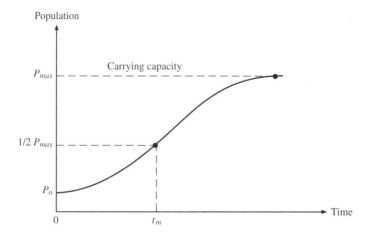

Figure 15.6   A logistic growth curve showing the characteristic S-shaped profile. P_o is the initial population and t_m is the time needed to reach half of the carrying capacity, P_{max}.

Now instead of population growth being proportional only to the current population, it depends also on the size of the current population relative to the carrying capacity, P_{max}. When the current population is small relative to the carrying capacity, the negative (resistance) term in Equation (15.4) is also small, and we have an exponential growth model as before. But as P gets larger and approaches P_{max}, the environmental resistance increases and the term in brackets approaches zero. The population growth rate (dP/dt) also then goes to zero. Between these two extremes the trend in total population transitions from an upward-bound curve to a horizontal asymptote, producing the S-shaped curve of a logistic growth model. Mathematically, the solution to Equation (15.4) is

$$P = \frac{P_{max}}{1 + e^{-r(t-t_m)}} \tag{15.5}$$

The growth rate, r, in this equation represents a composite growth rate over the sigmoidal shape of the logistic curve, so the value of r differs from that of the simple exponential growth model shown earlier. The two rates can be related if we define an initial exponential growth rate, r_o, associated with an initial population, P_o, at time $t = 0$. Then it can be shown that

$$r = \frac{r_o}{1 - \dfrac{P_o}{P_{max}}} \tag{15.6}$$

Similarly, we can show that the constant t_m in Equation (15.5) represents the time at which the population reaches *half* the carrying capacity (that is, the midpoint of the growth curve). Thus

$$\text{at } t = t_m, \ P = \tfrac{1}{2}P_{max} \tag{15.7}$$

By manipulating Equation (15.5) we can further show that t_m and r are related by

$$t_m = \frac{1}{r}\ln\left(\frac{P_{max}}{P_0} - 1\right)$$ (15.8)

A numerical example best illustrates the use of a logistic growth model for population projections.

Example 15.4

Estimating the world population in 2100.   Assume that world population growth can be described by a logistic growth model with a carrying capacity of 20 billion people. Estimate the global population in 2100 based on a current population of 6 billion in 2000 with an exponential growth rate of 1.5 percent. How would your answer differ if the carrying capacity were 15 billion people?

Solution:

The desired population can be found using Equation (15.5) with $t = 100$ years. But first we must find the logistic growth rate, r, and the midpoint time, t_m. From Equation (15.6),

$$r = \frac{r_0}{1 - \dfrac{P_0}{P_{max}}} = \frac{0.015}{1 - \dfrac{6.0}{20}} = 0.0214$$

From Equation (15.8),

$$t_m = \frac{1}{r}\ln\left(\frac{P_{max}}{P_0} - 1\right) = \frac{1}{0.0214}\ln\left(\frac{20}{6.0} - 1\right) = 39.5 \text{ years}$$

Substituting these results into Equation (15.5) gives

$$P = \frac{P_{max}}{1 + e^{-r(t-t_m)}} = \frac{20}{1 + e^{-0.0214(100-39.5)}} = 15.7 \text{ billion}$$

This is the projected population 100 years from now, assuming a carrying capacity of 20 billion people. The value of $t_m = 39.5$ years indicates that a population of 10 billion people (half the carrying capacity) will be reached about 40 years from now. Repeating the calculation for a global carrying capacity of 15 billion (instead of 20 billion), the estimated population in 2100 would be 13.4 billion people. By either estimate the world's population would more than double over the next 100 years based on these assumptions.

Logistic growth models are appealing because they reflect the type of long-term growth patterns that have actually been observed for microorganisms, insects, and other life forms. A logistic model further offers a simple way of representing a long-term limit to growth and a gradual stabilization of the population. For human populations, however, logistic models have been less successful in predicting carrying capacity and growth rates in the past. Rather, the value of carrying capacity seems to be a moving target that changes over time. The unique human capability for technological innovation has led to developments such as modern medicine and fertilizers for food production that continue to alter the apparent limit to global population. Thus a simple logistic model affords only a rough approximation based on assumptions about key parameters. Because of the availability of detailed data on actual population characteristics, other types of models are more commonly used to project future population growth, as discussed next.

15.4.4   Demographic Models

Demography is the study of the characteristics of human populations, including their size, age, gender, geographic distribution, and other statistics. The wealth of data on population characteristics allows much more detailed models to be developed for population projections in lieu of the models discussed so far.

Population statistics are most commonly collected, analyzed, and reported on a national basis by individual countries, private organizations, and international organizations like the United Nations. The basic data needed for population projections are current rates of births and deaths. The difference between these two rates gives an approximate measure of the overall population growth rate. In addition, a country may gain or lose population via migration. If more people regularly enter a country than leave, there is a net increase in the overall population growth rate. If the net immigration rate is negative (more people leaving than entering), the overall growth rate is reduced. In general, we can write

$$\text{Growth rate } = \text{ (Birth rate)} - \text{(Death rate)} + \text{(Immigration rate)} \quad \text{(15.9)}$$

The rates in Equation (15.9) are typically quantified in terms of the annual numbers of births, deaths, and immigrants per 1,000 people in the overall population. These overall statistics are referred to as the *crude rates,* which means they apply to the population as a whole, as opposed to specific segments of the population such as a particular age group.

Example 15.5

Estimating population growth rate.   A country with a total population of 50 million people has a crude birth rate of 20 births per year per 1,000 people, a crude death rate of 9 per 1,000, and a net immigration rate of 1 per 1,000. What is the net population growth rate expressed as a percentage of the total population?

Solution:

Because we are interested only in *rates,* the absolute size of the population does not enter this problem. Using Equation (15.9), we have

$$\text{Growth rate } = \text{ Birth rate} - \text{Death rate} + \text{Immigration rate}$$

$$= 20 - 9 + 1 \text{ (per 1,000 people/yr)}$$

$$= 12 \text{ per 1,000 people/yr}$$

On a percentage basis this annual growth rate is 1.2 percent of the population.

Age Structure of a Population   Data on overall birth and death rates, plus immigration statistics, allow us to construct an overall picture of population dynamics. Even more useful than the crude rates for the overall population are the *age-specific rates* by gender. Combining such data with information on the *age structure* of a population allows much more accurate projections of near-term population trends.

Figure 15.7 shows two examples of age-specific population distributions, illustrating the number of males and females in the population for various age intervals.* For clarity of presentation an age interval of five years is used in these figures, although a one-year interval is commonly used in population statistics. Notice the bulge in the United States in the middle years; this represents the baby boom cohort born after World War II. As this population ages, there will be an increasing percentage of people in the higher age brackets, and the age distribution profile will flatten. In contrast, the shape of the world population shows a predominantly younger population. This implies a substantial growth in future population as younger people enter their reproductive years.

Fertility Rates   A key factor in population projections is the *total fertility rate* of women in the population. A composite of the age-specific birth rates in a given year, this approximates the average number of children born to each woman during her lifetime. The higher the total fertility rate, the larger the future population is likely to be.

The *replacement* fertility rate is the average number of live births needed to replace each female in the current population with one female in the next generation. In modern industrialized countries this number is about 2.1, reflecting the slightly higher proportion of males that are born each year (it's not exactly 50–50, as seen by the higher number of males in the younger population in Figure 15.7) plus the number of females who die before childbirth. An important factor here is the number of newborns who do not survive the first year of life. In poorer societies with little access to the benefits of modern medicine and child care, the *infant mortality rate* historically has been high. Successful efforts to reduce infant mortality thus can significantly impact the replacement fertility rate and the total number of newborns who survive and contribute to the future population.

Fertility rates that exceed the replacement rate create a *population momentum* that leads to a sustained increase in population. The highest fertility rates in the world today are found among developing countries, whose populations are growing most rapidly (for example, fertility rates are above 7 in some African nations). In contrast, fertility rates in many industrialized countries (Western Europe) are currently about 1.5, well below the replacement rate. At this level the overall population will gradually decline over several decades (barring an increase in immigration).

Projecting Future Population   A simple example illustrates how population growth can be modeled using statistical data on age structure and age-specific birth rates and death rates. In order to simplify the arithmetic, the following example uses hypothetical data for a 10-year period rather than 1-year intervals.

---

* Most textbooks display the number of males and females in each age group side by side in the form of a *population tree* centered about the vertical axis. The graphical presentation in Figure 15.7, however, shows more clearly the differences in male and female populations in each age group.

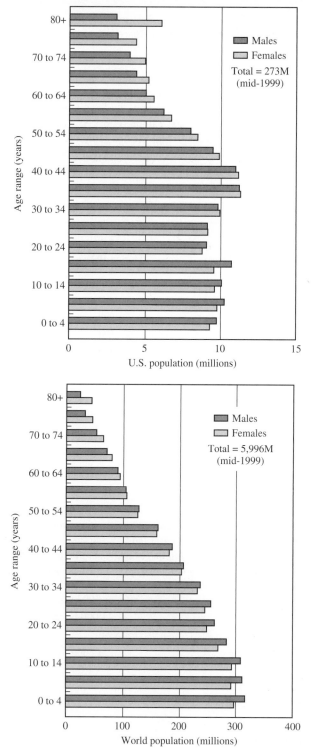

Figure 15.7 Age distribution of the U.S. and world populations in 1999. The U.S. population has a bulge in the middle age group, whereas the world population is markedly younger. The figure also shows that females outnumber males after age 30 in the U.S. population and after age 55 in the world population. (*Source:* Based on USDOC, 1999b and 1999c)

Example 15.6

A population projection based on age-specific data.   Table 15.1 shows the current age distribution for a hypothetical population of 500 million people. The age-specific birth rates and death rates also are shown based on data for the previous 10 years. Assuming these rates also apply for the next decade, calculate the total population and percentage of people in each age group expected 10 years from now. Assume immigration is negligible.

Table 15.1   Population statistics for Example 15.6.

Age Group (Years)	Current Population (Millions)	Births per 1,000 People (During 10 Years)	Deaths per 1,000 People (During 10 Years)
0–9	100	0	20
10–19	95	200	30
20–29	90	600	30
30–39	80	400	40
40–49	60	100	50
50–59	40	0	100
60–69	20	0	300
70–79	10	0	500
80–89	4	0	700
90–99	1	0	1,000
Total	500 million		

Solution:

First calculate the total number of births across all age groups over the 10-year period. This will determine the total number of children in the 0–9 age group in the next time period a decade from now. The reproductive ages in this example are the four groups between 10 and 49 years old. For simplicity, the rates are based on the total population of each age group rather than on the number of females. Thus

$$\text{Births to people aged 10–19} = 95 \times 10^6 \text{ people} \times \frac{200 \text{ births}}{1,000 \text{ people}} = 19 \times 10^6 \text{ births}$$

Similarly,

$$\text{Births to people aged 20–29} = (90 \times 10^6)\left(\frac{600}{1,000}\right) = 54 \times 10^6$$

$$\text{Births to people aged 30–39} = (80 \times 10^6)\left(\frac{400}{1,000}\right) = 32 \times 10^6$$

$$\text{Births to people aged 40–49} = (60 \times 10^6)\left(\frac{100}{1,000}\right) = 6 \times 10^6$$

Summing over all age groups, the total number of children born over the next 10 years would be $19 + 54 + 32 + 6 = 111$ million. This would be the population in the 0–9 age group 10 years from now (that is, children who have not yet reached their 10th birthday). Note that

Table 15.2 Summary of present and future population for Example 15.6.

Age Group (Years)	Population (in Millions)		Percentage of Total	
	Current	In 10 Years	Current	In 10 Years
0–9	100	111.0	20.0	19.2
10–19	95	98.0	19.0	17.0
20–29	90	92.2	18.0	16.0
30–39	80	87.3	16.0	15.1
40–49	60	76.8	12.0	13.3
50–59	40	57.0	8.0	9.9
60–69	20	36.0	4.0	6.2
70–79	10	14.0	2.0	2.4
80–89	4	5.0	0.8	0.9
90–99	1	1.2	0.2	0.2
Total	500	578.0	100.0	100.0

the infant mortality rate (for age 0 to 1 year) is not given separately in this example but rather is accounted for in the overall death rate for the 0- to 9-year-old population.

The remainder of the future population tree is quantified using the age-specific death rates given in Table 15.1. For instance, Table 15.1 shows that the 100 million children currently in the 0–9 age group will experience an average death rate of 20 per 1,000 (2.0 percent). This means that 98 percent of the 0–9 cohort will survive and subsequently enter the 10–19 age group. Thus

$$\begin{pmatrix} \text{Future population} \\ \text{aged 10–19} \end{pmatrix} = \begin{pmatrix} \text{Current population} \\ \text{aged 0–9} \end{pmatrix} \times (\text{Fraction surviving})$$

$$= (100 \times 10^6)\left(1 - \frac{20}{1,000}\right) = 98 \times 10^6 \text{ people}$$

Similar calculations apply to all other age groups. Notice that for the oldest group (90–99 years) the 10-year death rate in this example is 100 percent, meaning that no one in this age group survives past the 99th year.

Table 15.2 summarizes the results of these calculations. The table shows how each age group in the current population moves into the next age group 10 years from now. Also shown is the percentage of the total population in each age group now and in the future.

The result is a population of 578 million people 10 years from now. This represents a 16 percent increase over the current population. The percentage distributions also show an aging of the population, with nearly 10 percent over age 60 in the future compared to 7 percent currently.

The preceding example used a 10-year age interval to simplify the computations. Also, population migration was assumed to be negligible. A more refined analysis, done on a computer, would use one-year age categories and time steps, along with gender-specific and age-specific population figures, birth rates, and death rates. Examples of such projections appear in Figure 15.8(a), which shows the U.S. population projected to the year 2050, as estimated by the U.S. Census Bureau for three scenarios (labeled low, middle, and high series). Figure 15.8(b) shows how key parameters are assumed to change over time in the middle series Census Bureau projection. Different assumptions produced the higher and lower population projections.

Figure 15.8(c) shows the 2050 population distribution resulting from the middle series projections. Note that the profile is much more uniform than the 1999 profile in Figure 15.7(a). These projections imply a higher percentage of older people in the population than today: In the oldest groups the population over 80 will have roughly tripled. These results have important implications for public policy and the economy. For example, demand is likely to increase for health care services (as opposed to schools) and a smaller percentage of the population will be in the workforce. At the same time more

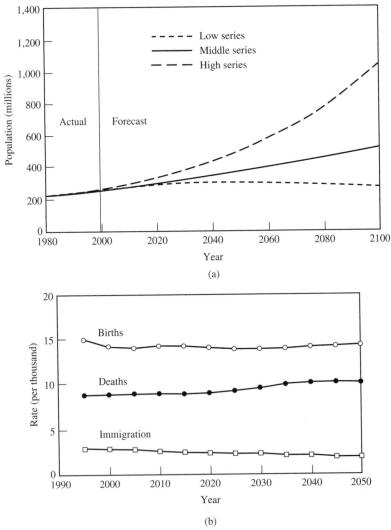

(a)

(b)

Figure 15.8   U.S. population projections to 2050, showing (a) the low, middle, and high estimates of the U.S. Census Bureau, (b) key assumptions behind the middle (best estimate) projection, and (c) the best-estimate age distribution in 2050. (*Source:* Based on USDOC, 1999b and 1999c)

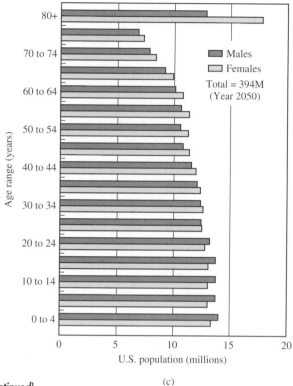

Figure 15.8 *(continued)*   (c)

senior citizens will be collecting Social Security benefits. The importance of population projections thus extends to a broad range of issues besides environmental impacts.

The procedure used in Example 15.6 to calculate the total future population based on age-specific birthrates and death rates can be expressed in general mathematical terms. A more detailed mathematical model also would divide the population into males and females and include separate data on the age-specific populations, fertility rates, and death rates by gender.

The writing of such equations is a bit tedious but essential for programming advanced models. A taste of this appears as an exercise for students in the problems at the end of this chapter.

Limitations of Demographic Models   Even when using detailed population data in sophisticated demographic models, we must make *assumptions* about how far into the future the current birth rates and death rates will prevail. In some cases mathematical models (including logistic models) have been proposed to estimate future changes in these parameters. But for the most part, assumptions about future fertility rates, death rates, and immigration patterns remain a matter of judgment because the future remains uncertain. Assumptions about demographic parameters are often linked to assumptions about future economic development and standards of living. For instance, fertility rates and infant mortality rates are substantially lower among wealthier populations than among poorer regions of the world.

Demographic models are thus limited in their ability to *forecast* long-term changes in population. On the other hand, they can be very useful for analyzing *scenarios* of future population trends under different conditions. The detailed population data of a demographic model also allow a richer set of "what if" questions to be asked, such as the effects of future changes in fertility rates and infant mortality rates. Such scenarios can reveal how policy actions might influence future population trends and hence environmental impacts. Demographic models also provide information on the size and age structure of the available labor force of the future, which provides an important link to economic projections, as we shall discuss shortly.

At the same time, demographic models are not always necessary or suitable for all types of environmental projections. Rather, in many situations the results of demographic projections can be used to estimate the parameter values needed for simpler models. Such a case is illustrated in the next example.

Example 15.7

Estimating growth rates from population data.   A demographic analysis projects the total population of a region to grow from 500 million to 772 million over the next 30 years. If this population increase were to be approximated by a simple exponential growth model, what is the implied annual growth rate?

Solution:

The exponential growth model was given earlier by Equation (15.2). In this case both the initial and final populations are known, and we are solving for the growth rate, r, over a period of 30 years. Thus

$$P(t) = P_o e^{rt}$$

$$772 = 500e^{r(30)}$$

$$e^{30r} = 1.544$$

$$r = 0.0145, \text{ or } 1.45\%$$

In most environmental forecasts or scenarios, the size of the future population is only one determinant of future environmental impacts. A second, closely related factor is the standard of living or affluence of a population. We look next at how the effects of economic development can be reflected in an analysis of environmental futures.

15.5   ECONOMIC GROWTH MODELS

Suppose you wanted to estimate the total mass of pollutant emissions from automobiles 25 years from now. One key factor in that analysis would be the number of cars in the future. That could depend on the future size of the population (the more people, the more cars), but it would also depend on how affordable a car is for the average citizen. In general, the more affluent the society, the more vehicles per capita as seen earlier in Chapter 3. Thus in the United States there are approximately 50 cars for every 1,000 people, whereas in China the ratio is about 1.5 cars per 1,000 people. So future improvements in standards of living that make cars more affordable could dramatically impact the future number of vehicles, especially in developing countries like China. This, in turn, could significantly increase the total level of future environmental emissions. (Of course, technology change also plays a big role here, as we will discuss later in the chapter.)

Economists try to understand the relationships between economic development and the resulting demands for goods and services in the context of different societies and cultures. Although the problem is extraordinarily complex, empirical evidence provides a basis for quantitative models that can help us peer into the not-too-distant future.

The most common measure of economic activity at the national level is the *gross domestic product* (*GDP*) of a country. This is the monetary value of all final goods and services produced during a year within a nation's borders. When normalized on the country's population, the *GDP per capita* offers a measure of the average affluence of the population. Thus one important indicator of overall economic growth is the change in GDP per capita over time.

Can we predict the future change in GDP of a country or region? And if so, can we relate changes in economic well-being to changes in environmental impacts? These are the two main questions of interest in this section.

The simple answer to both questions is that we cannot make *predictions* in these areas with any certainty. Rather, we can observe what has happened in the past and use these relationships to estimate what might happen in the future. Figure 15.9 shows one example of how the GDP of several countries has changed over time. Economic growth rates tend to be relatively high during the early stages of industrialization, then taper off as economic development raises the standard of living. Year-to-year changes in GDP also tend to be larger for developing countries than for industrialized countries, as depicted in Figure 15.9. Historical trends such as these

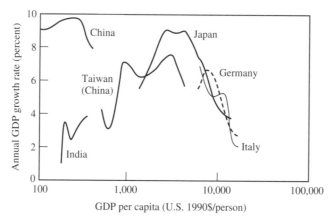

Figure 15.9   Historical economic growth rates for industrialized and developing countries. (*Source:* Nakicenovic et al., 1998)

provide guidelines for assumptions about future economic development. Such assumptions usually are expressed as an annual percentage increase in the overall GDP or in the factors that make up the GDP.

If the GDP grows faster than the total population, the GDP per capita also rises, which improves the standard of living. The question then becomes how to link changes in economic well-being to other measures of change that have environmental implications. Several approaches are outlined here, beginning with the simplest.

15.5.1   Activity Coefficients

Environmental forecasts often employ relatively simple assumptions or coefficients that act as surrogates for an economic growth model. For instance, to project the future number of automobiles in a region, an analyst might assume that the number of cars per 1,000 people increases from its current value to a higher ratio in the future. Combined with an estimate of the future population, this simple coefficient would yield the total number of cars in the future:

$$\text{Total cars} = \text{Cars/person} \times \text{Population}$$

An increase in the number of cars per person (or per 1,000 people) suggests an improvement in economic well-being that allows more people to own automobiles. The coefficient of cars per capita is not explicitly linked to a level of economic development; rather, a level of economic well-being is implicit in the assumed ratio of cars to people. For many types of environmental analyses this approach may be adequate, especially when analyzing a wide range of "what if" questions.

Example 15.8

A scenario of future automobile population.   What if economic development allowed automobile ownership in India to grow from its current level of 3 cars per 1,000 people to the current world average of 12 cars per 1,000 people? How many more cars would there be in India based on the current population of approximately 1 billion people? What if the future population were 1.3 billion?

Solution:

From the data given, the current number of cars in India is

$$\frac{3\,\text{cars}}{1{,}000\,\text{people}} \times 10^9\,\text{people} = 3 \times 10^6\,\text{cars (current)}$$

Based on the current population, an increase to the world average of 12 cars/1,000 people would give

$$\frac{12\,\text{cars}}{1{,}000\,\text{people}} \times 10^9\,\text{people} = 12 \times 10^6\,\text{cars (future)}$$

This would be an increase of 9 million cars. If the future population were 1.3 billion people, the total number of cars would be 15.6 million—more than five times the current number.

Simple coefficients of this type—we will call them *activity coefficients*—specify the *amount of a product or service per person* related to economic activity. Other examples are the per capita amounts of residential housing (m²/person), the number and type of appliances owned (such as refrigerators per household), and the amount of solid waste generated per year (kg/person-yr). All of these quantities directly or indirectly influence environmental impacts, and all are related to the level of economic development.

For environmental scenarios, such coefficients have the advantage of *transparency* because it is clear exactly what assumption lies behind the analysis. The major shortcoming of this approach is that the magnitude of activity coefficients is not explicitly coupled to economic assumptions or other relevant variables. One step toward addressing this shortcoming is to seek more explicit relationships, as in the next example.

15.5.2 Economic Growth and Energy Use

Researchers have attempted to correlate changes in GDP with other changes in the economy that have a bearing on environmental impacts. One widely studied factor is the relationship between economic development and energy use.

Figures 15.10 and 15.11 show two views of this relationship. In Figure 15.10 the per capita GDP of different countries is plotted as a function of the per capita energy use for a given year. This snapshot reveals a general trend toward higher per capita energy use as affluence increases. The solid line in the figure shows the curve that

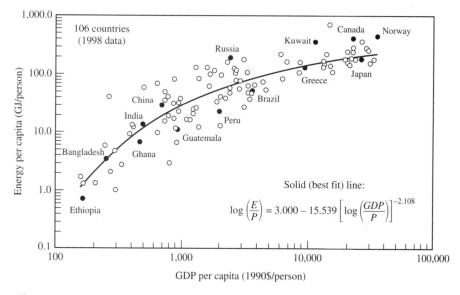

Figure 15.10 Per capita energy use as a function of GDP. Per capita energy consumption is highest in affluent countries with high per capita GDP. (*Source:* Based on 1998 data from USDOE, 2000)

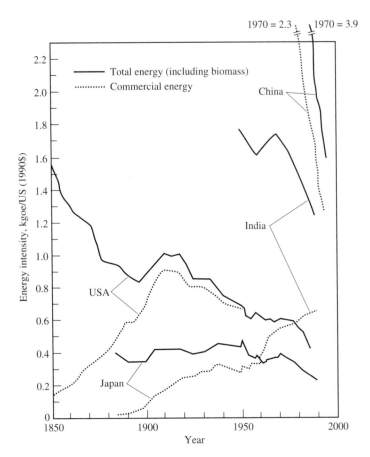

Figure 15.11   Trends in energy intensity. The total energy intensity of a country (including biomass fuels) tends to decrease over time with growing industrialization and economic development. However, the use of commercial energy tends to grow initially and then gradually decline over time. Trajectories for the United Kingdom, Germany, and France (not shown) lie between those of the United States and Japan. (*Source:* Nakicenovic et al., 1998)

best fits this data. This curve offers one way of estimating the expected increase in energy consumption associated with economic growth. Use of this relationship assumes that countries with low GDP per capita will develop along the same path as more affluent nations. However, the scatter in the data of Figure 15.10 indicates substantial variability in per capita energy use, even among countries with similar per capita GDP. This reflects differences in the makeup of the economies of different countries, differences in energy efficiency, and various other factors. It also underscores the uncertainty in forecasting the effects of economic development based on a cross-section of data from different countries.

An alternative approach is to use time series data for one particular country to model the relationship between economic growth and energy use. As an example,

Figure 15.11 shows a characteristic increase in the per capita consumption of commercial energy sources like oil, gas, and coal as different nations industrialize. However, if estimates of noncommercial energy (like local biomass) also are included, the energy intensity is initially much higher. Over time there is a gradual decline in energy intensity, reflecting improvements in energy efficiency as well as structural changes in the economy, such as a decline in energy-intensive industries.

Of course, historical relationships of this type may not necessarily hold in the future. Nonetheless, they can serve as a starting point for alternative scenarios about the effects of economic development on future energy use.

Example 15.9

Impact of economic growth on future energy use. Suppose that economic development raised the GDP per capita of a country from \$600/person to \$6,000/person. What would be the change in energy use per capita based on the best-fit curve of Figure 15.10?

Solution:

We can either read values directly from Figure 15.10 or use the equation shown in the figure (which gives greater accuracy). The best-fit relationship between GDP per capita (GDP/P) and total energy (GJ) per capita (E/P) is given by

$$\log\left(\frac{E}{P}\right) = 3.000 - 15.539\left(\log\frac{GDP}{P}\right)^{-2.108}$$

Thus for a GDP/capita of \$600/person we have

$$\log\left(\frac{E}{P}\right) = 3.000 - 15.539\,(\log 600)^{-2.108} = 1.197$$

$$\frac{E}{P} = 15.7 \text{ GJ/person}$$

Similarly, for (GDP/P) = \$6,000/person we find

$$\frac{E}{P} = 114 \text{ GJ/person}$$

So a 10-fold increase in GDP/person would raise energy consumption per capita by a factor of 114/15.7 = 7.3 (a 730 percent increase).

Note that because these figures are all on a per capita basis, an estimate of the total future population would be needed to calculate the total annual energy consumption.

To evaluate the environmental consequences of increased energy use, additional methods and models are required. Thus energy–economic growth models are frequently linked to other models that describe fuel and energy consumption in greater detail. This usually demands a more complete picture of economic activity.

The following sections discuss two methods used to model the structure of economic activity in greater depth. These methods provide further insight into how economic growth can affect future environmental impacts.

15.5.3   Input–Output Models

An economy is frequently described in terms of different *sectors*. In the United States, the U.S. Department of Commerce divides economic activity into 10 major areas, listed in Table 15.3. Each sector is further divided into subsectors. For instance, the manufacturing sector is divided into 20 different industry groups, as shown in Table 15.3. Additional subsectors are then defined within each industry.

To capture the interactions among different sectors and subsectors of an economy, economists sometimes use an *input–output model*. This type of model quantifies the value of goods and services that each sector requires from other sectors (the inputs) in order to make its own product (the output). For example, to produce an average automobile the auto manufacturing sector might require $2,000 worth of

Table 15.3   Principal sectors of the U.S. economy according to the Standard Industrial Classification (SIC) groups of the U.S. Department of Commerce.

SIC Code[a]	Major Economic Sectors
01–09	Agriculture, forestry, and fishing
10–14	Mining
15–17	Construction
20–39	Manufacturing
40–49	Transportation, communication, electric, gas, and sanitary services
50–51	Wholesale trade
52–59	Retail trade
60–69	Finance, insurance, and real estate
70–89	Services
90–98	Public administration

Detail of the Manufacturing Sector[b]

20	Food	30	Rubber and plastics
21	Tobacco	31	Leather
22	Textiles	32	Stone, clay, and glass
23	Apparel	33	Primary metals
24	Lumber and wood	34	Fabricated metals
25	Furniture	35	Machinery (excluding electrical)
26	Paper	36	Electrical and electronic equipment
27	Printing and publishing	37	Transportation equipment
28	Chemicals	38	Instruments
29	Petroleum and coal	39	Miscellaneous manufacturing

[a] SIC = Standard Industrial Classification. These are the first two digits of a code containing as many as eight digits to identify different subsectors. Some two-digit numbers are not used (e.g., 18, 19).

[b] Each two-digit industry code is further subdivided into three-digit categories, which are subdivided into four-digit categories, etc.

steel, $1,000 of plastics, $300 of aluminum, and additional quantities of many other products and services. These inputs represent a *supply chain* that draws on different sectors of the economy. In turn, each of these supply chain sectors has its own supply chain. Producing the $2,000 worth of steel requires the steel industry sector to purchase from other sectors certain amounts of iron, coal, electricity, transportation services, and all other inputs to its product.

By collecting data on the amount of goods and services each sector buys from other sectors in a given year, we can establish a picture of economywide interactions. In the United States, the U.S. Commerce Department collects and reports such data for 487 different subsectors. Other countries use a much smaller number of sectors, typically fewer than 100. These data provide the basis for an input–output model.

Table 15.4 illustrates an input–output model for the simple case of an economy with just three sectors, labeled 1, 2, and 3. For each producing sector, the coefficients along a row represent the fraction of the total direct inputs supplied by each of the other sectors. Thus b_{12} is the fraction of Sector 1 inputs purchased from Sector 2, and b_{13} is the fraction purchased from Sector 3. These fractions are based on the dollar value of all purchases, which is the easiest quantity to measure and report systematically. So if Sector 1 purchased a total of $100,000 worth of goods and services, of which $40,000 came from Sector 2 and $60,000 from Sector 3, b_{12} would be 0.40 and b_{13} would be 0.60. (In this case b_{11} is zero, though in general companies within a sector might purchase goods and services from other companies in the same sector.)

These input–output ratios allow us to analyze the overall effect of economic growth in any particular sector. Thus if Sector 1 increased its output by 10 percent, each of the other sectors that supply Sector 1 also would have to increase its output. The input–output matrix would quantify the magnitude of growth in other sectors required to supply the growth in Sector 1. A $10,000 (10 percent) increase in direct inputs to Sector 1 would require an additional $4,000 worth of output from Sector 2 and $6,000 from Sector 3. In turn, the supply chains of these two sectors would generate additional goods and services representing *indirect inputs* also attributed to the growth in Sector 1. The overall result is a "multiplier effect" in which the total economy grows because of increased activity along the supply chains of each sector.

What has all this got to do with environmental forecasting? The link is that any increase in economic activity also increases environmental emissions—at least in the near term, assuming no changes in production methods or environmental control

Table 15.4   An illustrative input–output model for a three-sector economy.

Producing Sector	Supply Sector		
	1	2	3
1	b_{11}	b_{12}	b_{13}
2	b_{21}	b_{22}	b_{23}
3	b_{31}	b_{32}	b_{33}

b_{ij} = fraction of producing sector i inputs supplied by sector j

requirements. Instead of looking only at the dollar values of inputs needed to make a product, we can also quantity the *environmental emissions per unit of production* based on data collected by government agencies. For instance, the U.S. Environmental Protection Agency annually reports the quantities of toxic releases from each of the manufacturing industries listed in Table 15.3. Adding such data to the input–output model allows the environmental impacts of economic growth in any sector to be evaluated in terms of increased emissions.

The main benefit of an input–output model is that it accounts for all of the indirect impacts along the supply chain, as well as the direct impacts of an activity. Thus, as shown in Table 15.5, the overall environmental impact of automobile production includes not only direct emissions from the auto production plants (where the steel, plastics, aluminum, and other materials are brought together to assemble a car), but also the indirect emissions from manufacturing the steel, plastics, aluminum, and all other components. Indirect impacts also include emissions from generating the electricity needed by each process, extracting the coal, ores, and other raw materials required, and shipping these products via ground transportation. Thus the input–output framework provides a comprehensive method for conducting environmental life cycle assessments of the sort discussed in Chapter 7. As Table 15.5 illustrates, the total indirect emissions are many times greater than the direct pollutant emissions from the sector being studied. Needless to say, the data requirements of this approach are formida-

Table 15.5   Environmental impacts of economic growth using an input–output model. Figures show the direct and indirect increases in emissions (in metric tons per year) for a 1 percent increase in U.S. auto production. Indirect emissions are shown for each of the top 10 suppliers of the auto production sector, with all other indirect sectors aggregated. All data are for 1992.

Sector Name	CO_2 (1,000s)	SO_2	CO	NO_2	VOC	Pb	PM_{10}
Direct Impacts							
Motor vehicles and passenger car bodies	64	73	98	95	645	0.0	11
Indirect Impacts							
Electric services (utilities)	605	3,691	116	1,689	13	0.1	76
Blast furnaces and steel mills	463	617	2,619	369	118	0.4	90
Crude petroleum and natural gas production	195	71	123	123	31	0.0	1
Coal production	114	2	1	2	0	0.0	7
Natural gas distribution	88	47	41	23	0.8	0.0	0.1
Trucking and courier services, except air	76	39	1,686	714	317	0.0	0
Air transportation	48	3	205	36	51	0.2	13
Motor vehicle parts and accessories	35	55	44	76	113	0.0	2
Wholesale trade	31	4	578	244	119	0.0	4
Industrial inorganic and organic chemicals	28	164	179	149	102	0.1	8
All other SIC sectors (476 subsectors)	263	1,650	3,940	1,619	853	5.3	480
Total National Impacts	2,010	6,416	9,632	5,138	2,364	6.1	693

Source: Based on CMU, 1999.

ble. Nonetheless, recent research programs have produced software that is now available for quantifying these life cycle impacts of economic growth (CMU, 1999).

A major limitation of the input–output framework is that all parameters of the model (both economic and environmental) reflect historical data for a particular year. This yields a static picture in which all production processes are fixed, and any growth in economic activity is assumed to produce proportional increases in environmental emissions. This assumption may be reasonable for relatively short time periods and small growth increments. Over longer periods, however, the structure of an economy and its environmental impacts will change. An input–output model based on today's economic and environmental data therefore may not be very useful for predicting impacts 10 years from now. For near-term analysis, however, this approach can be a useful way of assessing the overall environmental implications of economic growth.

15.5.4 Macroeconomic Models

Another type of economic model relevant to environmental analysis comes from the field of macroeconomics, which concerns the structure and performance of national economies and the effect of government policies on aggregate economic activity. To study these effects, mathematical models have been built using detailed economic data collected by firms and government agencies. These models provide deeper insights about how economic growth occurs and how environmental policies might affect the national economy. So let us take a brief look at these models and the types of questions they can (and cannot) address.

Macroeconomic models explain economic activity in terms of the behavior of three classes of economic actors: *firms, households,* and the *government.* These actors are assumed to trade with one another in various types of markets. Economists use a variety of labels to characterize these markets. Here we will use three categories known as the *labor* market, the *goods* market (which also includes services), and the *financial* market, which includes money in bank accounts, shares of stock, government bonds, and other financial instruments. In a market economy, private firms produce and sell the goods and services that are demanded. To do this, they must employ workers and invest in manufacturing equipment. The household sector supplies the necessary workers via the labor market, while other firms supply needed goods and equipment through trading in the goods market (as reflected in the input–output models discussed earlier in Section 15.5.3). The goods market also provides households and the government with the products and services they demand.

But not all of the income earned by households and firms goes into purchasing goods and services. Some of it goes to the government in the form of taxes, which are subsequently redistributed. Some is also saved and held in the form of financial assets such as cash, bank accounts, real estate, and other financial investments. The financial market is where these various types of assets are traded. So a firm in need of money to build a new production line would turn to the financial market for those resources.

If all of this sounds a bit complicated, rest assured it is! That is what makes economic activity so difficult to predict. Macroeconomic models use data about the various interactions shown in Figure 15.12 to explain overall changes in the economy. For instance, empirical studies of the U.S. goods market show that the level of GNP

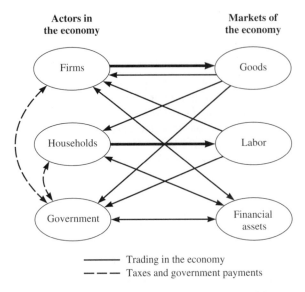

Figure 15.12   Macroeconomic interactions in a market economy. Bold arrows show firms supplying goods and households supplying labor. Other solid lines show trading in the economy. Dashed lines show the flows of taxes and government payments.

(gross national product)[2] during recent decades can be predicted reasonably well by this equation:

$$G = PK^{0.3} L^{0.7} \qquad (15.10)$$

where G is real GNP (billions of dollars),[3] K is the annual value of all capital goods used for production (billions of dollars), L is the size of the labor force (millions of workers), and P is the annual *productivity factor* for the economy, which is a measure of the overall efficiency of production.

This type of equation is known as a *production function*. It is a mathematical model of the factors that contribute to overall economic growth, as measured by GNP. According to this model, economic growth requires an increase in the amount of labor or capital used for production, or an increase in the productivity of those resources. In recent times the biggest gains in GNP have resulted from increases in productivity. Productivity improvements might come from such things as better management techniques or improvements in technology. Studies have shown that technology innovation has been the major contributor to productivity increases and is thus an important contributor to economic growth.

2 GNP differs slightly from GDP (gross domestic product). GDP reflects the value of all production within a nation's borders, regardless of whether done by domestic or foreign firms. GNP tracks the value of all production by domestically owned firms regardless of where it is performed. For most countries the two measures give very similar results (Abel and Bernanke, 1991).

3 Real GNP means adjusted for inflation. For a review of these terms, see Chapter 13.

Example 15.10

Factors contributing to economic growth. In 1989 the GNP of the United States was $4,118 billion (expressed in 1982 dollars), while total production capital was $3,960 billion and total labor was 117.3 million. Based on Equation (15.10), this yields a productivity factor of 12.21. According to Equation (15.10), how much would the overall economy grow if labor, capital, and productivity *each* increased by 2 percent in a given period?

Solution:

A 2 percent increase in each factor gives new values of

$$P = (12.21)(1.02) \quad = 12.45$$

$$K = (\$3,960)(1.02) = \$4,039 \text{ billion}$$

$$L = (117.3)(1.02) \quad = 119.7 \text{ million}$$

Substituting these values in Equation (15.10) gives

$$G = PK^{0.3}\,L^{0.7} = (12.45)(4,039)^{0.3}(119.7)^{0.7} = \$4,283 \text{ (billion 1982 dollars)}$$

This represents a 4.0 percent increase over the 1989 GNP of $4,118 billion. Half of this increase came from increased productivity, and half from capital and labor growth.

Macroeconomic models also use measures like household income, interest rates, taxes, and the prices of goods and services to explain the workings of market economies, including the balance between supply and demand for goods, labor, and financial assets. In the context of environmental analysis, these models are used mostly to predict the economic consequences of proposed environmental policy measures.

One current example is the case of global warming. Proposed measures to reduce carbon dioxide emissions are usually modeled by imposing a tax on the carbon content of fossil fuels. This raises the price of high-carbon fuels like coal, which causes less coal use and lower CO_2 emissions. But the higher cost of energy also drives up the cost of producing goods and services, which lowers consumer demand and slows economic growth. Macroeconomic models try to quantify such impacts. Figure 15.13 illustrates the predictions of several such models, expressed as a percentage loss in GDP from the imposition of a carbon tax in the United States (IPCC, 1996).

Some macroeconomic models, in conjunction with detailed models of specific energy demand sectors (residential, industrial, commercial, and transportation), predict how economic growth will affect the demand for specific fuels and energy supplies in different sectors of the economy. These predictions give a basis for estimating future environmental emissions from industrial processes, residential and commercial buildings, transportation systems, and electric power generation.

Of course, like all other types of models, macroeconomic models require *assumptions* about how key input parameters like oil prices or factor growth rates will change in the future. These assumptions underlie the resulting predictions of overall economic impacts. And because models are merely simplifications of complex human processes that also change over time, they are inherently limited in their predictive capabilities. Thus, although the use of macroeconomic models often can enrich an

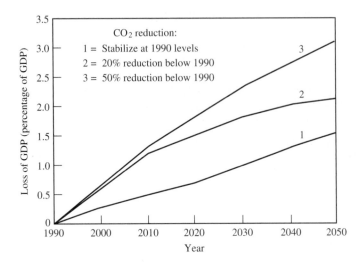

Figure 15.13   Macroeconomic model predictions of U.S. GDP loss from a tax on carbon emissions to control global warming. These average results across a large set of models predict that GDP losses will increase as CO_2 emissions are reduced. (*Source:* IPCC, 1996)

environmental analysis, considerable care and judgment are required to exercise and interpret the results of such models.

15.6   TECHNOLOGICAL CHANGE

We have already touched on the importance of technological change to economic growth. Here we look at some of the more direct ways that technology change can affect environmental forecasts or scenarios. Consider again, for example, the problem of estimating air pollutant emissions from automobiles 25 years from now. Or even 10 years from now. Not only will the number of cars on the road have changed (due to changes in population and standards of living), but vehicle designs also will have changed. Ten years from now a portion of the U.S. auto fleet is expected to be electric cars powered by batteries or fuel cells, which emit no air pollution directly. The design of conventional gasoline-powered vehicles also will have improved to emit fewer air pollutants than today's cars. Average energy consumption also might change significantly. Vehicles of the future might use more energy than today (from a continuing trend to large sports utility vehicles), or they might require less energy (from a transition to smaller, fuel-efficient vehicles).

These are examples of technological changes that can influence an environmental forecast or scenario. In this section we examine some of the ways that technological change can be considered analytically. The emphasis will be on relatively simple approaches that can be used easily in environmental analysis.

15.6.1 Types of Technology Change

Several types of technology changes can be important for environmental analysis:

Improvements to a current technology design. Incremental changes can reduce the environmental impacts of a current technology, typically via improvements in energy efficiency or a reduction in pollutant emission rates. An example would be an automobile with an improved catalyst or engine design emitting fewer hydrocarbons and nitrogen oxides per mile of travel.

Substitution of an alternative technology. Replacing a current technology with a different design often can provide the same basic service with reduced environmental emissions—for instance, replacing a gasoline-powered car with an electric vehicle, or an existing coal-fired power plant with an advanced gas-powered or wind-powered plant. However, the direct and indirect environmental impacts of the alternative technology must be carefully evaluated relative to the current technology design.

New classes of technology. This extension of the previous case encompasses technologies that offer a whole new way of doing things. For example, the automobile provided a new mode of personal transportation as an alternative to bicycles or horse-drawn buggies. Airplanes were a later example of an entirely new mode of transportation technology. The future environmental impacts of new classes of technology are inherently more difficult to evaluate than those of the technologies we know. Moreover, some indirect environmental impacts may be totally unforeseen (such as the extensive urban sprawl promoted by the growth of automobiles).

Change in technology utilization. Engineers are primarily concerned with the *design* of technology, but the *deployment* and *utilization* of technology determine its aggregate environmental impact. Environmental forecasts or scenarios must therefore consider how technology utilization might change in the future.

Technological innovation has changed the face of personal transportation in the 20th century. Imagine how things might look 50 or 100 years from now.

For instance, will the future use of an automobile (average distance driven per year) be the same or greater than today? Or might advances in air transportation or a growth in electronic commerce and telecommuting reduce the average usage of automobiles in the future?

The answers to such questions, and the importance of technological change, depend on the time frame of interest and the scope or objectives of the analysis. The further out in time we go, the more important these issues are likely to become. Next we discuss a few simple ways of incorporating technology change into environmental analysis.

15.6.2   Scenarios of Alternative Technologies

Perhaps the most direct way of modeling technological change is to postulate a transition from current technology to some improved or alternative technology. For instance, to characterize future emissions from automobiles, we could ask what might happen if x percent of future automobiles y years from now were battery-powered electric vehicles. What impact would this have on total air pollutant emissions and urban smog? A scenario of this type does not try to forecast the actual number of electric vehicles in the future. Rather, it asks a hypothetical question to assess the potential air quality benefits of an alternative technology.

The mathematical model in this case would quantitatively characterize the alternative technology. Key attributes of a technology for environmental analysis might include the types and quantities of air emissions, water pollutants, and solid wastes that are emitted; the fuel or energy consumption required for operation; and the natural resource requirements and materials needed for construction and operation of the technology. The environmental analysis also should capture any important indirect impacts of concern, such as emissions from the manufacturing or disposal of a new technology. Chapter 7 discussed this type of life cycle approach to environmental analysis, which applies to future technologies as well as to present-day systems.

Example 15.11

Future CO$_2$ reductions from a new technology.   Coal-burning power plants in many developing countries have an average efficiency of about 30 percent and emit approximately 1.1 kg of carbon dioxide (CO$_2$) for each kW-hr of electricity generated. CO$_2$ is a greenhouse gas that contributes to global warming. The amount of CO$_2$ released is directly proportional to the amount of coal burned. What if all future coal plants in these countries utilized advanced coal gasification combined cycle technology with an efficiency of 50 percent? How much would the CO$_2$ emission rate be reduced compared to plants using current technology?

Solution:

Recall that efficiency (η) is defined as the useful energy output of a process (in this case, electricity from the power plant) divided by the energy input (in this case, the fuel energy in coal). Thus

$$\text{Efficiency}(\eta) = \frac{\text{Electrical energy output}}{\text{Coal energy input}}$$

The higher efficiency of the advanced power plant technology means that less coal energy input is needed to achieve a given electrical output. Thus

$$\frac{(\text{Energy input})_{\text{advanced}}}{(\text{Energy input})_{\text{current}}} = \frac{\eta_{\text{current}}}{\eta_{\text{advanced}}} = \frac{30}{50} = 0.60$$

Because CO_2 emissions are proportional to the amount of coal burned, we also have

$$\frac{(CO_2)_{\text{advanced}}}{(CO_2)_{\text{current}}} = 0.60$$

Thus the advanced plant would emit 40 percent less CO_2 than the current plant design. Its CO_2 emission rate would be

$$(CO_2)_{\text{advanced}} = 0.60 \, (CO_2)_{\text{current}} = 0.60 \, (1.1 \text{ kg/kW-hr})$$
$$= 0.66 \text{ kg } CO_2/\text{kW-hr}$$

Scenarios like this can provide a simple way of estimating the potential environmental benefits of an advanced technology. If the results look interesting, a more sophisticated analysis would be needed to assess the feasibility of actually achieving such a result.

15.6.3   Rates of Technology Adoption

An important question in environmental forecasts is how long it takes for a new or improved technology to achieve widespread use. Chapter 6, for example, discussed the design of more energy-efficient refrigerators that eliminate the CFCs (chlorofluorocarbons) responsible for stratospheric ozone depletion. Such refrigerators first came on the market in the mid-1990s. But how long will it take until all U.S. households have these improved refrigerators? The answer is important for predicting atmospheric CFC levels, as well as energy-related environmental impacts.

The speed with which a new technology is adopted depends on many factors. Three of the most important are its price, its useful lifetime, and the number of competing options. High prices and many competing options inhibit the adoption of a new technology. So does a long useful lifetime because existing technologies are not quickly replaced. A number of methods are used to model the rate of adoption of new technology—some complex, others relatively simple. Three of these methods are highlighted here.

Specified Rate of Change   The most direct method of introducing a new technology is to specify its rate of adoption or diffusion into the economy. In general, that rate will depend on the growth of new markets for the technology, plus the opportunity to replace existing technologies at the end of their useful lives. The expected useful lifetime is thus an important parameter controlling the rate of adoption of a new technology. Table 15.6 shows the typical life of several technologies relevant to environmental projections.

Table 15.6   Typical technology lifetimes.

Technology	Typical Lifetime (years)[a]
Light bulbs	1–2
Personal computers	3–8
Automobiles	10–15
Refrigerators	15–20
Petrochemical plants	20–40
Power plants	30–50
Buildings	50–100

[a] Some internal components or subsystems may be replaced more frequently.

For the case of the household refrigerator, the expected useful life is about 18 years. If we assume a constant rate of replacement, this means that 1/18th (5.6 percent) of all current refrigerators must be replaced each year for the next 18 years. This defines the maximum rate of introduction for the improved refrigerator design discussed earlier—at least in the replacement market. The growth of new markets offers additional opportunities for adopting the improved technology; the size of this market depends mainly on population and economic growth.

Example 15.12

Adoption of CFC-free refrigerators.   Assume that all 120 million household refrigerators in the United States in 1995 are replaced with improved CFC-free refrigerators at the end of an 18-year lifetime. Assume further that population and economic growth increase the demand for new refrigerators by 2.0 million units per year over the next 10 years, and that all of these units are CFC-free. Estimate the percentage of all household refrigerators that are CFC-free in 2005.

Solution:

Since we are not given the age distribution of existing refrigerators, assume that each year 1/18th of all existing refrigerators die and are replaced with CFC-free models. Thus

$$\text{Replacement units/yr} = \frac{1}{18}(120 \text{ M}) = 6.67 \text{ million units/yr}$$

In addition, there is new demand of 2.0 million units/year from population and economic growth. Thus

$$\text{Total new units/yr} = \text{Replacement units/yr} + \text{New demand/yr}$$
$$= 6.67 \text{ M} + 2.0 \text{ M} = 8.67 \text{ M units/yr}$$

So after 10 years (in 2005) the total number of CFC-free units will be $8.67 \times 10 = 86.7$ M. The total number of refrigerators altogether will be

$$\text{Total units in 2005} = 120 \text{ M (as of 1995)} + 20 \text{ M (new growth)} = 140 \text{ M}$$

The fraction that are CFC-free in 2005 will be

$$\binom{\text{2005 fraction of}}{\text{CFC-free units}} = \frac{86.7 \text{ M}}{140 \text{ M}} = 0.62 \text{ or } 62\%$$

The fraction of CFC-free units will continue to grow for another eight years until it reaches 100 percent.

Specified Market Share   In Example 15.12 the only new refrigerator technology available after 1995 was the CFC-free design. In this case environmental laws actually prohibit the continued use of CFCs in new units. In general, though, a new or improved technology must compete with alternative options in the marketplace. In that case the adoption rate of a new technology depends not only on the size of the market, but also on its market share.

One way to model the diffusion of a new technology is to *specify the market share* at different points in time. For instance, one could *assume* that 10 percent of all new cars sold in 2005 will be electric vehicles. A more detailed specification might take the shape of a logistic curve like the ones illustrated in Figure 15.14. This type of S-shaped curve is frequently used to model the gradual diffusion of a new technology into the marketplace. The mathematical form of a logistic model was presented earlier in Section 15.4.3, where it was used to approximate population growth. Here we use it to represent the growth in market share of a new technology. All we have to do is to redefine the variable P as percentage market share rather than population. So if we define $P(t)$ as the percentage share of the market at time t, and P_{max} as the maximum market share (up to 100 percent), then

$$P(t) = \frac{P_{max}}{1 + e^{-r(t - t_m)}} \qquad (15.11)$$

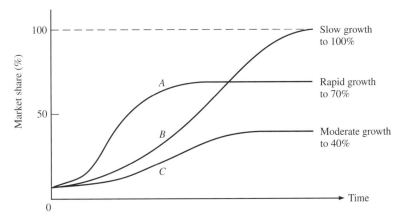

Figure 15.14   Logistic models of growth in technology market share for three scenarios involving different growth rates and final market share.

As before, t_m is the time needed to reach half the maximum value, and r is a composite growth rate given by

$$r = \frac{r_0}{1 - \dfrac{P_0}{P_{max}}} \tag{15.12}$$

where r_0 is the initial growth rate. In this case the characteristic time constants and growth rate for a logistic model would depend on the technology of interest.

Example 15.13

A logistic growth model for new technology adoption.   An auto industry analyst believes it will take 15 years for electric vehicles (EVs) to gain a 50 percent share of new auto sales once the initial share reaches 10 percent. Based on a logistic growth curve and an initial growth rate of 5 percent/year, how long would it take for EVs to gain 90 percent of the new car market?

Solution:

Assume the maximum possible market share is 100 percent. The logistic growth rate, r, from Equation (15.12) is

$$r = \frac{r_0}{1 - \dfrac{P_0}{P_{max}}} = \frac{0.05}{1 - \dfrac{10}{100}} = 0.0556$$

We want to find the time, t, at which the market share reaches 90 percent. Thus we use Equation (15.11) and solve for t based on $P(t) = 90$ and $t_m = 15$ years. Rearranging Equation (15.11) gives

$$1 + e^{-r(t - t_m)} = \frac{P_{max}}{P(t)}$$

$$1 + e^{-0.0556(t - 15)} = \frac{100}{90} = 1.111$$

$$e^{-0.0556(t - 15)} = 0.111$$

$$t = 55 \text{ years}$$

Remember that this means 55 years from the time the market share reaches 10 percent. (We are not told in this problem how long it will take to achieve that initial market share. That requires a separate analysis.)

Consumer Choice Models   Instead of directly specifying the market share or adoption rate of a new technology, some forecasting models introduce new technologies based on consumer preferences. Usually this is based on economic criteria. In such models the capital and operating costs of a new technology are specified along with those of all competing technologies. The model (typically a computer program) then selects the cheapest option. The cost of a new technology may change over time. A limit also may be imposed on its maximum market share to reflect the role of noneconomic factors in technology choice decisions. In some cases more sophisticated models of consumer preferences might be added to account for factors or features other than minimum cost.

Economic choice models are used frequently to analyze energy technologies and their environmental implications. One example is the National Energy Modeling System (NEMS) used by the U.S. Department of Energy for its annual outlook on future energy use (USDOE, 1999). This model couples a macroeconomic projection of the U.S. economy with detailed cost and performance assumptions for new and existing technologies in major sectors of the economy (industry, buildings, transportation, and electric power generation). The adoption rates and utilization of new technologies then emerge from complex calculations that determine the least expensive options over time. Chapter 5 showed examples of the growth in new power generation technologies based on the DOE modeling system. These projections provide a basis for estimating future environmental emissions.

An economics-based framework for modeling the adoption of new technology affords a rational, consistent basis for choosing among competing alternatives. One disadvantage, however, is the lack of transparency. Because many assumptions and relationships go into such models, it is often hard to know which assumptions most influence a particular outcome. Sometimes a simple assumption buried in a detailed calculation can affect key results without being recognized. And although cost is certainly a major factor in technology choice, many other considerations—including risk and personal preferences—also can be important. It thus bears repeating that any complex forecasting model must be exercised with care and understanding of its limitations as well as its strengths.

15.6.4 Rates of Technology Innovation

So far we have focused on ways of representing the adoption of new technologies with known characteristics (like higher efficiency and lower emissions). In many cases environmental forecasts assume that technology will continue to improve. Most energy–economic models, for instance, specify a rate of "autonomous energy efficiency improvements" that automatically reduces the energy needed each year to produce an increment of GDP. But how realistic is this assumption? What are the factors behind technology innovation, and how rapidly can we expect new innovations to materialize?

These are difficult questions. Social science research over the past few decades has improved our understanding of how factors like market forces, organizational structure, and government financing of research contribute to technology innovation. But we are still far from having any robust, mechanistic models that can predict technological innovations in the future. Rather, such projections typically are based on extrapolations of what has happened in the past. Such empirical models nonetheless provide an interesting and useful picture of what *could* happen in the future.

Figure 15.15 illustrates two examples of trends in technology innovation during the 20th century. The first shows advances in aircraft engine propulsion technology. In the early 1900s piston engines were the dominant technology, showing continued performance improvements through the 1940s. Then jet engines emerged as a new technology with greater performance capabilities. The overall result has been a 100,000-fold improvement over 90 years. The second example shows improvements

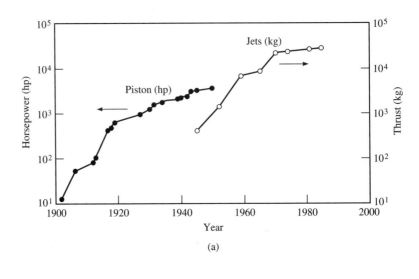

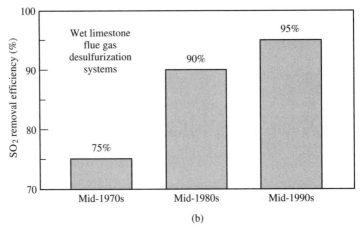

Figure 15.15   Examples of recent technology innovation trends showing increases in the (a) power of aircraft engines (*Source:* Ausubel, 1995) and (b) efficiency of new environmental control systems for sulfur dioxide capture at power plants.

in the pollutant removal efficiency of an environmental control technology used at electric power plants to control sulfur dioxide emissions. Since passage of the 1970 Clean Air Act, technology innovations have reduced pollutant emission rates by more than a factor of 10.

These examples illustrate innovation trends that have been documented for a wide array of technologies. Table 15.7 shows additional examples of energy efficiency improvements for a variety of new and existing technologies in the United States. But do these historical trends provide any way of estimating future changes in technology innovations, especially those with clear environmental implications?

Table 15.7 Annual rates of improvement in U.S. energy efficiency.

Technology and Measure	Time Period	Annual Rate[a]
Average New Technologies in the Marketplace		
New car fuel intensity normalized to vehicle weight (liters per 100 km and 100 kg)	1973–1983	−3.7%
Residential space heating intensity for new gas-heated houses (MJ per square meter and degree-day)	1954–1989	−1.6%
Electricity intensity of average refrigerator sold (kW-hr per year per cubic meter)	1972–1993	−2.0%
Electricity intensity of average room air conditioner sold (kW-hr per MJ)	1972–1993	−5.0%
Average of All Deployed Technologies		
Fuel intensity of electric utility fossil-fueled electricity generation (MJ per kW-hr)	1920–1960	−3.0%
Fuel intensity of all cars on the road (liters per 100 km for the fleet)	1973–1993	−2.1%
Energy intensity of space heating for all housing (MJ per square meter and year)	1973–1992	−2.6%
Electricity use of all refrigerators in households (kW-hr per refrigerator per year)	1973–1992	−1.2%
Energy intensity of steel production (GJ per tonne)	1970–1990	−1.4%
Energy intensity of all economic activity	1920–1970	−1.0%
(GJ per constant dollar of GDP)	1970–1990	−1.9%
Carbon intensity of all economic activity, corrected for structural change (grams C per constant dollar of GDP)	1970–1990	−1.9%

[a] Note that a rate of decline of 2 percent per year will, if it persists, halve the given measure in 35 years; a rate of decline of 4 percent per year will halve it in 18 years.
Source: PCAST, 1997.

A number of researchers have found that historical trends over long periods sometimes can be described rather well by logistic growth models. Two examples are shown in Figure 15.16. The first plots the trend in efficiency of different types of engines over 300 years, beginning with the earliest steam engines. The scale on the vertical axis employs a mathematical transformation (based on a log scale) that changes the S-shaped logistic curve to a straight line. This model suggests a technically plausible rate of continued technology innovation that would produce a high-efficiency fuel cell system within about 50 years.

Figure 15.16(b) shows another logistic model for the carbon intensity of the world's energy system. Over the past 150 years, a gradual shift from wood to coal to oil has reduced the carbon content per unit of energy consumed (Ausubel et al., 1998). Although total carbon emissions have increased because of growing energy consumption, the logistic model predicts a continued shift toward lower-carbon fuels such as natural gas (over the next 50 years or so) and ultimately to hydrogen (in 100 years or

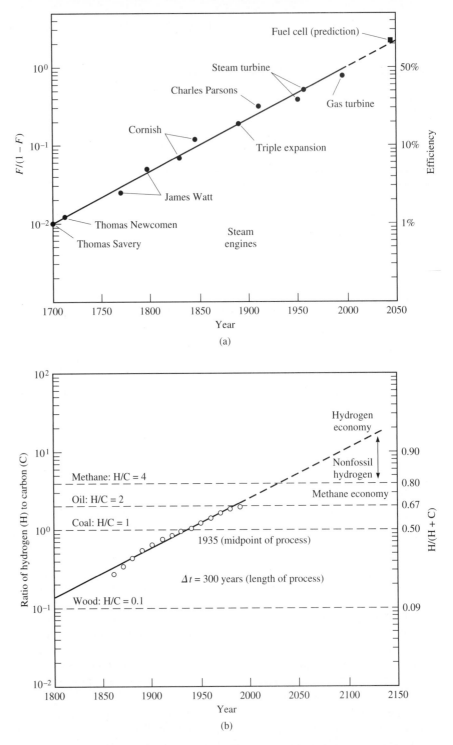

Figure 15.16 Logistic growth models for (a) efficiency improvements in engine technology and (b) carbon intensity of world energy use. The logarithmic scale used here transforms the S-shaped logistic curve to a straight line. The dashed lines show future predictions based on these models. (*Source:* Ausubel et al., 1998)

so). If such technology innovations actually materialize (the use of natural gas already is increasing), they could substantially ameliorate the problem of global warming.

Of course, any mathematical model based purely on historical data cannot be relied on to predict the future. But such models *can* suggest the types and rates of technology innovations that should be considered in the analysis of environmental scenarios. Unless one believes that technology innovations will no longer occur in the future, failure to consider technology trajectories (in conjunction with economic and population trends) could produce an incomplete or erroneous picture of future environmental change.

15.7 CONCLUSION

Addressing environmental problems often requires a forward-looking analysis to anticipate future problems and assess the merits of proposed actions or policies. The objective of this chapter has been to expose you to some of the challenges of environmental forecasting, as well as to some of the analytical methods that can help us peer into the future.

The material covered in this chapter represents only the tip of the iceberg. Ongoing research is continually improving our understanding of the drivers of environmental impacts, as well as our knowledge of the physical, chemical, and biological processes that govern those impacts. Although even the most sophisticated mathematical models cannot reliably predict the future of our world's environment, many of these tools, when used with wisdom and good judgment, can help us choose a path to a more sustainable environmental future.

15.8 REFERENCES

Abel, A. B., and B. S. Bernanke, 1991. *Macroeconomics.* Addison Wesley, Reading, MA.

Ausubel, J. H., 1995. "Technical Progress and Climatic Change." *Energy Policy,* 23(4/5), pp. 411–16.

Ausubel, J. H., C. Marchetti, and P. Meyer, 1998. "Toward Green Mobility: The Evolution of Transport." *European Review,* vol. 6, no. 2, pp. 137–56.

CMU, 1999. *Environmental Input–Output Life Cycle Assessment Model.* Green Design Initiative, Carnegie Mellon University, Pittsburgh, PA. http://www.eiolca.net.

IPCC, 1996. *Climate Change 1995: Economic and Social Dimensions of Climate Changes.* J. P. Bruce, H. Lee, and E. F. Haites (eds.), Intergovernmental Panel on Climate Change, Cambridge University Press.

Nakicenovic, N., A. Grubler, and A. McDonald, 1998. *Global Energy Perspectives.* International Institute for Applied Systems Analysis, Cambridge University Press.

PCAST, 1997. *Report to the President on Federal Energy Research and Development for the Challenges of the 21st Century.* President's Committee of Advisors on Science and Technology, Executive Office of the President, Washington, DC.

USDOC, 1999a. "Historical Estimates of World Population." Bureau of the Census, U.S. Department of Commerce, Washington, DC.

USDOC, 1999b. "Population Projections of the United States by Age, Sex, Race, and Hispanic Origin: 1995 to 2050." Bureau of the Census, U.S. Department of Commerce, Washington, DC.

USDOC, 1999c. International Database. Bureau of the Census, U.S. Department of Commerce, Washington, DC. http://www.census.gov/ipc/www/idbnew.htm.

USDOE, 1999. *Annual Energy Review 1998.* DOE/EIA-0484(98), Energy Information Administration, U.S. Department of Energy, Washington, DC.

USDOE, 2000. International Database. Energy Information Administration, U.S. Department of Energy, Washington, DC. http://www.eia.doe.gov/emeu/international.

USEPA, 1997. *National Air Pollutant Emission Trends Report, 1900–1996.* U.S. Environmental Protection Agency, Washington, DC.

USEPA, 1998. AIRS*Data.* Aerometric Information Retrieval System, U.S. Environmental Protection Agency, Washington, DC. http://www.epa.gov/airsweb/sources.htm.

15.9   PROBLEMS

15.1   Why are population growth and economic development important factors in projecting future environmental impacts? Under what circumstances might these factors be neglected in an environmental analysis?

15.2   A useful concept in mathematical growth models is that of the *doubling time,* defined as the time required for an initial value (say, of population) to double.

(a)   Derive a general expression that can be used to find the doubling time, $t_{2\times}$, for a given exponential growth rate, r.

(b)   Use this expression to find the growth rate that will cause a population to double in 10 years.

15.3   (a)   If the world population continues to grow at an exponential rate of 1.5 percent, how long will it take for the current population of 6 billion to grow to 12 billion?

(b)   If the exponential rate continued at 1.5 percent for 10 years, then dropped to 1.0 percent for 10 years, then decreased to 0.5 percent, when would the world population reach 12 billion?

15.4   Perform an uncertainty analysis for Example 15.2. Consider that the nominal population growth rate might vary by ± 0.5 percent/year (that is, from 2.5 percent to 3.5 percent/year) and that future per capita waste generation might be anywhere from 5 percent lower to 20 percent higher than the current value of 800 kg/person-yr. How would the uncertainty in these two assumptions affect the estimated mass of *additional* solid waste 15 years from now? Express your answer as

(a)   A range for the estimated mass of additional solid waste.

(b)   A percentage of the estimate from Example 15.2.

15.5   After studying recent demographic trends, the environmental planners in Pleasantville (population 100,000) conclude that the city's population could begin to grow more rapidly as new families are attracted there. Accordingly, the 15-year growth projection in Example 15.2 is revised to consider two time periods. Over the next five years, the recent 3 percent/year annual growth rate is assumed to continue. But over the following 10 years the annual population growth is assumed to increase to 5 percent/year. How do these new assumptions alter the previous estimates of additional population and additional solid waste generation 15 years from now?

15.6   Besides its environmental implications, population growth projection has implications for many other important public policy issues. One of these is Social Security payments. Assume, for example, that the U.S. government currently pays an average of $300/ month to each person aged 65 and over.

(a) Based on the 1999 population distributions in Figure 15.7, what is the total annual payment to Social Security recipients?

(b) Calculate the total annual payment based on the projected population distribution for 2050. Assume that the average monthly payment per person remains unchanged in real terms (that is, use constant 1999 dollars). By how much will the total annual payment increase, according to these projections? Give your answer both as an absolute dollar amount and as a percentage of the current (1999) value.

15.7 Suppose a logistic growth model had been used in the year 1900 to estimate future world population when the actual population was 1.6 billion with an approximate growth rate of 1.2 percent and an apparent carrying capacity of 10 billion. What world population would have been forecast for the year 2000? How does this compare to the actual value of 6 billion? Give your answer both in absolute terms (number of people) and as a percentage difference.

15.8 Develop a general mathematical model to project the future population of a country based on demographic data. Let $P(t)$ be the total population at time t, and let $P_i(t)$ be the age-specific population in any given age interval, i. The index i signifies each age group up to a total of N groups. Thus $P(t) = \sum_{i=1}^{N} P_i(t)$. If each age group is a one-year interval, N would equal the age of the oldest group. Finally, let b_i and d_i be the age-specific birth rate and death rate, respectively, expressed as fractions of the age-specific population at time t. Using these terms, write an expression for the total population, $P(t + 1)$, at the next time step, $(t + 1)$. For this problem assume the effects of immigration are negligible.

15.9 It has been proposed to increase the average fuel economy of new cars from the current U.S. standard of 27.5 miles per gallon (mpg) to 32.5 mpg in order to reduce energy consumption and lower the CO_2 emissions of the fleet by 15 percent. If the new mpg standard were enacted, but new car production continued to grow at a compound annual rate of 2.4 percent/year (with no change in the annual miles driven per vehicle), how many years would it take for the increased number of new cars to wipe out the CO_2 reduction achieved by the first 32.5 mpg fleet?

15.10 The average wealth of a country, as measured by GDP per capita, is an important indicator of environmental impacts. Suppose the per capita GDP of a country of 50 million people is just $500/person, and its population has an exponential growth rate of 2.3 percent. Based on the data in Figure 15.9, assume that the total GDP grows at a compound annual rate of 6 percent/year. If both of these rates persist, how long would it take for the per capita wealth to increase to an industrialized country level of $20,000/person? Assume all GDP amounts are in constant dollars, so you need not worry about the effect of inflation.

15.11 Based on the data in Figure 15.10, what would be the projected increase in total energy consumption for a country whose per capita GDP increases from $500 to $20,000/person? Give your answer both in absolute terms (energy/year) and as a percentage increase over the initial level.

(a) Use the solid curve in Figure 15.10 as the basis for your projection.

(b) Give an uncertainty estimate for the future energy use based on the scatter in data points at the higher value of GDP/capita.

15.12 Visit the website of the environmental input–output life cycle assessment model (http://www.eiolca.net). Use the most recent data in the model to assess the nationwide increase in environmental emissions from a 1 percent increase in U.S. automobile production. The name of that sector is "Motor Vehicles and Passenger Car Bodies" (SIC 590301). Production is measured in terms of the base year economic value given for each sector in the model. Compare your results to those given in Table 15.5, which were predicated on a 1 percent production increase for the base year 1992.

15.13 Repeat Problem 15.12 for a different economic sector of your choice. This time quantify the impacts of a 10 percent increase in economic activity.

(a) What major pollutants are emitted directly by the sector of interest?

(b) For these pollutants, how large are the indirect impacts from other sectors of the economy in the supply chain? What additional pollutants are important when indirect (supply chain) impacts are included?

15.14 Which of the three factors in the macroeconomic production function of Equation (15.10) contributes most to overall economic growth, assuming an equal percentage increase in each factor? Which factor contributes least?

15.15 Macroeconomic models predict that the overall economy as measured by GDP will suffer when money is spent to reduce pollution. Discuss some of the factors and assumptions that encourage such predictions. Also give your own view about the credibility of such projections and what additional factors would have to be considered to improve such forecasts.

15.16 Think about environmental problems that require (or would benefit from) an analysis of emissions or impacts at some time in the future.

(a) Give one example where considerations of technological change would be very important to such an analysis.

(b) Give another example where consideration of technological change would not be very important to the analysis.

15.17 You want to analyze the environmental consequences of future energy demands for residential space heating. You estimate that the effects of population growth will increase the total number of homes by an exponential rate of 1 percent/year. Furthermore, the average size (floor space) of each home will increase by 0.5 percent/ year as a result of increased affluence.

(a) Assuming no change in average weather conditions, how much larger will the heating energy demand be 20 years from now, assuming that the heating energy per square meter of floor space stays constant?

(b) Consider the additional effect of technological change. If the average energy intensity of space heating technology continues to improve at the historical rate given in Table 15.7, how large will the future heating energy demand be in 20 years relative to the current level?

(c) What percentage reduction in future heating energy demand is attributable to technological change?

(d) If there is no change in the fuel used for space heating, what are the likely environmental implications with and without technological change?

15.18 Assume that by 2015 the fraction of electricity generated from renewable energy sources will have increased to 20 percent, with a market share growth rate of 1.7 percent per year. If the market penetration of renewable technologies subsequently follows a logistic growth curve, and a 50 percent share is reached in 2050, in what year would the market share reach 95 percent?

APPENDIX

Table A.1 Atomic weight of selected elements

Element	Symbol	Atomic Weight	Element	Symbol	Atomic Weight
Aluminum	Al	26.98	Molybdenum	Mo	95.94
Antimony	Sb	121.75	Neon	Ne	20.18
Argon	Ar	39.95	Nickel	Ni	58.70
Arsenic	As	74.92	Nitrogen	N	14.01
Barium	Ba	137.33	Oxygen	O	16.00
Beryllium	Be	9.01	Palladium	Pd	106.4
Bismuth	Bi	208.98	Phosphorus	P	30.97
Boron	B	10.81	Platinum	Pt	195.09
Bromine	Br	79.90	Plutonium	Pu	244.
Cadmium	Cd	112.41	Potassium	K	39.09
Calcium	Ca	40.08	Radium	Ra	226.03
Carbon	C	12.01	Radon	Rn	222.
Cesium	Cs	132.90	Rhodium	Rh	102.91
Chlorine	Cl	35.45	Selenium	Se	78.96
Chromium	Cr	51.99	Silicon	Si	28.09
Cobalt	Co	58.93	Silver	Ag	107.89
Copper	Cu	63.55	Sodium	Na	22.99
Fluorine	F	19.00	Strontium	Sr	87.62
Gallium	Ga	69.72	Sulfur	S	32.06
Gold	Au	196.97	Thallium	Tl	204.37
Helium	He	4.00	Thorium	Th	232.04
Hydrogen	H	1.01	Tin	Sn	118.69
Iodine	I	126.90	Titanium	Ti	47.90
Iron	Fe	55.85	Tungsten	W	183.85
Krypton	Kr	83.80	Uranium	U	238.03
Lead	Pb	207.2	Vanadium	V	50.94
Lithium	Li	6.94	Xenon	Xe	131.30
Magnesium	Mg	24.31	Yttrium	Y	88.91
Manganese	Mn	54.94	Zinc	Zn	65.38
Mercury	Hg	200.59	Zirconium	Zr	91.22

Table A.2 SI unit prefixes

Quantity	Prefix	Symbol	Quantity	Prefix	Symbol
10^{-1}	deci	d	10^{1}	deka	da
10^{-2}	centi	c	10^{2}	hecto	h
10^{-3}	milli	m	10^{3}	kilo	k
10^{-6}	micro	μ	10^{6}	mega	M
10^{-9}	nano	n	10^{9}	giga	G
10^{-12}	pico	p	10^{12}	tera	T
10^{-15}	femto	f	10^{15}	peta	P
10^{-18}	atto	a	10^{18}	exa	E

Table A.3 Useful conversion factors

Quantity	Multiply	By	To Get
Length	inches (in)	2.540	centimeters (cm)
	feet (ft)	0.3048	meters (m)
	yards (yd)	0.9144	meters (m)
	miles (mi)	1.609	kilometers (km)
Area	in^2	6.452	cm^2
	ft^2	0.0929	m^2
	mi^2	2.590	km^2
	acres	4047	km^2
Volume	in^3	16.36	cm^3
	ft^3	0.02832	m^3
	std ft^3 (natural gas)	0.02679	normal m^3 (Nm^3)
	yd^3	0.7646	m^3
	gallons (U.S.)	3.785	liters (L)
Speed	ft/sec	0.3048	m/sec
	mi/hr (mph)	0.4470	m/sec
Mass	pounds (lb)	0.4536	kilograms (kg)
	tons (short)	0.9072	tons (metric)
Density	lb/ft^3	16.02	kg/m^3
	lb/gal	119.83	g/L
Force	pounds (lb_f)	4.448	newtons (N)
	kilograms (kg_f)	9.807	N
Pressure	lb_f/in^2 (psi)	6.895	kilopascals (kPa)
	lb_f/ft^2	0.04788	kPa
	atmospheres (atm)	101.325	kPa
	kg/m^2	9.807	Pa

(continued)

Table A.3 *(continued)*

Quantity	Multiply	By	To Get
Temperature	Rankine (°R)	5/9	kelvin (K)
	Farenheit (°F)	(5/9) (°F − 32)	Centigrade (°C)
Energy	foot-pounds (ft-lb$_f$)	1.356	joules (J)
	British thermal units (Btu)	1.055	kilojoules (kJ)
	Btu/lb (specific energy)	2.3244	kJ/kg
	kilocalories (mean)	4.190	kJ
	kilowatt-hours (kW-hr)	3600	kJ
Power	ft-lb$_f$/sec	1.356	watts (W)
	Btu/hr	0.2931	watts
	horsepower (hp)	0.746	kilowatts (kW)
Flow Rates	ft^3/min (cfm)	3.866×10^{-4}	m^3/sec
	gal/min (gpm)	0.06309	L/sec
	lbs/hr	0.1260	g/sec
	tons/day	10.50	g/sec
Heat Flux	Btu/hr-ft^2	3.155	W/m^2

INDEX

NAME & SUBJECT INDEX